Middle Cerebral Territory in Axial

W9-CXQ-253

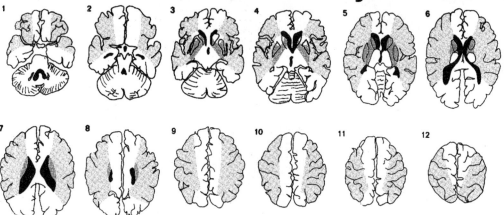

Cortical Branches

Cortical Functions

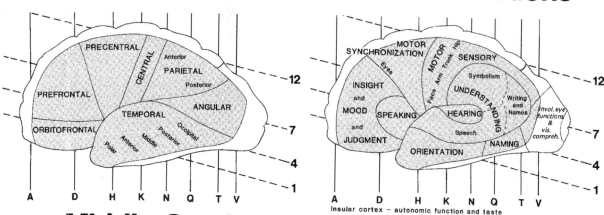

Insular cortex – autonomic function and taste

Middle Cerebral Territory in Coronal

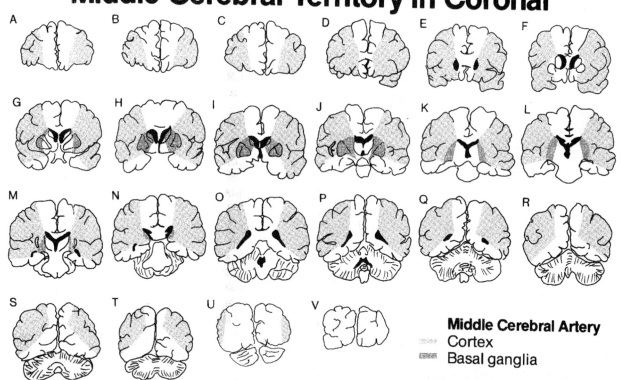

Middle Cerebral Artery
Cortex
Basal ganglia

Concise Text

OF

Neuroscience

ROBERT E. KINGSLEY, Ph.D.

South Bend Center for Medical Education
Indiana University School of Medicine

In Collaboration with

STEVEN R. GABLE, M.D.

T. R. KINGSLEY, Ph.D.

And the Assistance of

The Magnetic Resonance Imaging Center
The Saint Joseph Medical Center
South Bend, Indiana

Illustrated by

JACQUELINE SCHAFFER
MOLLIE DUNKER

Williams & Wilkins
A WAVERLY COMPANY

BALTIMORE • PHILADELPHIA • LONDON • PARIS • BANGKOK
BUENOS AIRES • HONG KONG • MUNICH • SYDNEY • TOKYO • WROCLAW

Editor: Timothy S. Satterfield
Managing Editor: Nancy Evans
Development Editor: Jude Berman
Production Coordinator: Carol Eckhart
Copy Editors: Thomas Lehr, Bonnie Montgomery
Designer: Norman W. Och
Illustration Planner: Ray Lowman
Cover Designer: Tom Scheuerman
Typesetter: Peirce Graphic Services
Printer: Rand McNally
Binder: Rand McNally

Copyright © 1996, Williams & Wilkins

351 West Camden Street
Baltimore, Maryland 21201–2436 USA

Rose Tree Corporate Center
1400 North Providence Road
Building II, Suite 5025
Media, Pennsylvania 19063-2043 USA

Accurate indications, adverse reactions, and dosage schedules for drugs are provided in this book, but it is possible that they may change. The reader is urged to review the package information data of the manufacturers of the medications mentioned.

Printed in the United States of America

Library of Congress Cataloging-in-Publication Data

Kingsley, Robert E.
    Concise text of neuroscience / Robert E. Kingsley : the assistence of the Magnetic Resonance Imaging Center, Saint Joseph Medical Center ; illustrated by Jacqueline Schaffer, Mollie Dunker.
      p. cm.
    Includes bibliographical references and index.
    ISBN 0-683-04621-7
    1. Neurosciences. I. Gable, Steven R. II. Kingsley, T. R. III. Saint Joseph Medical Center (South Bend, Ind.). Magnetic Resonance Imaging Center. IV. Title.
    [DNLM: 1. Neurosciences. WL 100 K55c 1995]
  RC341.K56   1995
612.8—dc20
DNLM/DLC
for Library of Congress   95-37148
                                             CIP

The publishers have made every effort to trace the copyright holders for borrowed material. If they have inadvertently overlooked any, they will be pleased to make the necessary arrangements at the first opportunity.

95 96 97 98 99
1 2 3 4 5 6 7 8 9 10

Reprints of chapters may be purchased from Williams & Wilkins in quantities of 100 or more. Call Isabella Wise, Special Sales Department, (800) 358-3583.

Dedicated to

W. v B. and R. D.,

*whose tragic deaths helped inspire and motivate me, and
also to the under-appreciated medical student, in whose hands
ultimately rests the future of medicine.*

PREFACE

The purpose of this book is to provide medical students with a concise presentation of brain anatomy and function in a form that is relevant to clinical practice. To do this, I feel it is necessary to present the subject matter of neuroanatomy, neurophysiology, and neurology together, in one place, integrated into a cohesive unitary structure—Neuroscience.

The size of the Neuroscience knowledge base precludes presenting a complete, systematic Neuroscience course to first-year medical students in the single semester typically allotted in medical curricula. This has necessarily placed severe limits on the amount and nature of the information that could be included in this book. Therefore, this is not intended to be a comprehensive work, but rather a practical textbook tailored specifically to the needs of the medical student. The greatest challenge in writing a Neuroscience textbook for medical students lies in selecting the subset of information such a book must contain. This particular book has been assembled with a close eye on the ultimate clinical usefulness of the information included. Of course, the selection of material has also been shaped by the special interests and knowledge of the author. Therefore, the final presentation of Neuroscience in this textbook represents my personal view of the core knowledge of Neuroscience appropriate to the practice of medicine.

It was with some concern that I chose to omit an extensive bibliography in this book. By omitting it I certainly do not wish to imply that the information presented here is original. It certainly is not. Nor do I wish to imply that factual statements cannot be documented in the original literature. They certainly can be. However, in keeping with the philosophy that this book is intended for medical students, who typically do not have the opportunity or the time to consult the original literature, extensive documentation and footnotes seemed to be more of a distraction than an aid. I have chosen instead to include a short list of secondary references that students may consult for further detail on the subjects presented. These sources also contain sufficient reference to the primary literature.

The size and rate of expansion of knowledge about the nervous system certainly precludes any one individual from having a complete or current understanding of this complex organ system. Writing a book of this nature makes one deeply aware of one's limitations and the remarkable contributions so many people have made to the field of Neuroscience. There are literally thousands of people whose sweat and toil in their research laboratories have made a book like this possible. Giving appropriate credit to all of them for their discoveries in any textbook is impossible; in a limited book such as this, entirely hopeless. Expressing the humility I feel in the presence of a research community that has so broadened our understanding of the human brain is the only acknowledgment I can extend. I do so with gratitude.

This book would not have been possible without the direct help and encouragement of many people. Foremost among these is T. R. Kingsley. Her unvarnished criticism, precise and detailed, and her very specific and useful suggestions have contributed immeasurably to the quality of this text. In addition, if she had not also volunteered to write the embryology appendix, that important topic would certainly be missing from these pages. Steve Gable read every word of this text and offered many important and useful contributions. In particular, he helped keep the focus of the text where it ultimately must be, on the patient suffering from neurological disease. Michael Hancock, Deloris Schroeder, and Jude Berman read the entire manuscript, made many important suggestions, and ferreted out errors. Arnold Hassen read several chapters and offered useful criticism and welcome encouragement. Mike Lannoo used a preliminary version of this work in the classroom; his comments and those of the freshman medical students at Muncie and South Bend were extremely valuable in shaping the final manuscript.

The inclusion of the clinical material found in this book would not have been possible without the help and cooperation of the Saint Joseph's Medical Center and the Magnetic Resonance Imaging Center, especially John Harding, Vic Jones, Tobin Mathews, and Barb Brown, who collected the material for me. Mary Donigan was very helpful in obtaining appropriate audiological material, while Bob King, Carl Marfurt, Mark Walsh, and Susan White kindly provided me with several cases. Former medical students C. L. Watts, J. L. Lackman, Janice Peterson, and Dennis Mishler collected most of the other cases.

The usefulness and beauty of this book have been materially enhanced by the superb illustrations of Jacqueline Schaffer and Mollie Dunker.

Several people offered for my use original prints of micrographs. I especially thank Richard Coggeshall, Steve Flieslor, Larry Squire, Gary Wright, and Jaime Dant, who generously provided outstanding examples of their work. The cover illustration was created by Caryl Erickson. The endsheets were kindly provided by the Eastman Kodak Co.

I would also like to acknowledge the assistance of Kathleen Drajus, who spent many hours obtaining and collecting copyright permissions and obtaining references for me from numerous libraries, and Marilyn Wacker of the Ruth Lily Library. Connie Gordon and Diane Huddlestun supported me in numerous ways, for which I am very grateful.

This book would not have been possible without David Burr, who first encouraged me to proceed with the project; Tim Satterfield, who was willing to take a chance on a new author; and Walter Daly, who arranged leave time that enabled me to complete the manuscript. Finally, and certainly not least, I want to extend my thanks and appreciation to the editorial staff of Williams & Wilkins, especially to Pat Coryell and Nancy Evans, who kept me and this project on track and on time. In spite of all of this help, errors of commission and omission must certainly have crept into the text, and for these I take total responsibility.

R. E. Kingsley
South Bend, 1995

CONTENTS

1 Gross Structure of the Nervous System

The mammalian nervous system is a control system that regulates all functions of the organism. In addition, at least in the human, it has unique functions that are independent of the other organ systems of the body. These independent functions allow us awareness of ourselves. The nervous system performs these functions as an information processing system. Information consists of internal representations of the external world in which the organism exists. The manner in which these internal representations of the world are created, transformed, and used to affect the behavior of the organism is the subject matter of **neuroscience.**

This book presents an introduction to a small segment of **neuroscience**—the part that is of particular concern to the physician. The principal topics that are addressed are the anatomy, physiology, and neurology (nervous system disease) of the human nervous system. The general plan is to present a description of the relevant anatomy of a functional system, followed by a discussion of its physiology. Then the neurology associated with that system is presented and, in most chapters, illustrated with case histories.

This chapter describes the surface features of the nervous system. The deep structures of the nuclei and tracts are dealt with separately in succeeding chapters. This description is not meant to be complete. It is a general overview for purposes of orientation and familiarization. It is essential that every physician know the structures described in this chapter, because they form the basis for one's visualization of the brain and one's ability to communicate about the brain with others. The purpose of this chapter is to lay a foundation that will enable one to develop a workable association between names and visualized structures, an association that is indispensable to understanding.

The central nervous system consists of the **brain,** the **brainstem,** the **cerebellum,** and the **spinal cord** (Fig. 1.1). The brain can be further divided into the **cerebral hemispheres** and **basal ganglia** (telencephalon), and the **thalamus** (diencephalon). The brainstem consists of the **midbrain** (mesencephalon), the **pons** (metencephalon without the cerebellum), and the **medulla** (myelencephalon).

The Cerebral Hemispheres and Basal Ganglia

The cerebral hemispheres form the largest part of the telencephalon, a paired structure that is the most prominent feature of the mammalian nervous system. The basal ganglia are a collection of closely associated nuclei that lie deep in the telencephalon and present no surface features. They are discussed in Chapter 8.

THE CEREBRAL HEMISPHERES

The visible, superficial portions of the telencephalon are the two cerebral hemispheres, separated by the **longitudinal cerebral fissure** and joined by a massive collection of axons, the **corpus callosum** [L. *corpus,* body, + L. *callosus,* hard] (Figs. 1.2 and 1.3). The cerebral hemispheres are the result of an extreme elaboration of the superficial embryonic telencephalon. This

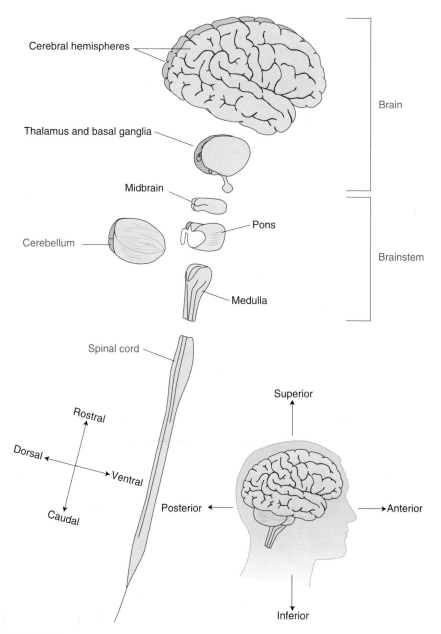

FIGURE 1.1.

Schematic drawing illustrating the relationships among the major divisions of the central nervous system. Note that the terms *superior, inferior, anterior,* and *posterior* are traditionally used to describe direction for the telencephalon, diencephalon, and cerebellum. The terms *dorsal, ventral, rostral,* and *caudal* are used for the remaining divisions. *Rostral* and *caudal* are frequently used for all parts of the central nervous system, rostral meaning "toward the head" and caudal, "toward the tail." Many authors use *posterior* and *anterior* interchangeably with *dorsal* and *ventral*. The choice is mostly a matter of style.

Longitudinal cerebral fissure

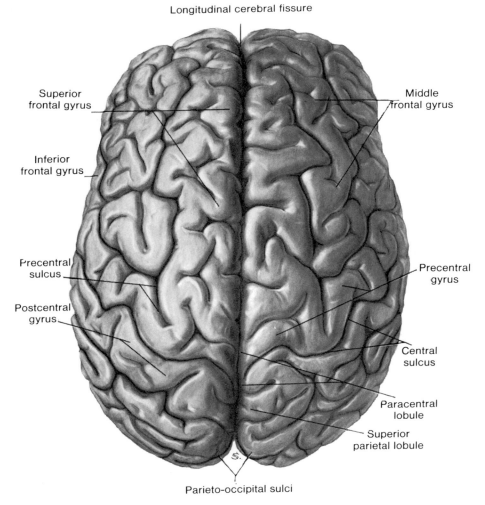

Superior
frontal gyrus

Middle
frontal gyrus

Inferior
frontal gyrus

Precentral
sulcus

Precentral
gyrus

Postcentral
gyrus

Central
sulcus

Paracentral
lobule

Superior
parietal lobule

Parieto-occipital sulci

FIGURE 1.2.

The superior surface of the cerebral hemispheres. They are separated by the longitudinal cerebral fissure. Note particularly the location of the central sulcus and the precentral and postcentral gyri. Note also the three frontal gyri.

elaboration is so extensive that the surface, the **cerebral cortex** [L. *cortex,* bark], becomes wrinkled during development. This wrinkling allows the cortical surface area to be greatly expanded while the volume enclosed remains at a minimum. A convex extension of the cortical surface is a **gyrus** [L. *gyros,* circle] and a concave fold, a **sulcus** [L. *sulcus,* furrow or ditch]. A particularly deep sulcus is frequently called a **fissure.**

The pattern of gyri on the surface of the hemispheres is not random. Within any mammalian species, the pattern of gyri and sulci remains re-

markably similar among individuals. This pattern establishes useful landmarks and should be memorized. Using these patterns, one can divide each hemisphere into five lobes (Figs. 1.3 and 1.4). The greatest of the wrinkles results, during fetal development, in part of the cerebral cortex, the **insular lobe,** being buried beneath cortical tissue that expands over it from two directions. The part of the cerebral cortex that covers the insular lobe is known as the **operculum** [L. *operculum,* a cover or lid]. Where these two sheets of cortex meet, a furrow is formed, the **lateral fis-**

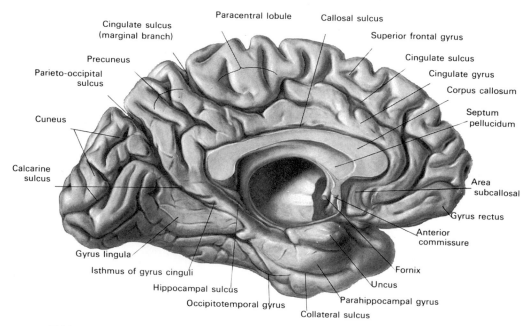

FIGURE 1.3.

Medial view of the left cerebral hemisphere. The corpus callosum has been cut in the midsagittal plane. If one follows the cingulate sulcus to the superior surface of the hemisphere, the postcentral gyrus lies immediately anterior to it. Moving anterior to the next sulcus, one finds the central sulcus. This is the most reliable method of identifying the central sulcus and is especially useful when interpreting magnetic resonance images. The parieto-occipital sulcus is easily identified on the medial surface of the hemisphere, and it meets the calcarine sulcus at nearly a right angle in most brains.

sure. This fissure separates the **temporal lobe** from the remaining cortical mass. Joining the lateral fissure at nearly right angles, the **central sulcus** separates the **frontal lobe** from the **parietal lobe.** The parietal lobe is separated from the **occipital lobe** by the **occipital-parietal fissure,** located on the medial wall of the hemisphere.

The Frontal Lobe

The largest of the five lobes, the frontal lobe is greatly expanded in the human compared to the other primates. On its superior-lateral surface (Figs. 1.2 and 1.5), the **precentral** gyrus spans across the hemisphere from the lateral fissure to the longitudinal cerebral fissure that separates the two hemispheres in the sagittal plane. Anterior to the precentral gyrus are three parallel gyri that are perpendicular to it, the **superior, middle,** and **inferior** frontal gyri. On the inferior surface (Fig. 1.6), at the most medial edge of

the hemisphere, lies the **gyrus rectus,** so named because it is unusually straight. The **olfactory tract** with its terminal **olfactory bulb** lies adjacent to the gyrus rectus in a depression known as the **olfactory groove.** Directly over the orbit lie several small gyri known collectively as the **orbital gyri.**

The Parietal Lobe

The anterior margin of the parietal lobe is denoted by the central sulcus, while the anterior portion of its inferior border is defined by the lateral fissure. Its most posterior regions are ill-defined on the lateral surface. One must view the hemisphere from the medial side (Fig. 1.3) to locate the definitive landmark that separates the parietal lobe from the occipital lobe, the **parieto-occipital sulcus.** This sulcus extends somewhat onto the superior surface of the hemisphere. To delineate the posterior boundary of

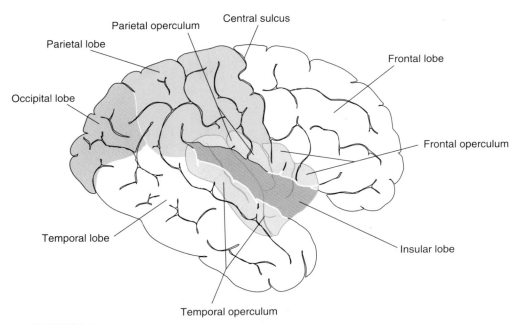

FIGURE 1.4.

Lateral view of the right cerebral hemisphere. The temporal lobe has been pulled away to reveal the insular lobe deep within the lateral fissure. The cortex that lies over the insular lobe is known as the operculum. Note the landmarks that indicate the remaining lobes. The boundaries of the occipital lobe are not obvious on the lateral view. The demarcation between the temporal lobe and the parietal lobe is not designated by any specific landmark.

the parietal lobe on the lateral surface, one must project imaginary extensions of the lateral fissure and the parieto-occipital sulcus. Anterior and superior to these lines lies the parietal lobe (Fig. 1.4).

On the superior-lateral surface of the parietal lobe (Fig. 1.5), the **postcentral gyrus** lies parallel to the central sulcus. Immediately posterior to this gyrus is the **supramarginal gyrus.** It is located immediately superior to the most posterior extent of the lateral sulcus. The **angular gyrus** lies posterior to the supramarginal gyrus.

The most prominent feature of the medial surface of the parietal and frontal lobes is the corpus callosum, a structure that is not a gyrus, but rather a massive collection of axons interconnecting the two hemispheres. This structure extends the entire length of the parietal lobe and well into the frontal lobe (Fig. 1.3). From posterior to anterior, it is divided into the **splenium** [G. *splenion,* a bandage], the **body,** and the **genu** [L. *genu,*

knee]. Along its superior margin lies the **callosal sulcus,** and superior to it is the **cingulate gyrus.** The cingulate gyrus is defined on its superior margin by the **cingulate sulcus** [L. *cingulum,* girdle, from *cingo,* to surround].

The Occipital Lobe

Lying posterior to the parieto-occipital sulcus is the occipital lobe, which is relatively small on its lateral aspect (Fig. 1.4). It is divided nearly in two on its medial surface by the **calcarine sulcus** [L., from *calcar,* spur-shaped] (Fig. 1.3). This sulcus joins the parieto-occipital sulcus and then extends along the medial wall of the temporal lobe. In fresh material, if one slices a cross-section perpendicular to the calcarine sulcus, one can see a white stripe in the cortex, following the sulcus and the gyrus on either side. This is the **line of Gennari,** a unique feature of this part of the cerebral cortex, conferring upon it its special name, the **striate cortex.**

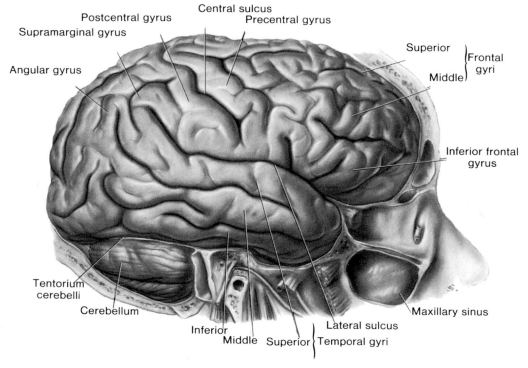

FIGURE 1.5.

The lateral surface of the right cerebral hemisphere. Note how the superior temporal sulcus terminates in the parietal lobe with the angular gyrus wrapped around it. The supramarginal gyrus lies immediately anterior to the angular gyrus.

The Temporal Lobe

Inferior to the lateral sulcus and its imaginary extension and anterior to the parieto-occipital sulcus and its imaginary extension is the temporal lobe (Fig. 1.4). Running approximately in the anterior-posterior direction and parallel with the lateral fissure are three gyri, the **superior, middle,** and **inferior temporal gyri** (Fig. 1.5). The **superior temporal sulcus** separates the superior temporal gyrus from the middle temporal gyrus and extends into the parietal lobe, terminating at the angular gyrus.

The medial surface of the temporal lobe contains several important structures. Most prominent is the **uncus** [L. *uncus,* a hook] (Figs. 1.3 and 1.6), a medial protrusion of the temporal lobe. Deep to it lie the **amygdaloid nuclei** [G. *amygdale,* almond]. Inferior to the uncus is the **parahippocampal sulcus** [G.

hippocampos, seahorse] and adjacent to it, the **hippocampal gyrus.** Deep to the hippocampal gyrus lies the **hippocampal formation,** an important structure of the temporal lobe that cannot be seen without cutting into it. The parahippocampal gyrus extends along the superior-medial aspect of the temporal lobe, wraps around the splenium of the corpus callosum, and becomes continuous with the cingulate gyrus (Fig. 1.3). At the posterior region, the calcarine sulcus lies inferior to the parahippocampal gyrus.

The Insular Lobe

If one were to widen the lateral fissure, the insula would be brought into view. This small cortical area is completely covered by the operculum of the overlying frontal, parietal, and temporal lobes (Fig. 1.4). Little is known about

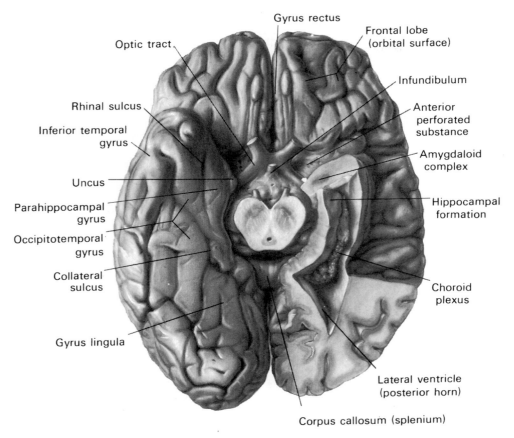

FIGURE 1.6.

A view of the inferior surface of the brain and brainstem.

the functions of the insula and little is to be gained from learning the names of its gyri.

Cortical Maps

At about the turn of the century, a number of neuroanatomists proposed systems by which the cerebral cortex could be subdivided into anatomically distinguishable areas. Most of these systems were based on histological distinctions. Brodmann's map (Fig. 1.7) delineates more than 50 regions. Although the cortical areas were originally differentiated by their histological features, many of the areas have subsequently been found to correlate well with neurophysiological and neurological function. Therefore, Brodmann's map has retained a certain popularity, and many areas of the cerebral

cortex are simply referred to by their Brodmann number. In subsequent chapters the Brodmann number of cortical areas will be mentioned when the numbered area is associated with specific neurological functions. For most of the Brodmann areas, however, there is no clear correlation and thus there is no need to memorize the entire map.

THE INTERNAL CAPSULE

The basal ganglia and thalamus, when separated from the cerebral cortex, appear as a large bulbous structure. Dividing it approximately in two is a massive fiber tract, the **internal capsule,** the principal tract connecting the brain with more caudal structures. The internal capsule is broad and fan-shaped as it leaves the

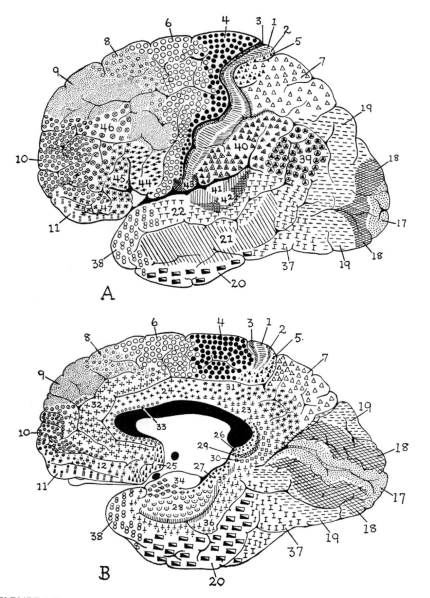

FIGURE 1.7.

Brodmann's cytoarchitectural map of the human cerebral cortex identifies more than 50 areas based on their histological differences. Since some of these areas have distinct neurological functions associated with them, many people refer to them simply by their Brodmann number. These regions are pointed out in the text.

cerebral cortex (Fig. 1.8). Here it is known as the **corona radiata** or "radiating crown." Descending between the thalamus and basal ganglia, the internal capsule loses many fibers to these structures, becomes smaller and more compact, and finally emerges at the rostral end of the mesen-

cephalon as a compact, round structure known as the **cerebral peduncle** [L., from *pes*, foot].

THE BASAL GANGLIA

Medial to the internal capsule lies the head and body of the **caudate nucleus** [L. *cauda*, tail]

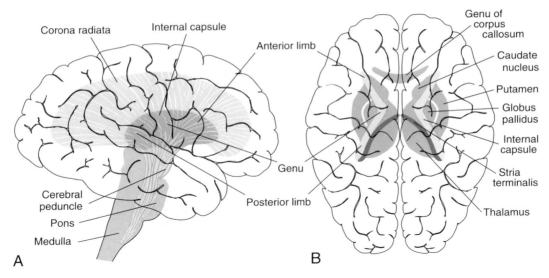

FIGURE 1.8.

A, The corona radiata is the narrow band of axons leaving the cerebral hemisphere. Within the cortical matter, as in this illustration, the structure takes on the appearance of a crown, hence its name. Deeper within the cortex, these same axons form a more compact bundle and are named the internal capsule at the level of the basal ganglion and thalamus. The fibers that continue caudally beyond the thalamus become the cerebral peduncles just before they enter the pons. **B,** Viewed from above, with the superior half of the cerebral hemispheres removed, the internal capsule is shaped like a V. Three divisions are usually identified: the anterior and posterior limbs with the intervening genu. Note how the putamen and globus pallidus are separated from the caudate nucleus by the anterior limb and from the thalamus by the posterior limb.

and the body of the **thalamus** [G. *thalamos,* a bedroom]. Lateral to it is the **putamen** [L. *puto,* to prune] and the **globus pallidus** [L. *globus,* a round body + L. *pallidus,* pale] (Fig. 1.8). The caudate nucleus, the putamen, and the globus pallidus are the major nuclei of the **basal ganglia.** The details of their internal structure are considered in Chapter 8.

The Thalamus

The thalamus is the adult manifestation of the embryonic diencephalon. It is located rostral to the brainstem and lies in close association with the basal ganglia. The thalamus is a paired structure that contains several nuclei. Based on its embryological development, it is divided into three regions: the principal body of the **thalamus,** the **hypothalamus,** and the **epithalamus.** Although there is much anatomical detail asso-

ciated with the thalamus, the discussion here will be limited to the external features that represent orientation landmarks.

When viewed from the dorsal aspect (Fig. 1.9), the principal body of the thalamus can be distinguished from the head of the caudate nucleus by a ridge that courses diagonally across it and medial to the internal capsule. This ridge is a fiber tract, the **stria terminalis** [L. *stria,* furrow + L. *terminalis,* terminal]. Medial and ventral to it lies the thalamus; lateral and dorsal is the caudate nucleus.

The adult structures derived from the epithalamus are the **habenula** [L. *habenula,* strap], the **stria medullaris,** and the **pineal body.** The habenula is located in the midline, at the posterior limit of the thalamus. It is a thin layer of tissue forming a roof over the posterior part of the third ventricle. Leaving the habenula is a ridge of fibers known as the stria medullaris, a tract that courses along the medial wall of the thalamus where it terminates in the septal nuclei, parts of the hypothalamus, and the anterior nuclear group of the

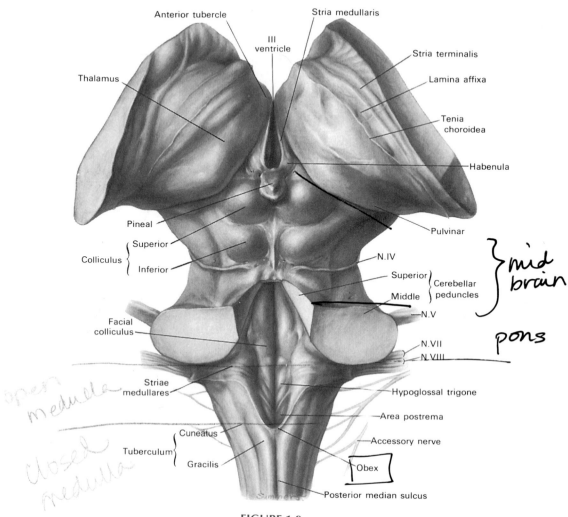

FIGURE 1.9.

A, Dorsal view of the brainstem.

principal thalamus. These nuclear structures are discussed elsewhere (see Chapter 12). In the midline, and merging with the habenula on either side, is the **pineal** body [L. *pineus,* like a pine].

The principal body of the thalamus is divided into a large number of nuclei serving sensory, motor, and cognitive functions. These nuclei are discussed in the various chapters dealing with these systems (see Chapters 5, 8, and 12). Two nuclei, the **medial geniculate** and the **lateral geniculate** nuclei, sometimes called the metathalamus, can be seen in gross dissection as bulges on the lateral surface of the thalamus (Fig. 1.9).

Viewed from the lateral aspect of the thalamus, one can see the **optic tract** (CN II) wrapped around the side of the cerebral peduncle and terminating at the lateral geniculate nucleus (Fig. 1.9). From the ventral aspect (Fig. 1.6), one can follow the left and right **optic tracts** rostrally to the point of their union in the **optic chiasm** [G. *chiasma,* two crossing lines (as in the Greek letter χ)]. From the optic chiasm, the **optic nerves** extend to the eyes. It is customary in neuroanatomy to describe collections of *axons in the central nervous system as tracts, fascicles, peduncles, or stria.* In the *pe-*

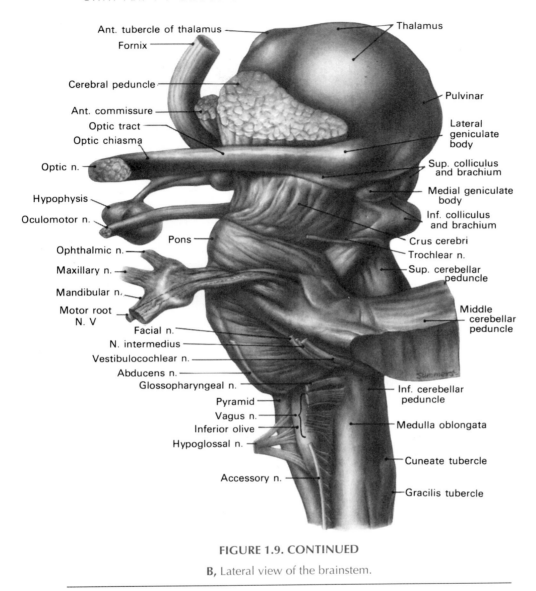

Ant. tubercle of thalamus
Fornix
Cerebral peduncle
Ant. commissure
Optic tract
Optic chiasma
Optic n.
Hypophysis
Oculomotor n.
Pons
Ophthalmic n.
Maxillary n.
Mandibular n.
Motor root N. V
Facial n.
N. intermedius
Vestibulocochlear n.
Abducens n.
Glossopharyngeal n.
Pyramid
Vagus n.
Inferior olive
Hypoglossal n.
Accessory n.

Thalamus
Pulvinar
Lateral geniculate body
Sup. colliculus and brachium
Medial geniculate body
Inf. colliculus and brachium
Crus cerebri
Trochlear n.
Sup. cerebellar peduncle
Middle cerebellar peduncle
Inf. cerebellar peduncle
Medulla oblongata
Cuneate tubercle
Gracilis tubercle

FIGURE 1.9. CONTINUED

B, Lateral view of the brainstem.

ripheral nervous system, collections of axons are called nerves. However, in the case of the axons coursing between the eye and the brain, a confusing exception to this rule exists. By ancient convention, the axons between the eye and the optic chiasm are called the optic nerve, and from the optic chiasm to the lateral geniculate body the same axons are called the optic tract. These axons in both locations are truly part of the central nervous system, but the convention is so convenient that it is thoroughly entrenched in the nomenclature. The details of the visual pathways are given in Chapter 11.

The most inferior portion of the thalamus is the hypothalamus, a diencephalic derivative that is separated from the principal body of the thalamus by the **hypothalamic sulcus** (Fig. 1.10). Its small size belies its importance ; it is discussed in detail in Chapter 12. The **infundibulum** [L. *infundibulum,* funnel] is a protuberance that extends out of the ventral surface of the hypothalamus. The infundibulum is a piece of tissue that

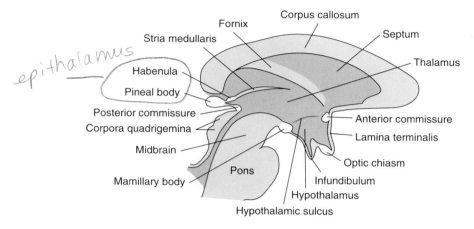

epithalamus

FIGURE 1.10.

Medial view of the brainstem sectioned in the midsagittal plane. This section reveals the medial wall of the third ventricle.

---

connects the hypothalamus with the **pituitary gland** [L. *pituita,* a thick nasal secretion] (Fig. 1.9), an endocrine organ of some significance. Caudal to these structures one can identify the paired structures known as the **mamillary bodies** (Figs. 1.6 and 1.10). The mamillary bodies mark the most caudal extent of the hypothalamus, while the optic chiasm marks its rostral border.

The Brainstem

The embryonic mesencephalon, metencephalon (without the cerebellum), and myelencephalon together make up the **brainstem,** a long cylindrical structure that lies at the base of the brain. It is a phylogenetically older part of the nervous system serving diverse functions. Many of these are very primitive and independent of conscious control, such as breathing, vomiting, and regulating temperature and blood pressure. The brainstem is also the entry point for all the cranial nerves except the olfactory, optic, and accessory nerves. Sensory information from the spinal cord and cranial nerves, and motor commands from the cerebral cortex and cerebellum must all pass through the brainstem. Much of this information is modified by brainstem structures. Therefore, this region is of great clinical significance.

Only the ventral surface of the brainstem can be seen in the intact brain because the cerebral hemispheres and cerebellum lie over the dorsal surface. Therefore, when studying the external features of the brainstem, it is convenient to remove these structures, as has been done in the illustrations that appear in the following section.

THE MIDBRAIN

The midbrain or mesencephalon is most easily appreciated from its dorsal aspect, where four bumps, the **corpora quadrigemina** [L. *corpus,* body + L. *quadri,* four + L. geminus, twin], are readily identified (Fig. 1.9). The rostral pair are known as the **superior colliculi** [L. *collis,* hill] and the caudal pair, the **inferior colliculi.** Immediately caudal to the inferior colliculi, and very close to the midline, one can identify the **trochlear nerves** [G. *trochileia,* a pulley] (CN IV) leaving the brainstem on its dorsal surface and wrapping around the lateral aspect of the cerebral peduncles. From the ventral surface of the brainstem (Figs. 1.6 and 1.11), the most prominent features of the mesencephalon are the cerebral peduncles. There is a deep space between them known as the **interpeduncular fossa.** Emanating from the interpeduncular fossa are the **oculomotor nerves** (CN III).

THE PONS

The pons forms a bridge between the medulla and the midbrain. It is dominated by the hemi-

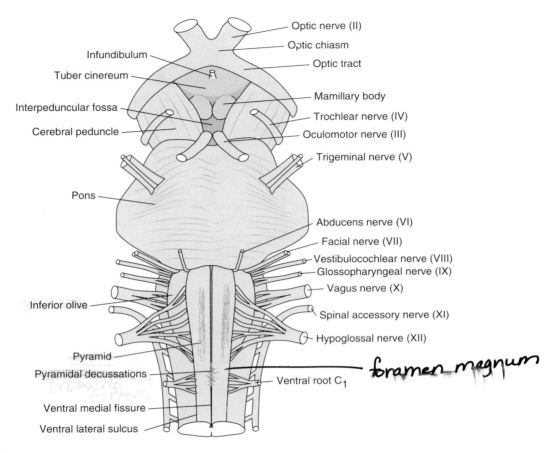

FIGURE 1.11.

Ventral surface of the brainstem. All of the cranial nerves are identified, as well as some other prominent landmarks.

spheres of the cerebellum, which will be described separately. The most conspicuous external feature of the pons other than the cerebellum is the great bulbous body that forms its ventral surface. Known as the **basis of the pons,** this structure is composed mostly of a great mass of axons entering the cerebellum (Fig. 1.11). At the lateral margins of the pons these fibers are known as the **middle cerebellar peduncle.**

Viewing the ventral surface of the pons, one can identify several pairs of cranial nerves. The largest and most conspicuous is the **trigeminal nerve** (CN V). This nerve leaves the middle cerebellar peduncle at the lateral edge near, but not at, the rostral border. The remaining cranial nerves associated with the pons are all located at the **pontomedullary** junction, the ridge in the basis of the pons that marks its most caudal extent. The most medial pair of nerves is the **abducens nerve** (CN VI). More lateral are the **facial** and the **vestibulocochlear nerves.** Of these two, the facial nerve leaves the brainstem from the ventral aspect and is the most medial, while the vestibulocochlear nerve leaves from the dorsolateral surface and is more lateral.

THE MEDULLA

The **medulla** is the most caudal division of the brainstem. On its ventral surface (Fig. 1.11) one can identify in the midline a deep furrow, the **ventral median fissure,** running parallel to the long axis of the medulla and another, the **ventral lateral sulcus,** parallel to it. The **medullary pyra-**

mid is a conspicuous fiber tract that resides between these two landmarks. The ventral median fissure terminates rostrally at the basis of the pons. At approximately the junction between the medulla and the spinal cord, it is less distinct, its cavity being partly filled with fibers. This is the area of the **pyramidal decussation,** where the two pyramidal tracts cross the midline. Although it is more clearly seen in histologic cross sections, it is an important gross landmark too, marking the transition between the spinal cord and the brainstem.

Lateral to the pyramidal tract is a bump on the side of the medulla, the **inferior olive** (Fig. 1.9). The **hypoglossal nerve** (CN XII) leaves the medulla between the pyramidal tract and the inferior olive. The **vagus** and the **glossopharyngeal nerves** (CN X and CN IX) can be found leaving the medulla on the dorsal margin of the inferior olive.

On the dorsal aspect of the medulla, one can discern two pairs of ridges running parallel to the long axis of the medulla separated by the **dorsal intermediate sulcus** (Fig. 1.9). Under these ridges lie important sensory fiber tracts known as the **fasciculus gracilis** (medial) and the **fasciculus cuneatus** (lateral). The gracile fasciculi are separated at the midline by the **dorsal median sulcus.** At the rostral extension of the fasciculus gracilis, the **area postrema** forms a slight ridge that looks like a **V** across the surface of the medulla. The midline origin of this eminence is known at the **obex** [L. *obex,* barrier]. The obex is a rough point of demarcation between the spinal cord and the medulla, being slightly more rostral than the pyramidal decussation.

The Cerebellum

Located on its dorsal surface, the cerebellum is the most conspicuous feature of the metencephalon (Fig. 1.1). To remove the cerebellum from the main body of the pons, one must sever three pairs of fiber tracts known as the **cerebellar peduncles** (Fig. 1.12). The cerebellar peduncles are named according to their anatomic location: **superior, middle,** and **inferior.** The superior peduncles connect the cerebellum with the mesen-

cephalon, the middle peduncles attach the cerebellum to the pons, and the inferior peduncles communicate with the medulla and spinal cord. The cerebellar peduncles are best appreciated in histologic sections. All information entering and leaving the cerebellum must pass through these fiber tracts.

The body of the cerebellum is divided into two **hemispheres** and a midline structure called the **vermis** (Fig. 1.12). The surface is elaborated into a series of deep ridges or folds known as **folia.** As with the cerebral cortex, this anatomical expediency greatly increases the surface area of the organ without increasing its volume. The deepest of the folia are known as fissures, and many authors divide the cerebellum into lobules according to these fissures. The appropriateness of these divisions has been the subject of lively debates among anatomists, physiologists, and physicians. Since the clinically relevant functional organization of the cerebellum does not follow the physical divisions based on the fissures, they will not be discussed or elaborated here. Memorizing their names is unnecessary. The **tonsil,** however, is a lobule that all physicians should know because it may be squeezed into the foramen magnum under conditions that cause the intracranial contents to shift. This situation is discussed further below.

Of greater interest, however, is the phylogenetically oldest part of the cerebellum, called the **flocculonodular lobe** (Fig. 1.12). The **flocculus** is a small lobe of the cerebellum, located immediately adjacent to the inferior cerebellar peduncle. The **nodulus** forms the midline structure between the pair of floccular lobes and is the most caudal segment of the vermis. During embryogenesis, the separation of the flocculonodular lobe is the first division of the cerebellum, and this fact attests to its ancient origin. The lobe is noteworthy because injury to it produces unique clinical symptoms. These symptoms are discussed in Chapter 8.

The Spinal Cord

The characteristic physical feature of the spinal cord is that in its adult form, it retains the segmental organization with which it was endowed during embryonic development. There-

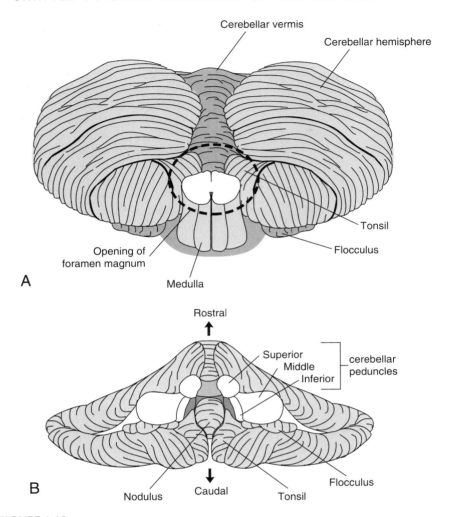

FIGURE 1.12.

A, Inferior view of the cerebellum seen as if one were looking up through the foramen magnum. Note the position of the tonsil, a structure that can herniate into the foramen magnum. Under such conditions, the brainstem is compressed, effectively separating it from the spinal cord. **B,** A view of the ventral surface of the cerebellum illustrating the relationship of the cerebellar peduncles with the cerebellar hemispheres. Note the lateral portion of the flocculus.

fore, the spinal cord is composed of 31 segments that, although not identical, are essentially similar in structure (Fig. 1.13).

At the spinal cord, the **dorsal median septum** and the **ventral median fissure** are continuous with those structures in the medulla. Lateral to the dorsal median septum is the **dorsal root entry zone,** a continuous line along the spinal cord from which small fascicles or bun-

dles of axons exit (Fig. 1.14). A number of these fascicles, collected together, form a **dorsal root.** The cells that give rise to the dorsal root axons are **sensory, pseudounipolar** (see Chapter 2) neurons. With only one exception,[1] the cell bodies of sensory pseudounipolar neurons lie in dense collections of cells called peripheral gan-

---

[1]The mesencephalic root of CN V; see Chapter 9.

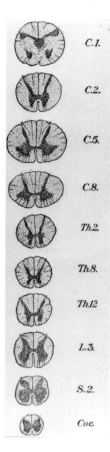

C.1.

C.2.

C.5.

C.8.

Th.2.

Th.8.

Th.12

L.3.

S.2.

Coc.

FIGURE 1.13.

Cross sections of the spinal cord all drawn to the same scale, illustrating the relative size of the spinal cord at various levels as well as the differences in the shape of the gray matter.

glia. Each dorsal root has associated with it such a ganglion, known as the **dorsal root ganglion.**

In a similar fashion, the **ventral root entry zone** lies lateral to the ventral median fissure, and the fascicles that emanate from it collect together to form the **ventral root.** The cells from which the axons of the ventral root originate are **multipolar motor neurons** (see Chapter 2). These cell bodies lie within the spinal cord; there is no ganglion associated with the ventral roots.

The ventral root and the dorsal root of each segment anastomose to form a single **spinal nerve.** Each pair of spinal nerves leaves the protection of the spinal column through an **intervertebral foramen.** Since each pair of spinal nerves is associated with a vertebra (Fig. 1.15), it is convenient to name the nerve according to that vertebra. An exception is made in the cervical region because there is an extra pair of spinal nerves, the first pair being anomalous, having only ventral roots. Therefore, by convention, the first eight pairs of spinal nerves are called **cervical;** the next twelve, **thoracic;** the next ten are divided equally between **lumbar** and **sacral;** and finally the last pair are named **coccygeal.** Consequently, in the cervical region, the first seven pairs of spinal nerves exit rostral to the vertebra for which they are named. All others exit caudal to the vertebra for which they are named, the C8 pair being an exception. At the cervical and lumbar enlargements, spinal nerves anastomose with adjacent spinal nerves, divide and re-anastomose, forming a complicated **plexus** [L. *plexus,* a braid] of nerves. The **peripheral nerves** emerge after the last anastomosis (Fig. 1.16).

One should appreciate that the *dorsal root is a pure sensory nerve* and the *ventral root is a pure motor nerve.*[2] When the dorsal and ventral roots merge to form the spinal nerve, it and all of its derivatives contain both afferent (sensory) and efferent (motor) axons. Thus any trauma that affects a spinal nerve, the brachial or lumbar plexuses, or a peripheral nerve produce complementary sensory and motor findings. On the other hand, pure lesions to the dorsal or ventral roots produce pure sensory or motor findings, respectively. This is a very important diagnostic principle.

The spinal cord itself is shorter than the vertebral column that protects it because the spinal cord stops its embryologic extension sooner than does the vertebral column. As the vertebral column continues to develop, it carries the spinal nerves with it, causing the lumbar and sacral roots to be greatly elongated within the central vertebral foramen (Fig. 1.15). Since it resembles a horse's tail, this extension of the lumbar and sacral spinal nerves is frequently called the **cauda equina.** In the adult, the spinal cord extends only to about the level of the L2 vertebra. The **filum terminalis,** a fine strand of connective tissue, extends from the last segment of the spinal cord to the coccyx, where it fuses with the periosteum.

---

[2]Ventral root afferents have been described, but their clinical significance has yet to be established.

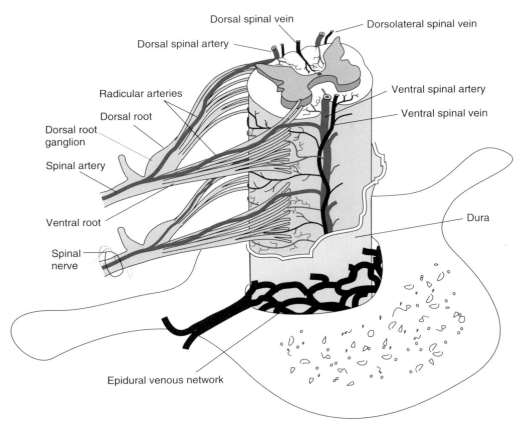

FIGURE 1.14.

Oblique view of the spinal cord with the dura removed. Note that the dorsal and ventral roots enter the spinal cord as a continuous series of rootlets. The radicular arteries supply each segment of the spinal cord, but the arteries are not equally robust at every segment. The dorsal spinal arteries and the ventral spinal artery provide intersegmental anastomoses. Venous drainage follows the arteries. In addition, there is a large epidural venous network on the ventral side of the spinal cord.

The spinal cord is also characterized by two swellings, or enlargements. The **cervical enlargement** and the **lumbar enlargement** correspond to the areas of the spinal cord that innervate the extremities (Fig. 1.15). Not surprisingly, more neuronal support is required to manage the sensorimotor requirements of the extremities than is required for the trunk, and this requirement is reflected in an increased size of the spinal cord serving the extremities. Also, the rostral portion of the spinal cord is larger than the caudal region because all the axons serving the entire spinal cord must pass through the cervical regions, whereas only those needed for the

lumbosacral regions are present at more caudal levels. For these reasons, the spinal cord is very closely confined in the central foramen of the cervical vertebrae.

The Peripheral Nervous System

The peripheral nervous system is limited to the parts of the nervous system that are outside the dura mater, or, stated somewhat differently, the parts of the nervous system that are ensheathed by Schwann cells (see Chapter 2). It is

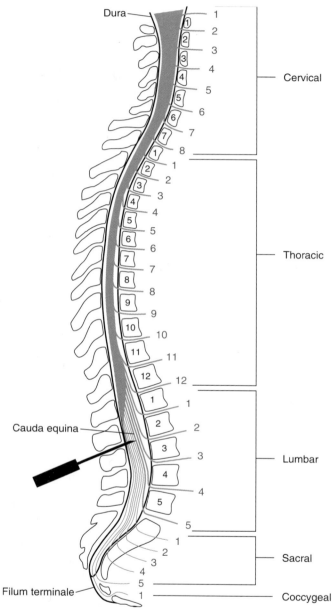

FIGURE 1.15.

The relationship between the spinal cord and the vertebral column. Note that in the region of the cervical enlargement, the spinal cord completely fills the vertebral foramen. Since only the cauda equina exists below approximately the L2 vertebra, a needle can be safely inserted into the lumbar cistern below the L2 vertebra as shown.

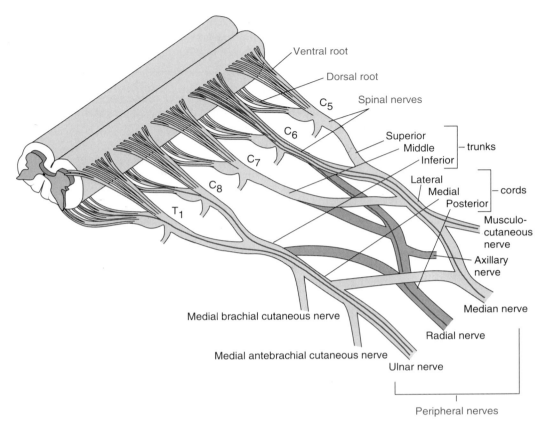

FIGURE 1.16.

Spinal roots are either sensory (dorsal roots) or motor (ventral roots). Spinal nerves are mixed, serving both motor and sensory functions. The axons from individual spinal nerves separate from and join axons from other spinal nerves as they pass through the plexuses of the brachial and lumbosacral regions. Therefore, peripheral nerves not only are mixed, but can contain axons from several spinal roots. Illustrated here, spinal segment C6 supplies axons to the median, musculocutaneous, and radial nerves. Segments C8 and T1 both supply axons to the ulnar nerve. For simplicity, examples of mixed motor and sensory innervation are not shown.

composed of three parts: (*a*) the somatosensory or **afferent** neurons, (*b*) the motor or **efferent** neurons, and (*c*) the **autonomic** neurons. The autonomic neurons are discussed in Chapter 12. The peripheral nervous system is distinguished from the central nervous system by several other, less precise but useful criteria. For example, unlike the central nervous system, it is not surrounded by bone and is therefore quite susceptible to injury. When injured, however, it is more capable than the CNS of repairing itself. In addition, the peripheral nervous system is subject to

certain diseases that cannot affect the central nervous system, and vice versa.

FASCICULAR ORGANIZATION

In the peripheral nerve, individual axons are enveloped in a loose connective tissue, the **endoneurium** (Fig. 1.17). Small groups of axons are closely associated within a bundle called a **nerve fascicle.** The fascicle is defined by a sheath of connective tissue known as the **perineurium.** The perineurium gives mechanical strength to the peripheral nerve and also provides the surgeon a material that can hold sutures without tearing. In addition to its mechanical strength, the perineurium serves as a diffusion barrier, isolating the endoneurial space around the axons from the surrounding tissue. This barrier helps to preserve the ionic milieu of the axon. The fascicles are collected together to form the peripheral nerve that is embedded in loose connective tissue called the **epineurium.**

It is worth noting that the fascicles are not continuous throughout the course of the peripheral nerve. They divide and anastomose with one another as frequently as every few millimeters (Fig. 1.18). However, axons within a small set of adjacent bundles redistribute themselves within the same set of bundles such that the axons remain in approximately the same quadrant of the nerve for several centimeters. This is of practical concern to the microsurgeon faced with suturing a severed nerve. If the cut is clean, one may be able to suture individual fascicles together. Anastomosed in this way, there is a high probability that the distal segment of nerves synapsing with muscles will be sutured to the central stump of motor axons, and similarly for sensory axons. In such cases, good functional recovery is possible. If a short segment of the nerve is missing, the fascicles in the two stumps may no longer correspond with one another. However, if the surgeon can maintain the original axial orientation, good functional recovery may still be possible since the axons maintain themselves within the same nerve quadrant over long distances. However, if the nerve is badly mangled, good axial alignment may not be possible. In such cases functional recovery is greatly compromised.

BLOOD SUPPLY

The peripheral nerve is supplied with a rich vasculature (Fig. 1.17). Vessels extrinsic to the nerve run parallel to it and extend branches that penetrate the epineurium every few millimeters. These epineural vessels branch and anastomose forming a rich plexus of vessels that in turn send branches into the perineurium of the fascicles. The perineurial plexus in turn sends penetrating branches into the endoneurium. Within the endoneurial space, an extensively anastomosed microvascular network exists that consists of arterioles, capillaries, and venules. This highly collateralized network affords a wide safety margin against vascular insult.

The External Coverings of the CNS

The tissue of the nervous system is soft and fragile. Although it can hold its shape outside the body and does not need skeletal support, it does require skeletal protection from external insults. Therefore, the brain and spinal cord are contained within the rigid framework of the skull and spinal column. The latter is modified to allow some flexibility, but on the whole, the central nervous system is encased in a formidable suit of armor. If for any reason the cranial or spinal contents increase in volume, as for example due to the growth of a tumor, this protective covering becomes a liability. Since the skull cannot expand, the intracranial pressure rapidly increases, causing specific and important neurological signs (see below).

Within the bony covering, the central nervous system is further enclosed in a series of three membranes (Fig. 1.19). The most external of these is the **dura mater** [L. *dura mater,* hard mother] a tough and resilient sheet of connective tissue that encloses the brain and spinal cord. Though it is a continuous structure, it is important to note that the cranial portion of the dura mater differs from the spinal portion. The dura mater encloses the entire central nervous system but is perforated in many places to allow for the entrance of cranial nerves and blood vessels. At each of these perforations, however, the dura adheres to the penetrating object, forming a tight seal. The **pia mater** [L. *pia mater,* tender mother] is the most internal layer that closely adheres to the surface of the CNS. The **arachnoid** [G. *arachne,* spider] lies between them.

THE CEREBRAL DURA MATER

The cerebral dura mater consists of two layers, an inner **meningeal** and an outer **endosteal** layer. The endosteal layer adheres closely to the inner surface of the skull and serves as its inner periosteum (Fig. 1.20). It is attached to the skull by small strands of fibrous connective tissue.

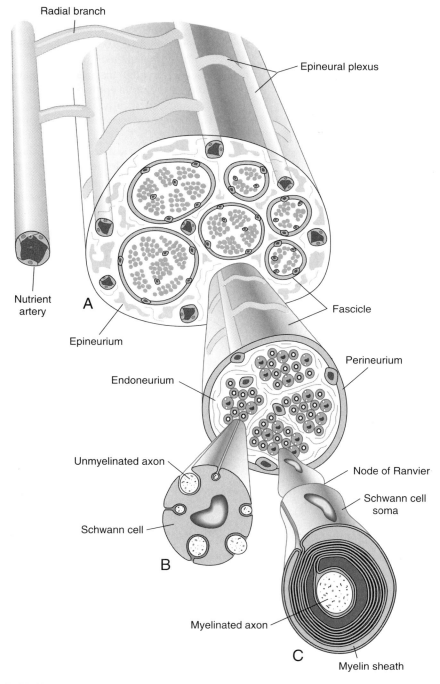

Radial branch

Epineural plexus

Nutrient
artery

A

Fascicle

Epineurium

Perineurium

Endoneurium

Unmyelinated axon

Node of Ranvier

Schwann cell
soma

Schwann cell

B

Myelinated axon

C

Myelin sheath

FIGURE 1.17.

   A, Peripheral nerves are divided into fascicles by the perineurium, a tough connective tissue that isolates the fascicle both physically and chemically. Nutrient blood vessels supply peripheral nerves via a large number of radial branches along their length. These radial branches supply an extensively collateralized network of vessels within the nerve that further collateralize and penetrate the individual fascicles. **B,** Several unmyelinated axons within a fascicle are enveloped in a single Schwann cell. **C,** One Schwann cell forms the myelin for a single myelinated axon. Myelin structure is discussed in Chapter 2.

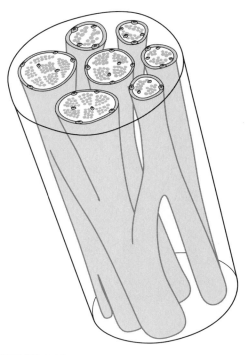

FIGURE 1.18.

Individual fascicles divide and re-anastomose along the length of a peripheral nerve. Over a short distance, the fascicles remain in approximately the same quadrant of the nerve. Over larger distances, considerable reorganization may occur.

Many small blood vessels also pass from the dura mater into the bones of the skull. If the dura is violently separated from the skull, these blood vessels bleed into the space created by the separation, forming an **epidural hematoma.** This occurs most frequently after skull fractures. If the fracture should *lacerate one of the meningeal arteries* that lie between the dura and the bone, death can occur within minutes. The more common **subarachnoid hematoma** is formed when bleeding occurs between the brain and the dura. This may follow a severe blow to the head that displaces the brain within the cranium, but does not fracture the skull. The displacement may *rupture one or more of the veins* lying on the surface of the brain, resulting in bleeding. The bleed may continue for days to months and lead to neurological symptoms so far removed in time from the initial injury that diagnosis may be difficult.

Throughout most of the dura, the endosteal and meningeal layers are tightly attached to one another and form one continuous membrane. At certain places, the layers separate to form a cavity or sinus (Fig. 1.20). The walls of the V-shaped sinus are formed from the meningeal layers, while the roof is composed of the endosteal layer. These sinuses form part of the venous drainage system of the brain (to be described later).

At the inferior margin of a sinus, the two

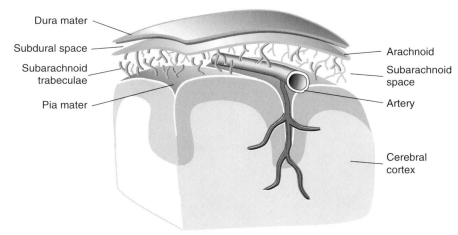

Dura mater

Subdural space

Subarachnoid trabeculae

Pia mater

Arachnoid

Subarachnoid space

Artery

Cerebral cortex

FIGURE 1.19.

Relationships between the three meningeal layers. Note that the pia mater adheres closely to the brain surface and even follows the blood vessels deep into the tissue of the brain. The arachnoid and the dura mater, normally closely adherent to one another, are shown separated here for clarity.

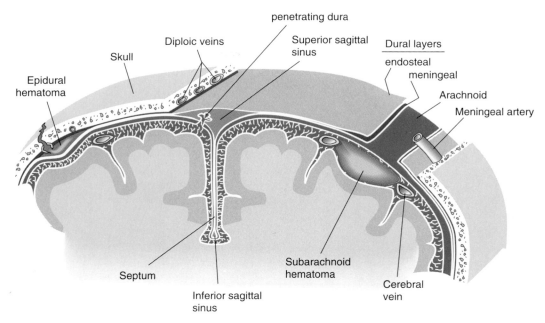

FIGURE 1.20.

Relationships between the dura mater and the vasculature of the brain. Note that the meningeal and endosteal layers separate to form the sinuses. Two meningeal layers come together to form the septa. Arachnoid villi are sites where CSF is returned to the vascular space. Hematomas can develop between the arachnoid and the pia mater (subarachnoid hematoma) or between the skull and the dura mater (epidural hematoma). The latter rapidly expand because they are the result of arterial bleeding. The former, caused by venous bleeding, expand slowly over a period of days to months.

meningeal layers of the dura fuse to form a two-layered dural fold or **septum** that separates the cranium into major compartments (Figs. 1.20 and 1.21). The most conspicuous of these septa is the **falx cerebri** [L. *falx,* sickle], which lies between the two cerebral hemispheres in the midsagittal plane. At the anterior margin, it is attached to the ethmoid bone. It arches posteriorly until it meets another septum, the **tentorium cerebelli** [L. *tentorium,* a tent], at the occipital pole of the skull. The inferior border of the falx remains unattached, leaving a space between it and the floor of the skull. The diencephalon and corpus callosum lie in this space.

The tentorium separates the cerebellum from the cerebral hemispheres. It is attached to the skull at the margin of the petrous bone and along the inferior part of the occipital bone. The medial or central edge of this septum is unattached and joins the free edge of the falx. This border of the tentorium circumscribes a space through

which the brainstem passes, the **tentorial incisure.** The falx and the tentorium together separate the interior of the cranium into three compartments. The space inferior to the tentorium is called the **posterior fossa.** Frequently one will also refer to the supratentorial or the infratentorial space.

THE SPINAL DURA MATER

At the level of the spinal cord, the meningeal layer of the dura separates from the endosteal layer. The latter blends into the periosteum at the foramen magnum. The meningeal layer does not adhere to the vertebra, creating an **epidural space** between it and the bone (Fig. 1.14).

The spinal dura, like the cerebral dura, envelopes the nerve roots as they leave the spinal cord, forming a tight seal at the intervertebral foramina. At the caudal end of the spinal cord, the dura is closely adherent to the filum terminale.

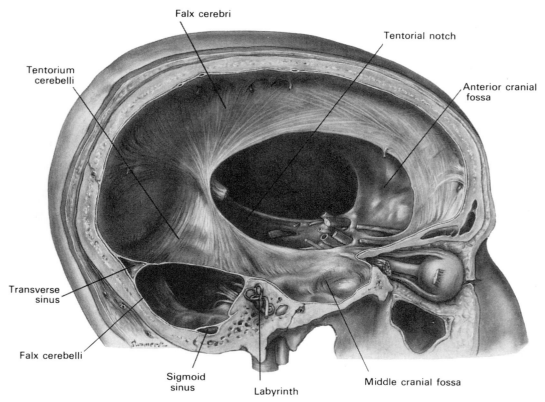

Falx cerebri

Tentorial notch

Tentorium
cerebelli

Anterior cranial
fossa

Transverse
sinus

Falx cerebelli

Sigmoid
sinus

Labyrinth

Middle cranial fossa

FIGURE 1.21.

  Sagittal view of the head with part of the skull and all of the brain removed, showing
the position and relationships of the various septa. The tentorium cerebelli isolates the
posterior fossa. The tentorial incisure is its medial margin and marks the border through
which the brainstem must pass.

Together they attach to the coccyx and merge
with the periosteum.

THE PIA MATER

  The surfaces of the central nervous system are
lined with a connective tissue membrane, the pia
mater. In contrast to the dura and arachnoid, the
pia closely follows all the surface undulations of
the nervous tissue (Fig. 1.19). In the spinal cord,
special extensions of the pia form triangular connections between the lateral wall of the spinal
cord and the adjacent dura. These **dentate ligaments** anchor the spinal cord to the dura.

THE ARACHNOID

  The inner surface of the dura is lined with the
arachnoid. Between the arachnoid membrane

and the pia mater is a delicate lacework of connective tissue that resembles a spider's web, the
arachnoid trabeculae (Fig. 1.19). This space,
the **subarachnoid space,** is filled with **cerebrospinal fluid (CSF).** Although narrow over
most of the cerebrum, this space is wider over
the spinal cord. In some places the subarachnoid
space is so large that these areas are called **cisterns** (Fig. 1.22). The cisterns are in communication with one another, allowing the free flow
of cerebrospinal fluid over the entire external
surface of the brain, brainstem, and spinal cord.

  The **lumbar cistern** is noteworthy because it
provides a convenient location for sampling
cerebrospinal fluid. Other than CSF, this cistern
contains only the cauda equina, and it is easily
reached by inserting a needle between the lumbar vertebrae (Fig. 1.15). Once the needle is in

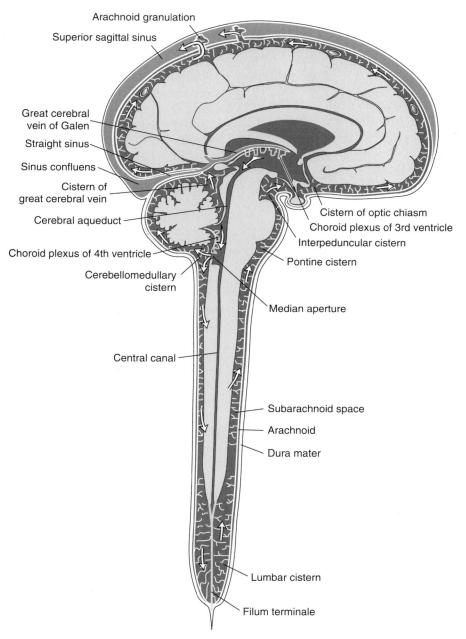

FIGURE 1.22.

Sagittal view of the central nervous system showing the location of the various cisterns. *Arrows* indicate the direction of CSF flow from the areas of secretion in the choroid plexuses to the areas of reabsorption through the arachnoid granulations. CSF passes from the ventricular system into the cisterns at the two lateral apertures (not shown) and the median aperture in the fourth ventricle. Within the ventricular compartment, the two intraventricular foramina and the cerebral aqueduct are narrow passages that can be easily occluded, producing hydrocephalus.

place, CSF can be sampled and its pressure measured. Contrast material can be injected to replace the withdrawn CSF, a process that makes the outlines of the dural sac visible on x-ray, creating a **myelogram.** This procedure is now frequently replaced by noninvasive magnetic resonance imaging (MRI). However, the myelogram is still very useful in emergency situations and for patients who cannot undergo MRI.

The Neurovasculature

It is essential that the physician have a good understanding of the vasculature of the nervous system, because many neurological difficulties are based on problems of blood supply. A good knowledge of the brain's vasculature will help one to evaluate various differential diagnoses and to localize lesions. Furthermore, many neurovascular problems are treatable. Appropriate diagnostic evaluation is essential in such cases if optimum results are to be achieved.

THE CAROTID CIRCULATION

The brain is supplied by only two pairs of arteries, the **carotid arteries** and the **vertebral arteries.** The carotid artery enters the base of the skull and divides into the **middle cerebral artery** and the **anterior cerebral artery** (Fig. 1.23). The middle cerebral artery ascends to the lateral surface of the hemisphere through the lateral sulcus. From the lateral sulcus branches of the middle cerebral artery spread across the lateral aspect of the frontal, parietal, and temporal lobes. These branches do not extend to the margins of these lobes, and a definite "watershed" circumscribes an arc around the lateral surface of the hemisphere (Fig. 1.24).

The anterior cerebral artery follows a course on the medial surface of the hemisphere, arcing around the corpus callosum (Fig. 1.24). Branches of the anterior cerebral artery ascend the medial surface of the hemisphere. They extend over the medial ridge, onto the most superior aspect of the lateral surface of the frontal and parietal lobes and descend to meet, but not to anastomose, with the ascending branches of the medial cerebral artery arising from below.

THE VERTEBROBASILAR CIRCULATION

The two **vertebral arteries** enter the skull separately through the foramen magnum and course along the ventrolateral aspect of the medulla (Fig. 1.23). They merge, forming the **basilar artery** at approximately the pontomedullary junction. The basilar artery follows the midline of the pons to its rostral border, where it divides into the two **posterior cerebral arteries.** They send branches to the inferior surface of the temporal lobe and to the occipital lobe.

The basilar artery gives rise to many important branches that perfuse the basis of the pons via the **paramedian** and **circumferential** branches. A separate branch, the **labyrinthine artery,** supplies the inner ear. In addition, the **superior cerebellar artery** and the **anterior inferior cerebellar artery (AICA)** are branches of the basilar artery that supply most of the superior region of the cerebellum. The **posterior inferior cerebellar artery (PICA)** is derived, in most individuals, from the vertebral artery and perfuses the remaining part of the cerebellum and the lateral wall of the caudal medulla (Fig. 1.23). The superior cerebellar artery and the anterior inferior cerebellar artery also supply portions of the pons.

The thalamus and basal ganglia are supplied by branches of several major cranial arteries (Fig. 1.25). The thalamus is supplied principally by **thalamic** branches of the posterior cerebral artery. The adjacent caudate nucleus and globus pallidus are supplied by the **striate branches** from the middle cerebral artery. The choroid plexus of the third ventricle is supplied by the **choroidal artery,** derived from the posterior cerebral artery. These branches are small relative to the main vessel, which they leave at right angles. They commonly rupture at the junction, producing an intracerebral hemorrhage.

The rostral spinal cord derives part of its circulation from the vertebral arteries, which give caudally directed branches that soon fuse with one another to form the **ventral spinal artery** (Fig. 1.23). Since the fusion takes place near the pontomedullary junction, the ventral spinal artery

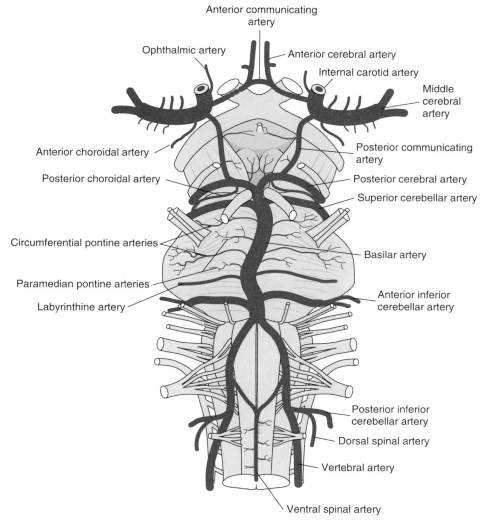

FIGURE 1.23.

The blood supply to the brainstem and cerebellum is derived from the vertebral arteries and the basilar artery. These arteries give rise to a number of penetrating branches, the most important of which are illustrated here.

courses nearly the full length of the medulla and the entire length of the spinal cord. The most medial structures of the medulla are supplied by the anterior spinal artery, while the more lateral paramedian structures are supplied by branches of the vertebral arteries. The posterior inferior cerebellar artery and the dorsal spinal artery supply the dorsal and lateral portions of the medulla.

The **dorsal spinal arteries** are also branches of the vertebral arteries, arising near the lateral border of the medulla. Each wraps around the medulla to reach the dorsal surface, and then travels caudally, in parallel with its mate from the opposite side, along the dorsal surface of the medulla and spinal cord. The dorsal spinal arteries are most conspicuous at medullary and high cervical levels, merging with one another at low cervical levels. After fusion, the dorsal spinal

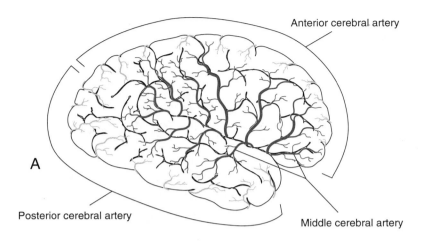

Anterior cerebral artery

Posterior cerebral artery

Middle cerebral artery

A

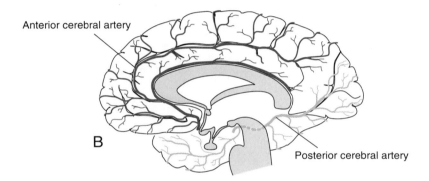

Anterior cerebral artery

Posterior cerebral artery

B

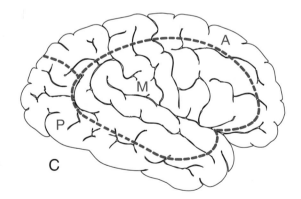

C

FIGURE 1.24.

  A, The principal blood supply to the lateral portion of the cerebral hemisphere is derived from the middle cerebral artery. The superior, posterior, and inferior boundaries
are supplied by the anterior and posterior cerebral arteries. These systems do not anastomose, and a definite "watershed" exists at their margins (**C**). **B,** The medial portion of
the cerebral hemisphere is supplied by the anterior and posterior cerebral arteries,
whose distribution is illustrated here. Note that the posterior cerebral artery supplies the
medial and inferior surface of the temporal lobe. **C,** The "watershed" boundaries superimposed on a lateral view of the hemisphere.

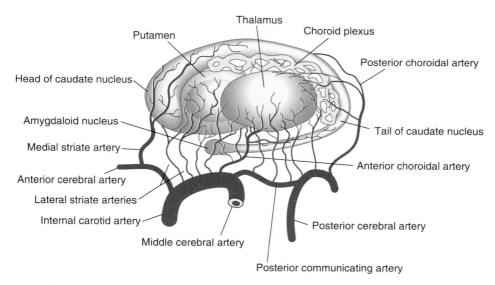

Thalamus

Putamen

Choroid plexus

Head of caudate nucleus

Posterior choroidal artery

Amygdaloid nucleus

Posterior choroidal artery

Medial striate artery

Tail of caudate nucleus

Anterior cerebral artery

Anterior choroidal artery

Lateral striate arteries

Internal carotid artery

Posterior cerebral artery

Middle cerebral artery

Posterior communicating artery

FIGURE 1.25.

The blood supply to the choroid plexus, thalamus, and basal ganglia is derived from short, penetrating branches of the posterior and middle cerebral arteries and posterior communicating artery as depicted here.

artery usually does not descend the entire length of the spinal cord, but peters out at the midthoracic level (Fig. 1.23).

THE CIRCLE OF WILLIS

In most individuals, the posterior circulation derived from the vertebrobasilar system and the anterior circulation derived from the carotid system anastomose through the pair of **posterior communicating arteries.** The two anterior cerebral arteries are linked by the single **anterior communicating artery.** This creates a circular pathway of arteries known as the **circle of Willis** (Fig. 1.23).

There is evidence that little blood flows through the communicating arteries, as one would expect, if normal blood pressure exists in all of the vessels supplying the circle of Willis. The communicating arteries are, in fact, normally small and cannot immediately deliver the amount of blood required if one of the major arteries were to be suddenly occluded. However, in cases of slowly developing stenosis of one of the four major nutrient arteries, such as the development of an atherosclerotic plaque at the carotid bifurcation, the communicating arteries

can gradually expand and establish a sufficient collateral circulation to maintain normal brain function.

THE SPINAL CORD CIRCULATION

Besides blood supplied to the cervical spinal cord by branches of the vertebral arteries, there is a segmental blood supply to the spinal cord. In theory, every spinal nerve carries with it a spinal artery that divides into a **dorsal** and a **ventral radicular artery** that follows the dorsal and ventral roots, respectively (Fig. 1.14). The radicular system of arteries is the major source of blood to the spinal cord. These radicular arteries anastomose with the dorsal or ventral spinal arteries, as appropriate, forming a rich collateral network. However, as a practical matter, there is a great deal of variation in both the number of radicular arteries in the adult and in their distribution. In particular, only about half the dorsal radicular arteries fully develop and anastomose with the dorsal spinal artery, and even fewer of the ventral radicular arteries join the ventral spinal artery.

Given the capriciousness of the development of this arterial network, it is not surprising that certain areas of the spinal cord are particularly vulnerable to the loss of a single spinal artery. In general this vulnerability is greatest at the thoracic levels because the intercostal arteries that give rise to the spinal arteries are themselves poorly anastomosed. Infarction of the thoracic spinal cord is not uncommonly associated with dissecting aortic **aneurysms**[3] [G. *aneurysma,* wide] or surgery that occludes a single intercostal artery.

VENOUS DRAINAGE OF THE SPINAL CORD

In the spinal cord, the pattern of veins closely follows the pattern of arteries, there being **ventral** and **dorsal spinal veins,** and **radicular** veins. In addition to these veins found in the subarachnoid space, there is an extensive **epidural venous network** that receives blood from the radicular veins (Fig. 1.14). This rete extends the entire length of the spinal cord and continues to the clivus under

---

[3]An aneurysm is any widening, or ballooning, of an artery. In some cases, particularly with the larger arteries, the intima may separate from the arterial wall, a condition described as a dissecting aneurysm.

the pons, where it joins the cavernous sinus (see below). In the spinal cord it drains into the segmental veins associated with the vertebra, and from these, finally into the vena cava.

THE CEREBRAL VENOUS SINUSES

In the brain, the penetrating veins of the cerebral cortex also closely parallel the penetrating arteries. However, once on the surface of the brain, these veins empty into a series of **sinuses** that are formed by a separation of the two dural layers (Fig. 1.20). The venous sinus system drains from the **superior sagittal sinus** into a pair of **transverse sinuses.** The transverse sinuses are formed at the junction of the tentorium cerebelli with the superficial dura mater. Each follows the margin of the occipital bone to the petrous bone, where it becomes the **sigmoid sinus.** The sigmoid sinus follows an S-shaped medial course to the point where it exits the skull as the **internal jugular vein** (Fig. 1.26).

Several other sinuses are tributaries to this system. Most notable is the **inferior sagittal sinus.** This sinus is formed at the free, inferior edge of the falx cerebri and drains in an anterior-to-poste-

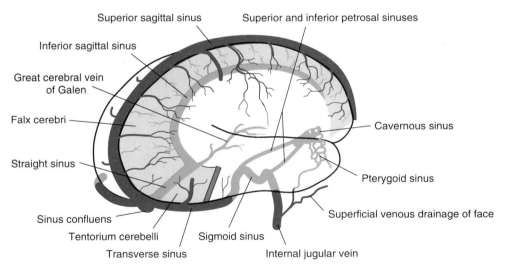

Superior sagittal sinus

Superior and inferior petrosal sinuses

Inferior sagittal sinus

Great cerebral vein of Galen

Falx cerebri

Straight sinus

Cavernous sinus

Pterygoid sinus

Superficial venous drainage of face

Sinus confluens

Tentorium cerebelli

Transverse sinus

Sigmoid sinus

Internal jugular vein

FIGURE 1.26.

The superficial veins of the brain drain into the various interconnected sinuses, as depicted here. The venous blood flows to the transverse sinus at the posterior of the brain. After passing through the sigmoid sinus, the blood finally leaves the cranium through the internal jugular vein. An alternate pathway exists through the cavernous and pterygoid sinus.

rior direction, forming the **straight sinus** at the tentorial incisure. It passes directly posterior to join the superior sagittal sinus and the two transverse sinuses at the **sinus confluens.**

The straight sinus, in addition to receiving blood from the inferior sagittal sinus, also receives blood from the deep cerebral veins that drain the basal ganglia, thalamus, and related structures. These deep veins converge to form the **great cerebral vein of Galen,** which flows directly into the straight sinus.

The **cavernous sinus** lies at the base of the brain in the sella turcica. It envelopes several important structures, among them the pituitary gland, the internal carotid artery, and several cranial nerves, including the oculomotor, the trochlear, the abducens, and the maxillary division of the trigeminal. It receives blood from the anterior-inferior portion of the brain, the ophthalmic veins, and the sphenoparietal sinuses. The cavernous sinus drains toward the posterior into the transverse sinus, through the **superior** and **inferior petrosal sinuses.** It can drain inferiorly as well, into the extracranial pterygoid venous plexus and from there into facial veins and finally the internal jugular vein.

ABNORMALITIES OF THE NEUROVASCULATURE

Among normal individuals, there is considerable variation in the arrangement of the major arteries at the base of the brain. The point where the basilar artery is formed from the union of the vertebral arteries can lie almost anywhere along the base of the medulla and pons. In cases where this junction is more rostral than normal, arteries that normally arise from the basilar artery simply arise from one of the vertebral arteries instead. Also worth noting is that, in some individuals, the posterior communicating arteries can be as large as the basilar artery or, conversely, can be missing altogether. These anomalies usually occur only on one side.

One should recall from the study of gross anatomy that the circulation from the external carotid artery can potentially reach the distal circulation of the internal carotid artery in the brain. This is accomplished by reversal of the blood flow through the ophthalmic artery, the first intracranial branch of the internal carotid artery. Normally the ophthalmic artery is supplied from the internal carotid artery. However, if the internal carotid artery is occluded between the carotid bifurcation and the origin of the ophthalmic artery, blood from the facial, maxillary, and superficial temporal arteries can reach the ophthalmic artery. Through it, blood can flow backwards into the cranium, reaching the middle and anterior cerebral arteries (Fig. 1.27).

Arteriovenous Malformations

Under certain circumstances, the arteries and veins in nerve tissue develop with direct shunts between them. Known as an **arteriovenous malformation** or **AVM,** these developmental abnormalities are composed of a tangle of vessels with abnormally thin walls that shunt blood from the arterial to the venous circulation without passing through a capillary bed (Fig. 1.28). Because of the delicate nature of the vessel walls, they are frequently a site of intracerebral hemorrhage. They can also be the source of focal seizures (see Chapter 13). Arteriovenous malformations lie undetected in the brain until their presence causes the development of some neurological sign. Although treating arteriovenous malformations is a dangerous and complicated procedure, they can frequently be isolated by applying vascular clips or can be occluded by injecting them with beads or polymers.

Berry Aneurysms

Intracranial bleeding can also be caused by leaking or ruptured aneurysms. A localized improper development of the arterial wall leaves a point of weakness in the artery that balloons under pressure (Fig. 1.29). Named for their berry-like appearance, in the brain they are usually (more than 90% of the time) seen in the arteries associated with the anterior circle of Willis. They may lie dormant for years before leaking or rupturing, causing an intracranial hemorrhage that is frequently fatal. If the patient survives the original bleeding episode, the neck of the aneurysm can often be occluded with a small steel clip.

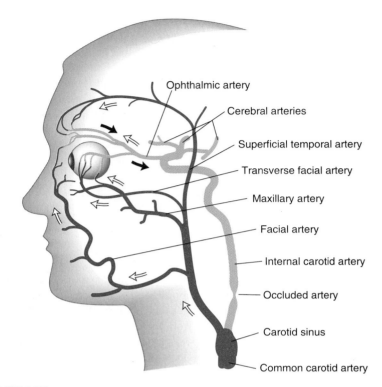

Ophthalmic artery

Cerebral arteries

Superficial temporal artery

Transverse facial artery

Maxillary artery

Facial artery

Internal carotid artery

Occluded artery

Carotid sinus

Common carotid artery

FIGURE 1.27.

The ophthalmic artery is the first intracranial branch of the carotid artery. In cases of gradual stenosis of the internal carotid artery, the flow of blood through the ophthalmic artery can be reversed (solid arrows), making it a nutritive artery of the brain. The collateral circulation in such cases is derived from the facial artery, the maxillary artery, and the superficial temporal artery. Normal blood flow is indicated by open arrows.

The Ventricular System

The nervous system is a hollow organ. There are four cavities within the brain, called ventricles, that form a continuous space (Fig. 1.30). The largest are the paired **lateral ventricles,** located in the cerebral hemispheres. They are an internal reflection of the superficial form of the cerebrum, hence there is an **anterior horn** that is found within the frontal lobe, a **posterior horn** that lies within the occipital lobe, and an **inferior horn** that penetrates the core of the temporal lobe. The **body of the lateral ventricle** connects the anterior and posterior horns and lies deep to the frontal and parietal lobes.

The **third ventricle** lies in the midline between the two halves of the diencephalon. It is a thin, narrow structure that communicates with the lateral ventricles through a pair of passages, the **intraventricular foramina.** Toward the posterior, the third ventricle narrows at the beginning of the mesencephalon to form the very small and relatively long **cerebral aqueduct** that opens into the **fourth ventricle.**

The fourth ventricle is a large space lying dorsal to the pons and medulla and ventral to the cerebellum. A pair of **lateral recesses** in the fourth ventricle slide under the cerebellum and communicate with the subarachnoid space through the **lateral apertures.** In the midline the fourth ventricle opens into the cisterna magna through the single **median aperture.** Hence the ventricular system and the subarachnoid space are continuous.

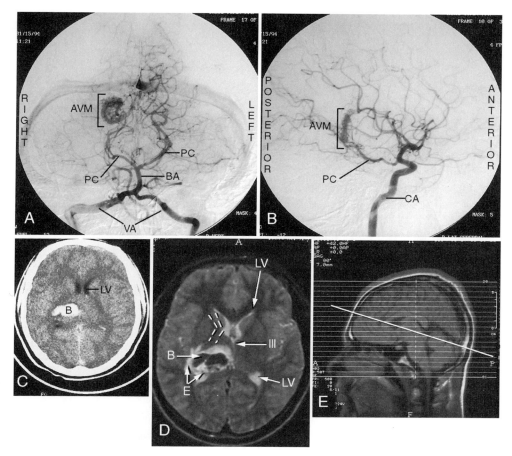

FIGURE 1.28.

A, An anterior-posterior view arteriogram of an arteriovenous malformation (*AVM*). The AVM is derived from the posterior cerebral (*PC*) artery. The contrast material used to opacify the blood vessels was injected into the left vertebral artery (*VA$_l$*). It was transported into the basilar artery (*BA*) and the two posterior cerebral arteries. Some contrast material also refluxed into the right VA. **B,** A lateral view arteriogram from the same patient shown in **A.** The dye was injected into the right carotid artery (*CA*). **C,** A CT from the same patient illustrating the appearance of fresh blood in the brain (*B*) after a leak from the AVM. The anterior horns of the lateral ventricles (*LV*) are just visible in this section. Note that CSF appears dark in CT. **D,** A T2-weighted MRI from the same patient illustrating the appearance of the same fresh blood (*B*) shown in **C** and also surrounding edema (*E*) that is not visible in the CT scan. The lateral ventricles (*LV*) and the third ventricle (*III*) are clearly shown. The internal capsule is outlined by *dashed lines* on the left. **E,** A T1-weighted MRI sagittal view of the brain. The horizontal *white lines* represent the levels at which various images were taken in this patient. The *diagonal line* illustrates the plane of the CT section shown in **C.** Ordinary CT sections of the brain are parallel to this plane.

Cerebrospinal Fluid

The ventricles and subarachnoid space are filled with a clear, colorless fluid known as **cere-** **brospinal fluid,** or **CSF.** The cerebrospinal fluid is an actively secreted product of the **choroid plexus,** a highly vascularized tissue that partially lines the ventricles. CSF is a simple fluid that resembles plasma. However, it differs from plasma in several important respects.

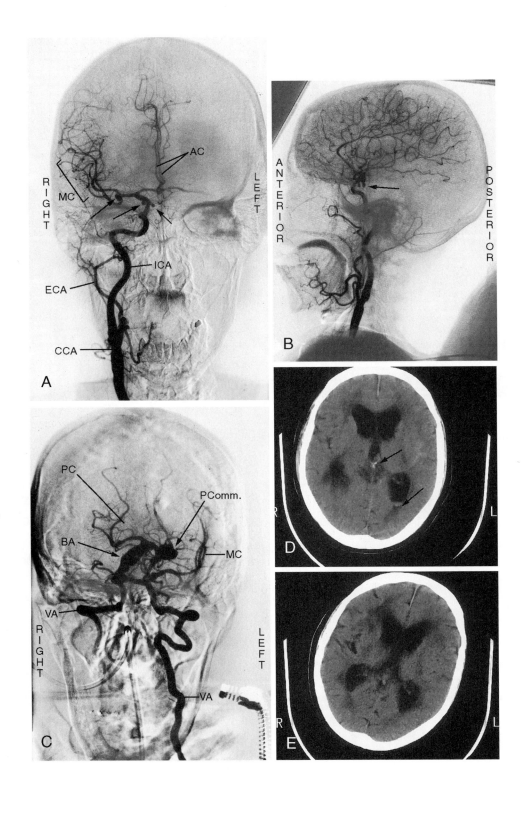

Table 1.1 summarizes the comparison between plasma and CSF. Specifically, every physician should remember that *CSF normally contains less than 50 mg protein per 100 ml* and *no more than 5 cells per milliliter*. Normal CSF pressure at the lumbar cistern is less than *200 mm H₂O* when the patient is in the recumbent position.

PRODUCTION AND CIRCULATION OF CSF

CSF is produced at a relatively constant rate of 400 to 500 ml/day and must therefore be reabsorbed at that same constant rate. This is accomplished through **arachnoid villi,** small structures that communicate between the subarachnoid space and the dural sinuses (Fig. 1.20). They are particularly prominent in the sagittal sinus. CSF can freely pass through the arachnoid granulations into the venous blood if CSF pressure is slightly higher than venous pressure. Blood is unable to pass into the subarachnoid space when the opposite is true.

Since CSF reabsorption occurs from the subarachnoid space into the venous sinuses, and CSF is produced in the ventricles, it should be quite apparent that if reabsorption is to equal production, free circulation between all of the ventricles and from the ventricular system into the subarachnoid space must be maintained. When this communication is disrupted, production of CSF exceeds reabsorption in the part of the ventricular system between the blockage and the choroid plexus. In such cases, an **increase in intracranial pressure** follows, because the brain is nearly incompressible and it is enclosed within the skull, which is not expandable. One should also understand that overproduction of CSF with normal reabsorption or poor reabsorption with normal production can cause increased intracranial pressure without blockage to CSF flow.

HYDROCEPHALUS

Aberrations in CSF fluid dynamics are called **hydrocephalus.** Hydrocephalus may be associated with an increase in intracranial pressure. In such cases it is designated **communicating** or **noncommunicating hydrocephalus,** reflecting the presence or absence of a blockage to the flow of CSF.

Hydrocephalus is not always associated with an increase in intracranial pressure. For example, newborn infants with hydrocephalus have enlarged ventricles and an enlarged cranium because, in the infant, the skull is not yet a rigid structure. The cranial sutures have not closed so the skull expands, allowing an increase in ventricle size without a dramatic increase in CSF pressure (see Appendix A).

Unfortunately the dilatation of the ventricles causes neuronal damage by thinning the cerebral cortex, and these children become neurologically defective. To prevent or minimize the damage, a tube is placed in the ventricular system and through it CSF is shunted into the jugular vein or the peritoneal cavity, where it can be reabsorbed. These shunts are fitted with a one-way valve that also serves as a pressure regulator.

FIGURE 1.29. A, An anterior-posterior view arteriogram following dye injection into the right common carotid artery (*CCA*). Three berry aneurysms (*arrows*) are revealed. *ICA,* internal carotid artery; *ECA,* external carotid artery; *MC,* arteries derived from the middle cerebral artery, colloquially known as the "candelabra"; *AC,* anterior cerebral arteries (note filling of both right and left arteries). **B,** A lateral view arteriogram from the same patient following dye injection into the right common carotid artery. The *arrow* points to two berry aneurysms. The middle cerebral candelabra is well demonstrated above the aneurysms. **C,** An anterior-posterior view arteriogram of a different patient illustrating a fusiform aneurysm of the basilar artery (*BA*) and another aneurysm of the posterior communicating artery (*Pcomm*). The dye was injected into the left vertebral artery (*VA*) and crossed over into the right VA, partially filling it. The right posterior cerebral (*PC*) and right middle cerebral arteries (*MC*) are also filled. **D,** A CT image of the same patient illustrated in **C,** showing enlarged ventricles and fresh blood (*arrows*). Note the blood in the posterior horn of the left ventricle has collected in the posterior crevice because the patient had been lying on his back for some time, allowing the red cells to settle due to gravity. **E,** CT of the same patient taken 2 days later, still showing enlarged ventricles but no fresh blood.

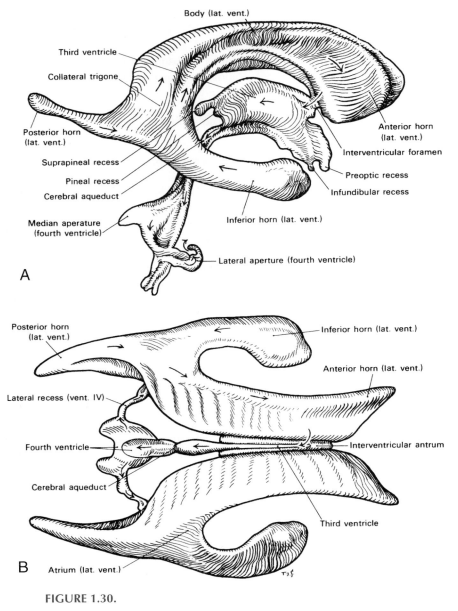

FIGURE 1.30.

The ventricles of the human brain shown in lateral (**A**) and superior (**B**) views.

Normal-pressure hydrocephalus is also seen in adults, usually in the aged. In these cases, the ventricles enlarge but CSF pressure remains relatively normal. Impaired reabsorption is believed to be the cause of this syndrome, but the etiology is uncertain. It is noteworthy because it is a treatable form of dementia, urinary incontinence, and/or gait disturbance in the elderly; these patients sometimes respond well to CSF shunts.

Increased intracranial pressure can also be caused by the development of intracranial masses such as subarachnoid hematomas or tumors. As these masses expand, the brain becomes distorted and some structures may be displaced. Initially some of the pressure can be relieved by an increased reabsorption of CSF and a corresponding collapse of the ventricles. This strategy soon fails and pressure increases rapidly, presenting com-

Table 1.1.   Comparison Between CSF and Plasma

Component	CSF	Serum
Protein	**35 mg/dL**	**7000 mg/dL**
Glucose	**60 mg/dL**	**90 mg/dL**
Red cells	**0/mL**	**NA**
White cells	**<5/mL**	**NA**
Na$^+$	138 mEq/L	138 mEq/L
K$^+$	2.8 mEq/L	4.5 mEq/L
Ca^{2+}	2.1 mEq/L	4.8 mEq/L
Mg^{2+}	0.3 mEq/L	1.7 mEq/L
Cl$^-$	119 mEq/L	102 mEq/L
pH	7.33	7.41

[a]The major constituents of CSF and serum are listed here along with their normal values. The most important components are listed in boldface.

plications in patient management. Perhaps the greatest difficulty involves the decrease in the perfusion of the brain. It is intuitive that cerebral perfusion pressure must equal mean arterial pressure less intracranial pressure; therefore, cerebral blood flow decreases as intracranial pressure increases. As cerebral perfusion decreases, clouding of consciousness, coma, and eventually death follow (see Chapter 13).

If the CSF pressure is seriously elevated, and one relieves that pressure by inserting a needle into the lumbar cistern, there is great danger that the cerebral contents will shift. In such cases, it is possible that part of the uncus of the temporal lobe will be squeezed through the incisure and place pressure on the cerebral peduncles and the mesencephalon. Or, more likely, the cerebellar tonsils may be thrust into the foramen magnum, squeezing the medulla and cervical spinal cord.

Both situations are catastrophic to the patient; frequently they are fatal. Obviously, before sampling CSF, the physician is obliged to rule out all causes of increased intracranial pressure.

The Blood-Brain-CSF Compartments

The nervous system consists of three fluid compartments: the **vascular** compartment, the **CSF** compartment, and the **neuronal** or **extracellular** compartment (Fig. 1.31). These compartments do not freely communicate with one another. The interface between the vascular system and the extracellular compartment has traditionally been called the **blood-brain barrier.** The other two interfaces, the **blood-CSF bar-**

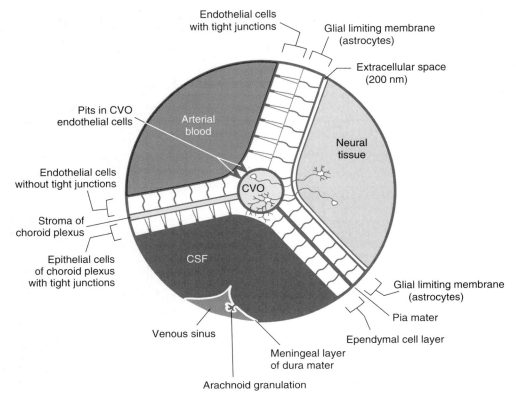

FIGURE 1.31.

This schematic illustrates the three chemical compartments of the brain and the various barrier systems that separate them. Note that specialized tight-junction endothelial cells are the principal barrier between the blood compartment and the CSF and neural tissue. There is essentially free passage between the CSF and the neural tissue. In the center of the figure, a circumventricular organ (CVO) is depicted that has relatively free exchange with the vascular compartment. Neurons provide transport of materials to and from CVOs and the neural tissue and the CSF.

rier and the **CSF-brain barrier** have, until recently, received little attention.

BLOOD-BRAIN INTERFACE

Endothelial cells found in most capillary systems have open, 10-nm spaces between the cells. However, the capillaries of the CNS are formed from unusual endothelial cells that have tight junctions between them. Large macromolecules that ordinarily would be delivered to the extracellular compartment by diffusion between the endothelial cells are almost completely excluded from the CNS. Even ionic species can be excluded by these specialized endothelial cells.

Ordinary endothelial cells transport molecules by transcellular mechanisms. They may engulf extracellular materials nonspecifically, transport the resulting vacuole across the cell, and then expel it into another extracellular space, a process known as **fluid-phase endocytosis.** Or they may transport specific molecules by a process known as **receptor-mediated endocytosis.** Brain endothelial cells do not have these capabilities.

Some molecules do pass from the blood into the brain extracellular space. In almost every instance, there exists a special transport mechanism for these molecules. For example, amino acids have three specific transport systems designed for acidic, neutral, and basic amino acids. D-glucose also has a specific transporter. Some other molecules can cross the endothelial cell based on their lipid solubility.

Astrocytes (see Chapter 2) line the external surface of the CNS capillaries and play a special role in maintaining the extracellular milieu. Their most important function in this respect seems to be related to their ability to regulate the extracellular concentration of K⁺ ions, rather than preventing the entry of molecules from the capillaries.

BLOOD-CSF INTERFACE

The potential for communication between the blood and CSF compartments also exists in the choroid plexus. Here the endothelial walls of the capillaries do not come in direct contact with the CSF compartment. They are a part of the stroma of the choroid plexus, and the endothelial cells have standard fenestrated junctions. However, the epithelium of the choroid plexus itself does contain tight junctions, and they serve as the molecular barrier. The result is that the CSF compartment is isolated from the blood compartment in a manner that is nearly identical to the way the neuronal extracellular space is isolated.

CSF-BRAIN INTERFACE

The ventricles are lined with ependymal cells. There are no tight junctions between these cells, nor does the pial membrane present a significant diffusion barrier. Therefore, there is essentially free molecular communication between the CSF and the intercellular space of the CNS.

The Circumventricular Organs

The barrier that exists between the blood stream and the neurons of the central nervous system is circumvented at specific places in the brain. These sites, known as **circumventricular organs** (CVOs), exist in close association with the CSF. Seven sites are currently recognized, the **area postrema,** the **pineal gland,** the **subcommissural organ,** the **subfornical organ,** the **organum vasculosum of the lamina terminalis,** the **median eminence,** and the **neurohypophysis** (Fig. 1.32). CVOs contain neurons but have some characteristics of endocrine glands. For example, ordinary nervous tissue is only able to utilize glucose as an energy source via the tricarboxylic acid cycle. CVOs, like endocrine glands, are able to utilize free fatty acids in addition to glucose as energy sources. Furthermore, they can metabolize these precursors through the hexose monophosphate shunt, a pathway not available to neurons.

Circumventricular organs have a higher capillary density than the surrounding neural tissue (Fig. 1.33), and these capillaries are structurally

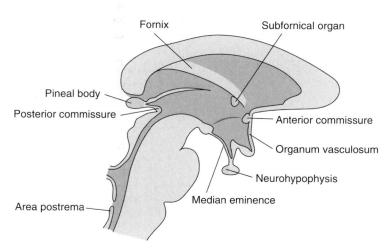

FIGURE 1.32.

Sagittal view of the third and fourth ventricles showing the locations of the circumventricular organs.

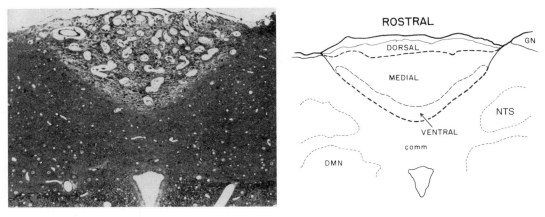

FIGURE 1.33.

A, The area postrema of the rat, shown here, is clearly distinguished from the surrounding neural tissue by the density of its capillary bed (×45). **B**, Schematic drawing of the same area identifying major structures: *NTS*, nucleus of the tractus solitarius; *DMN*, dorsal motor nucleus of the vagus; *GN*, gracilis nucleus. (These structures are discussed in subsequent chapters.)

quite different from the capillaries of the brain. Unlike the capillaries in ordinary neural tissue, CVO capillaries have a pericapillary space, fenestrations between the endothelial cells, and surface pits (Fig. 1.34). There is evidence that specialized receptors line the surface of the pits, where they bind circulating hormones and transduce the chemical signal into a neuronal signal. Therefore, the brain is subject to hormonal regulation via the CVOs that act at the interface between the endocrine system and the nervous system (see Chapter 12).

The brain is also an endocrine organ, releasing into the blood substances that affect remote targets. The best documented role of the brain as an endocrine organ is its release of oxytocin and vasopressin from neurosecretory granules into the perivascular space of the neurohypophysis (one of the CVOs), where these hormones diffuse into the capillaries through fenestrated endothelial cells. Another CVO, the median eminence, is an interface between the nervous system and the adenohypophysis. The median eminence not only deposits hormonal signals (releasing factors) into the hypophyseal portal blood stream, it receives feedback signals in the form of circulating hormones. These systems are discussed in detail in Chapter 12.

Many high-molecular weight neuroactive peptides that cannot cross the blood-brain interface are nevertheless found in the CSF. They appear to be secreted into the CSF and/or detected by one or more of the circumventricular organs. Therefore it seems likely that there may be a humoral regulatory system within the CNS that uses the CSF as its transport vehicle, much as the endocrine system uses the blood stream to transport hormones to distant targets. Since the CSF is isolated from the blood stream, these neuroactive peptides are restricted to CNS targets.

SUMMARY

The central nervous system is divided into the cerebral hemispheres, the basal ganglia, the brainstem, the cerebellum, and the spinal cord. The brainstem is subdivided into the thalamus, the midbrain, the pons, and the medulla. The cerebral hemispheres are subdivided into the frontal, the parietal, the occipital, the insular, and the temporal lobes. The cerebral hemispheres are wrinkled into gyri and sulci. The cerebellar hemispheres are wrinkled into folia. Both schemes serve to greatly increase the surface area without greatly increasing the volume of the organ.

The brainstem receives all the cranial nerves except the olfactory and accessory nerves. The optic nerve enters the thalamus. The oculomotor and trochlear nerves enter the midbrain, while the trigeminal nerve enters the pons. The abducens,

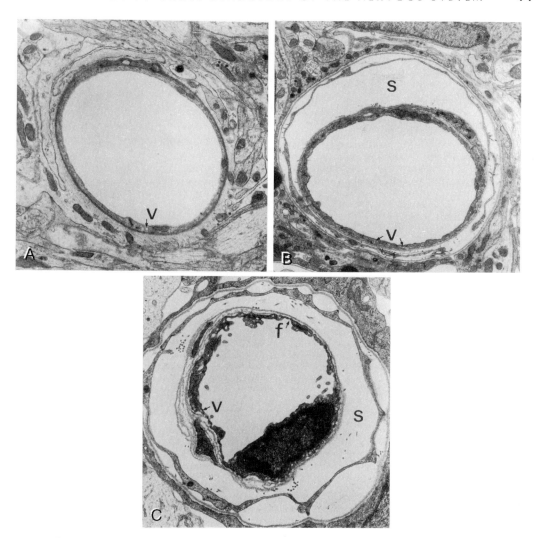

FIGURE 1.34.

Electron micrographs of three types of capillaries present in CVOs. **A,** Typical non-fenestrated capillaries are rarely seen in CVOs. These are typical of ordinary neural tissue. **B,** The prominent pericapillary space (*S*) and vesicles or pits (*V*) are commonly found in CVO capillaries. **C,** This type of CVO capillary not only has pits and a pericapillary space, but also has fenestrations between the endothelial cells (*f*).

facial, and vestibulocochlear nerves all enter at the pontomedullary junction. The remaining nerves, except the accessory, enter the medulla; the accessory nerve, being a component of the spinal cord, has an aberrant course into the cranium.

There are 31 pairs of spinal nerves. Each nerve consists of a merged dorsal and ventral root; the dorsal root has a sensory function, and the ventral root has a motor function. Sensory nerves, both spinal and cranial, are mostly pseudounipolar and

have ganglia associated with their roots. There are exceptions, those being the olfactory and optic nerves and the mesencephalic branch of the trigeminal nerve (all to be discussed later).

The central nervous system is enclosed in a bony suit of armor that affords it considerable protection from external forces. It is further covered, from the outside in, by the dura mater, the arachnoid, and the pia mater. The dura mater also forms several septa that divide the brain into compartments. The falx cerebri is the largest, lying in the sagittal plane and separating the two cerebral hemispheres. The tentorium cerebelli separates the cerebral hemispheres from the cerebellum. The tentorial incisure is a hole in the union of these septi through which the brainstem passes. It can be a site of entrapment of the temporal pole under conditions that bring about the shifting of the intracranial contents.

The arachnoid creates a space between the dura and the pia that contains cerebral spinal fluid. Where this space is very large, it is called a cistern. For the physician, the most useful cistern is the lumbar, from which CSF is sampled.

The intracranial nervous system is supplied by only two pairs of arteries, the vertebral and the internal carotid arteries. The carotid circulation gives rise to the middle cerebral and anterior cerebral arteries. Together they supply the major portion of the hemispheres. The vertebral arteries form the basilar artery of the pons that in turn splits to form the posterior cerebral arteries. These supply the occipital pole of the brain and the inferior portion of the temporal lobes. The cerebellum is supplied by three arteries. The superior and the anterior-inferior cerebellar arteries arise from the basilar artery, and the posterior-inferior cerebellar artery arises from the vertebral. The spinal circulation is derived from the dorsal and ventral spinal arteries as well as several radicular arteries. The spinal circulation is quite variable, but is richly anastomosed except in the thoracic region.

Blood is drained from the central nervous system by a network of veins. In the spinal cord this system finds its way to the radicular veins in a straightforward manner. The intracranial veins collect as sinuses, specialized structures formed at the intersections of the external dura and the septa. The principal sinus, the superior sagittal, divides to form the two transverse sinuses that eventually find their way to the internal jugular veins. Other sinuses drain other parts of the intracranial nervous system.

The brain is a hollow organ, having four internal cavities known as ventricles: the two lateral ventricles (one on each side of the midline), and the third and fourth ventricles. The ventricles are filled with CSF that is secreted from the choroid plexus lining many of the ventricles. The CSF flows out of the ventricular system through two lateral foramina and one middle foramen in the fourth ventricle into the subarachnoid space. Eventually the CSF is reabsorbed into the blood stream through the arachnoid villi, found mostly in the superior sagittal sinus. If CSF production exceeds reabsorption, or if circulation among any of the CSF spaces is occluded, intracranial pressure increases (hydrocephalus). Pressure may also be increased by expanding intracranial masses such as tumors or the seepage of blood into the epidural or subarachnoid space (hematoma).

There are three biochemical compartments in the brain: the blood, the CSF, and the extracellular space of the CNS. There is greatly restricted access from the blood to either of the other two spaces, but the CSF and the intercellular CNS space freely exchange molecular material. The blood-brain interface is specifically absent at certain regions of the brain known as circumventricular organs. At the circumventricular organs, hormones and large peptides can be exchanged between the blood, the brain, and the CSF.

FOR FURTHER READING

Brodal, A. *Neurological Anatomy in Relation to Clinical Medicine*. New York: Oxford University Press, 1981.

Carpenter, M. B. *Core Text of Neuroanatomy*. Baltimore: Williams & Wilkins, 1991.

Haines, D. *Neuroanatomy: An Atlas of Structures, Sections, and Systems*. Baltimore: Urban & Schwarzenberg, 1991.

Lundborg, G. *Nerve Injury and Repair*. Philadelphia: Churchill Livingstone, 1988.

Montemurro, D., and Bruni, J. E. *The Human Brain in Dissection*. New York: Oxford University Press, 1988.

Netter, F. H. *The Ciba Collection of Medical Illustrations. Vol 1*. West Caldwell, NJ: Ciba, 1983.

Netter, F. H. *The Ciba Collection of Medical Illustrations. Vol 2*. West Caldwell, NJ: Ciba, 1983.

Nieuwenhuys, R., Voogd, J., and van Huijzen, C. *The Human Central Nervous System; A Synopsis and Atlas*. Berlin: Springer-Verlag, 1981.

Waddington, M. *Atlas of Cerebral Angiography with Anatomic Correlation*. Boston: Little, Brown, 1974.

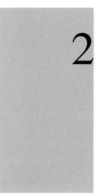

2 Microstructure of the Nervous System

The human nervous system is perhaps the most complicated structure ever to evolve on Earth. Yet its structural components are quite simple. Excluding the meninges and vasculature, the nervous system consists only of three cell types: **neurons, glia,** and **Schwann cells.** While there are a large number of morphologically different neuron cell types, the differences among them are derived from simple variations of cellular phenotype, not from major differences in cell metabolism, structure, or function. Furthermore, neurons communicate among themselves through contacts called **synapses.** Although there are billions of synapses in the mammalian nervous system, most of these synapses share common chemical, structural, and functional features. It is important to understand that *the nervous system accomplishes its functions not by increasing the diversity of its component parts but by replicating a small number of simple structures and modifying the specificity with which those elements are interconnected with one another.* This microcosm of cells is organized into a basic macrostructure.

The primary division in the neural macrostructure is between the **central nervous system (CNS)** and the **peripheral nervous system (PNS).** The CNS is contained within the meninges, while the PNS is distributed outside the meninges, within the structure of the body. Neurons are found in all parts of the nervous system, while the glia are found only in the CNS. Schwann cells are found only in the PNS.

Within the central nervous system the neuron cell bodies are organized into **nuclei,** and their processes are collected together into organized bundles called **tracts.** The glia are cells in the CNS that provide structural and nutritive support for the neurons. **Ganglia** are collections of neuron cell bodies in the peripheral nervous system. The axons of those neurons along with the axons from nuclei that leave the CNS form the **peripheral nerves.** The only supporting cell in the PNS is the Schwann cell.

This chapter describes the cells that make up the nervous system, emphasizing the morphological and functional characteristics that establish their unique role in neuronal function. The electrical properties of neurons are discussed in Chapter **3.**

The Neuron

Estimates place the number of neurons in the human brain at about 10^{11} nerve cells. A number that large is hard to comprehend. To help put it into perspective, assume for a moment that the average diameter of the neurons is 10 μm (a conservative estimate). If these neurons were placed side by side, they would form a chain 10^3 km long. Traveling at approximately the speed of sound ($\sim$1000 km/hr), it would take more than 1 hour to traverse this chain from end to end.

Neurons are cells and therefore have the characteristics common to all cells (Fig. 2.1). They are bound by a cell membrane. They have a nucleus and cytoplasm. Within the cytoplasm is the usual complement of intracellular organelles: endoplasmic reticulum, Golgi bodies, mitochondria, peroxisomes, free ribosomes, and inclusion bodies such as lysosomes, lipofuscin bodies, and vesicles.

Typical neurons consist of three anatomically distinguishable regions: the **soma** or cell body, **dendrites** [G. *dendrites,* relating to a tree], and a single **axon** [G. *axon,* axis]. The axon branches, and each tip terminates as a specialized structure known as a **synaptic bouton.** (Fig. 2.2)

Neurons are commonly classified according to the number of processes they possess. According to this traditional scheme, neurons with a single process are called **unipolar,** those with two processes are called **bipolar,** and those with more than two processes are called **multipolar.** Most neurons are multipolar. In humans, the only truly bipolar neurons are located in the ganglia of the first, second, and eighth cranial nerves. The neurons of the remaining ganglia have a single process that almost immediately divides into two. These neurons are called **pseudounipolar** (Fig. 2.3). This classification system affords a succinct description that is easily applied to light microscopic preparations.

SOMA

The soma of most neurons is compact and globular. It contains the ultrastructural organelles needed to carry out the metabolic functions of the cell. *Neurons are secretory cells* that produce an extensive variety of proteins. Consequently, they have a large centrally placed **nucleus** with a very prominent **nucleolus.** Within the nucleus, most of the DNA exists in the extended form, making it readily available for transcription. It has been estimated that more than 20,000 mRNA sequences are expressed by neurons. This is 10 to 20 times more sequences than

are expressed by the "typical" secretory cells of the liver.

Protein synthesis occurs on ribosomes. In keeping with the extraordinary amount and diversity of protein synthesis that occurs in neurons, the soma contains large numbers of ribosomes and a complex set of internal membranes (Fig. 2.4). This internal membrane complex is divided into several compartments: the **cell membrane,** the **nuclear membrane,** the **rough endoplasmic reticulum** (rER), the **smooth endoplasmic reticulum** (sER), the **Golgi apparatus,** and **membrane-limited secretory granules** such as **lysosomes** and **endosomes.** This membrane system is (at least potentially) continuous, the lumen of the endoplasmic reticulum and the Golgi apparatus being continuous with the extracellular space.

One portion of the major membrane system, the rough endoplasmic reticulum, is so named because the cytoplasmic side of its membranes is decorated with ribosomes. The rough endoplasmic reticulum of neurons is so extensive that it is easily visualized with light microscopy when the cells are stained with basic dyes. Displayed in this way, the rough endoplasmic reticulum of the neuron is called **Nissl substance,** named after the 19th century cytologist who first described it.

Protein synthesis occurs on both the free ribosomes and those bound to the membranes of the rough endoplasmic reticulum. Soluble proteins remain within the cell and are assembled on the free ribosomes in the cytosol. Secretory proteins are destined to leave the cell and are assembled on the ribosomes of the rough endoplasmic reticulum. As they are assembled, they are extruded through the membrane into the lumen of the endoplasmic reticulum. Further elaboration of proteins occurs in the Golgi apparatus. Eventually, membrane-bound vesicles con-

FIGURE 2.1.

Electron micrograph (×15,000) of a typical neuron illustrating the various intracellular organelles mentioned in the text. Note the well-organized endoplasmic reticulum (*ER*) that with the light microscope is seen as Nissl substance. The Golgi apparatus (*G*) is well developed in neurons, and there are numerous mitochondria (*mit*). On the neuron is a synapse (*ba*) and adjacent to it numerous dendrites from other cells in the neuropil. (*m*, microtubules; *ncl*, nucleolus; *Nuc*, nucleus.)

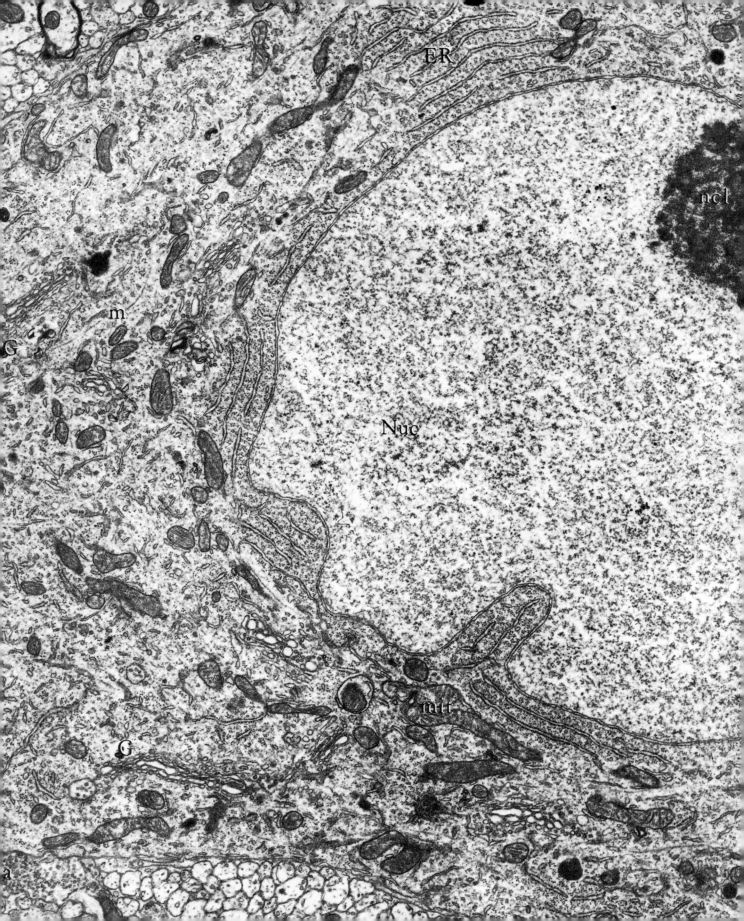

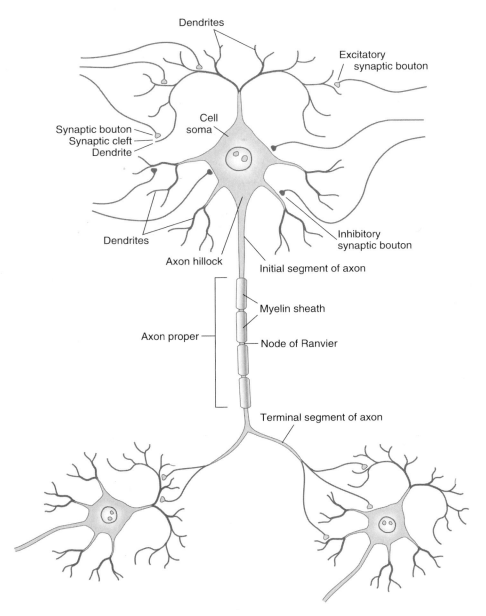

FIGURE 2.2.

The major features and relationships of neurons in the CNS of mammals. Most CNS neurons have a dendritic field that receives synapses from other neurons. Synapses may also be located on the soma. Some synapses are excitatory and others inhibitory. Neurons have one axon that leaves the soma at the axon hillock and may or may not be myelinated. Before the axon terminates, it divides and contacts numerous other neurons. One of the basic principles of neuroanatomy is illustrated here; information from multiple sources **converges** on a single neuron that in turn distributes its information to multiple **divergent** targets.

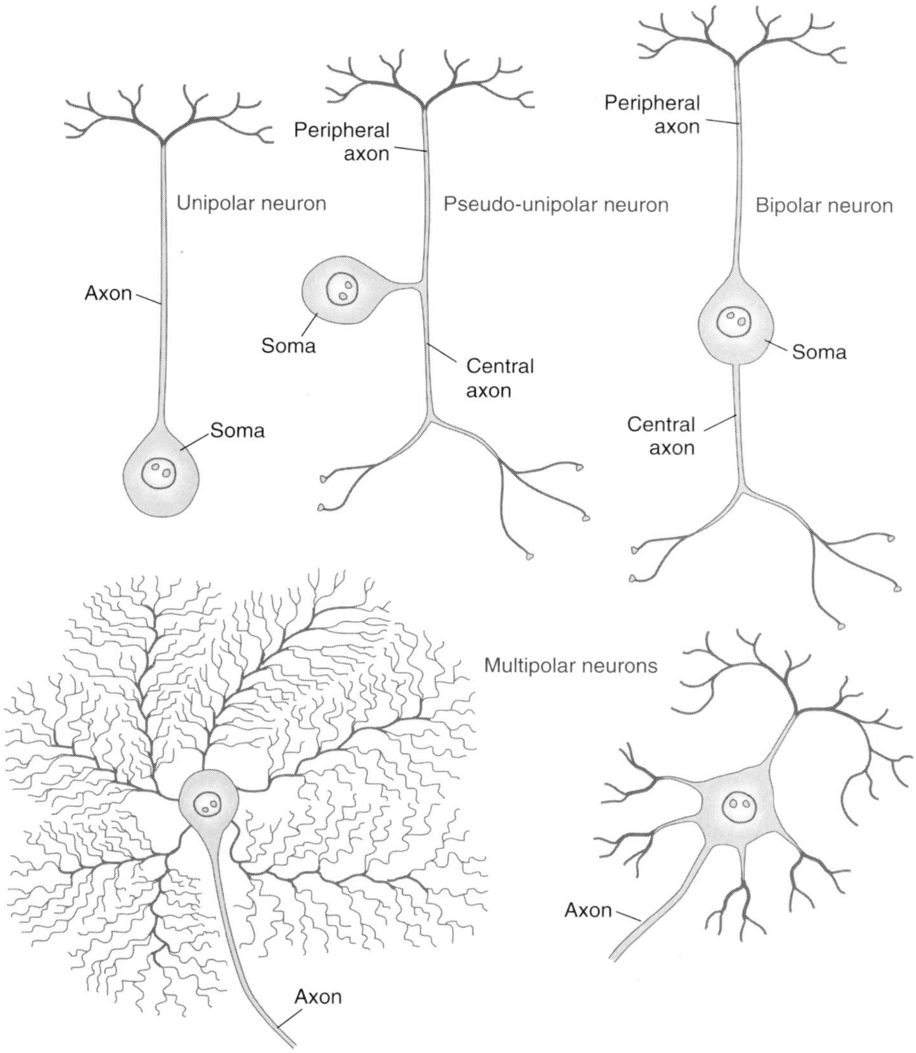

FIGURE 2.3.

The traditional histologic differentiation of neurons is by the number of processes they possess. True unipolar neurons are present in mammals only during development. Pseudounipolar and bipolar cells are sensory neurons found in peripheral ganglia. All other neurons in the mammalian nervous system are multipolar. As shown here, multi-polar neurons can assume a number of different forms.

taining the proteins and enzymes are pinched off from the Golgi membrane. These vesicles are then transported to various locations within the cell, eventually to be secreted into the intercellular space by exocytosis (Fig. 2.4).

The cytoplasm of neurons contains a number of polymeric molecules that constitute the cyto-skeleton. The principal structures formed by these molecules are **microtubules, neurofilaments, and microfilaments.** Each of these skeletal structures is constructed from a different class of proteins. Together they form an intracellular matrix that determines the shape of the cell, gives it some stiffness, and also provides the mecha-

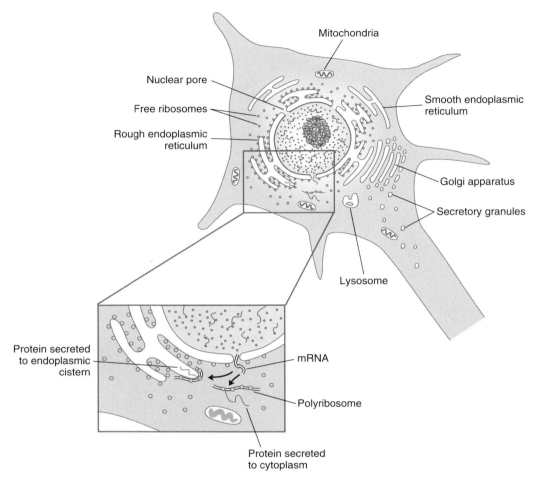

FIGURE 2.4.

The internal membrane system of cells is, in essence, an extension of the extracellular space. The nuclear membrane and the endoplasmic reticulum are a continuous network that forms a giant cistern within the cell. RNA in the form of polyribosomes decorates the surface of the rough ER. Proteins produced on the ribosomes enter either the cytosol or the endoplasmic cistern. Portions of the cistern become isolated as the Golgi apparatus, within which proteins are further elaborated. Secretory granules bud off from the Golgi apparatus and migrate to the cell surface, where they may be secreted or, in the case of neurons, transported down the axon.

nisms by which molecules are transported within the cell.

The largest of the cytoskeletal elements, the microtubules (Fig. 2.5), are not simple structures. Each microtubule is formed by a circular arrangement of 13 protofilaments. The protofilaments are long assemblages of two alternating molecular subunits, α- and β-**tubulin.** The tubulin subunits are asymmetrical, so the protofilaments formed

from them are polar molecules. The "head" has chemical properties that are different from those of the "tail." The assembled microtubule is about 25 nm in diameter and can be as long as 100 μm.

There are several **microtubule-associated proteins (MAPs)** that control the assembly of microtubules from the tubulin protofilaments. Microtubules, like all the cytoskeletal elements, are dynamic structures in a continuous state of assembly and disassembly. The degree of phosphorylation of the MAPs determines the state of polymerization of the microtubules.

Microtubules can cross-link with neurofilaments and contribute to the overall stiffness and shape of the cytosol. They are also essential in the transport of macromolecules and membrane-bound structures throughout the soma and along the axon.

Neurofilaments are approximately 10 nm in diameter. They are polymerized bundles of four strands of proteins called **cytokeratins** (Fig. 2.6). These are the most numerous cytoskeletal element in neurons. Normally neurofilaments, like all cytoskeletal elements, are aligned in orderly parallel arrays. They are cross-linked with each other and with microtubules. Under some circumstances this regular neurofibrillar organization becomes disrupted. For example, neurofilaments increase in number in cells that have been exposed to aluminum salts and they sometimes form dense, disorganized filamentous tangles. These knots of neurofilaments superficially resemble the neurofibrillary tangles that are one characteristic of Alzheimer's disease (see Chapter **13**). Similar disorganized neurofibrillary tangles are seen in neurons of patients with Down's syndrome and in some types of parkinsonism (see Chapter **8**), suggesting that cytoskeletal disorders may play an important role in some neurodegenerative diseases.

The smallest of the cytoskeletal elements are the microfilaments, 3 to 5 nm in diameter. These protein filaments are composed of two twisted strands of polymerized **actin** (Fig. 2.7). Microfilaments are closely associated with the cell membrane and can serve as bridges to anchor proteins to it. They are particularly abundant in the tip of growing axons, the neural **growth cones,** where they apparently play an important role in the mobility and plasticity of this structure (see Chapter **4**).

All of the cytoskeletal elements are dynamic polymers. Monomers can be added or removed at either end of the polymerized molecule (Fig. 2.8). Depending on the balance between the addition and removal of the monomers, the cytoskeletal strand can rapidly extend or collapse in length. The cytoskeletal elements are polar molecules, that is to say, their longitudinal structure is asymmetrical. Under some circumstances, new monomers can be added at the "head" of the molecule while other monomers are removed from the "tail," a phenomenon known as **treadmilling.** Such a treadmilling molecule can be a source of

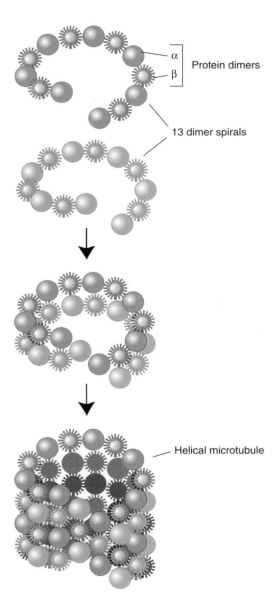

α
β   Protein dimers

13 dimer spirals

Helical microtubule

FIGURE 2.5.

Microtubules are formed from a protein dimer (the α and β subunits) that is assembled as a 13-segment coil from which the hollow spiral structure of the microtubule is formed.

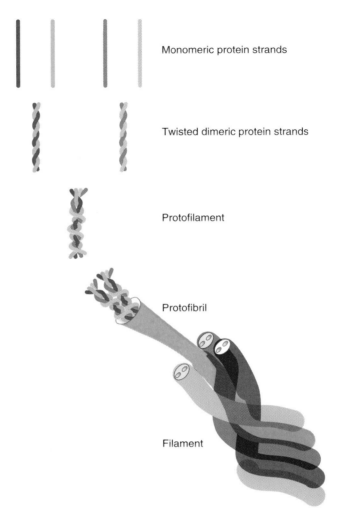

Monomeric protein strands

Twisted dimeric protein strands

Protofilament

Protofibril

Filament

FIGURE 2.6.

Neurofilaments are complex structures formed from four protein strands. Pairs of the strands twist together to form larger strands that in turn twist together in pairs to form the protofilament. Two protofilaments combine to form the protofibril, four of which combine to make the final, 10-nm-diameter neurofilament.

mechanical energy. These dynamic properties are important in cell motility. During growth and regeneration of the nervous system, for example, they account for the extension of the growing axon. They also account for the intracellular transport of macromolecules and organelles.

DENDRITES

Most neurons display a number of processes. The most numerous, the **dendrites,** are exten-

FIGURE 2.7.

Actin filaments are composed of two strands assembled from protein subunits. The subunits are asymmetrical and give the entire actin filament asymmetry that is important in determining its motile characteristics.

sions of the cell body. Beginning with a relatively large trunk at their base on the soma, dendrites repeatedly divide into ever narrower branches, forming a tree-like structure whose shape differs depending on the class of neuron (Fig. 2.3).

At the ultrastructural level, the dendrites contain many of the structures commonly found in the soma. At the base of the dendrites, one usually finds rough and smooth endoplasmic reticulum. All but the smallest dendritic branches have free ribosomes, mitochondria, microtubules, and neurofilaments (Fig. 2.9).

The principal function of the dendrites is to increase the surface area of the cell. Accordingly, most of the synaptic connections received by a neuron occur on the dendrites. This principle is carried to an extreme in some cells, in which mushroom-like protrusions, or **dendritic spines,** exist over the entire dendritic surface (Fig. 2.10). Numerous synapses occur on these

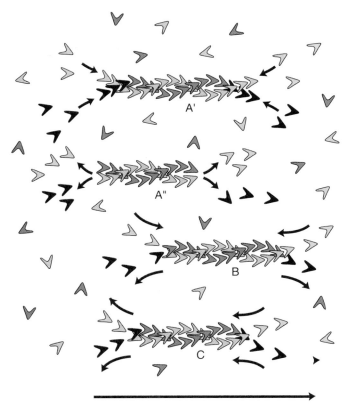

FIGURE 2.8.

The cytoskeletal structural proteins—neurofilaments, microfilaments, and microtubules—are polymeric structures built from monomers. The state of polymerization can be in flux, and this leads to three recognized states. **A,** Instability occurs when monomer separation and attachment are not balanced. This results in the rapid expansion (*A'*) or contraction (*A"*) of the parent molecule. **B,** True equilibrium is achieved when the rate of monomer separation from the parent molecule is equal to the rate of monomer attachment. The separation/attachment may both occur at either end of the parent molecule, as shown here. **C,** Treadmilling occurs when monomer separation takes place at one end of the parent molecule while monomer attachment takes place simultaneously at the other end. The parent molecule maintains approximately the same size during the process but can move through the cytosol (indicated by *long arrow*).

spines. The spines are so important in certain cells of the cerebral cortex that mental function is closely correlated with the number of spines, and hence synapses, present on the neurons.

The most characteristic morphologic feature of a neuron is the arrangement of its dendrites. Based on these patterns, scores of families of neurons have been described, but classification into a useful systematic nomenclature has not been possible. When one considers that the dendrites provide almost all of the synaptic contact the neuron has with other neurons, it then becomes apparent that the extent and diversity of these connections will be determined to a large degree by the physical arrangement of the dendrites (Fig. 2.11). If a neuron has a small, compact, and sparsely branched array of dendrites, then the number of synaptic contacts it can receive will necessarily be limited. The number of cells that can establish contact with such a sparse tree will also be limited. Neurons with large, extensively branched dendritic arborization not only receive more synaptic contacts because of the large surface areas, but also receive contacts from a larger number of different cells due to the spatial arrangement of the tree.

Some cells have unusually shaped dendritic fields. For example, the dendritic arborization of the Purkinje cell of the cerebellum, although large in cross section, is thin (resembling a palm leaf), so the cell receives contact only from axons traveling perpendicular to it. Some cells, such as those found in the sensory ganglia, do not have dendrites at all. The synaptic contacts made by such neurons are extremely limited, being constrained to specialized axo-axonic synapses (see below).

AXONS

Neurons have a single specialized process called an **axon.** Generally this process is thinner than the dendrites and considerably longer. Most neurons have quite short axons (less than 100 μm), but certain cells, such as motor neurons in whales, may have axons as long as 10 meters. A few highly specialized neurons do not have an axon, but these are exceptional in mammals. The axon consists of three regions: the transition zone from the soma to the axon proper, the **initial segment;** the main extent of the axon or the **axon proper;** and the swollen tip that constitutes the terminus, the **synaptic bouton** (Fig. 2.2).

Like dendrites, axons branch extensively, but unlike dendrites, the axons branch primarily near their distal end, just before they terminate. Extremely distal profuse branching allows one axon to establish many synaptic contacts with a single cell or a small group of propinquitous cells (Fig. 2.11). Many axons also divide into several major branches as they approach their targets. Each of these major divisions extends to a different group of cells before exhibiting the profuse branching prior to making the final synaptic contacts (Fig. 2.12).

Because of their great length, axons contain a significant proportion of the cytoplasm of the cell. For example, a typical motor neuron soma with its dendrites might have an average radius of 50 μm. Its axon would be approximately 10 μm in diameter and (in the human lumbar spinal cord) be about 500 cm long. For such a neuron, the volume of the soma would be 523×10^3 μm^3 and the axonal volume, assuming minimal branching, 39×10^6 μm^3, or 75 times the total cell volume. In extreme cases the axonal volume can be as much as 10,000 times the cell volume.

The Initial Segment

In large neurons, the cone of cytoplasm from which the axon originates is devoid of Nissl sub-

---

FIGURE 2.9.

The ultrastructure of dendrites is shown in this electron micrograph (×44,000) that shows three dendrites, two in longitudinal section (*Den₁* and *Den₂*) and one in cross section (*Den₃*). In all three dendrites, neurofilaments (*nf*), microtubules (*m*), smooth endoplasmic reticulum (*SR*), and mitochondria are clearly shown. Free ribosomes (*r*) are also seen in *Den₁* and *Den₂*. Two synapses with active zones make contact with *Den₃;* the upper contains small spherical vesicles and the lower, small oval vesicles. (*As,* astrocyte process.)

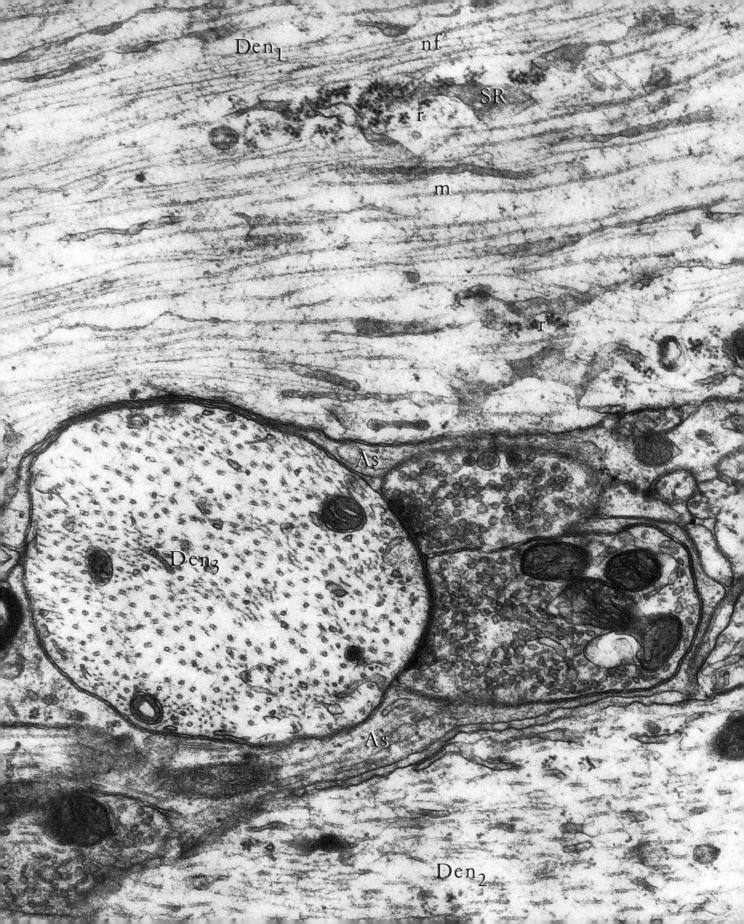

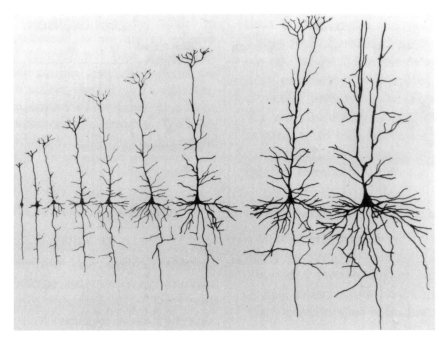

FIGURE 2.10.

The number of synapses impinging on a neuron is critical to its function. This is made clear by the fact that dendritic arborization becomes more complex during development. Here, pyramidal cells from humans of various ages are arranged in order, with fetal cells on the left and, at the exreme right, an adult pyramidal neuron. Note that the number of branches increases dramatically during development. Not shown here are the dendritic spines that also increase dramatically during development.

stance and gives a clear appearance in light microscopic preparations. This area is known as the **axon hillock.** In this region, neurotubules and neurofilaments coalesce into long parallel bundles before entering the axon. They remain collected into fascicles for the first 20 to 50 μm of the axon that is known as the initial segment (Fig. 2.13).

A special feature of the initial segment is a dense membrane-associated undercoating of osmophilic material. This undercoating may be the anatomical correlate for the voltage-sensitive sodium gates (see Chapter 3) that are abundant at the initial segment and the internodal region of myelinated axons (see below).

The Axon Proper

The main body of the axon contains the usual collection of organelles such as mitochondria, microtubules, neurofilaments, and smooth endoplasmic reticulum. Conspicuously absent from the axoplasm of the main shaft are free ribosomes and rough endoplasmic reticulum (Fig. 2.13). Thus *the axon does not have the necessary intracellular organelles required to manufacture proteins.* These macromolecules must be continuously resupplied from the soma.

Many axons in both the central and peripheral nervous systems are covered with **myelin** [G. *myelos,* marrow]. Myelin is a dense structure consisting almost entirely of cell membranes layered one upon another with the intervening cytoplasm removed. The glia produce myelin in the CNS, while it is formed by the Schwann cells in the PNS. Not all axons are myelinated. However, separation between unmyelinated axons is maintained by intervening glia or Schwann cells.

The Synaptic Bouton

Neurons communicate with one another through structures known as **synapses** [G.

synapto, to join], which are found at the junction of two neurons (Fig. 2.9). The synapse is a specialized structure involving both the presynaptic and the postsynaptic cell. In most synapses, the synaptic bouton of the axon is the presynaptic element of the synapse. It contains secretory vesicles that contain a specific chemical (the neurotransmitter) that is released by exocytosis into the extracellular space under certain carefully regulated conditions.

Although most synapses in the mammalian nervous system occur between axons and dendrites, synaptic contact can occur at any region of the neuron. For example, nearly all cells in the central nervous system receive synapses on their somas from axons. These are, appropriately enough, called **axo-somatic** synapses. Other arrangements are possible, the most common of which are the **axo-dendritic** (axon to dendrite) and **axo-axonic** (between axons) types. Even dendro-dendritic, dendro-somatic,

and somato-dendritic synapses have been described, although their number and distribution seem to be quite limited (Fig. 2.14).

It is worth noting that, while the chemical synapses described above are the most common type found in mammalian systems, electrical synapses do occur in vertebrates and they are quite common in invertebrates. However, their neurological importance has yet to be established, and they are not discussed here. The mechanisms of chemical synaptic transmission are discussed in Chapter **3.**

TRANSPORT OF PROTEINS

Since the axon does not contain the intracellular organelles necessary for protein synthesis, nearly all of the products of metabolism necessary for cell maintenance must be transported down the axon from the soma. In addition, materials from the environment are taken up at

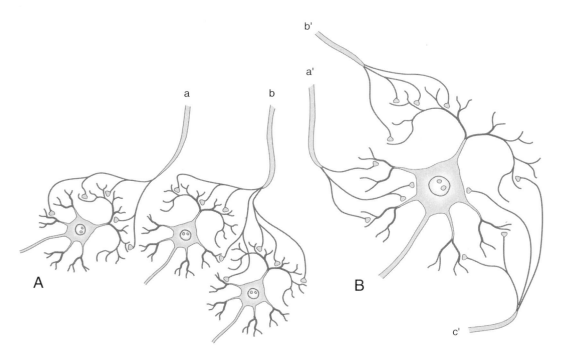

FIGURE 2.11.

The information a neuron has available to it is determined to a large degree by the size and shape of its dendritic field. **A,** Several small neurons receive tightly focused information from only a few afferent axons (*a* and *b*). **B,** A large neuron receives convergent input from several axons (*a'*, *b'*, and *c'*).

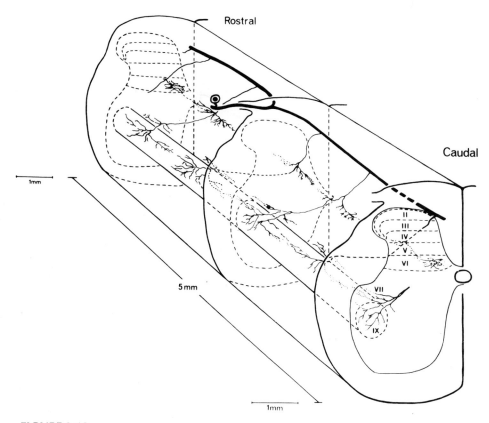

FIGURE 2.12.

Illustration showing how a single primary afferent neuron is distributed within the spinal cord. The entering axon bifurcates into a long ascending and descending branch. Each long branch gives off several short penetrating branches that divide extensively, sending terminal branches among the dendrites of the neurons with which they eventually make contact via synapses.

FIGURE 2.13.

The axon hillock (*AH*) of the neuron is an area that appears pale in the light microscope due to a lack of Nissl substance. Here one can see that a small amount of rough endoplasmic reticulum (*ER*) does remain in the axon hillock, but it does not extend into the initial segment of the axon. Ribosomes (*r*) are also plentiful in the perikaryon, but few enter the initial segment (r_1). Microtubules (*m*), neurofilaments, and mitochondria (*mit*) are prominent components. In this example from a pyramidal cell, a number of synaptic terminals can be seen making contact with the initial segment (At_1, At_2) (×23,000).

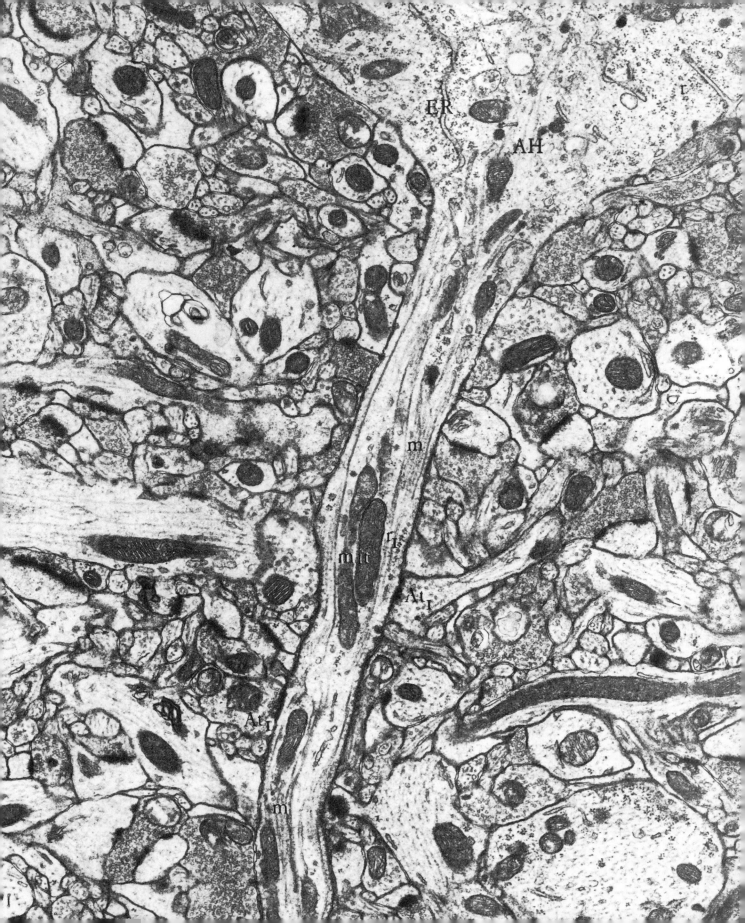

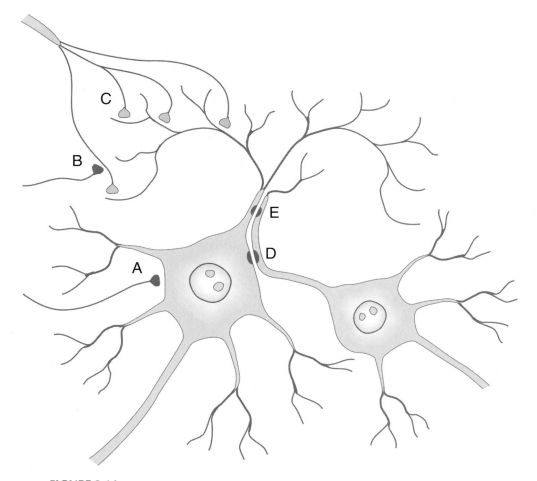

FIGURE 2.14.

Synapses can be formed between almost any neuronal structures. The most common sites for synaptic contact are between axon boutons and neuron somas (**A**), between two axons (**B**), and between axon boutons and dendrites (**C**). Much less common are synapses between dendrites and somas (**D**) or between two dendrites (**E**).

axon terminals and must be transported back to the soma, where they can participate in the regulation of cellular metabolism.

There are two transport mechanisms in axons, one fast and one slow (Table 2.1). Cytoskeletal and secretory proteins are carried from the soma to the boutons (anterograde) [L. *antero,* in front + L. *gradior,* to go] by **slow anterograde axoplasmic flow** at a rate of 0.1 to 10 mm/day. Membrane-bound organelles are carried by **fast anterograde axoplasmic flow** at rates from 50 to 500 mm/day. Materials are carried from the boutons to the soma (retrograde) [L. *retro,* in back + L. *gradior,* to go] by **retrograde axoplasmic flow** at 200 to 300 mm/day.

Slow Axoplasmic Flow

There are two components to the slow rate of axoplasmic flow, slow component A (SC_A) and slow component B (SC_B). The slower, SC_A, transports materials at about 0.1 to 1 mm/day. Somewhat faster, SC_B operates at about 1 to 10 mm/day. The principal cytoskeletal proteins, α- and β-tubulin, neurofilament proteins, and MAPs make up about 80% of the materials

Table 2.1. Major Components of the Axoplasmic Transport Systems

Transport Component	Transport Rate (mm/day)	Substances Transported
Slow Transport		
Anterograde SC_A	0.1–1.0	Cytoskeletal molecules
Anterograde SC_B	1–10	Soluble proteins and enzymes
Fast Transport		
Anterograde	50–500	Membrane-limited vesicles; mitochondria
Retrograde	200–300	Lysosomes, enzymes

transported by SC_A. Actin, myosin, calmodulin, clathrin, and about 200 soluble proteins, including many cytosolic enzymes, are carried at the faster SC_B rate.

The cytosol of the axon is structurally dynamic; its constituents are continuously being replenished by cytoskeletal materials freshly synthesized and assembled in the soma and transported along the axon by slow axoplasmic flow. This is especially evident during the growth and regeneration of axons that proceeds at about 1 mm/day, a rate that roughly corresponds to the fastest SC_A rate of axoplasmic transport. Since actin and myosin are carried in this compartment, it is tempting to speculate that the mobility of the growing axon tip is based on these well-known motor proteins, although the actual mechanisms remain unknown.

Fast Axoplasmic Flow

Fast transport mechanisms carry the intracellular organelles that have membranes. These include the mitochondria, secretory granules, and other membrane-bound vesicles. The mecha- nism of fast transport involves two "motor" proteins, **kinesin** and **dynein,** both of which are able to hydrolyze ATP. One end of these molecules apparently locks onto the organelle membrane. The other end forms temporary bridges to the tubulin of the microtubules and, by sequentially making and breaking these bonds, is able to propel the organelle along the microtubule tracks (Fig. 2.15). Interestingly, the kinesin molecule can only move toward the "head" end of the tubulin, and dynein can move only toward the "tail." Most tubulin is oriented with the "head" toward the axon terminal, making kinesin the anterograde motor. Dynein is responsible for retrograde flow.

Supporting Cells of the CNS— The Glia

The supporting cells of the central nervous system are collectively known as **glia.** Originally, the glia were not recognized as cells, but were thought to be an amorphous intercellular

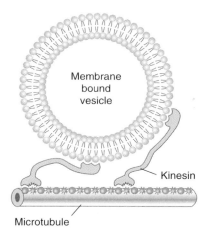

FIGURE 2.15.

Schematic drawing showing the presumed relationship between the transported organelle, kinesin, and the microtubules. It is thought that the kinesin molecule flexes at a central hinge point and ratchets along the microtubule, which acts like a track.

matrix in which the neurons were suspended. The early histologists coined the term glia, which means glue, because it seemed that this matrix "glued" the neurons together. The development of silver stains and the electron microscope has clearly shown the glia to be individual cells. In general the glia cells are small but numerous, being approximately 10 times more abundant than neurons.

Three glial cell types are recognized (Fig. 2.16 *A* and *B*). The most numerous glia are the **astrocytes** and **oligodendrocytes.** The **microglia** are relatively rare. Because the microglia develop from an embryonic tissue type different from the rest of the nervous system (see Appendix **A**), some neuroscientists do not classify them as glia. However, since the microglia play an important physiological role in the nervous system, it is appropriate to consider them as one of the glial elements.

ASTROCYTES

The astrocytes [G. *astron,* a star + G. *kytos,* hollow] are, in the light microscope, star-shaped cells that fill the interneuronal spaces in the CNS. Two forms are recognized, the **protoplas-**

mic and the **fibrous** astrocytes (Fig. 2.16 *A* and *B*). The former occur in the gray matter, while the latter are found in the white matter. Their names are derived from the presence or absence of cytoplasmic fibers that can be discerned with light microscopy in appropriately stained material. Electron microscopy reveals that both types of astrocytes contain numerous 8- to 9-nm fibrils. These fibrils, though similar in appearance to neurofilaments, are composed of polymerized strands of **glia fibrillary acidic protein (GFAP),** not cytokeratins. At the ultrastructural level, the difference in the amount of GFAP fibrils present in the two types of cells is a matter of degree (Figs. 2.17 and 2.18).

Astrocytes fill virtually the entire extraneuronal space in the CNS of mammals. Their processes are long and sinuous, wrapping around boutons, dendrites, and neuronal cell bodies, filling the tiniest crevasses between neuronal elements (Fig. 2.18). Where astrocytes are in apposition with one another or with oligodendrocytes, they frequently form **gap junctions.** These gap junctions are identical to the nexuses of epithelial cells and smooth and cardiac muscle. Astrocytes also envelop capillaries and establish a boundary at the surface of the CNS. At the surface of the CNS, the astrocytes form a layer several micrometers thick between the pia mater and the neuronal elements, where there are also a basal lamina and collagen (Fig. 2.19). This layer of astrocytes is frequently referred to as the **glial limiting membrane.**

A number of functions have been proposed for astrocytes. One of the most persistent ideas recognizes that they provide gross support for the CNS since the fibrils within the astrocytes makes them fairly rigid cells and renders some stiffness to the otherwise malleable neurons. Furthermore, except at the points of synaptic contact, they separate neurons from one another. Astrocytes also serve as reservoirs for potassium ions. The significance of this will be made apparent in the next chapter.

In the embryo, radial glial cells appear to play a critical role in early CNS development by providing a pathway along which neurons can migrate. In the neural tube, for example, radial glial cells form slender processes that extend from the

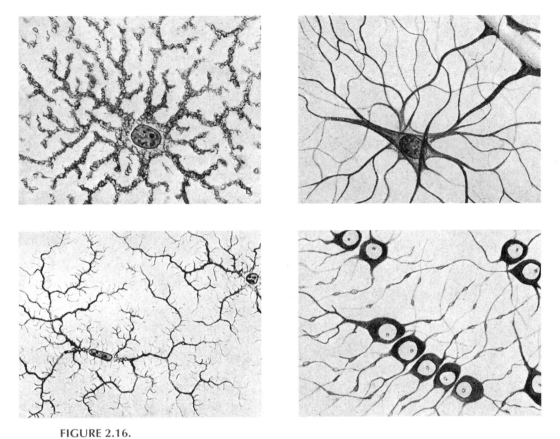

FIGURE 2.16.

The neuroglia as they appear in silver stained material in the light microscope. **A,** Protoplasmic astrocyte; **B,** Fibrous astrocyte; **C,** Microglia; **D,** Oligodendrocyte.

central canal to the surface. They form a network along which the neuroblasts migrate. Once the neuroblasts are in place, the radial glial cells differentiate into mature astrocytes (Fig. 2.20).

The pathfinding role that astrocytes play during development seems to be lost in the adult mammal. For example, after CNS injury, astrocytes proliferate. They probably participate in the phagocytosis of cellular debris but, more importantly, they form a "glial scar" by filling in and occupying the space created by the destruction of CNS tissue. As a further response to injury, damaged neurons attempt to reestablish their former connections by elaborating new axons. These regenerating axons are unable to penetrate this glial scar, and ultimately CNS regeneration fails (see Chapter **4**). However, if an embryonic CNS graft is placed into an adult CNS wound, the embryonic neurons frequently do survive and establish functional synapses. The embryonic astrocytes seem to play a critical role in this process, but the mechanisms are far from being understood. What is clear is that *embryonic astrocytes are different from mature astrocytes* and that that difference is critical to neuronal growth, development, and regeneration.

OLIGODENDROCYTES

Oligodendrocytes [G. *oligos,* few + G. *dendron,* tree] are small glial cells with, as their name implies, few processes (Fig. 2.16*D*). They are most commonly found in the white matter of the CNS, but they do exist in the gray matter as well. The oligodendrocytes produce myelin in the central nervous system.

MICROGLIA

The microglia are small cells with a few spindle-shaped processes (Fig. 2.16C). They are found in both white and gray matter, but in normal tissue are not abundant in either place. Only about 5 to 10% of the total glia population is composed of the microglia.

The microglia are normally quiescent, but become active in response to brain injury or inflammation. Once activated they proliferate and migrate to the site of injury, where they engulf and phagocytose cellular debris, myelin fragments, and injured neurons. When vascular injury or inflammation accompanies brain injury, macrophages from the bloodstream and pericytes also invade the neural tissue and participate in the phagocytotic activity. Under these conditions, microglia and macrophages are indistinguishable. To acknowledge this ambiguity, the term microglia is usually reserved for the resting microglial cells that are a normal component of the ensemble of cells forming the central nervous system. The cells that respond to brain injury, whether they be microglia or macrophages, are simply called **phagocytes.**

Supporting Cells of the PNS— The Schwann Cell

In the peripheral nervous system, there is only one type of supporting cell, the **Schwann cell.** In the PNS Schwann cells perform all of the basic functions that are accomplished by the three types of glia cells in the CNS. For example, Schwann cells enclose unmyelinated axons and separate them from each other, much like the astrocytes do in the CNS (Fig. 2.21). Like astrocytes in the CNS, they reside in the interneuronal space between neuron somas. Like the microglia, they can become phagocytes under conditions of inflammation or peripheral nerve injury. And like the oligodendrocyte in the CNS, they produce myelin for the axons.

Schwann cells secrete **laminin, fibronectin, and collagen,** proteins that are the principal components of the neuronal **basal lamina** and **extracellular matrix.** Unlike the glia, the Schwann cell secretes a basement membrane, or basal lamina, that surrounds the cell membrane. The extracellular matrix is found between the axons in nerve trunks, where it is a principal constituent of the epineurium (see Chapter **4**).

Neurons in the peripheral ganglia are surrounded by **satellite cells.** Originally thought to be an independent cell type, satellite cells are now recognized to be indistinguishable from Schwann cells. In the ganglion, a single layer of Schwann cells (formerly called satellite cells) surrounds the ganglion cell. Much like the astrocytes in the CNS, the Schwann cells effectively isolate the neuron from the general extracellular environment. The intercellular space between the neuron and Schwann cell is reduced to about 20 nm.

This investment of the neuron by the Schwann cells continues outside the ganglion and includes the entire axon. Most of the larger axons are enclosed by myelin, a product of Schwann cells. The enclosure of unmyelinated axons is simply accomplished by a series of Schwann cells nestling axons into a trough of membrane-bound cytoplasm around their circumference (Fig. 2.21). Through these mechanisms, both the myelinated and the unmyelinated axons are isolated from the extracellular environment over most of their surface.

FIGURE 2.17.

The large, mottled nuclei of two fibrous astrocytes (*Nuc*) are illustrated here in contrast to the dark nucleus of the oligodendrocyte (*O*). The relationship between the astrocyte processes (*AsP*) and the bundles of myelinated axons is apparent. Within the cytoplasm of the astrocytes, numerous fibrils are the most prominent feature (×17,000).

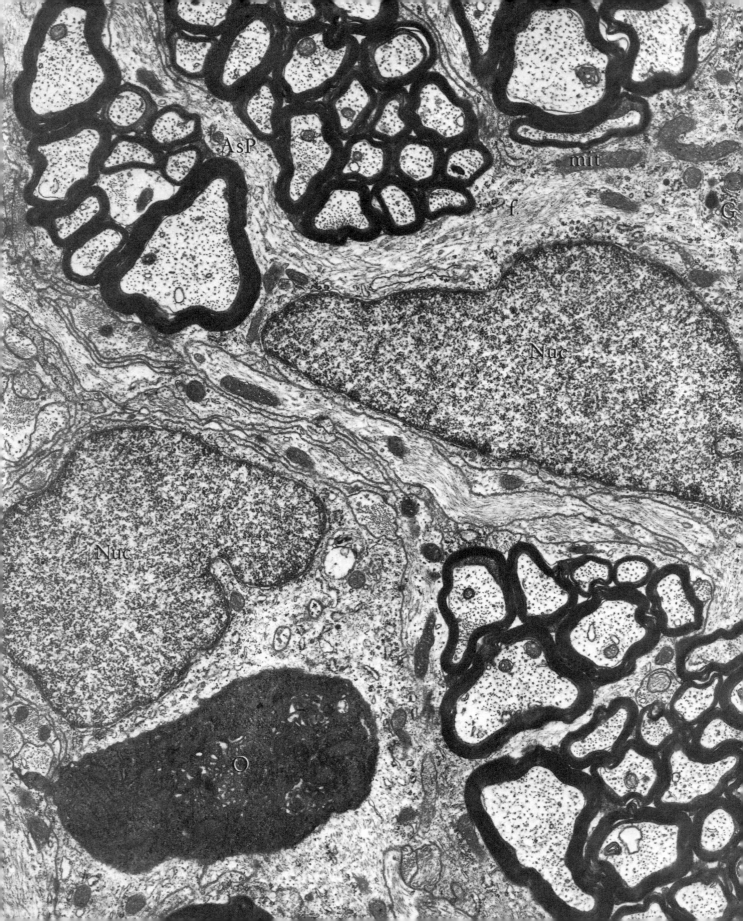

Myelin

Myelin is produced in the central nervous system by the oligodendrocytes and in the peripheral nervous system by the Schwann cells. Both cells produce myelin by wrapping themselves around the axon to form a dense laminated structure consisting of lipids and membrane proteins. Myelin is arranged in segments along the length of an axon, the segments being separated longitudinally by **nodes of Ranvier** (Fig. 2.22). There are important differences between central and peripheral myelin based on the differences in how they are produced.

MYELIN FORMATION

Peripheral myelin is produced early in development when Schwann cells become associated with bundles of developing axons in the nerve. Individual axons become nestled in furrows along the margin of Schwann cells (Fig. 2.21). As the axons elongate, the Schwann cells associated with them proliferate by mitotic division to maintain a continuous covering for the axons.

As the individual axons mature, many of them increase in diameter. When an axon exceeds approximately 1 μm in diameter, it leaves the bundle of immature, smaller axons and attracts a number of individual Schwann cells, which begin to envelop it. This results in a continuous series of Schwann cells becoming associated with a single axon along its entire length. As these Schwann cells mature, each develops a single myelin segment for the axon. *An individual Schwann cell produces one segment of myelin for a single axon.*

Myelination is initiated when the axon becomes nestled into a crevice of the Schwann cell and becomes completely enclosed (Fig. 2.23). The apposition of the outer layers of the Schwann cell membrane where the two edges of the Schwann cell converge is called a **mesaxon** [G. *mesos,* middle + G. *axon,* axis]. One of these edges slides under the other, forming an inner and an outer lip. The outer lip contains most of the cytoplasm and the nucleus of the Schwann cell. The best current evidence now supports the theory that the inner cytoplasmic lip lengthens and slides under and around the outer, stationary, cytoplasmic lip. As the migration of the inner lip proceeds, the mesaxon necessarily elongates, eventually becoming layered upon itself. The cytoplasm is squeezed out of the diminishing intercellular space between the spiraled mesaxon membranes until only a compact membranous structure remains around the axon. Ultimately, the Schwann cell cytoplasm and nucleus are relegated to a small peripheral band lying around the external perimeter and to cytoplasmic pockets at the edges of the myelin segments (Fig. 2.22).

The characteristic laminated appearance of myelin is determined by the asymmetrical nature of the cell membrane. Where the *outer surfaces* of the apposing Schwann cell membranes come in contact with one another at the mesaxon, an **intraperiod line** is formed. As the membranes spiral around the axon and the membranes meet again, the *inner surfaces* of the membranes meet and form a **major dense line.** The intraperiod line (outer surface apposition) appears *lighter* than the major dense line (inner surface apposition) in electron micrographs, giving myelin is characteristic laminated appearance (Fig. 2.24).

In the central nervous system, the oligodendrocytes do not individually attach themselves to

---

FIGURE 2.18.

The nucleus of a protoplasmic astrocyte (*Nuc*) is characterized by a mottled appearance with condensed nuclear material lining its external limiting membrane. The protoplasmic astrocyte contains numerous filaments (*f*), mitochondria (*mit*), endoplasmic reticulum (*ER*), and occasional lysosomes (*Ly*). The cell insinuates itself between the other cellular elements of the CNS. In this way, astrocytes occupy almost the entire extracellular space. The approximate outline of this astrocyte (thin line and *arrowheads*) illustrates this remarkable characteristic (×20,000).

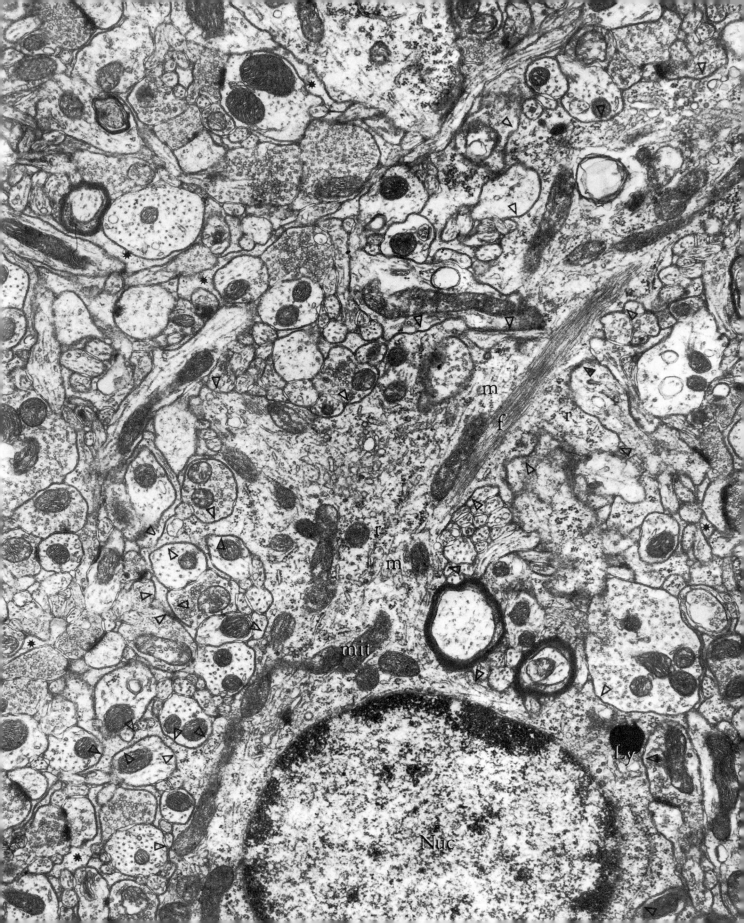

single axons during development. Rather, a single oligodendrocyte extends processes to several axons. Each process envelops a different axon and forms one segment of myelin (Fig. 2.25). *A single oligodendrocyte forms one myelin segment on 20 to 60 different axons.*

Once an oligodendrocyte process has become associated with an axon, the axon becomes enveloped and a mesaxon is formed in essentially the same manner as in the peripheral nervous system. However, axons of the CNS always have fewer spirals of myelin than peripheral axons of the same diameter.

In the CNS the cytoplasm of the oligodendrocyte is more completely extruded from the myelin structure than in the peripheral nervous system. There is no enveloping cytoplasmic blanket surrounding the myelin in the CNS; the remaining cytoplasm is reduced to a thin ridge running along the surface and to pockets at the edges of the segment.

Axons are myelinated in a segmental fashion. The segments meet but do not join one another at **nodes of Ranvier** (Figs. 2.22 and 2.26). At the nodes, the major dense lines separate and cytoplasmic pockets form the edge of the myelin segment. These pockets are continuous and themselves spiral from the inner layer near the axon to the surface, where they meet with the surface cytoplasm of the Schwann cell or oligodendrocyte. Thus cytoplasmic continuity is maintained throughout the cell (Fig. 2.23*F*).

At the center of the nodal space, the axon is exposed more or less directly to the extracellular environment. One can detect an osmophilic dense undercoating along the axon membrane at the node, similar to that seen at the initial segment of the axon (Fig. 2.26). Although the significance of this observation is not yet clear, it is tempting to speculate that the density is a manifestation of a concentration of voltage-sensitive gates that are characteristic of excitable cells (see Chapter **3**).

THE CHEMICAL STRUCTURE OF MYELIN

Myelin is a chemically complicated structure. Approximately 70% of myelin is lipid, with the remainder being composed of several different proteins. Most of the lipid is cholesterol and cerebroside. Central and peripheral myelin do not appear to have important differences in lipid content.

There are several different proteins that are important in myelin structure. To date, seven genes have been cloned that either produce the structural proteins of myelin itself or are involved in regulating its formation. Three proteins are particularly important: **myelin basic protein (MBP), proteolipid protein (PLP), and protein zero (P_0).**

MBP is common to both central and peripheral myelin. It constitutes about 30% of the total protein in CNS myelin but no more than 18% in PNS myelin. It is a membrane-bound protein that resides entirely on the cytoplasmic face (Fig. 2.28). The MBP molecule is formed from seven different protein cassette exons, encoded by a single gene. The final MBP molecule is determined by pre-mRNA splicing when one or more of the cassettes is deleted. The final MBP molecule may be assembled in any of a number of different forms depending on which cassettes are deleted (Fig. 2.27). The functional significance of this complicated mechanism is not clear.

MBP is essential to adhesion of the CNS myelin structure at the major dense line (Fig. 2.28). This has been aptly demonstrated in certain mutant mice that synthesize defective MBP. The small amount of CNS myelin produced in these animals is lacking major dense lines, although the intraperiod lines appear normal. Interestingly, the peripheral myelin appears to be

---

FIGURE 2.19.

The processes of numerous astrocytes form the glial limiting membrane. The wall of an arteriole (*A*) lies at the top of the figure, with the pia mater (*PM*) immediately beneath it. A basal lamina (*B*) lies over the most extreme layer of astrocytes (*As*) that are conspicuously filled with numerous filaments (*f*). At the bottom of the figure are a number of small axons (*Ax*) (×15,000).

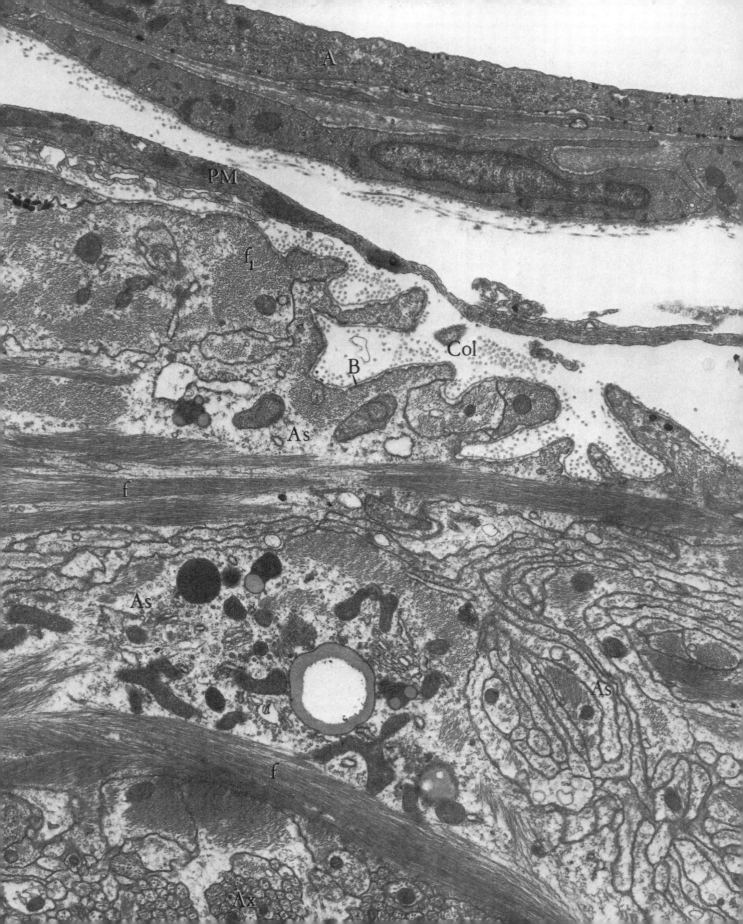

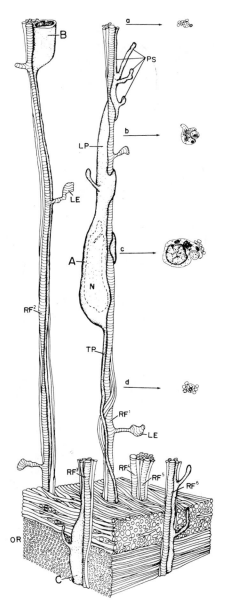

FIGURE 2.20.

Radial glia serve as tracks along which neuroblasts migrate during development. Two radial glia are shown here with neuroblasts in different stages of migration. The leading process (*LP*) follows the radial glial fiber (*RF₁*) by sending out exploratory pseudopodia (*PS*). Note the trailing process (*TP*), which will become the axon.

normal in these animals in spite of the fact that it contains the same defective MBP.

PLP is a membrane protein found only in central myelin, accounting for about 50% of the myelin protein. It exists mostly on the extracellular side of the membrane, but has both cytoplasmic and intercellular domains (Fig. 2.28). PLP is essential for the adhesion of myelin at the intraperiod line. It links the two outer layers together by linking with another PLP molecule from an adjacent membrane.

The PLP gene is also subject to a number of mutations seen in various mammalian species, including the human, where it is responsible for the rare **Pelizaeus-Merzbacher disease.** This disease is characterized (at autopsy) by an almost complete absence of CNS myelin with normal peripheral myelin. Genetic analysis has shown that a similar disease in mice is caused by a single base mutation that results in a single amino acid substitution from among the 276 amino acids in the PLP molecule.

P_0 is a membrane glycoprotein found only in peripheral myelin, and it accounts for about half of the protein in peripheral myelin. Although it assumes the same role in peripheral myelin that PLP plays in central myelin, its structure is entirely different. Like PLP, P_0 is a membrane-spanning protein, but most of the molecule is on the extracellular side (Fig. 2.28). It is a homophilic cell adhesion protein and plays a critical role in attaching the myelin membranes at the intraperiod line by linking to itself. However, unlike PLP it also apparently plays a role in forming the major dense line, since peripheral myelin appears to be normal in the absence of functional MBP. There is no recognized human genetic disease associated with mutations of the P_0 gene.

In addition to Pelizaeus-Merzbacher disease,

FIGURE 2.21.

In the peripheral nerve, a single Schwann cell forms one node of myelin around a single axon (*Ax₁*). A single Schwann cell (*SC*) will envelop a large number of unmyelinated axons, as shown here. Note the basal lamina (*B*) that surrounds the external surface of the Schwann cell. In cases where the Schwann cell cytoplasm does not fully enclose an axon (*Ax₂*), the basal lamina extends over it.

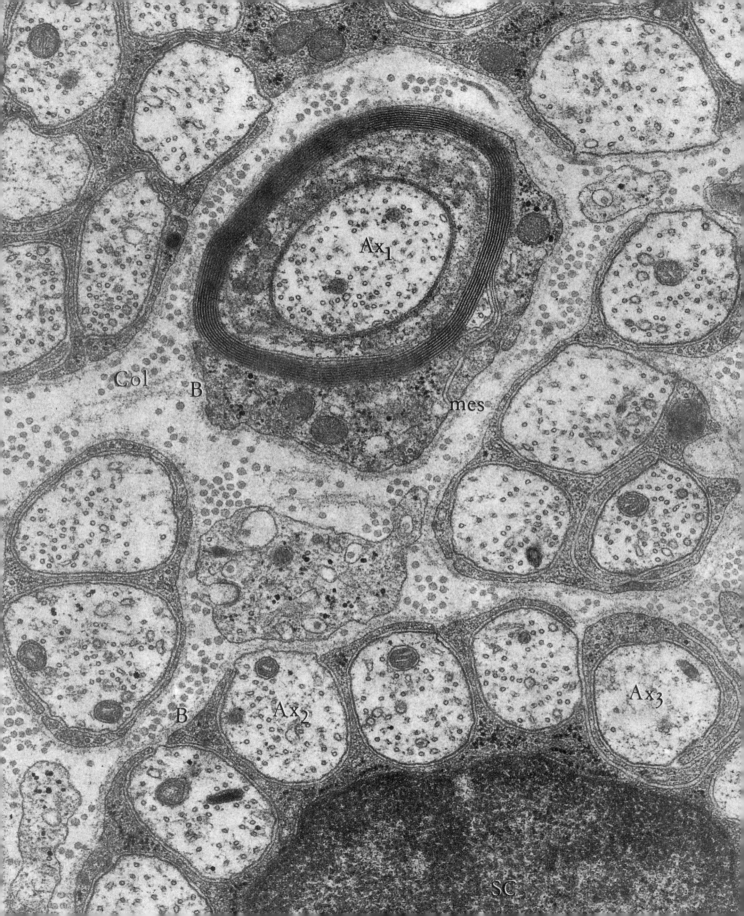

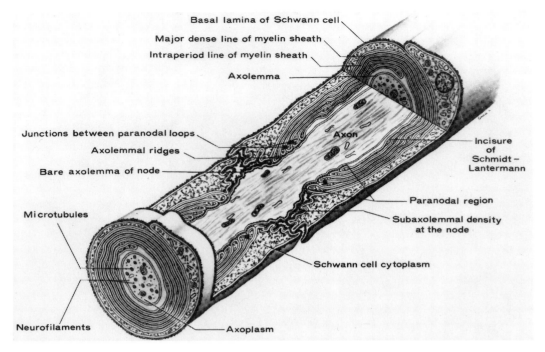

FIGURE 2.22.

The principal structures of peripheral myelin are illustrated here. Note particularly how, in the paranodal region, the Schwann cell cytoplasm is squeezed to the edges of the laminae, forming the paranodal loops. The cytoplasmic pockets in these loops are continuous, spiraling toward the surface, where they merge with the main cytoplasmic mass of the cell.

there are a number of genetically determined diseases of myelin, grouped together simply as **leukodystrophies,** that are characterized as progressive neurologic diseases, caused by metabolic errors of myelin. Some, like Pelizaeus-Merzbacher disease, are clearly genetic with a recognized inheritance pattern and a known gene error. Most, however, are simply "unclassified" leukodystrophies because their precise etiology is unknown. All are relentlessly progressive and ultimately fatal.

MYELIN AS AN ANTIGEN

The myelin proteins are potent antigens. In rats, two important autoimmune diseases can be initiated by immunizing the animals to their own myelin. If central myelin is used as the antigen, **experimental allergic encephalomyelitis (EAE)** develops. This disease is characterized by chronic, relapsing episodes of CNS demyelina-

tion. Peripheral myelin is spared. Conversely, peripheral myelin can be used to induce a similar disease, **experimental allergic neuritis (EAN),** that attacks only myelin associated with the peripheral nerves, leaving the central myelin unscathed. These two experimental diseases, EAE and EAN, resemble to a remarkable degree two human diseases, **multiple sclerosis** and the **Guillain-Barré syndrome,** respectively. Although an autoimmune etiology has not been conclusively established for either of these human disease, immunological attack is clearly a major component of the disease process.

SUMMARY

The nervous system is a complex structure composed of only three cell types: neurons, glia, and Schwann cells. The nervous system of mammals is separated into central and peripheral divisions. The central nervous system (CNS) lies within the meninges and consists of neurons and

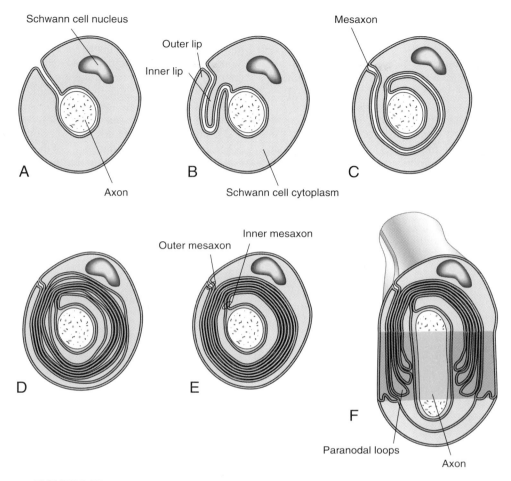

FIGURE 2.23.

The various stages in the formation of myelin (see the text for a more complete description). **A,** The mesaxon is formed by the apposition of two outer layers of the Schwann (or oligodendrocyte) cell membrane. **B,** The inner lip begins to wrap itself around the axon by invaginating the cytoplasm of the glial support cell. **C** and **D,** Further elaboration of the myelin structure. **E,** The final form is achieved (for Schwann cells). **F,** A cross section and longitudinal section exposing the paranodal loops.

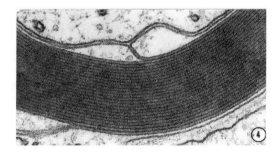

FIGURE 2.24.

The relationships between the various components of myelin are clearly shown in this electron micrograph. The inner and outer layers of the cell membrane are particularly well preserved. Note that within the structure of the myelin, the dark, major dense line is formed by the apposition of the *cytoplasmic* side of the cell membrane, while the intraperiod line is formed by the apposition of the *extracellular* side of the membrane.

glia. The peripheral nervous system (PNS) is outside the meninges and consists of neurons and Schwann cells. In the CNS, neuron cell bodies are collected together as nuclei and their axons form tracts. In the PNS, neuron cell bodies are found in ganglia. Their axons and the axons of certain CNS neurons form peripheral nerves.

Neurons are highly energy-dependent secretory cells. In the neuron soma there is a well-developed major membrane system within which the proteins are produced. In neurons the rough

endoplasmic reticulum is called Nissl substance. Neurons have a cytoskeleton that is composed of microtubules, neurofilaments, and microfilaments. The cytoskeleton determines the shape of the cell, provides an intercellular transport mechanism for organelles and macromolecules, and renders some stiffness to the cells.

Neurons have two types of processes, dendrites and a single axon. Dendrites, whose primary purpose is to increase the surface area of the cell, are essentially extensions of the soma. In some cells

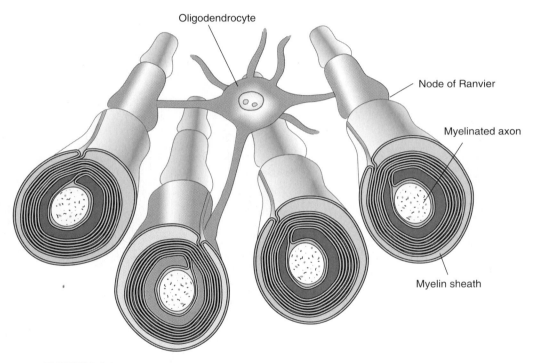

Oligodendrocyte

Node of Ranvier

Myelinated axon

Myelin sheath

FIGURE 2.25.

The relationship between a single oligodendrocyte and the many axons it myelinates can be seen in this schematic view. Note that, like the Schwann cell, the oligodendrocyte myelinates only a single node on any given axon, but unlike the Schwann cell the oligodendrocyte myelinates segments on many axons. The myelin maintains its contact with the parent cell body by a long, thin process.

FIGURE 2.26.

At the node of Ranvier, the myelin becomes progressively thinner as the inner layers terminate in the paranodal loops of cytoplasm (*P*). Between segments of myelin there is a short distance where the axon is exposed to the extracellular space. Here the dense undercoating (*D*) is apparent ($\times 60,000$; inset, $\times 100,000$). (*SR*, smooth endoplasmic reticulum; *m*, microtubule; *nf*, neurofilament; m_1, microtubules in cross section within paranodal loops; *Al*, axolemma.)

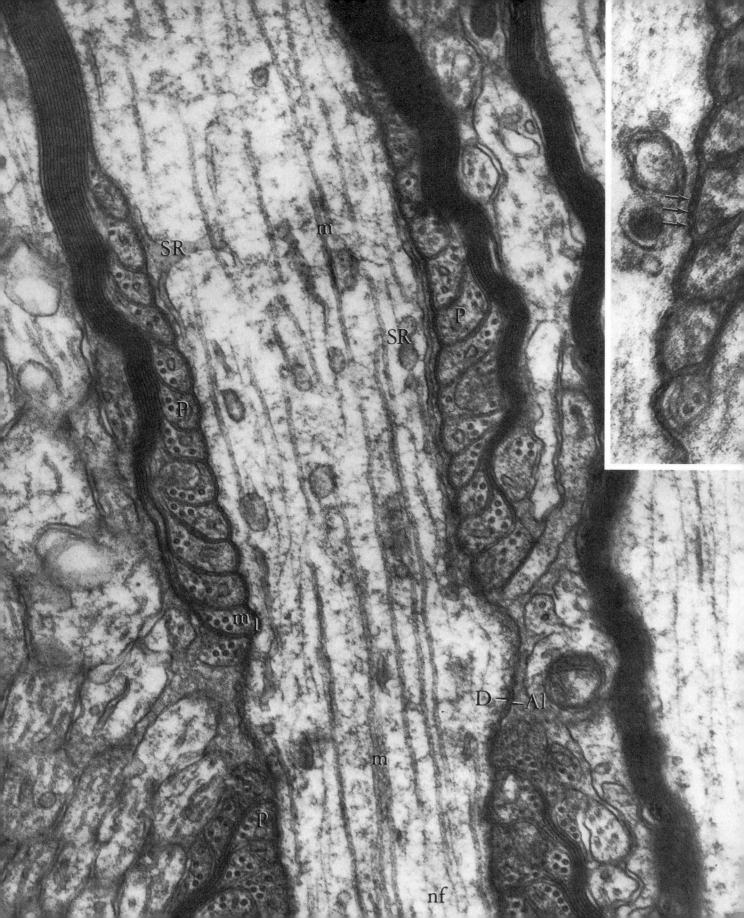

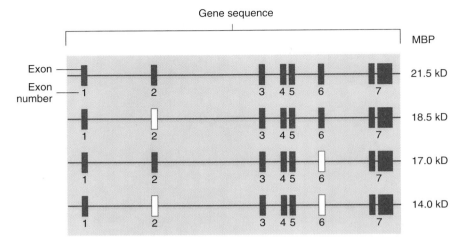

FIGURE 2.27.

Myelin basic protein (MBP) is not a single protein, but a family of closely related proteins encoded by a single gene. The various forms of MBP are assembled by exon deletion (*empty boxes*) at the pre-mRNA stage of protein synthesis. Here, a schematic representation of the assembly of the various cassettes that make up various forms of MBP, each with a different apparent molecular weight, is shown.

dendrites have spines that further increase the surface area. Most synaptic contact is made on dendrites. Axons are thinner, but longer than dendrites. They have three regions, the initial segment, the axon proper, and the terminal bouton. Axons are branched, but unlike dendrites that branch along their entire length, axonal branching takes place near the area of termination. The initial segment of the axon has specialized voltage-sensitive gates that are essential for the initiation of the action potential. The axon proper contains most of the cytoplasm of the neuron. The tip of the axon, the synaptic bouton, contains specialized structures necessary for chemical synaptic communication with other cells.

Neurons have elaborated intracellular transport mechanisms to a high degree to convey organelles and macromolecules along the great length of the axon, a process known as axoplasmic flow. There are two types of axoplasmic flow, slow and fast. The fast system transports materials both anterograde and retrograde; the slow system is only anterograde.

Three types of supporting cells, or glia, are found in the central nervous system: astrocytes, oligodendrocytes, and microglia. Astrocytes separate and isolate neurons from one another and help regulate the extracellular milieu. The oligodendrocytes form myelin in the central nervous system. Microglia become phagocytes under conditions of inflammation or CNS infection.

The Schwann cells are the only supporting cells in the peripheral nervous system. They perform all the same functions as the three types of glia cells.

Many axons in both the central and peripheral nervous system are invested with myelin. Myelin is made by Schwann cells and oligodendrocytes. Each Schwann cell produces a single segment of myelin for a single axon. The oligodendrocyte also produces a single segment of myelin for a given axon, but it is capable of myelinating as many as 60 different axons.

Myelin is formed from lipids and several different proteins. The most important of these are myelin basic protein (MBP), proteolipid protein (PLP), and protein zero (P_0). MBP is common to both central and peripheral myelin. It is the major molecule responsible for the adhesion of the cytoplasmic layers of the cell membranes (the major dense line). PLP is a component only of central myelin, where it attaches the intercellular side of the myelin membranes together, forming the intraperiod line. P_0 is the corresponding protein for peripheral myelin.

Myelin production is subject to a number of genetically determined disorders. At autopsy these genetic diseases are characterized by severe hypomyelination and are classified as leukodystrophies. Either the lipid or the protein components of myelin may be affected.

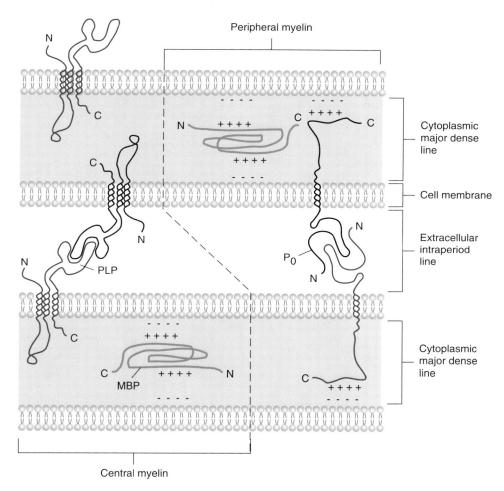

FIGURE 2.28.

Schematic representation of the molecular composition of central and peripheral myelin (see text for details). Myelin basic protein (MBP) is common to both peripheral and central myelin and is the major adherent molecule attaching the cytoplasmic surfaces of the membranes together. PLP is found only in central myelin. Multiple PLP molecules hold the extracellular surfaces together by attaching to sister molecules. P_0 is the peripheral analog to PLP, holding the membrane together at the intraperiod line by self-adhesion. P_0 seems also to play a prominent role in membrane attachment at the major dense line.

The proteins in myelin are potent antigens, and the differences in the chemical composition of central versus peripheral myelin are important for understanding their immunogenic potential. Multiple sclerosis appears to be an autoimmune disease directed against central myelin, while the Guillain-Barré syndrome seems also to be an autoimmune response, but against peripheral myelin.

FOR FURTHER READING

Fawcett, D. and Raviola E. *A Textbook of Histology.* Philadelphia: Chapman and Hall, 1994.

Hall, Z. W. *An Introduction to Molecular Neurobiology.* Sunderland, MA: Sinauer Associates, 1992.

Matthews, J. L. and Martin, J. H. *Atlas of Human Histology and Ultrastructure.* Philadelphia: Lea & Febiger, 1971.

Pappas, G. and Purpura, D. *Structure and Function of Synapses.* New York: Raven Press, 1972.

Peters, A., Palay, S., and Webster, H. *The Fine Structure of the Nervous System.* New York: Oxford University Press, 1991.

Siegel, G., Agranoff, B., Albers, R. W., and Molinoff, P. *Basic Neurochemistry.* New York: Raven Press, 1989.

Smith, C. U. M. *Elements of Molecular Neurobiology.* New York: John Wiley & Sons, 1989.

Watson, J., Tooze, J., and Kurtz, D. *Recombinant DNA: A Short Course.* New York: W. H. Freeman, 1983.

3 Electrochemical Signaling Systems

Signaling within and between cells is essential for the development and survival of multicellular organisms. In complex organisms, the signaling mechanisms are predominantly chemical. Within cells, signaling mechanisms direct the transcription of genes, the translation of proteins, and the distribution of proteins within the nucleus, within the cytosol, or to the cell surface where they may be secreted. Intercellular signaling mechanisms may regulate the activity of adjacent cells through chemical synapses or distant cells by means of hormones. The behavior of entire organisms may be manipulated chemically at great distances by pheromones. Chemicals are probably the most omnipresent and diverse signaling mechanisms used by living systems.

The nervous system is the most complex signaling system to evolve from living matter. Neurons, unlike other cells, use two types of signaling mechanisms, chemical and electrical. The interchange of information between these two mechanisms within the same organ system allows for complex transformations of information that would not otherwise be possible. Information transformations are the means by which the higher-order functions of the nervous system are accomplished. The basic mechanisms of electrical and chemical signaling within and among neurons are discussed in this chapter.

Electrochemical Potentials

In the mammalian nervous system, *electrical signals are represented by changes in potential electrical energy across the neuron cell membrane*. The source of the potential electrical energy is derived from the electromagnetic force[1] associated with **electrical charge**. Charge, as Benjamin Franklin discovered, exists in two forms, labeled for convenience positive (+) and negative (−). The electromagnetic force acts on charges, causing like charges to be repelled and opposite charges to be attracted. Therefore, if one physically separates opposite charges, energy is expended and *work is performed on the system*. If one releases these charges, they will come together, returning the energy that was expended in separating them. Therefore, *separated charge represents potential electrical energy*. This simple relationship is expressed as:

$$\text{Energy} = \frac{\text{Work}}{\text{Charge}} \qquad (3.1)$$

or, expressed in units of measure:

$$\text{Volts} = \frac{\text{Joules}}{\text{Coulombs}} \qquad (3.2)$$

MEMBRANE PERMEABILITY

In living systems, charges are separated by the cell membrane. The hydrophobic lipid bilayer structure of the membrane is quite resistant to penetration by any hydrophilic substance, including ions. The hydrophobic nature of the membrane is greatly modified, however, by membrane-

---

[1]In physics, four forces are recognized: the strong force, the weak force, the electromagnetic force, and the gravitational force. The interactions of these forces and the five basic physical quantities—temperature, length, mass, time, and charge—represent, for scientific empiricism, the definition of reality.

spanning proteins. Composed of long chains of amino acids, these proteins twist and bend to form a three-dimensional shape that contains a central channel or pore that allows passage of ions through the lipid backbone (Fig. 3.1). The three-dimensional, or tertiary, structure of these molecules is very specific. Depending on the tertiary structure of the protein, the channels can be very selective with regard to which ions can pass, making the cell membrane a **selectively permeable** barrier.

SINGLE-ION SYSTEMS

To visualize how a selectively permeable membrane allows charge separation to develop in a cell, imagine a hypothetical pure lipid membrane that has no channels; it is impermeable. Suppose that a solution containing 10 mEq of KCl bathes one side of the membrane and a solution containing 100 mEq of KCl bathes the other (Fig. 3.2A). Since there are no channels for either ion, there is no ion flux across the membrane (assume in this hypothetical example that

the lipid backbone is strictly impermeable). In each compartment all ions are paired with an electrically opposite partner. Electrical neutrality is maintained on both sides of the membrane, and thus there is no charge separation and consequently no membrane potential.

If nonselective holes were placed in the membrane, *pairs* of Cl^- and K^+ ions would diffuse down their concentration gradient until equilibrium was established. Electrical neutrality, however, would still exist at all times. There would be no charge separation and no membrane potential (Fig. 3.2B).

If, however, one were to place channels in the membrane that would allow only K^+ to pass, excluding Cl^-, an entirely different situation would exist at equilibrium (Fig. 3.2C). As the concentration gradient caused K^+ to diffuse through the membrane, a separation of charge would be established because, in our hypothetical environment, Cl^- could not diffuse with the K^+. This establishes a separation of charge across the membrane. One side of the system becomes neg-

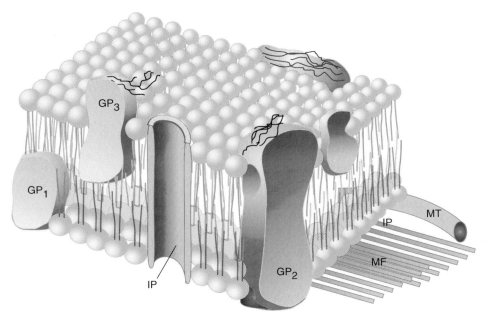

FIGURE 3.1.

The cell membrane's lipid bilayer serves as a matrix within which a number of other structures, mainly glycoproteins (*GP*), reside. Some of these proteins are membrane spanning (*GP$_2$*), while others reside primarily on the external (*GP$_3$*) or the internal (*GP$_1$*) side. Ionophores (*IP*) are membrane-spanning proteins that form ion-selective channels through the membrane. Microfilaments (*MF*) are numerous at the internal surface, along with a few microtubules (*MT*).

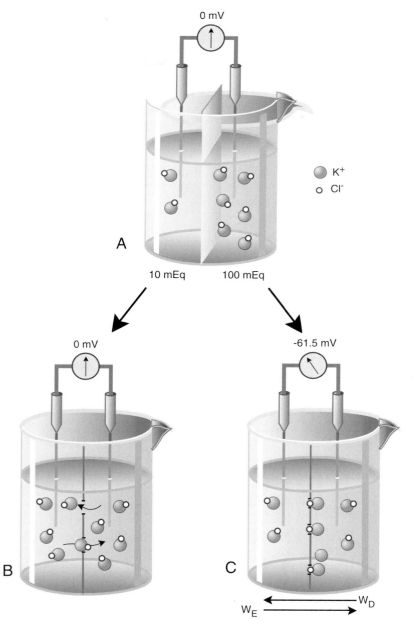

FIGURE 3.2.

A, In this hypothetical situation, two solutions of KCl are separated by an impermeable membrane. Note that the K$^+$ and the Cl$^-$ are associated and migrate together. The
voltmeter, connected by electrodes to the solutions on either side of the membrane, indicates no electrical potential across the membrane. **B**, In this situation the separating
membrane is permeable, but not selective. Charge separation does not occur because
the paired ions migrate together. No electrical potential is established across the membrane. **C**, If the separating membrane is permeable but selective, then charge separation
will occur and an electrical potential can be established. Note that the ions remain
paired but oriented across the membrane. Although the principle is valid, this drawing
is very schematic. In real membranes the ions retain a great deal of mobility and are not
really "stuck" in the membrane as shown.

ative with respect to the other side due to this excess Cl^-.

The work of diffusion separates the charge and establishes an electrical force that draws the K^+ against the diffusion gradient. When equilibrium is established, the **work of diffusion** is just balanced by **electrical work**. *At equilibrium a residual concentration gradient and a residual charge separation exists across the membrane.* Potassium ion fluxes remain, but there is no *net* movement of ions across the membrane.

This phenomenon can be expressed mathematically. The work of diffusion is related to the diffusion gradient by the following expression:

$$W_D = RT \ln \frac{[K^+]_{OUT}}{[K^+]_{IN}} \qquad (3.3)$$

where $[K^+]_{OUT}/[K^+]_{IN}$ is the concentration gradient for potassium ions, **R** is the gas constant (8.3143 joules/mole-degree), and **T** is the absolute temperature. Similarly, the electrical work related to the separation of charge is expressed as:

$$W_E = V_m F Z \qquad (3.4)$$

where V_m is the membrane potential measured in volts, **F** is Faraday's constant (96,487 coulombs/mole), and **Z** is the ionic valence.

At equilibrium, the two forces are balanced so Equations (3.3) and (3.4) can be equated.

$$V_m F Z = R T \ln \frac{[K^+]_{OUT}}{[K^+]_{IN}} \qquad (3.5)$$

After rearrangement Equation (3.5) becomes:

$$V_m = \frac{RT}{FZ} \ln \frac{[K^+]_{OUT}}{[K^+]_{IN}} \qquad (3.6)$$

which is commonly known as the **Nernst equation**. A potential calculated in this way is referred to as the **equilibrium potential** or **Nernst potential**. This equation applies to simple systems that are selectively permeable to a single ion species. At 37°C, **RT/FZ** reduces to the constant 0.0267 for univalent ions. In our example, the natural log of 10 mEq/100 mEq is −2.302, which, multiplied by the constant, predicts an equilibrium potential of −61.5 mV.

If, in our hypothetical cell, the K^+Cl^- on each side of the membrane were replaced with Na^+Cl^- of equal concentration, there would be no transmembrane potential because the sodium ions could not pass through a membrane that contains only potassium-selective channels. Electrical neutrality would exist on both sides of the membrane. To develop a membrane potential with the sodium ions, one must add sodium-selective channels. Then, sodium ions would pass down their concentration gradient until enough charge separation developed to balance the diffusion force for sodium.

MULTI-ION SYSTEMS

In physiological systems, the interior and exterior milieus contain many ions, some of which are permeable to the membrane. In neurons, the most relevant ions are Na^+ and K^+. Each of these ions, when considered independently, has an equilibrium potential associated with it, and each contributes to the resting membrane potential. Chloride ions are not relevant in this context because they behave as if they were freely permeable to the membrane. Therefore they are distributed passively across the cell membrane in response to whatever membrane potential is generated by other ions.

Table 3.1 summarizes the distribution of the physiologically important ions in mammalian skeletal muscle and the calculated equilibrium potential for each ion. The actual membrane potential recorded from a cell reflects the potential generated simultaneously by all of these ions. However, all the ions do not contribute equally to the membrane potential because they do not all penetrate the membrane with equal facility.

Consider a hypothetical cell (Fig. 3.3A) with, initially, only K^+ channels inserted into the membrane; Cl^- is freely permeable, and the ionic environment is as given in Table 3.1. Potassium ions will be driven out of this cell due to the concentration gradient. The expulsion of K^+ will be opposed by the excess negative charge carried by the cytoplasmic proteins. An equilibrium will develop when enough potassium has been expelled to establish a $^-$97 mV membrane potential as predicted by the Nernst equation. Cl^- will also be driven out of the cell until a chloride concentration gradient is established that just balances the $^-$97 mV membrane potential.

If sodium channels were also inserted into the membrane, sodium ions would be driven

Table 3.1. Equilibrium potentials for selected ions

Equilibrium potentials for the most important ions in mammalian neurons. Data are typical values for cat motor neurons. The concentration for internal calcium is an estimate of the free CA^{2+}, since most intracellular calcium is sequestered in the smooth endoplasmic reticulum.

Ion	[Out] Concentration, mMoles	[In] Concentration, mMoles	Equilibrium potential (mV)
K^+	5.5	150	-88.3[a]
Na^+	150	15	$+61.5$
Cl^-	125	9	-70.3
Ca^{2+}	100	0.0001	$+184$
A^-	—	385	b

[a] The equilibrium potentials were calculated at 37°C.
[b] No equilibrium potential is associated with intracellular anions because they are impermeable.

into the cell both by the concentration gradient for sodium and by the electrical negativity inside the cell (Fig. 3.3B). If sodium ions were able to cross the membrane as easily as potassium ions, then there would be a nearly equal exchange of sodium for potassium ions and the membrane potential would approach 0 mV. Normally, however, the membrane is about 100 times more permeable to K^+ than to Na^+, and therefore sodium ions have only a small, but measurable, effect on the membrane potential.

The effects on the membrane potential of variations in ionic permeability in multi-ionic systems can be calculated with the **Goldman-Hodgkin-Katz constant field equation**:

$$V_m = \frac{R\,T}{F} \ln \frac{p_K[K^+]_{OUT} + p_{Na}[Na^+]_{OUT} + p_{Cl}[Cl^-]_{IN}}{p_K[K^+]_{IN} + p_{Na}[Na^+]_{IN} + p_{Cl}[CL^-]_{OUT}} \quad (3.7)$$

where the different ionic permeability coefficients are represented by p_K, p_{Na}, p_{Cl}, etc. Note that this equation is essentially an expansion of the Nernst equation (Equation (3.6)) for univalent ions, with the permeability coefficients accounting for the fraction of the membrane potential contributed by each ion species. In mammalian muscle the measured membrane po-

tential is about -90 mV; close to the potassium equilibrium potential of -97 mV and quite close to the value predicted by the G-H-K equation (-89.5 mV) if sodium and potassium are both permeable at a ratio of 1:100 (chloride is ignored because it is passively distributed in response to the membrane potential).

THE NA$^+$/K$^+$ TRANSPORTER

Unlike the hypothetical single-ion situation, in the real-world multi-ion environment a steady inward flow of sodium exists because the membrane is not perfectly sealed and sodium is driven into the cell by a large concentration gradient and a substantial electrical gradient (66 mV + -90 mV = 156 mV gradient), both acting to draw sodium into the cell. Similarly, a steady outward potassium flow exists because the outward potassium diffusion gradient is not exactly balanced by the negative electrical potential inside the cell. Although these currents are small, if allowed to continue indefinitely, they would eventually deplete the concentration gradients and consequently the transmembrane potential.

To counteract this slow leakage and therefore

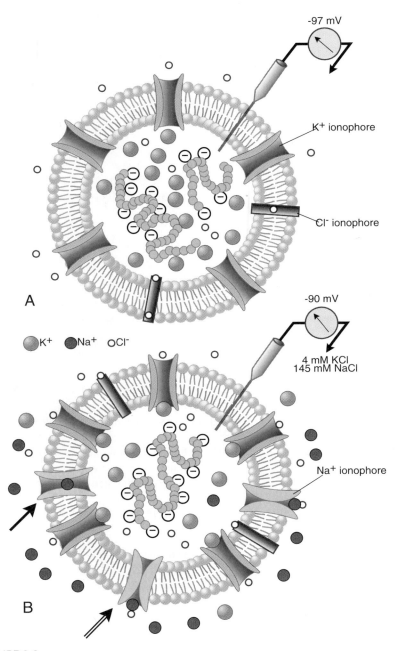

FIGURE 3.3.

A, In a typical cell with permeability to K^+ and Cl^-, potassium ions are driven out of the cell because of the concentration gradient. Anions attached to large impermeable proteins remain inside and make the interior of the cell negative. Chloride distributes itself passively across the membrane in response to the membrane potential because chloride channels offer essentially no resistance to the flow of Cl^-. **B**, The addition of sodium-selective pores to the membrane has little effect on the membrane potential if the pores offer a high resistance to the flow of sodium ions (*double arrow*). The membrane is not totally impermeable to Na^+ (*single arrow*). In the case of typical cells, the ratio of sodium to potassium flow is about 1:100.

to establish a steady state, a special transporter mechanism exists in the neuronal cell membrane. This transporter, the Na/K-ATPase "pump" is a transmembrane molecule. For every ATP molecule it hydrolyzes, three Na^+ ions are removed from the cell and two K^+ ions are inserted. This pumping action exactly counterbalances the leakage of sodium and potassium. The 3:2 ratio of ionic transport also makes the pump **electrogenic**, since more positive charge is removed from the cell than is returned to it. This property effectively adds about -3 mV to the membrane potential.

The absolute number of ions needed to establish the membrane potential is very small. The amount of charge separation needed to establish the membrane potential for an average-size neuron is only about 10^{-17} moles of univalent ions. An average neuron contains approximately 10^{-12} moles of potassium. This represents about 100,000 times more charge than is necessary to maintain the membrane potential. With such an excess of charge to draw upon, the small potassium and sodium currents associated with membrane potentials do not substantially affect the ionic concentrations within the cell.

Local Membrane Currents

The relatively large potential differences across the membrane and along the length of the axon generate electrical currents. The energy they represent is dissipated over short distances, and thus these currents remain close to their source. For these reasons they are called **local currents**. The membrane properties that affect the local currents are called **passive properties** because they do not involve any structural or metabolic alterations in the membrane. These passive membrane properties can be modeled with simple resistors, capacitors, and batteries.

In passive systems, the relationship between electrical potential energy and current flow is expressed by Ohm's law:

$$E = I \cdot R \qquad (3.8)$$

where the potential electrical energy, **E**, is measured in volts, electrical current, **I**, is measured in amperes and the resistance to current flow, **R**, is measured in ohms.[2] Resistance to current flow is a property of matter and varies from one substance to another.

THE TIME CONSTANT

Stimuli that generate rapid changes in membrane current evoke changes in membrane potential that occur more slowly than the stimulus. This observation seems to conflict with Ohm's law, which states that current and voltage are linearly related. However, membranes have a property known as **capacitance**. A capacitor is simply two conductors separated by an insulator. In an electrical circuit, charge accumulates on one of the conductive plates of a capacitor in proportion to the size of the plate, the applied current, and the resistance of the circuit. Opposite charge is attracted to the other plate. Larger plates can accumulate more charge than smaller ones, and circuits with low resistance can deliver charge to the plates faster than circuits with high resistance. Therefore the **time** required to fully charge a capacitor is dependent on both the capacitance (i.e., the size of the plates, which affects how much charge is needed) and the resistance of the circuit (which affects how rapidly the charge can be delivered). This relationship can be described by the **time constant** τ and is related to membrane capacitance, C_m, and membrane resistance, R_m, by the following simple expression:

$$\tau = R_m C_m \qquad (3.9)$$

Note particularly that τ *is independent of the applied stimulus current.* The time constant can be measured as the time required for the membrane potential to change by 63% $(1 - [1/e]$, where $e = 2.718)$, or to 37% of the original steady-state value (Fig. 3.4*C*).

The concept of the time constant is best understood by considering an experiment in which one isolates an unmyelinated axon in a chamber filled with a balanced salt solution. At the midpoint along the axon, two electrodes are inserted into the axon; one electrode is used to measure

---

[2]Expressed in this way Ohm's Law is in alphabetical order and easily remembered.

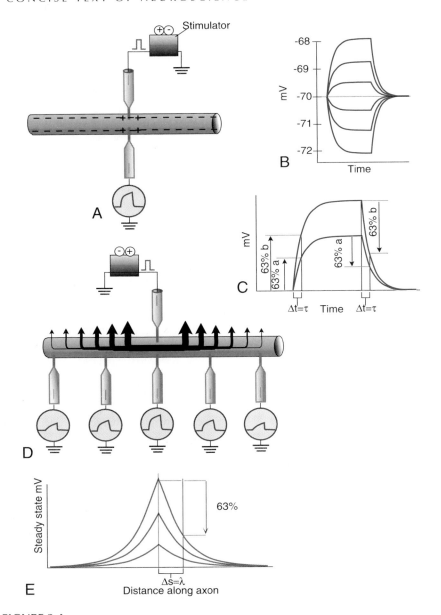

FIGURE 3.4.

A, The method by which the time constant of a membrane can be measured. A small, rectangular current pulse is injected into the axon with one electrode while the change in membrane potential is recorded with a different electrode. With good equipment, the leading and trailing edges of the stimulus pulse change nearly instantaneously. **B,** The membrane response to stimuli of various strengths is illustrated. Note that the rise and fall times of the membrane response are not as fast as the stimulus (shown in A). However, the shapes of the responses are fully symmetrical for stimuli of different intensities and polarities as long as the stimuli are sub-threshold for the voltage-gated sodium channels. **C,** The time constant is the time it takes for the response to change by $1-1/e$ (63%). The time constant, τ, is not affected by the strength or polarity of different stimuli. The time constant varies only with the physical properties of the membrane that affect membrane capacitance and resistance. **D,** The apparatus used to measure the space constant

the transmembrane potential and the other, to inject current into the axon (Fig. 3.4A). If the axon is at rest, a membrane potential of approximately −70 mV is recorded.[3] When positive charge is injected instantaneously (in picoseconds) into the axon, the membrane will **hypopolarize**[4] (the membrane potential will become less negative).

A recording of the change in the membrane potential versus time reveals that the membrane potential changes more slowly than the stimulus but eventually reaches a stable plateau. When the stimulus is released instantaneously (again within picoseconds), the membrane potential slowly returns to its original resting level (Fig. 3.4B). If several recordings are made, each with a different stimulus strength, a series of symmetrical membrane responses can be recorded, each showing a characteristic exponential response at the leading and trailing edges of the stimulus, culminating in a steady state. The symmetry of the response is maintained for both hypopolarizing and **hyperpolarizing** stimuli. One can demonstrate the independence of the time constant from the stimulus by measuring the time it takes to effect a 67% change in membrane potential for each stimulus (Fig. 3.4C).

---

[3]This is a reasonable value for the membrane potential of neurons. The actual membrane potential of a specific neuron is determined by a large number of variables and therefore is not readily predicted. Most neurons have a resting membrane potential between −60 and −80 mV.

[4]Many authors use the term **depolarization** in this situation. Strictly speaking, a depolarization of the cell membrane implies no separation of charge and hence no voltage across the membrane. This is rarely the case. I prefer to use the strictly correct term **hypopolarization** because it precisely describes the actual event, a **decrease** in the membrane potential. The word depolarization should be reserved for the rare situations when membrane polarity is lost.

THE SPACE CONSTANT

If we expand our experiment by inserting a series of recording electrodes into the axon at equally spaced intervals on either side of the stimulating electrode, we can measure the membrane potential simultaneously at various distances from the stimulus (Fig. 3.4D). Superficially, the voltage changes recorded at the various locations along the length of the axon all appear to be roughly the same as the response recorded close to the stimulus. However, upon further inspection one can observe that the farther the recording electrode is from the stimulus, the smaller the amplitude of the steady-state response and the slower the rise-time (Fig. 3.4D). If one plots the amplitude of the steady-state response versus distance from the stimulus, an exponential decline in the membrane steady-state response is revealed (Fig. 3.4E).

The decrease in the amplitude of the steady-state potential along the axon is due to the fact that some of the injected current leaks across the membrane close to the stimulus, more current leaks out a little farther away, and so on (Fig. 3.4D). There is less current available at a distant recording site than at a more proximal site. Some of the electrical energy is also dissipated by the resistance of the axoplasm. According to Ohm's law, if less current is available, a smaller voltage will be observed across a specific resistance. Since the transmembrane resistance is, for all practical purposes, the same along all parts of the unmyelinated axon, the effect of injecting current at one point will be a steady decline in transmembrane voltage

---

is the same as for measuring the time constant. However, instead of recording the membrane potential at a single point, several recordings are made along the axon as illustrated. While superficially similar, the recordings diminish in steady-state amplitude as the recording site becomes further removed from the stimulus site and the rise and fall times become extended, because τ increases with increasing distance from the stimulus site. This is due to the accumulation of membrane capacitance along the axon. **E,** The space constant is measured from a graph such as the one illustrated here, where the steady-state amplitude of each response is plotted against distance from the stimulus site. Connecting the data points produces the curves illustrated. Each curve represents the steady-state membrane potential to stimuli of different intensities. The space constant, λ, is defined as the distance from the stimulus where the steady-state membrane potential has fallen by $1-1/e$ of the steady-state membrane potential at the stimulus site. Note that this distance is the same for all three stimulus intensities.

change the farther one gets from the source of the current.

The measured distance from the current source to the point where the transmembrane voltage has decayed by $1 - (1/e)$ of its peak value is λ, **the space constant** (Fig. 3.4*E*). From the previous discussion, it should be apparent that if the membrane resistance is increased, less current will leak across the membrane, and therefore the effect of the stimulus will extend farther along the axon. The space constant will be lengthened. Similarly, reduction in the axoplasmic resistance, for example by increasing the diameter of the axon, will also increase the space constant. The relationship between the space constant λ, the membrane resistance $\mathbf{R_m}$, and the axoplasmic resistance $\mathbf{R_a}$ is approximately:

$$\lambda \simeq \sqrt{\frac{R_m}{R_a}} \qquad (3.10)$$

The Action Potential

The discussion up to this point has centered on the channels in the membrane that have a constant permeability to a single ionic species. These channels convey the passive membrane currents just discussed. Other channels, however, are more complex. Many of them can alter their structure, and hence their permeability to ions, in response to external events. Electrical signaling depends on a class of ion-selective channels that alter their specific permeability as a function of the transmembrane potential. Such channels are called **voltage-gated channels**. Specific voltage-gated channels have been described for Na^+, K^+, and Ca^{2+}. Cells with voltage-gated channels can modulate the transmembrane potential of neurons, and *this modulation is the electrical signal that represents information in the nervous system.*

There are two types of electrical signals that result from the action of voltage-gated ion channels, the **action potential** and the **postsynaptic potential**. *The action potential is a brief reversal of the membrane potential caused by a modulation of the membrane permeability to sodium and potassium ions* (see V_m, Fig. 3.5*A*). The nature of the action potential is discussed here; postsynaptic potentials are discussed later in the chapter.

VOLTAGE-GATED ION CHANNELS

Voltage-gated ion channels are membrane proteins whose tertiary structure is determined, in part, by the transmembrane voltage. The ion channel through the protein randomly snaps open and shut (Fig. 3.6). *The state of the channel, open or closed, is a probability function related to the transmembrane potential.* At any given membrane voltage, there is a certain probability that the channel will snap open; once it has opened, a different probability determines the length of time the channel will remain open before snapping shut again.

FIGURE 3.5.

A, In 1952, Hodgkin and Huxley published a mathematical model of the action potential. This graph is a computer-generated solution of their model. It illustrates the changes in the membrane potential (V_m) and conductance changes for sodium (g_{Na}) and potassium (g_K) ions and the relative state of sodium and potassium gates during an action potential generated by a single hypopolarizing pulse (bottom, S_1). **B,** This computer simulation illustrates the membrane response to two stimuli delivered 10 msec apart (at S_1 and S_2), the second stimulus being 5 times stronger than the first. No action potential is generated because the membrane is absolutely refractory. Although the sodium activation gate begins to open in response to the stimulus, sodium current (I_{Na}) does not increase substantially, it remains much smaller than the potassium current (I_K), because the sodium inactivation gate is still nearly completely closed. **C,** After a stimulus interval of 15 msec, however, the stronger pulse (S_2) is able to elicit another action potential. The membrane is relatively refractory at this point. **D–F,** This series of simulations is similar to those shown in **B** and **C**, except both stimuli are the same magnitude. A second action potential can be elicited only after a delay of 25 msec.

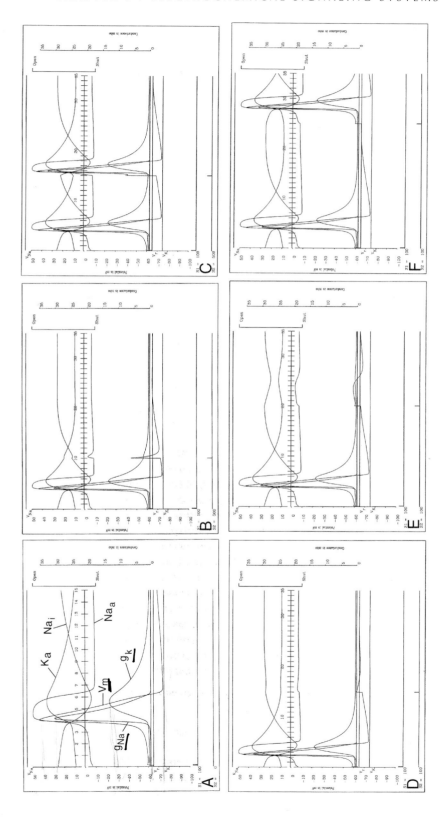

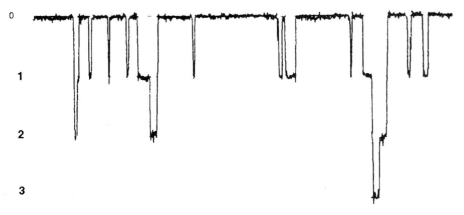

FIGURE 3.6.

This recording illustrates current flowing through three ion channels. Note that the channels open and close independently and that the channels snap open and shut almost instantaneously; there are no intermediate states. When more than one channel is open, the currents sum, giving the larger steps seen here. At any given membrane potential, a channel flickers between the open and shut state. The time it remains open or shut is determined by a probability function that favors the open state as the transmembrane potential decreases.

The opening and closing of the channel is nearly instantaneous, occurring in 10 μs or less, and is discontinuous. In other words, the opened or closed state is an **all-or-nothing** phenomenon; there are no intermediate conditions. At the normal cell resting potential for neurons, about −70 mV, the sodium and potassium channels are closed most of the time. When the channel is open, ions for which the channel is selective can cross the membrane and generate an electrical current. Each channel can pass only about 10^7 ions per second, so the electrical current associated with each channel is small (about 2 pA).

SODIUM CURRENT

The sequence of events that results in an action potential begins with any stimulus that causes a hypopolarization of the neuron. If the hypopolarization is large enough, the membrane potential rapidly diminishes to zero, reverses, and reaches a magnitude of about +30 mV. It then rapidly returns to its original resting potential. The process continues at a slower pace, causing a prolonged hyperpolarization that slowly returns to the original resting level.

This transient alteration in membrane potential is initiated by the initial hypopolarization.

This increases the probability that individual ion channels will open and also increases the average duration that they will remain open. When the sodium channels open, the inward sodium current increases, driven inward both by the sodium concentration gradient and the voltage gradient (see g_{Na} in Fig. 3.5A). This **inward sodium current** brings positive charge into the cell and hypopolarizes the cell further, reinforcing the probability that voltage-gated channels will open, further increasing the inward sodium current. *The process becomes irreversible when inward sodium current is greater than the outward potassium current* (*arrow,* Fig. 3.5A). The membrane potential at which the hypopolarization becomes irreversible is called **threshold**. Once that point has been reached, the inward sodium current continues to increase rapidly until the membrane potential reverses and becomes inwardly positive (see V_m in Fig. 3.5A).

The opening of the sodium channels, called **sodium activation**, is a *positive-feedback system*. As in all positive-feedback systems, the stimulus evokes a response that acts to reinforce the stimulus. Positive-feedback systems must be terminated by mechanisms that are outside the feedback loop. For example, chemical explosions are positive-feedback systems terminated by the consumption of the explosive material. Sodium activation is terminated by a separate process, **sodium inactivation**.

A voltage-activated sodium channel behaves as if it consists of two gates, an activation gate and an inactivation gate. The gates are arranged in series, so both must be open for sodium to pass through the channel (Figs. 3.5 and 3.7). At the normal resting potential, the probability functions that determine the conformation of the ionophore protein favor the closed state for the activation gate and the open state for the inactivation gate. When a stimulus causes hypopolarization of the cell, the probability function associated with the activation gate *rapidly* changes to favor the open state, while the probability function associated with the inactivation gate *slowly* changes to favor the closed state. Briefly both gates are open and sodium flows freely, further hypopolarizing the membrane. The inactivation gate probability function continues to change until it favors the closed state. This shuts off the flow of sodium into the cell, which repolarizes it. The repolarization of the cell changes the gate probabilities back toward their resting state. First the probabilities associated with the activation gate change to favor the closed state. Then, more slowly, the probabilities associated with the inactivation gate change to favor the open state. This restores the sodium channel to its normal resting state.

Within the protein that constitutes the voltage-gated channel, the domain that represents the inactivation gate is physically separate from the domain that represents the activation gate. When the membrane is hypopolarized, the conformational change in the protein that opens the sodium channel initiates a secondary, more slowly progressing change that closes the sodium channel. The opening and closing actions are separate but linked processes. They do not act independently. Consequently, the process that represents activation cannot be reset until the process that represents inactivation is completed.

POTASSIUM CURRENT

Hypopolarization also activates voltage-gated potassium channels. The probability functions for the potassium channels are such that, statistically, they open more slowly than the sodium channels and they remain open longer. This re-

sults in a prolonged **outward potassium current** that removes positive charge from the cell, resulting in a **hyperpolarization** of the membrane (see g_K, Fig. 3.5A). Potassium channels are simpler than the sodium channels, as there is no mechanism associated with potassium channels that corresponds with sodium inactivation.

Because of the difference in the time course of the sodium and potassium currents, most of the increase in potassium current occurs after sodium inactivation has reduced the sodium current to resting levels. During the later stage of the action potential, the increased potassium current is unopposed, and it drives the membrane potential beyond (more negative than) its normal resting potential. This **after-hyperpolarization** may last for several hundred milliseconds in some cells.

THE REFRACTORY PERIOD

Once hypopolarization has reached threshold it is more difficult to initiate an action potential until the membrane has been fully restored to its resting state. This is the **refractory period** (Fig. 3.5B and C). Immediately after threshold has been reached, the activation of sodium becomes irreversible, and further stimulation of the membrane cannot affect the generation of the action potential. The membrane is **absolutely refractory** to further stimulation. After the peak of the action potential, the membrane is **relatively refractory**; an action potential can be initiated, but only with a supranormal stimulus. Two mechanisms account for the decreased sensitivity of the membrane to hypopolarizing stimuli during the refractory periods.

First, during sodium inactivation, the sodium activation probability favors the open state while the inactivation probability slowly changes to the closed state, preventing sodium from flowing through the channel. Until the activation probability resets, no amount of hypopolarization can increase sodium permeability further, and the membrane will be **absolutely refractory**. Even after the activation probability has reset, the inactivation probability favors the closed state for about 2 msec, making any stimulation during this time ineffective (Figs. 3.5B and 3.7).

Second, the refractory period is also caused in part by the increased potassium current that is just reaching its peak as the sodium current is declin-

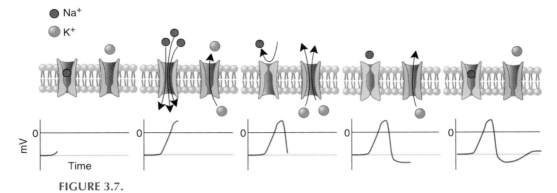

FIGURE 3.7.

    The developing action potential is shown below schematic drawings of the gating of the sodium (*left*) and potassium (*right*) channels. Note particularly the sequencing of the sodium activation (*bottom*) and inactivation (*top*) gates. Compare this schematic with the graphs in Figure 3.5 **A–F**.

ing (Fig. 3.5*A*). Recall that threshold is achieved when inward current equals outward current. During the after-hyperpolarization, the outward flowing potassium current is greater than normal, and consequently a greater than normal inward, hypopolarizing current is required to overcome it and hypopolarize the membrane to threshold. This **relative refractory period** lasts as long as the potassium after-hypopolarization.

Propagation of the Action Potential

    The rapid modulation of the membrane potential, the action potential, is limited both in time and in space. Temporally, an action potential exists for only a few milliseconds and at any given instant is localized to a part of the axon that measures only a few micrometers. For it to be an effective signaling device, the action potential must travel from one part of the axon to another. Therefore the action potential must be capable of regenerating itself, micrometer by micrometer, in a continuously repetitive manner along the entire axonal length.

CONVERSION OF LOCAL CURRENTS INTO ACTION POTENTIALS

    At the peak of the action potential, the membrane potential is reversed. This reversal exists over only a small segment of the axon and re-

sults in a separation of charge along the length of the axon on either side of the region where the membrane potential is reversed. This longitudinal charge separation generates local currents along the axon (Fig. 3.8).

    The local currents generated by the action potential are hypopolarizing because internally positive charge is being delivered to the resting region of the membrane. At the proximal region of the axon, these hypopolarizing local currents have very little practical effect since that area is refractory. However, at the distal portion ahead of the action potential, these hypopolarizing currents bring that region to threshold and initiate another action potential.

    In this manner, the action potential regenerates itself. Since the action potential is regenerated de novo and continuously along the entire axonal membrane, *there is no attenuation of the amplitude of the action potential with distance.* Each patch of membrane sustains the complete action potential at that location and generates local currents that serve as the stimulus to initiate the action potential at the adjacent location.

FACTORS AFFECTING CONDUCTION VELOCITY

    The continuous regeneration of the action potential is a relatively slow process. Action potentials can be propagated at rates up to 120 m/sec. In contrast, electrical currents in copper wires travel at approximately 300,000 m/sec.

But unlike electrical currents in wires, the conduction velocity of action potentials is greatly affected by the physical parameters of the axon.

Adjusting these parameters in any way that lengthens the space constant will extend the effective spread of the local currents along the axon. Consequently, the region of hypopolarization is also extended, bringing a longer segment of the axon to threshold and expanding the distance consumed by the action potential. Since each regeneration of the action potential consumes a greater length of the axon, the conduction velocity is increased. For example, increasing the diameter of the axon reduces the longitudinal resistance of the axon. This lengthens the space constant and increases the conduction velocity of the action potential (Fig. 3.8*A* and *B*). The relationship between axon diameter and conduction velocity for unmyelinated axons is approximately 0.5 m/sec/μm.

Many animals—particularly the invertebrates,

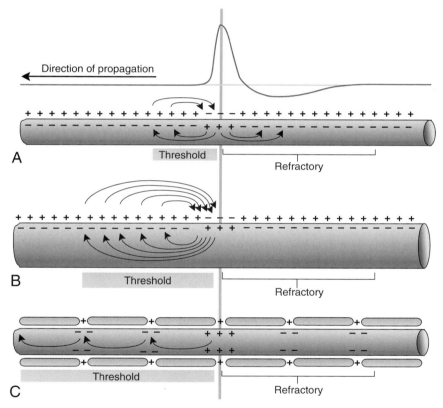

FIGURE 3.8.

Action potentials generate local currents that serve to hypopolarize adjacent regions of the membrane, bringing it to threshold. Here the local currents associated with the action potential are depicted for three different types of axons. **A** and **B** are unmyelinated axons of different diameters. The local currents extend farther down the larger axon and are able to hypopolarize to threshold more of the axon membrane than the local currents in the smaller axon because of the reduced axoplasmic resistance in the larger axon. The portion of the axon brought to threshold by the local currents is indicated by the *bar*. The axon shown in **C** is myelinated and the same diameter as **A**. For all practical purposes, the local currents can only leave the axon at the nodes of Ranvier. Therefore the effective transmembrane resistance is increased in myelinated axons, causing the hypopolarizing local currents to extend farther along the myelinated axon. The effect of extending the influence of the local currents is to bring larger regions of axonal membrane to threshold and thereby increasing the velocity of propagation of the action potential.

in which axons are frequently several hundred micrometers in diameter—use large, giant axons to increase conduction velocity. This is not without cost, however, for the efficiency gained in velocity is offset by the inefficiency in volume consumed by these giant axons. As a nervous system becomes more complex, more axons are needed to meet the increased demand for information carriers, and volume efficiency becomes a critical factor. The most complex nervous systems are those found in mammals, and these are noted for their small axons. Mammalian systems typically have axons that range from 0.1 to about 10 μm in diameter. To achieve relatively fast conduction velocities, the mammals lengthen the space constant by increasing the axon membrane resistance. They do this by wrapping the axon with insulation (myelin).

The myelin sheath found around most mammalian axons increases the membrane resistance so much that current leakage across the membrane is reduced to negligible levels. In effect, the only place where current can leave the axon is at the nodes of Ranvier (Fig. 3.8C). Furthermore, unlike unmyelinated axons, in which the voltage-gated channels are strewn along the entire axon, in myelinated axons sodium channels exist only at the nodes, the only locus where action potentials can exist.

An action potential at a single node of Ranvier establishes local currents that extend along the axon. Typically the space constant is long enough to include several adjacent nodes, which are all brought to threshold by the hypopolarizing local currents. Since action potentials are generated only at the nodes, the internodal segments of the axon are simply skipped. This phenomenon is frequently referred to as **saltatory conduction**, saltatory meaning to dance by leaping. Myelinated axons conduct at approximately 6 m/sec/μm.

The Synapse

Neurons exchange information at specialized contact points called **synapses**. The physiological role of the synapse is to *convert an electrical signal from the presynaptic cell into a chemical*[5] *signal that can be transferred to the postsynap-*

tic cell. This conversion is not a simple process, for information is not simply handed from one cell to the next, like the baton in a relay race. During synaptic transmission, information is transformed such that signals leaving the postsynaptic cell are never the same as those entering it. This process of **information transformation** is central to the function of the nervous system. Synaptic interactions are necessary for all nervous system functions. Without functional synapses, the nervous system fails, causing, among other things, failure of perception, failure of motor acts, and loss of consciousness.

STRUCTURE OF CHEMICAL SYNAPSES

Axons terminate as **synaptic boutons** that form the presynaptic portion of a complex structure known as the synapse. Specializations in the neuronal membrane of the postsynaptic neuron opposite the bouton complete the structure (Fig. 3.9).

Synaptic Vesicles

The synaptic bouton contains large numbers of membrane-limited structures, the **synaptic vesicles**. Synaptic vesicles are derived by endocytotic invagination from the bouton membrane and contain a unique set of membrane proteins[6] that appear to be common to all synaptic vesicles and are not found in other intracellular membranes. The origin of synaptic vesicles from the bouton membrane and the special proteins contained within their membranes differentiate them from the secretory vesicles that are derived from the endoplasmic reticular–Golgi body secretory system in the soma.

The synaptic vesicles contain a **chemical messenger** that is released during the signaling process. Chemical messengers can be differentiated into two types based on the physiological effect they have on the postsynaptic cell. Chemical messengers that, when combined with a receptor protein, directly open or close ion channels or initiate second messenger cascades are usually

---

[5]There are two types of synapses, electrical and chemical. Electrical synapses are rare in mammals and are of no clinical significance. They will not be discussed.

[6]Four proteins—SV2, synaptophysin, synaptotagmin, and synaptobrevin—have been described, but their exact physiological roles have yet to be determined.

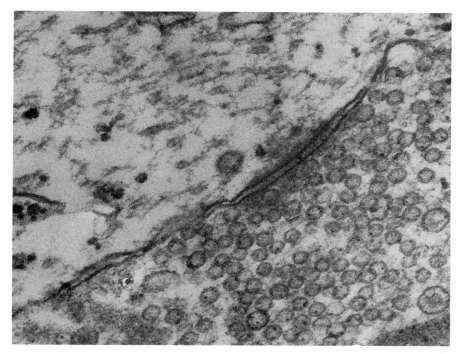

FIGURE 3.9.

This electron micrograph illustrates some of the important ultrastructural features of chemical synapses. Note particularly the active zone. Here the presynaptic thickening is quite evident, although its hexagonal structure is not shown with the staining used in this preparation. Note also the membrane-limited vesicles. Within the synaptic cleft, one can discern structural elements, the function of which is not certain. The postsynaptic thickening is also visible. Its significance is also not known.

termed **neurotransmitters**. Chemical messengers that modify the function of receptor proteins are usually called **neuromodulators**. Most chemical messengers are synthesized in the synaptic bouton and subsequently packaged in the synaptic vesicle.

To a limited degree, the type of chemical messenger contained within the vesicle is revealed by its shape and by the electron density of its contents (Table 3.2). For example, most vesicles are spherical, about 40 to 60 nm in diameter, with clear centers. They are considered **excitatory** since the chemical messengers they contain usually cause a *hypopolarization* in the postsynaptic cell. Other clear core vesicles appear ellipsoidal, being about 30 by 50 nm. Although the ellipsoidal morphology is probably an artifact of aldehyde fixation, it is a useful one since these vesicles are usually associated with

hyperpolarizing postsynaptic events and are therefore **inhibitory**. Most synapses in the mammalian nervous system have small clear-core vesicles. However, boutons associated with the autonomic nervous system and other systems that use monoamines as chemical messengers contain small (40- to 60-nm) vesicles with dark centers. Neuroactive peptides are packaged in large, dark-centered vesicles that range in size from about 80 to 150 nm. Most boutons contain only one type of vesicle and release only one type of chemical messenger. However, many boutons co-release a neuroactive peptide and either an excitatory or an inhibitory transmitter. These boutons contain the two types of vesicles associated with those chemical messengers.

Neurotransmitters are chemical messengers that convey information from one neuron to another. They are contained within the vesicles

Table 3.2. Classification of synaptic vesicles based on ultrastructure

Size (nm)	Shape	Appearance	Contents
40–60	Spherical	Clear	Various
30 × 50	Ellipsoidal	Clear	Various
40–60	Spherical	Dark	Monoamines
80–150	Spherical	Dark	Neuropeptides

and released from the presynaptic bouton to combine with a specific receptor at the postsynaptic membrane. The transmitter-receptor complex alters either the electrical or the biochemical state of the postsynaptic neuron.

Identifying neurotransmitters has been difficult. Many substances that are not normally used by the nervous system can affect neuron function and mimic the action of true neurotransmitters. The following criteria are frequently used to qualify a substance as a neurotransmitter:

1. Neurotransmitters are *synthesized* by neurons and are *present* in presynaptic terminals;
2. Neurotransmitters are *released* at the terminal as the result of electrical activity;
3. Neurotransmitters *bind to a specific receptor* on the postsynaptic cell; and
4. Neurotransmitters are associated with a *specific mechanism of inactivation* at the synapse.

Docking Complex

Only a small portion of the synaptic bouton participates in synaptic transmission. On the presynaptic side of the synapse, small patches of membrane appear irregular and dark in electron micrographs (Fig. 3.9). These densities, although seemingly amorphous, are shown with special stains to be a highly structured complex that forms a hexagonal grid. The size of the grid is matched to the size of the vesicles such that the vesicles are nestled within this structure and held close to but not touching the membrane of the bouton

(Fig. 3.10). Because of the physical relationship between the vesicles and the grid, it is known as a **docking complex**, and it appears to be present in all chemical synapses.

Synaptic Cleft

Between the synaptic bouton and the postsynaptic cell is a 10- to 20-nm space, the **synaptic cleft** (Figs. 3.9 and 3.10). This cleft is filled with 4- to 6-nm filaments that bridge the gap and help bind the two cells together. Both structural proteins and mucopolysaccharides have been identified within the cleft, but their function is uncertain. It is assumed that their presence is useful in cell recognition during development and helps determine the specificity of synaptic connections.

Postsynaptic Density

The postsynaptic membrane frequently shows a thickening directly opposite the presynaptic docking complex (Figs. 3.9 and 3.10). This material is composed primarily of structural proteins such as actin filaments and tubulin. Its thickness is not consistent in all synapses, and many authors describe synapses as either "symmetric" or "asymmetric" depending on the relative thickness of the presynaptic and postsynaptic membrane specializations.[7] It has not been possible to consistently correlate the relative thickness of

---

[7]Sometimes referred to as the Gray type I and type II synapses.

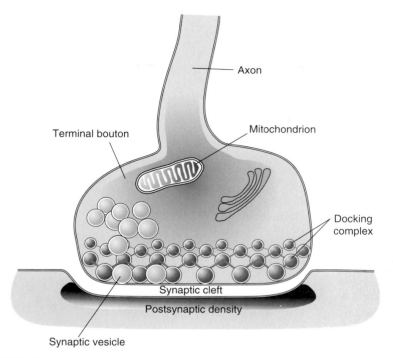

Axon

Terminal bouton

Mitochondrion

Docking
complex

Synaptic cleft

Postsynaptic density

Synaptic vesicle

FIGURE 3.10.

Schematic drawing of the bouton, showing the vesicles locked into the docking complex but not fused with the presynaptic membrane.

these densities with differences in physiological function. In fact, there is a continuum of relative thicknesses between symmetric and asymmetric synapses that blurs the distinction. However, the membrane densities are useful markers, indicating the place on the membrane where information transfer between cells takes place. Therefore the densities are called the **active zone**. A single bouton may have several active zones.

The active zone of the postsynaptic membrane also contains membrane-spanning proteins called receptors. Some of these proteins are similar to the voltage-gated channels found in the axon; others are quite different. All of the receptor proteins at the postsynaptic active zone have specialized domains in the extracellular region that can bind the chemical messenger.

SYNAPTIC TRANSMISSION

Chemical synaptic transmission, the passing of information from one cell to another, is a complex process. First, as the action potential envelops the bouton, the synaptic vesicles attach to the bouton membrane at the active zone. Once attached they release their chemical messenger into the synaptic cleft where it diffuses to the postsynaptic membrane. There the chemical messenger combines with a receptor protein, a process that alters the ionic conductance and consequently the membrane potential of the postsynaptic cell. This may initiate an action potential in the postsynaptic neuron.

Vesicle Binding

In addition to the Na^+ and K^+ voltage-gated channels found on the axon, synaptic boutons contain **voltage-gated calcium channels**. These calcium channels exist near the active zones, possibly as a structural part of the docking complex (Fig. 3.11). When an action potential envelops a bouton, the voltage-gated calcium channels open, allowing Ca^{2+} to enter the bouton very close to the docked vesicles. In the presence of Ca^{2+}, the synaptic vesicles already locked

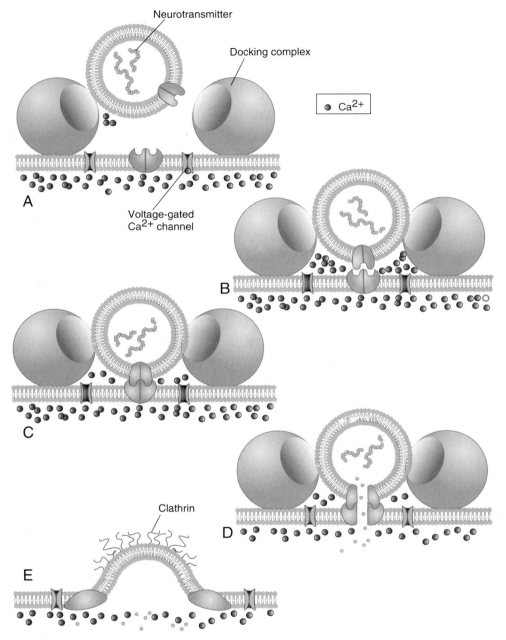

FIGURE 3.11.

The steps involved in the binding of synaptic vesicles with the presynaptic membrane are schematically illustrated. **A**, The vesicle must become attached to the docking complex where it is held in place, awaiting the arrival of an action potential. This attachment is facilitated by Ca^{2+}. **B**, The action potential opens voltage-sensitive Ca^{2+} gates. The increased concentration of Ca^{2+} facilitates the attachment of vesicles with the bouton membrane. **C,** Once attached, the vesicular and bouton membranes fuse. This releases the neurotransmitter into the synaptic cleft (**D**). This step cannot occur in the absence of Ca^{2+} and may also involve specialized membrane proteins as depicted here. **E**, A patch of the bouton membrane invaginates and becomes coated with clathrin, a protein that helps recover membrane from the bouton and recycle it into new synaptic vesicles.

into the docking sites fuse with the bouton membrane. Calcium ions probably serve as a bridging mechanism between the proteins in the vesicular membrane and proteins in the bouton membrane. The exact mechanism is not completely understood, but fusion cannot occur in the absence of Ca^{2+}. Once the vesicle has fused with the bouton membrane, its contents (the chemical messenger) empty into the synaptic cleft and diffuse onto the postsynaptic membrane. Simultaneously, the vesicle membrane becomes incorporated into the bouton membrane.

The continued presence of Ca^{2+} in the region of the active zone facilitates the docking of new vesicles. However, calcium must be rapidly sequestered or the newly docked vesicles would immediately fuse with the bouton membrane. The inactivation of Ca^{2+} in the bouton is a complex, multi-stage process (Fig. 3.12). Initially, Ca^{2+} can be rapidly sequestered by proteins in the cytosol, effectively removing Ca^{2+} from an active role in vesicle release. Additionally, special calcium storage cisterns in the bouton have transporter molecules in their membrane that rapidly pump Ca^{2+} out of the cytosol. Finally, a Ca^{2+}/Na^+ exchange transporter exists in the bouton membrane that transports Ca^{2+} out of the bouton cytosol into the interstitial space. Ca^{2+} ions are exchanged among these systems until the normal intracellular concentration is restored.

The sequestering and removal of Ca^{2+} is a slow process. Under some circumstances, action potentials can invade the bouton at such a rapid rate that the entry of Ca^{2+} is greater than the rate of sequestering and removal. This can cause a temporary increase in the Ca^{2+} concentration in the bouton. Subsequent action potentials initiate a greater than normal release of the chemical messenger because the docking-fusion mechanism is facilitated by this increase in the intracellular concentration of Ca^{2+}. This process, **post-tetanic potentiation**, may be an important physiological regulator of synaptic function.

The fusion of the synaptic vesicle with the bouton is balanced by a system that recovers the vesicular membrane from the synaptic bouton (Fig. 3.12). The first stage in this process occurs when a patch of the bouton membrane invaginates. **Clathrin**, a cytoplasmic protein, is essential to the process and coats the invaginating membrane as it separates, forming a coated vesicle. Later the clathrin coat is shed when the coated vesicle fuses with a cistern of the smooth endoplasmic reticulum (SER) in the bouton. Within the cistern, the chemical messenger is synthesized and inserted into pockets of membrane that are pinched off, forming new synaptic vesicles.

Receptors

Broadly speaking, there are two classes of receptors for neurotransmitters. The simplest are **ligand-gated ion channels**. When this kind of receptor is bound to its specific neurotransmitter, the main structure of the protein is altered in a way that modifies the channel's permeability to ions. The resulting altered flow of ions generates local currents, postsynaptic potentials (PSPs), that alter the membrane potential of the postsynaptic neuron. Because the receptor itself is the active agent of change, the action of these receptors is *direct*.

Other membrane-spanning proteins in the postsynaptic membrane have specialized receptor domains, but they are not ionophores. These proteins, often called **collision-coupled second-messenger systems**, initiate a cascade of biochemical reactions that can modify cell metabolism, gene expression, or the ion selectivity of certain membrane proteins. The receptor is an *indirect* agent of change in these systems, being only the initiator of a long series of biochemical events.

The action of a neurotransmitter on the postsynaptic cell depends on the nature of the receptor. Some neurotransmitters bind to more than one kind of receptor. In such cases a single transmitter can initiate a number of different actions, depending on the receptor with which it interacts. Consequently, *transmitters may have different effects at different synapses*. For example, acetylcholine (ACh) is an excitatory transmitter when acting through direct ionophore receptors on skeletal muscle, but it is an inhibitory transmitter in cardiac muscle and the central nervous system when operating through second-messenger indirect receptors. Glutamate, a common excitatory neurotransmitter in the central nervous system, binds to at least four different receptors.

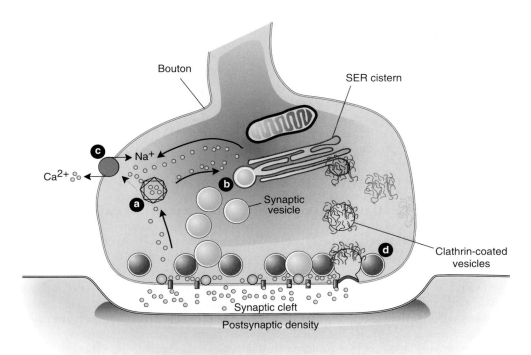

FIGURE 3.12.

Calcium must be removed from the synaptic bouton or vesicle binding and fusion will continue unabated. The three mechanisms that account for this are shown here. First, cytosolic proteins (*a*) can temporarily sequester Ca^{2+}. Second, special transporters exist on the smooth endoplasmic reticulum (SER) that remove Ca^{2+} from the cytosol so that it can be isolated within the SER (*b*). Ca^{2+} ions are stored in the SER where they can be made rapidly available (see text). Ultimately, Ca^{2+} must be returned to the extracellular space; special membrane transporters (*c*) exist that move Ca^{2+} ions into the extracellular environment. Calcium can be exchanged between all three compartments. In addition, vesicular membrane must be recycled. After a vesicle has fused with the bouton membrane, an invagination forms that is coated with an electron-dense material, forming a coated pit (*d*). This invagination enlarges and separates as a clathrin-coated vesicle. This fuses with a cistern of the SER, within which a new transmitter is assembled. Synaptic vesicles separate from the SER and eventually migrate to the docking complex.

Three receptors are selective ion channels directly affected by glutamate, and one initiates a second-messenger cascade.

Neuromessenger Inactivation

Once released, the chemical messenger must be inactivated to limit the duration of its influence. One characteristic of synaptic systems is the presence of mechanisms for the removal of the messenger (Fig. 3.13). The simplest of these mechanisms is **diffusion**. This passive mecha-

nism is augmented in some cases by a specific **reuptake transporter** either in the synaptic bouton or in the adjacent astroglia. Reuptake has been specifically described for the catecholamines, glutamate, glycine, and γ-amino butyric acid (GABA). Alternatively, the messenger may be destroyed. For example, ACh is rapidly inactivated by **enzymatic degradation** by the enzyme acetylcholinesterase. This enzyme breaks ACh into the inactive substances choline and acetate. Choline is subsequently returned to the bouton by a specific transporter molecule.

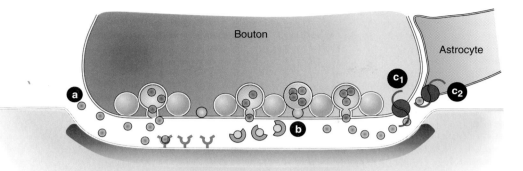

FIGURE 3.13.

Neurotransmitters are inactivated in three ways: by simple diffusion of the transmitter away from the receptors into the intracellular space (*a*), by enzymatic degradation (*b*), and by the action of specific membrane transporters that pump the transmitter back into the bouton (c_1) or into adjacent astroglia cells (c_2).

Ligand-Gated Ion Channels

In the presence of its neurotransmitter, the ligand-gated ion channel receptor protein opens its ion-selective channel. While the ligand is bound, the channel flickers between the open and the closed state (Fig. 3.14), remaining predominantly in the open state. The binding of the ligand to the receptor is a tenuous one with a high probability of dissociation. Once the ligand-receptor complex dissociates, the channel closes. The channel remains open for only a few milliseconds. While the channel is open, ions enter or leave the postsynaptic cell in response to the electrical and diffusional forces upon them, generating local currents that are recorded as postsynaptic potentials.

The release of transmitter into the synaptic cleft is a quantum phenomenon. Each vesicle, with its cargo of transmitter is the minimal, quantum unit of excitation or inhibition and each produces one quantum of postsynaptic current. The integration of multiple quantum hypopolarizing events, represented by the release of transmitter in numerous synaptic vesicles, produces the postsynaptic potential (PSP), a local current that extends throughout the entire postsynaptic cell. This PSP is a relatively brief event, lasting about 15 msec. Its *amplitude,* unlike the action potential, is *variable,* depending on the number of quanta, or packets of transmitter, released at the synapse.

The dynamic behavior of ligand-gated ion channels is more complex than the dynamic behavior of voltage-gated ion channels. The time the channel remains in the open or the closed state is determined by separate rate constants for association and dissociation of the ligand-receptor complex. Each rate constant is subject to modification by a number of environmental factors, including other molecules that can attach to the receptor (i.e., neuromodulators). Therefore the generation of PSPs is more complicated than the generation of the action potential.

EXCITATORY SYNAPSES

One class of ligand-gated ion channels selectively passes cations, (Na^+, K^+, and Ca^{2+}). Postsynaptic neurons are excited (hypopolarized) by the activation of cationic channels because local currents are generated that hypopolarize the postsynaptic cell. The local currents, known as **excitatory postsynaptic potentials (EPSPs)**, are the result of an increase in the transmembrane **conductance**[8] for Na^+, K^+, and sometimes Ca^{2+} ions.

---

[8]Conductance is simply the reciprocal of resistance: $g = (1/R)$. The unit of conductance is the siemens.

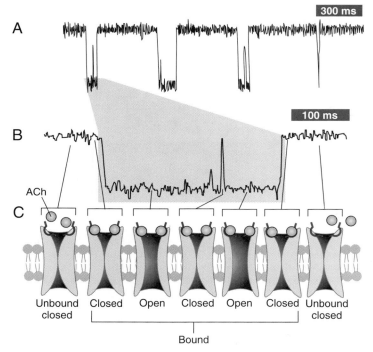

FIGURE 3.14.

Many ligand-gated channels, when bound with their transmitter, flicker between the opened and the closed state. **A**, Four separate receptor-ligand associations are recorded from a single channel in the presence of ACh. **B**, During the first association, which lasted about 250 ms, the channel briefly closed, even though the ligand was still associated with the receptor. **C**, A schematic representation of the state of the receptor during the 250 ms that it was associated with ACh.

The cationic channels are only partly selective. When open, they allow both Na^+ and K^+ (and sometimes Ca^{2+}) to pass through the same pore (Fig. 3.15). If the conductance of Na^+ and K^+ were to increase identically, there would be no alteration in the membrane potential because the inward (Na^+) and outward (K^+) current flows would balance. However, potassium is close to its equilibrium potential and so there are only moderate forces driving it out of the cell. Sodium, on the other hand, is driven into the cell both by a strong diffusion gradient and by a substantial electrical gradient. Consequently, sodium conductance increases about 7.5 times more than the increase in potassium conductance. The result is that more positive charge enters the cell than leaves it.

INHIBITORY SYNAPSES

Ligand-gated channels that are selective for Cl^- also generate local currents but these currents have the opposite effect of EPSPs (Fig. 3.15). The inward chloride current, the **inhibitory postsynaptic potential (IPSP)**, has the effect of neutralizing any positive charge in the cell, making it more difficult to hypopolarize or excite it.

In most neurons the resting potential is about -60 mV and the chloride equilibrium potential is about -70 mV. In these neurons, opening chloride channels will cause an inward flow of Cl^-, since chloride will be driven into the cell both by its concentration gradient and the chloride equilibrium potential (Fig. 3.16A). This inward chloride current will hyperpolarize the cell.

If the resting potential of the neuron were equal to the chloride equilibrium potential of -70 mV—and in many neurons this is the case—opening the chloride channels will cause *no increase in the inward flow of Cl^-* (Fig. 3.16B). This situation is still inhibitory, however, because the *potential* to increase chloride conductance ex-

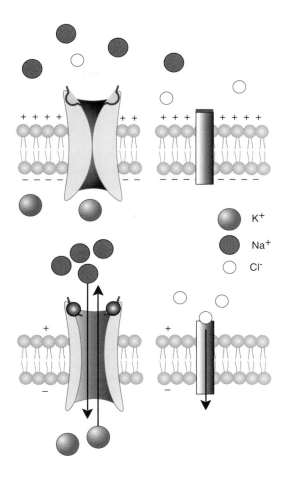

FIGURE 3.15.

Typical ligand-gated cationic receptor channels are not as selective as the voltage-gated ion channels. When bound to their transmitter, these proteins open a channel that allows Na+ and K+ to pass simultaneously through the same pore (*left*). Because the driving forces are greater for sodium than for potassium, approximately seven Na+ ions enter for every K+ ion that leaves the cell (see text). A chloride-specific channel is also shown (*right*).

ists. For example, any hypopolarizing event, such as an EPSP, will allow positive charge to enter the cell, making the membrane potential more positive. As the membrane potential becomes more positive, chloride ions will immediately be drawn into the cell through the open chloride channels. These negative chloride ions will neutralize the positive charge flowing to produce the EPSP. Chloride will enter the cell as long as the membrane potential is more positive than the chloride equilibrium potential. As the membrane potential approaches the chloride equilibrium potential, the net chloride current approaches zero.

The membrane is said to be "stabilized" around the chloride equilibrium potential.

The IPSP can even be a hypopolarizing potential. If the resting potential of the neuron were −75 mV, a voltage quite within the normal physiological range, opening chloride-selective channels will cause an *outward* flow of Cl⁻ (Fig. 3.16C). This drives the membrane potential more positive, but as the membrane potential approaches the chloride equilibrium potential, the net chloride current approaches zero. As in the above example, the membrane potential is "stabilized" at the chloride equilibrium potential.

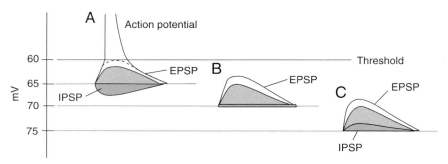

FIGURE 3.16.

The shape of postsynaptic potentials is affected by the resting membrane potential and its relationship to the equilibrium potential of the ions participating in the PSP. The EPSP is not much affected, since the resting membrane potential is usually quite far from the combined equilibrium potential for K^+ and Na^+ (about -8 mV). The effect is dramatic in the case of the IPSP because the equilibrium potential of chloride is close to the resting potential of most neurons. **A,** An EPSP (*black line* and *dashed line*) will generate an action potential if it exceeds threshold (*solid black line*). If the resting membrane potential is more positive than the equilibrium potential of chloride, then the IPSP (*red area*) will be a hyperpolarizing event. The temporal summation of an EPSP and an IPSP (*black area*) will decrease the amplitude of the EPSP over what it would be in the absence of the IPSP, in this case preventing the occurrence of the action potential. **B,** If the resting potential is equal to the equilibrium potential of chloride, then the IPSP will cause no change in the membrane potential (*red line*), but it will still affect the amplitude of an EPSP (*black line*), reducing its amplitude (*black area*). **C,** If the resting potential is more negative than the chloride equilibrium potential, then the IPSP will be a hypopolarizing event (*red area*). Even so, the IPSP will affect the amplitude of a simultaneous EPSP (*black line*), reducing its amplitude (*black area*).

INITIATION OF THE POSTSYNAPTIC ACTION POTENTIAL

Postsynaptic potentials are local currents that spread throughout the neuron. The current generated at a synapse, like all local currents, diminishes with distance from the source. Consequently, the amplitude of postsynaptic potentials at the soma is determined by the *electrical distance* from the bouton and is best measured in space constants rather than metric units.

The electrical distance that separates a synapse from the soma dramatically affects both the amplitude and the shape of the postsynaptic potential *as recorded at the soma*. The *amplitude* of the postsynaptic potential is diminished in proportion to the space constant, λ, due to the leakage of current out of the cell. The *shape* is affected in proportion to the time constant, τ. The time constant increases with greater distance between bouton and soma because of the greater membrane surface area, which increases membrane capacitance. The changes in both

shape and amplitude of an EPSP with increasing distance are illustrated in Figure 3.17. The amount of synaptic current that reaches the soma has important consequences for the initiation of an action potential in the postsynaptic neuron.

Action potentials, of course, require the presence of voltage-gated channels. With few exceptions, however, the dendrites of neurons have very few voltage-gated channels embedded in their membranes. Instead, these channels are concentrated at the initial segment of the axon. Therefore, to be an effective trigger of action potentials, synaptic currents must be of sufficient amplitude *at the initial segment* to affect the voltage-gated channels located there. If hypopolarization at the initial segment is above threshold for the voltage-gated channels, an action potential will be initiated. This action potential, once initiated, will be conducted along the axon as previously discussed (Fig. 3.18*A* and *B*).

Most individual EPSPs are quite small at the initial segment (Fig. 3.17) because it is several

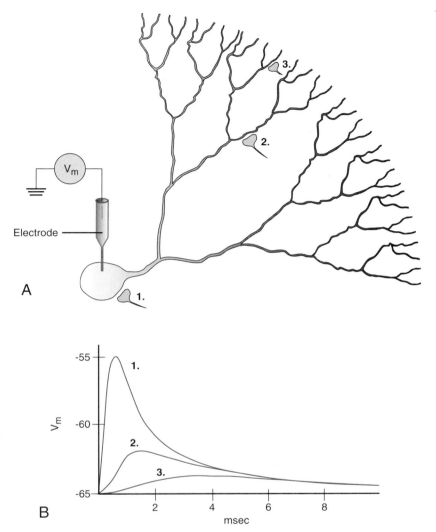

FIGURE 3.17.

A, Schematic diagram of a hypothetical dendritic tree with a simplified branching pattern. A recording site on the soma is indicated by the electrode. **B**, Recordings of EPSPs are based on computer simulations. The numbers correspond with the numbered boutons. The model calculated the resulting EPSP from a brief hypopolarizing event at the boutons indicated. Note that the amplitude and time course of the EPSP, as recorded in the soma, vary considerably, depending on the location of the active synapse.

space constants away from the distal dendrites (Table 3.3). A single EPSP, generated at a remote synapse, generally cannot bring the neuron to threshold. However, the current from multiple EPSPs combine at the initial segment such that many active synapses, acting in concert, can augment the amplitude of the hypopolarization (Fig. 3.18*C* and *D*). The **spatial** and **temporal summation** of hundreds, if not thousands, of EPSPs,

from as many individual synapses, is required to generate enough current to provoke a neuron into transmitting a **single** action potential.

INTERACTION OF EPSPs AND IPSPs

Local currents like IPSPs and EPSPs sum. The ebb and flow of synaptic currents follows the combined forces of the ionic equilibrium poten-

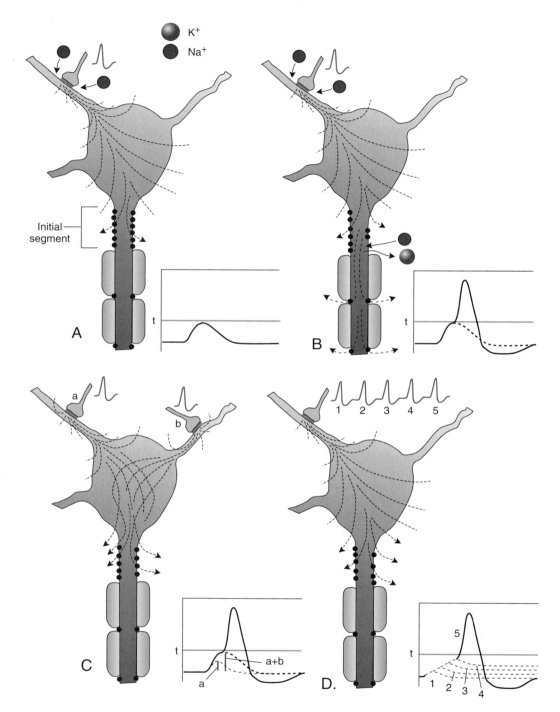

FIGURE 3.18.

The conversion of local postsynaptic currents into action potentials occurs at the initial segment. **A,** The hypopolarizing currents spread from the active bouton throughout the cell (*arrows*). At the initial segment there are voltage-gated sodium channels (*black dots*) that can cause a secondary increase in inward current if the EPSP is above thresh-

tials, modifying the neuron membrane potential continuously in response to synaptic activity. At the initial segment, this activity is translated into action potentials if the membrane potential reaches threshold. Therefore, any mechanism that can control the amount of current reaching the initial segment is important in controlling the activity of the neuron.

Synapses that produce IPSPs tend to be concentrated at the base of large dendrites and on the soma of neurons. The excitatory synapses are, generally speaking, located more distally. Consequently, proximal inhibitory synapses will have more influence on the membrane potential at the initial segment than the distal excitatory synapses. By providing a low-resistance pathway, or "sink," for electrical current to leave the cell,[9] the IPSP can short-circuit the positive currents generated by EPSPs (Fig. 3.19). An active IPSP at the base of a dendrite collapses the space constant for that dendrite by reducing its effective membrane resistance. *Inhibitory synapses, in effect, regulate the space constant of dendrites.* In this manner, a few well-placed inhibitory boutons can control the excitatory effectiveness of hundreds of excitatory boutons.

The smooth integration of excitatory and inhibitory events over time and space ultimately determines whether or not the postsynaptic cell

---

[9]It is customary, when referring to current flow, to define its direction in terms of the flow of positive charge. Therefore the inward negative chloride current associated with the IPSP is referred to as an outward (positive) current.

will initiate an action potential. This integration brings together information from diverse sources at a single moment and transforms that collective set of information into a single postsynaptic event, the action potential. *It is the summation of synaptic currents, both excitatory and inhibitory, at the initial segment of the axon that is the fundamental decision making process of the nervous system.*

PRESYNAPTIC INHIBITION

The postsynaptic interplay between hyperpolarizing and hypopolarizing currents at the initial segment is not the only way that the excitability of a neuron can be regulated. Presynaptic mechanisms exist that regulate the release of neurotransmitter and thus can modulate the amplitude of individual EPSPs and IPSPs. Presynaptic regulation is very specific, affecting only a single bouton, in contrast to postsynaptic regulation, which affects an entire neuron and all of the synapses on it. While both presynaptic inhibition and facilitation have been described in various species, only the presynaptic regulation of the EPSP, **presynaptic inhibition**, is of great significance in mammals.

Presynaptic regulation occurs when an axon synapses on the terminal bouton of another axon, which in turn forms a synaptic contact with another neuron (Fig. 3.20). The presynaptic bouton (*c* in Fig. 3.20) contains vesicles, releases transmitter, and triggers ligand-gated events very much like other chemical synapses. Most presyn-

---

old. A subthreshold EPSP is shown below. **B**, If threshed is exceeded, the voltage-gated sodium channels open and the regenerative process known as the action potential is initiated. This, in turn, creates new local currents that spread down the axon, allowing the action potential to be propagated. Below, the action potential is shown; the *dotted line* indicates the continuing EPSP, which is obscured by the action potential. **C**, Spatial summation occurs when two nearly simultaneous EPSPs are produced by different synapses, here at *a* and *b.* Independently each EPSP is subthreshold, but if they occur within an appropriate time frame, their currents can sum at the initial segment and exceed threshold (depicted below). The active boutons may be derived from collateral branches of a single axon or from different neurons. **D**, Temporal summation occurs when sequential action potentials invade a bouton. Each postsynaptic potential produced occurs before the effects of the previous postsynaptic potentials have subsided. Therefore their currents can sum, as shown below. Under physiological conditions, temporal summation and spatial summation usually occur simultaneously and involve multiple EPSPs and IPSPs generated at numerous active boutons.

Table 3.3. Equilibrium potentials for multi-ion systems

Equilibrium potentials computed using the Goldman-Hodgkin-Katz equation (3.7) with the permeability coefficients shown. Potentials are shown to the right of the permeability coefficients at 0, 2, and 4 space constants (Equation (3.10)) removed from the hypothetical single point origin of the potential. Permeability coefficients taken from C.U.M. Smith.

	P_{Na}	P_K	P_{Cl}	V_m		
				$0\,\lambda$	$2\,\lambda$	$4\,\lambda$
Rest	1.0^{-8}	1.0^{-7}	1.0^{-8}	-54	na	na
Peak of AP	1.0^{-6}	1.0^{-7}	1.0^{-8}	42	$^-16$	-39
After hypopolarization	1.0^{-8}	5.0^{-7}	1.0^{-8}	-77	na	na
Astrocytes	1.0^{-8}	1.0^{-6}	1.0^{-8}	-82	na	na
IPSP	1.0^{-8}	1.0^{-7}	1.0^{-6}	-68	$^-60$	-56
EPSP	7.5^{-7}	1.0^{-6}	1.0^{-8}	$^-8$	$^-36$	-47

aptic terminals in mammals secrete GABA, a chemical messenger that, when bound to the receptor, opens ligand-gated Cl^- channels in the postsynaptic bouton (*a* in Fig. 3.20). The chloride channels remain open for 100 to 150 msec, considerably longer than the 15 msec of a typical PSP. Since the chloride equilibrium potential is close to the axon resting potential, there is very little change in the bouton membrane potential.

The open chloride channels in the postsynaptic bouton, however, provide a large sink for positive current that diminishes the amplitude of any action potentials that arrive at the postsynaptic bouton. The diminution of the amplitude of the invading action potential decreases the probability that voltage-gated Ca^{2+} channels will open. Thus less calcium will enter the postsynaptic bouton and fewer vesicles will fuse with the bouton membrane, resulting in fewer quanta of transmitter being released into the synaptic cleft. This ultimately results in a diminished EPSP in the postsynaptic cell (*b* in Fig. 3.20).

Collision-Coupled Receptor Systems

The second method by which neurotransmitters interact with postsynaptic cells is mediated through G protein-coupled, second-messenger systems. Although these systems can affect ionophores, the principal advantage is their ability to affect cell metabolism and to modulate the function of the receptor proteins themselves. In this latter role they are acting as neuromodulators.

ACTIVATION OF G PROTEINS

The general pattern for second messenger-mediated synaptic responses is the same for all of the systems, regardless of their effects. In these systems, three membrane elements work together. The key molecule is a **G protein** (guanosine nucleotide binding protein), which is freely float-

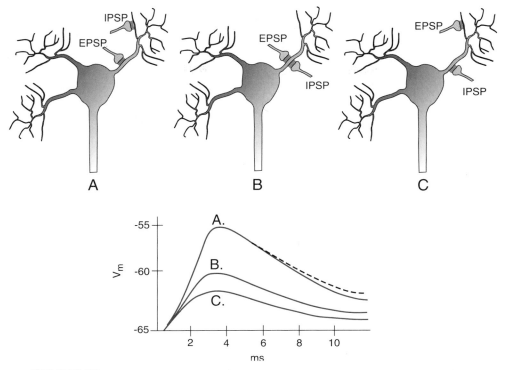

FIGURE 3.19.

   The physical relationship between boutons that produce EPSPs and IPSPs affects the amplitude and time course of the postsynaptic potential as recorded at the soma of the neuron. In this illustration, hypothetical, computer-generated EPSPs at the base of a dendrite were calculated for three different physical arrangements. **A,** An EPSP generated at the indicated site produces a response at the base of the dendrite as seen in the *dashed black line.* A simultaneous IPSP applied at a site distal to the EPSP hardly affects the amplitude or shape of the EPSP (*red line*). **B,** If the EPSP and the IPSP are simultaneously generated at the same physical position along the dendrite, the amplitude of the EPSP is diminished by about half and its time course is slowed. **C,** If the IPSP-generating bouton is located between the recording site and the EPSP-generating bouton, the inhibitory effect is even more pronounced.

ing within the lipid structure of the membrane. It randomly bumps into the receptor and other membrane proteins but cannot, under normal circumstances, react with them. However, when the neurotransmitter binds to the cytoplasmic side of the receptor molecule, the receptor is altered so that it can bind a G protein if one happens to bump into it (Fig. 3.21). Once bound to the receptor, the G protein is activated by the replacement of its GDP with GTP. The activated G protein is subsequently released from the receptor and again freely floats around within the structure of the membrane. Eventually, this activated G protein will bump into and combine with an appropriate enzyme (e.g., adenylate cyclase). This complex can catalyze the formation of the second messenger. This cumbersome process has become known as a **collision-coupled mechanism**.

SECOND MESSENGERS

   While there are at least 12 G proteins, there are only three second-messenger systems that have been described in neurons. Each system utilizes reactions induced by a different intermediate enzyme. The three enzymes are **cAMP, IP_3-DAG,** and **arachidonic acid** (Fig. 3.22). Each of these systems initiates a cascade of reactions that results in the phosphorylation of proteins (cAMP system), the liberation of Ca^{2+} into the cytosol (IP_3-DAG system), or the production of arachidonic acid metabolites (arachidonic acid system).

The neurotransmitters that operate indirectly through second messengers have the potential to affect a much larger variety of cell functions than the direct ionophore systems. The later mechanisms can only manipulate the cell's membrane potential and consequently participate only in the fast intercellular signaling that is manifested by the traffic of action potentials. Second-messenger systems can affect the way neurons transform information by interacting with cells in four ways: (a) by modulating the conductance of ion-selective channels, (b) by regulating the concentration of intracellular Ca^{2+}, (c) by modulating the properties of neurotransmitter receptors, and (d) by altering gene expression.

For example, ACh in the CNS acts through a muscarinic receptor to hyperpolarize the cell.

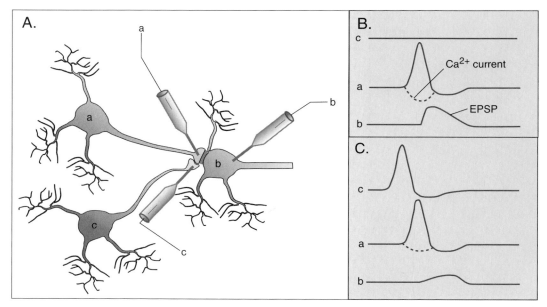

FIGURE 3.20.

A, The physical arrangement between neurons that results in presynaptic inhibition. A bouton from neuron *c* terminates on the bouton of another neuron, *a,* that in turn synapses with neuron *b.* Neuron *a* is excitatory to *b.* **B** and **C,** Recordings from each cell correspond with the labeled electrodes. **B,** In the absence of activity from neuron *c,* an action potential enveloping bouton *a* opens calcium channels (*red dotted line*) that in turn results in the release of neurotransmitter causing an EPSP in neuron *b.* **C,** An action potential from neuron *c* diminishes the calcium current (*red dotted line*) generated by a subsequent action potential arriving at bouton *a.* This in turn causes fewer vesicles to bind and consequently a diminished EPSP in neuron *b.* (See text for details.)

FIGURE 3.21.

Collision-coupled systems, like the G protein second-messenger system illustrated here, are loosely coupled systems in which one or more components freely float within the matrix of the cell membrane. **A,** Initially the G protein and receptor are dissociated. **B,** When a specific ligand (i.e., a neurotransmitter [*NT*]) is bound to the receptor, the receptor conformation changes, making it receptive to the G protein. **C,** The G protein at-

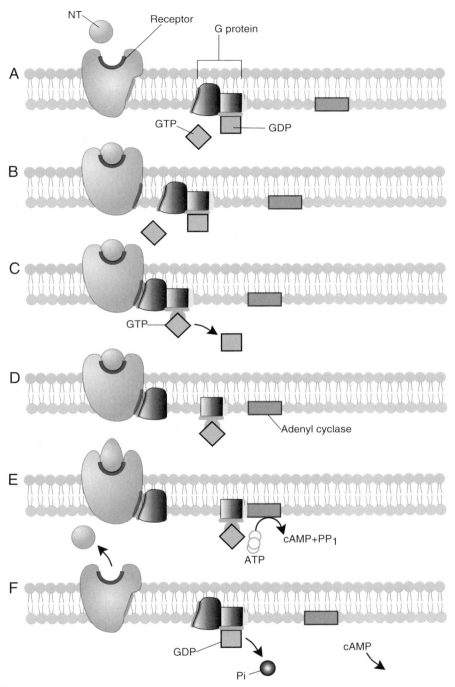

taches to the receptor when it bumps into it. This allows GTP to replace GDP on the G protein. **D**, A subunit of the G protein separates from the complex, becoming mobile within the membrane matrix. **E**, The subunit catalyzes the production of the second messenger with an appropriate enzyme (here adenyl cyclase). **F**, With the dissociation of the ligand from the receptor, the entire G protein is reunited. In this state it cannot catalyze adenyl cyclase.

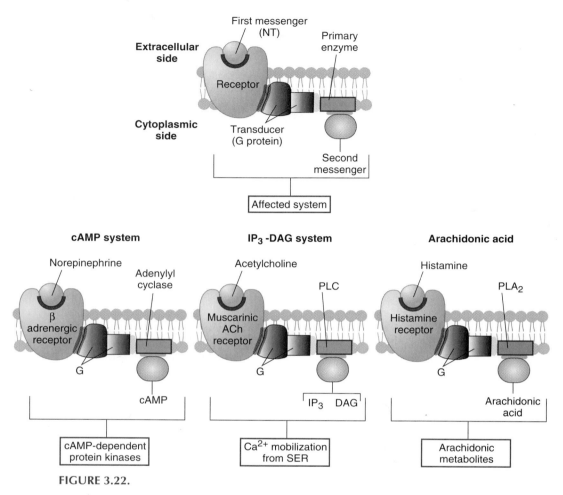

FIGURE 3.22.

G protein-coupled systems fall into three classes that are defined by the nature of the enzyme that is catalyzed by the active form of the G protein. These three systems, along with specific examples, are schematically illustrated.

This is accomplished through an interaction of the receptor–G protein–GTP complex with a potassium ionophore. This unusual ionophore is neither voltage gated nor ligand gated. It is a second messenger-gated ionophore. When coupled with the G-protein complex, the potassium ionophore opens, increasing the conductance to potassium, which hyperpolarizes the cell (Fig. 3.23). This is not as straightforward as it seems, however, because at least four different muscarinic receptors have been uncovered in the CNS, and all seem to respond somewhat differently to ACh.

Cell metabolism can be regulated by second-messenger systems also. The inositol phosphates (IPs) form a large and complex set of second messengers that are activated by phospholipase C (PLC). Many of the reactions initiated by the IPs liberate Ca^{2+} from intracellular stores, particularly from the endoplasmic reticulum. Calcium, in turn, has a profound effect on the activation of protein kinases and is therefore a critical component in the regulation of most cell functions.

Finally, it is possible for neurotransmitters to regulate the expression of the genes themselves. This is accomplished through phosphorylation of transcription regulatory proteins. Genes have a coding and a regulatory region with a proximal promoter section and a distal enhancer section. Attached to these sections are various regulatory proteins that, when phosphorylated, control the binding of RNA polymerase to the gene. Acting through second messengers, neurotransmitters can regulate the rate of protein transcription by phosphorylating the regulatory proteins. For example, in some neurons, the availability of the transmitter norepinephrine is regulated by the transcription of *tyrosine hydroxylase,* the enzyme that controls the first step in its synthesis. Synaptic activity on the neuron,

acting through second-messenger systems, causes the phosphorylation of the gene regulatory proteins that in turn control the transcription of tyrosine hydroxylase and ultimately the availability of norepinephrine.

In addition to providing great variety to the ensemble of synaptic interactions that can occur between cells, second-messenger systems are extremely sensitive to low levels of transmitter because the system greatly amplifies the signal.

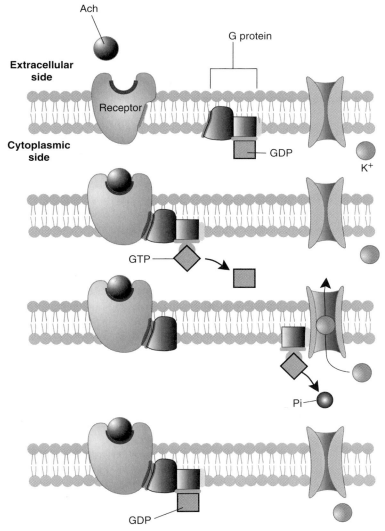

FIGURE 3.23.

   Some potassium channels are neither voltage gated nor ligand gated. As in the muscarinic system illustrated here, the permeability of the ion channel is controlled by a collision-coupled system.

For example, one transmitter molecule can activate about 10 G proteins, each of which in turn can activate about 10 adenyl cyclase molecules, an amplification factor of 100.

This variety and sensitivity of function is accomplished, however, at the expense of speed. Direct ligand-gated ionophores respond within about 0.5 msec, and the response is usually complete in about 15 msec. Indirectly regulated systems may not respond for hundreds of milliseconds and may continue their effects for hundreds of seconds. The effects related to gene regulation may persist for days.

The chemical details of these second-messenger systems are beyond the scope of this book, but may be found in standard biochemical textbooks.

Common Neurotransmitters

Neurotransmitters can be divided into several broad classes (Table 3.4). The best studied are the **small-molecule transmitters** such as acetylcholine (ACh), glycine, glutamate, and γ-aminobutyric acid (GABA). Most neurons in the nervous system secrete one of these molecules. The **catecholamines** are the principal neurotransmitters of the autonomic nervous system. They are also found in several important systems in the CNS. Recently, **neuroactive peptides** have been recognized as true neurotransmitters.

SMALL-MOLECULE NEUROTRANSMITTERS

The small-molecule neurotransmitters are the most commonly used transmitters in the nervous system. Within this group, **glutamate** is the most pervasive excitatory neurotransmitter in the mammalian brain. It binds with four receptors (Fig. 3.24), the simplest of which, kainate, opens a channel for Na^+ and K^+. A different glutamate receptor, NMDA, when bound opens a channel that passes not only Na^+ and K^+, but Ca^{2+} also. The action of this receptor is regulated by separate binding sites for Zn^{2+}, Mg^{2+}, glycine, and PCP (the hallucinogen commonly known as "angel dust"), which act as neuromodulators. These additional sites modulate the sensitivity of the ion channel when it is bound to glutamate. The complexity of the glutamate receptors offers a rich milieu for synaptic regulation, a richness that should dispel any notions that synapses are simple relay mechanisms designed to transfer action potentials from one cell to another.

Glutamate is also a potent neurotoxin and its effects are widespread because nearly every CNS neuron has glutamate receptors. The mechanisms of glutamate neurotoxicity are not known, but one hypothesis proposes that an excessive extracellular concentration of glutamate prolongs the conductance of Ca^{2+} into neurons.

Table 3.4. Common Neurotransmitters

Classes of neurotransmitters and the most common examples in each class. Substances listed in parentheses are probably neurotransmitters, but they have not rigorously met all the criteria (see text).

Small Molecule	Catecholamines	Neuroactive Peptides
Glutamate	Dopamine	Substance P
GABA	Norepinephrine	(CGRP)
Glycine	Serotonin	Enkephalin
Acetylcholine	(Histamine)	Endorphin

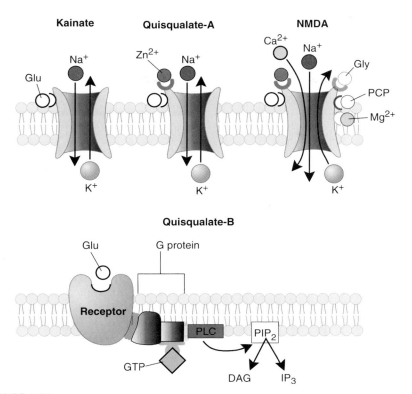

FIGURE 3.24.

Glutamate, perhaps the most pervasive of the neurotransmitters, interacts with at least four different receptors. The various receptors differ in several respects. The kainate receptor is a simple cationic channel with no apparent regulatory sites. The quisqualate-A receptor is similar, but has a Zn^{2+} regulatory site. The NMDA receptor not only has several regulatory sites, but also is permissive to the flow of Ca^{2+}. These are all direct ionophores that, when open, allow both Na^+ and K^+ ions to pass. In contrast, the quisqualate-B receptor initiates a G protein–IP_3-DAG second-messenger cascade.

Since calcium is a potent regulator of intracellular metabolism, excessive amounts may inappropriately activate proteases that can lyse the cell. Glutamate neurointoxication can occur under a number of conditions. For example, prolonged seizures (see Chapter **13**) may cause an excessive liberation of glutamate at the synapses of hyperactive cells and subsequently lead to neuron death.

γ-**Aminobutyric Acid (GABA)** and **glycine** are the major inhibitory neurotransmitters in the mammalian CNS. Both transmitters operate through receptors that are selectively permeable to Cl^- and act to inhibit postsynaptic cells. The GABA receptor has additional domains that are sensitive to the benzodiazepines and barbiturates, and its function can be modulated by these drugs. Both kinds of drugs increase the Cl^- permeability of the ionophore in the presence of GABA but have little effect in its absence. The benzodiazepines are a class of common tranquilizers (e.g., diazepam [Valium] and chlordiazepoxide [Librium]), and the barbiturates are well-known anesthetics.

Acetylcholine (ACh) is a neurotransmitter that can bind to two kinds of receptors, the **nicotinic** and the **muscarinic**. These names are historical, representing the first agonists that were used to differentiate the two receptors. ACh, acting through the nicotinic receptor, is best known in its direct excitatory role at the neuromuscular junction where the bound complex opens an ion-selective channel that increases the permeability to both Na^+ and K^+. In cardiac muscle cells, ACh operates through muscarinic receptors that open only K^+ channels through a second-messenger G protein.

CATECHOLAMINES

Another large group of neurotransmitters are the **catecholamines**. Norepinephrine, serotonin, and dopamine are the most important molecules in this class. Although there is little doubt that they function as transmitters, their exact role is not well understood. The catecholamines are a group of neurotransmitters that exert their effects through the cAMP system. They are widely distributed in the nervous system but, unlike the omnipresent glutamate, are restricted to specific neuron systems. For example, **dopamine** is located predominantly in the nigrostriatal pathways, while **norepinephrine** is principally located in the locus ceruleus pathways and the autonomic systems. **Serotonin** is largely limited to the cells of the raphe nuclei. Unlike the small-molecule transmitters, the catecholamines are expressed by relatively few neurons, but the axon processes of these neurons extend widely throughout the brain.

NEUROACTIVE PEPTIDES

Neuroactive peptides form a large class of neurotransmitters. While only a few peptides have been accepted as neurotransmitters, more than 100 have been shown to have effects on the nervous system. None of the neuroactive peptides appear to work through receptors that have direct effects on ion channels but exert their influences indirectly through second-messenger systems.

Substance P is a small polypeptide. There is very good evidence that it acts as a neurotransmitter. Substance P, like most of the neuroactive peptides, has effects that are separate from its role as a transmitter. For example, substance P is able to induce cell division and is therefore a mitogen. As such it plays a role in wound healing.

Another neuroactive peptide, **CGRP** (calcitonin gene-related peptide), acting through the cAMP system can phosphorylate an ACh receptor, making it less responsive to the specific transmitter ACh. In this role, CGRP acts as a **neuromodulator** rather than a neurotransmitter, because it regulates the function of the receptor protein.

Disturbances of Neural Function

The complexity of chemical signaling between neurons creates many possibilities for malfunction. Consider the four criteria for transmitter identification; each also represents an opportunity for synaptic failure. For example, genetic or metabolic dysfunction can lead to failure to synthesize the transmitter. Disturbances with the voltage-gated Ca^{2+} channels lead to diminished capacity to release the transmitter. False ligands can bind to the receptor in a way that blocks its physiological function, and methods of eliminating the transmitter can be foiled. Many important neurological diseases can be understood in terms of the failure of synaptic transmission.

MYASTHENIA GRAVIS

Myasthenia gravis is a common disease with a prevalence of about 3/100,000. It is characterized by a general muscle weakness that waxes and wanes throughout the day and also over longer periods of time. This weakness almost always involves the ocular muscles and the muscles of the face and of the mouth. **Diplopia** (double vision), **ptosis** (eyelid droop), **dysphagia** (swallowing disorders), and **dysarthria** (speech slurring) are common symptoms. To varying degrees, all the other muscle groups of the body may be involved. Frequently, but not necessarily, episodes of weakness are correlated with exercise. The weakness improves with rest and improves dramatically if cholinergic agonist drugs are administered. It is now possible to show that antibodies to the ACh receptor protein are present in the victim (see Case History below).

ACh receptor antibodies interfere with synaptic transmission at the neuromuscular junction in two ways. *The primary problem is the immunological destruction of the receptor.* In normal people, the ACh receptors normally degrade and consequently are continuously replaced (Fig. 3.25). In the myasthenic patient the receptors are destroyed faster than they can be replaced. The normal population of receptors becomes depleted. Additionally, while bound to the receptor, the antibody interferes with the receptor's gating properties. Therefore, the patient has a subnormal number of functional ACh receptors and ionophore channels at the motor endplate. Second, there is complement-mediated lysis of the endplate membrane itself that reduces its surface area.

Together these deficiencies lead to a dramatic reduction in the effectiveness of the released transmitter and subsequently to an abnormally small muscle endplate potential (EPSP of the muscle). Many endplate potentials do not reach threshold for the muscle, so fewer muscle fibers are activated and the compound muscle action potential is smaller than normal. Fewer muscle fibers participate in the contraction, and the patient experiences weakness.

Clinically, **electromyography** is used to record the compound muscle action potential (Fig. 3.25). Low-frequency repetitive stimulation (typically 3 Hz for 3 seconds) of the motor nerve produces a *decrementing response* in the amplitude of the compound muscle action potential in patients with myasthenia gravis (Fig. 3.25*B*), whereas in normal patients all the muscle action potentials are of the same amplitude (Fig. 3.25*A*). This same decrementing response can be recorded from patients that have been given *d*-tubocurarine[10], a drug that blocks ACh receptors. In both cases, the release and enzymatic removal of transmitter is unaffected.

EATON-LAMBERT SYNDROME

The Eaton-Lambert syndrome is also an autoimmune disease. It is associated with carcinoma of the bronchus and is only rarely seen in its absence. The syndrome is characterized by abnormal voltage-regulated Ca^{2+} channels in the motor nerve terminals. The disease affects both muscarinic and nicotinic receptors. Most evidence suggests that the channels are being im-

munologically attacked. The defective channels allow fewer Ca^{2+} ions to enter the bouton with each stimulus. This results in an abnormally small number of vesicles binding to the active site, and consequently less transmitter is released. This leads to a smaller endplate potential and weakness of muscle contraction. In contrast to myasthenia gravis, repetitive stimulation of the motor nerve results in an *incrementing response* of the compound muscle action potential. This is probably caused by an accumulation of intracellular Ca^{2+} in the presynaptic terminal. As the Ca^{2+} builds up with each succeeding stimulus, more vesicles are mobilized and fuse with the preterminal membrane. Therefore, more transmitter is released with each stimulus (Fig. 3.25*C*).

NEUROTOXINS

Botulinum toxin, produced by the spores of *Clostridium botulinum,* also acts on muscarinic and nicotinic muscle synapses by interfering with a calcium dependent-mechanism for the release of transmitter. The exact mechanism is unknown, but since it interferes with other known calcium-dependent processes, it is assumed that calcium channels are either destroyed or blocked. Its mode of action is similar to **tetanus toxin**, another potent microbial neurotoxin. The effects of botulinum toxin are very similar to those seen in Eaton-Lambert syndrome, including the incremental response to repetitive stimulation. Botulinum toxin is one of the most potent poisons, being lethal at 10^{-12} g/kg (mouse data; extrapolated to a human, a lethal dose would require only 800 pg).

Black widow spider venom (α-latrotoxin) is also a presynaptic poison. This unusual toxin binds to specific receptors in the presynaptic motor nerve terminal where it is incorporated into the membrane. It then acts as a continuously open ionophore that allows Na+, K+, and Ca^{2+} to enter the cell. The result is both membrane hypopolarization and a massive influx of Ca^{2+} that leads to inappropriate vesicle binding and rapid depletion of transmitter. Since the incorporation of the toxin into the membrane is essentially irreversible, vesicle recycling is impossible because the intracellular calcium concentration can never be restored. Clinically, victims of this spi-

---

[10]Curare is a natural toxin used by certain Indian groups to coat their poison darts. The drug paralyses their prey by blocking ACh receptors at the neuromuscular junction.

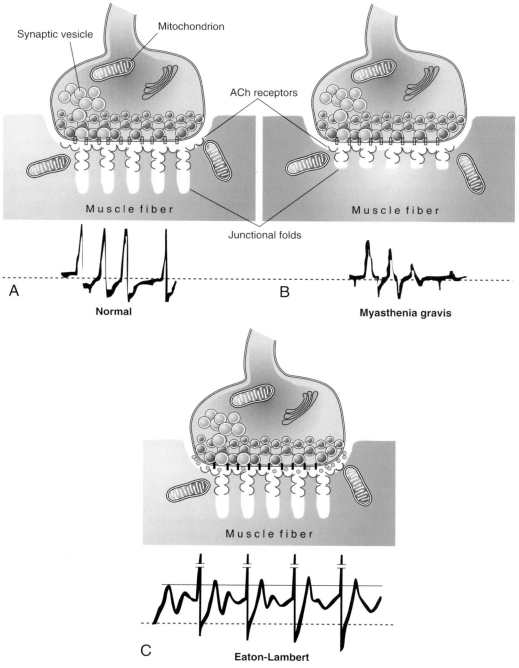

FIGURE 3.25.

    The myoneural junction is depicted with EMG recordings below. **A**, The normal junction is characterized by an end plate with deep folds with the ACh receptors located at the top. Repetitive electrical stimulation of the nerve produces muscle action potentials of equal amplitude. **B**, The disease process of myasthenia gravis not only reduces the number of ACh receptors but also diminishes the size of the postsynaptic folds in the motor end plate. In response to repetitive nerve stimulation, the EMG recording shows a decrease in the amplitude of successive muscle action potentials. **C**, Eaton-Lambert disease affects the calcium ionophores in the presynaptic bouton, not the myoneural junction. The EMG in this case increments with repetitive stimulation (truncated spikes are shock artifacts produced by the stimulus).

der's bite first manifest spasms and rigidity, followed by flaccid paralysis. Death is uncommon (about 4%), probably because this small spider is unable to inject a large amount of venom.

CASE HISTORY

Ms. S. R.

S. R. is a 14-year-old female who was in good neurological health until July of 1980. At that time, she began to have difficulty with speaking, swallowing, and vision and was experiencing weakness in her arms. The first problem to become apparent to her was a difficulty with speaking that would occur late in the day, generally just before bedtime. Her speech at this time became characterized by a nasal tone. This progressed over a period of about 10 days. As it progressed, her difficulties occurred earlier and earlier in the day, until it was a persistent problem. About this time, she began to have difficulty swallowing solid foods, but had no difficulty with liquids. To the patient, it seemed as if the solid foods would stick in her throat. Also about this time, the patient stated that she began to have visual disturbances late in the day, but she denied having double vision. In the evening, her eyelids would droop, and she had to arch her head backward to watch television. She also noted that she had difficulty with her arms, particularly after moving them for any length of time. She also had noted difficulty with smiling, spitting, and sucking while drinking through a straw.

There was no family history of similar difficulties, and the patient had no history of any other neurological or immunological disorders.

NEUROLOGICAL EXAMINATION, September 11, 1980:

I.   MENTAL STATUS: Normal

II. CRANIAL NERVES:

    A. **Facial nerve:**   there was bilateral drooping of the mouth and the smile was not full.

    B. **Vagus:**   the voice was somewhat nasal, but no apparent fatigue was noted during the interview.

    C. The other cranial nerves were normal. In particular, there was no ptosis, and there were no visual disturbances.

III. MOTOR SYSTEMS:

    A. Gait, casual walk, walking on toes, heels, and tandem walk were all normal.

    B. The patient could hop well on either foot and performed deep knee bends without difficulty.

IV. SENSORY SYSTEMS: Normal

V.   CEREBELLAR FUNCTIONS: Normal

VI. REFLEXES: All normal

Comment: At this time, the patient's neurological examination was fairly undramatic and not sharply defined. Nevertheless, one is clearly required to entertain the diagnosis of myasthenia gravis at this time based on the history. One hallmark of myasthenia gravis is increasing weakness with exercise and a restoration of strength with rest. This is illustrated in the case of S.R., since her problems began late in the day but would resolve by morning. One should review the history and note how weakness in various muscle groups accounts for each of the specific symptoms of weakness reported by the patient.

SUBSEQUENT COURSE

After the examination described above, the following laboratory evaluations were performed. To demonstrate that the patient's symptoms are caused by failure of neuromuscular transmission, **edrophonium,** an anti-cholinesterase drug, was injected intravenously. This drug prevents the destruction of ACh by the enzyme acetylcholinesterase, thereby prolonging the action of ACh. Ms. S. R.'s strength dramatically improved within 30 seconds of the injection. Her weakness returned about 5 minutes later, because the drug is very short acting. A dramatic response such as this to edrophonium is considered essential for the diagnosis of myasthenia gravis because it proves that the site of the pathology is limited to the neuromuscular junction. Subsequently, a CT scan was performed to determine if S. R. had a thymoma, a nonmalignant hyperplasia of the thymus gland that is present in about 15% of patients with myasthenia gravis. S. R.'s CT scan was normal.

At this time, the patient was placed on a long-acting anticholinesterase drug, pyridostigmine. In spite of the therapy, her general weakness progressed. She was examined four more times in the course of 1980 in an effort to establish the proper drug dosage. These examinations revealed a relentless progression of the disease and the inability of the drug to control the symptoms. At this time, unless she took two pyridostigmine pills in the morning, she could not get out of bed unaided. She was able to walk while at school, but her speech remained poor even when the medication was most effective.

S. R. was admitted on July 6, 1981, for a thymectomy in an attempt to bring about a remission of her disease. Removal of the thymus gland brings the disease into remission in about 80% of the cases.

FURTHER APPLICATIONS

- Contrast the clinical picture given here of a patient suffering with myasthenia gravis with what you would expect to observe in a patient suffering from the Eaton-Lambert syndrome. Should you be able to differentiate between these two diseases based only on the history? How? What would you emphasize during the neurological examination to draw out the differences?

- Give possible explanations for the fact that patients with myasthenia gravis have involvement of only the ACh receptor sites at the neuromuscular junction?

- From other resources, learn about plasmapheresis. How can it be of benefit to patients with myasthenia gravis? What complications would you expect from this procedure?

- Consider the difference in the meaning of the terms hypertrophy and hyperplasia. Is this distinction important in interpreting the pathology report in this case? Is hyperplastic tissue necessarily malignant? What is the rationale for removing the thymus gland in patients such as Ms. S. R.?

SUMMARY

The nervous system uses both chemical and electrical processes to generate and transmit signals. The energy for electrical signaling comes from ion imbalances that exist across a selectively permeable cell membrane. The ion imbalances cause ions to move through the membrane, down their concentration gradient. If the membrane is impermeable to one species of charged molecules, charge will be separated and an electrical gradient established. When the force of diffusion is just balanced by the electrical force represented by the separated charge, an equilibrium exists. The electrical potential at equilibrium is the equilibrium potential or Nernst potential. If a membrane has several ion channels that select different ions, each ion contributes to the equilibrium potential in proportion to its relative permeability. The membrane potential of a cell is the equilibrium potential established by several ions.

The membrane potential can be manipulated by altering the permeability coefficients for the various ions. Neuron membranes contain membrane-spanning proteins that alter their permeability in response to the transmembrane potential. These voltage-gated ion channels create the action potential, a brief modulation of the membrane potential, by altering the membrane permeability to Na^+ and K^+ in response to changes in membrane potential.

In contrast to action potentials, local currents are not associated with dynamic changes in membrane permeability but are simply the passive distribution of current throughout the neuron. The potential associated with local currents diminishes with distance from the current source in proportion to the membrane resistance and the cytoplasmic resistance. Local currents associated with action potentials are distributed along axons and serve as the hypopolarizing stimulus needed to propagate the action potential along the axon. Local currents associated with synaptic events are distributed along the dendrite and soma, and they serve as the hypopolarizing stimulus at the initial segment where action potentials are generated.

Axons terminate as complex structures known as synapses. At most synapses in the mammalian nervous system, presynaptic action potentials are converted to chemical signals that, in turn, are converted into electrical signals at the postsynaptic cell. The presynaptic action potential causes voltage-gated calcium channels to open, which allows Ca^{2+} to enter the synaptic bouton. In the presence of Ca^{2+}, vesicles containing the neurotransmitter bind with the bouton membrane and release their contents into the subsynaptic space. The transmitter diffuses across the space and combines with a membrane-spanning receptor protein on the postsynaptic cell.

There are two recognized types of receptor proteins, those that are ion-selective ionophores and those that initiate collision-coupled second messenger biochemical cascades. The association of transmitter and receptor with the former group opens ion channels in the postsynaptic cell. The postsynaptic potentials that are associated with the ionic currents resulting from these channel openings can be either hyperpolarizing or hypopolarizing. At the initial segment, the forces of all postsynaptic potentials are integrated in proportion to their electrical distance from each synapse, and action potentials are generated if threshold is exceeded.

Receptor proteins that initiate collision-coupled reactions operate through G proteins that activate one of three second-messenger systems. Cell metabolism can be altered by cAMP systems acting through protein phosphorylation or by IP_3-DAG systems that regulate intracellular Ca^{2+}. A third second messenger, arachidonic acid, can also be invoked by neurotransmitters acting through specific receptors. Second-messenger systems regulate the metabolic function of the cell by regulating intracellular Ca^{2+} levels, the phosphorylation of proteins, and the regulation of gene expression.

FOR FURTHER READING

Hall, Z. W. *An Introduction to Molecular Neurobiology.* Sunderland, MA: Sinauer Associates, 1992.

Kandel, E. Transmitter Release. In: *Principles of Neural Science,* edited by E. Kandel, J. Schwartz and T. Jessell. New York: Elsevier, 1991, pp. 194–212.

Kandel, E., Siegelbaum, S., and Schwartz, J. Synaptic Transmission. In: *Principles of Neural Science,* edited by E. Kandel, J. Schwartz, and T. Jessell. New York: Elsevier, 1991, pp. 122–134.

Kandel, E., and Schwartz, J. Directly Gated Transmission at Central Synapses. In: *Principles of Neural Science,* edited by E. Kandel, J. Schwartz, and T. Jessell. New York: Elsevier, 1991, pp. 153–172.

Kelner, K., and Koshland, D. J. *Molecules to Models: Advances in Neuroscience.* Washington DC: American Association for the Advancement of Science, 1989.

Koester, J. Membrane Potential. In: *Principles of Neural Science,* edited by E. Kandel, J. Schwartz, and T. Jessell. New York: Elsevier, 1991a, pp. 81–94.

Koester, J. Passive Membrane Properties of the Neuron. In: *Principles of Neural Science,* edited by E. Kandel, J. Schwartz, and T. Jessell. New York: Elsevier, 1991b, pp. 95–103.

Koester, J. Voltage-Gated Ion Channels and the Generation of the Action Potential. In: *Principles of Neural Science,* edited by E. Kandel, J. Schwartz, and T. Jessell. New York: Elsevier, 1991c, pp. 104–119.

Partridge, L. and Partridge, L. D. *The Nervous System: Its Function and Its Interaction with the World.* Cambridge, MA: The MIT Press, 1993.

Rowland, L. *Merritt's Textbook of Neurology.* Philadelphia: Lea & Febiger, 1989.

Rowland, L. Diseases of Chemical Transmission at the Nerve-Muscle Synapse: Myasthenia Gravis. In: *Principles of Neural Science,* edited by E. Kandel, J. Schwartz, and T. Jessell. New York: Elsevier, 1991, pp. 235–243.

Schwartz, J. Ion Channels. In: *Principles of Neural Science,* edited by E. Kandel, J. Schwartz, and T. Jessell. New York: Elsevier, 1991a, pp. 66–80.

Schwartz, J. Chemical Messengers: Small Molecules and Peptides. In: *Principles of Neural Science,* edited by E. Kandel, J. Schwartz, and T. Jessell. New York: Elsevier, 1991b, pp. 213–224.

Schwartz, J. Synaptic Vesicles. In: *Principles of Neural Science,* edited by E. Kandel, J. Schwartz, and T. Jessell. New York: Elsevier, 1991c, pp. 225–234.

Schwartz, J. and Kandel, E. Synaptic Transmission Mediated by Second Messengers. In: *Principles of Neural Science,* edited by E. Kandel, J. Schwartz, and T. Jessell. New York: Elsevier, 1991, pp. 173–193.

Shepherd, G. *The Synaptic Organization of the Brain.* New York: Oxford University Press, 1979.

Shepherd, G. *Neurobiology.* New York: Oxford University Press, 1994.

Smith, C. U. M. *Elements of Molecular Neurobiology.* New York: John Wiley & Sons, 1989.

4 Principles of Sensory Transduction

The mammalian nervous system is an information processing organ. Information by itself is sterile unless given meaning by association with objects and processes that are significant to the organism. The process of converting energy from external world events into an internal representation creates meaningful information. In this chapter we will see how elements of the external world are transcribed and represented as local currents and action potentials in the internal world of our brains.

There is an important distinction to be made between the **physical stimuli** that the external world presents to our senses and the **perception** our conscious awareness has of these stimuli. In our everyday lives we behave as if the relationship between stimulus and perception is both accurate and tightly linked. This is not always the case. Everyone is familiar with optical illusions (Fig. 4.1). These are familiar situations in which our perceptions do not match reality. All of our senses are capable of deceiving us.

To understand the relationship between **physical stimuli** (i.e., sensations) and conscious **perception**, one must understand the nature of **transduction** and **representation** in the nervous system. Transduction is the process by which the energy of a physical stimulus is detected and converted into a form of energy used by the nervous system. Representation is the way information is encoded and organized as an **analog** of the stimulus within the structure of the nervous system.

Transduction

The process of transforming the energy content of an environmental stimulus into coded action potentials is called **sensory transduction**. Two steps are involved. First the physical stimulus is converted into a **receptor potential**. Second, the receptor potential is converted into **action potentials.** The receptor potential is a change in the membrane potential produced by a transducer mechanism. Receptor potentials are local, graded responses that diminish in amplitude with distance from the receptor membrane. Therefore, for the information represented by the receptor potential to be transported to the central nervous system, it must be converted into action potentials.

IONIC MECHANISMS

Receptor potentials are produced by changes in membrane permeability in a manner that is similar to the mechanisms associated with voltage-gated and ligand-gated systems. The process of initiating a receptor potential is best illustrated in certain nerve endings associated with some **mechanoreceptors** that are incorporated into the terminal membrane of certain sensory axons. Mechanoreceptors convert mechanical energy into receptor potentials. The current hypothesis suggests that ion channels specifically sensitive to

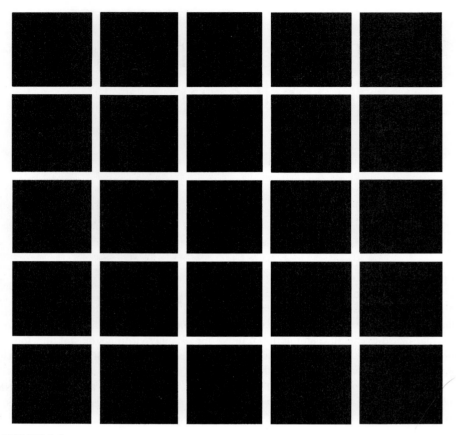

FIGURE 4.1.

The gray spots seen at the intersections of the white lines are not "real." The true physical stimulus at the intersections is the same as at the middle of the white lines. The perception of gray is caused by neuronal interactions occurring in the retina.

mechanical deformation are incorporated in the terminal axon membrane (Fig. 4.2). Unlike voltage-gated or ligand-gated channels, these channels are linked to the cytoskeleton in such a way that mechanical distortions of the receptor membrane cause the channels to open. These channels are selective for Na^+ and K^+, so when they open, the axon terminal becomes hypopolarized.

The hypopolarization caused by the opening of the channels is the receptor potential. *The receptor potential is simply a local current* and displays all the properties of local currents; its *amplitude* is proportional to the number of open ion channels, i.e., it is a graded potential. Its *duration* lasts approximately as long as the stimulus. These properties make the receptor potential

an **analog** [G. *analogos,* according to ratio, proportionate] of the stimulus (Fig. 4.3). Being an analog of the stimulus, the receptor potential follows the stimulus, more or less accurately, in intensity and duration.

INFORMATION CODING

At the receptor, the energy content or **intensity** of the stimulus is represented by the *amplitude* of the receptor potential (Fig. 4.3). The degree to which mechanical energy deforms the receptor membrane determines the number of ion channels that open, which in turn determines the amount of hypopolarization of the nerve ending. Since the amplitude of the receptor potential varies in time

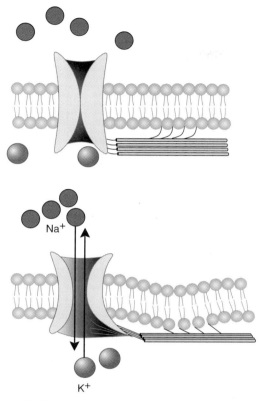

FIGURE 4.2.

Mechanoreceptor membranes are thought to contain mechanically coupled ionophores. They are normally closed (*top*), but if the membrane is deformed (*bottom*) they open due to their linkages with the cytoskeleton. When open, they allow both Na^+ and K^+ to cross the membrane.

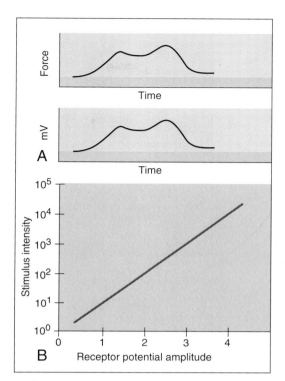

FIGURE 4.3.

A perfect transducer would create an analog of the stimulus that is identical to it in every respect. **A**, The figure illustrates the idealized receptor potential shown to represent a voltage analog of a mechanical stimulus. The stimulus is plotted as force versus time, while the corresponding idealized receptor potential is plotted as millivolts versus time. **B**, An idealized graph illustrating a log relationship between the amplitude of the stimulus and the amplitude of the receptor potential.

as the stimulus intensity varies, the receptor potential is an **amplitude-modulated (AM)** signal.

The receptor potential, like all local currents, degrades in amplitude with distance from the current source. This degradation is in proportion to the space constant, λ, of the membrane. For the information content of the receptor potential to be transmitted to the central nervous system, it must be transformed into action potentials. Action potentials are all of the same amplitude; they are, in effect, a binary signal, being either present or absent. Since stimulus intensity cannot be represented by action potential amplitude, it must be represented by the interval between action potentials, which is usually expressed as **frequency;** i.e., the number of action potentials occurring per

second (Fig. 4.4). The variation in the frequency of the action potentials correlates with the intensity of the stimulus. In this form, the internal representation of the stimulus intensity is **frequency modulated (FM)**.

Action potentials are, of course, produced by voltage-sensitive channels in the membrane. However, there are no voltage-sensitive gates in the receptor membrane; they first appear a short distance away from the receptor membrane. In myelinated axons this is at the first node of Ranvier. The conversion of the amplitude-modulated receptor potential into frequency-modulated action potentials is similar to the conversion of local currents into action potentials when an ac-

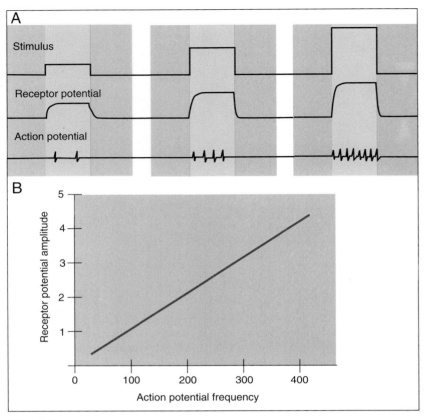

FIGURE 4.4.

Frequency modulation is a form of representation. **A**, A receptor potential is plotted versus time below the stimulus that produced it. Below it are shown the resulting action potentials. Note that the inter-spike interval decreases with increasing amplitude of the receptor potential. **B**, The idealized relationship between the amplitude of the receptor potential and the frequency of action potentials is shown.

tion potential is propagated along the axon. The principal difference is that receptor potentials vary in amplitude. If the receptor potential amplitude is large, it will continue to hypopolarize the voltage-sensitive membrane above threshold after the initial action potential has recovered. This means that another action potential can be initiated after the absolute refractory period subsides. The interval between the first and the second action potential is determined by the interaction between the amplitude of the receptor potential and the relative refractory period of the action potentials (Fig. 4.5). The larger the amplitude of the receptor potential, the shorter the interval between succeeding action potentials.

The frequency of these action potentials is the inverse of the interspike interval:

$$f_{Hz} = \frac{1}{\Delta t_{sec}} \qquad (4.1)$$

where f is the frequency in hertz and Δt is the interspike interval.

AM-FM CONVERSIONS

The conversion of intensity information from AM to FM occurs in several critical places in the nervous system. The initial transduction of external stimuli is into an AM signal (the receptor potential). This is immediately transformed into an FM signal (action potentials) at the distal portion

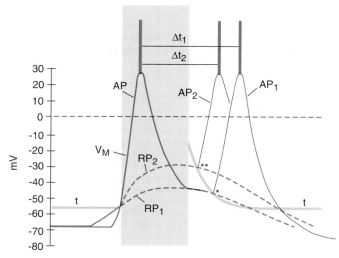

FIGURE 4.5.

The relationship between the receptor potential amplitude and the inter-spike interval of the generated action potentials is illustrated. A low-amplitude receptor potential (RP_1) reduces the membrane potential to threshold (t), at which time an action potential (AP) is produced. From this point until some time during the repolarization of the action potential, the membrane is absolutely refractory (*red shaded area*). During the subsequent relative refractory period, threshold is high but declining (t after spike). When threshold declines and meets the recovering membrane potential (at *), a second action potential is initiated (AP_1). Given a larger receptor potential (RP_2) that initiates an identical initial action potential (AP), the recovering membrane potential will meet the declining threshold level sooner (at **) than in the previous example, initiating a second action potential (AP_2). Note that the interspike interval Δt_1 is longer than Δt_2. In both examples, during the action potential the receptor potentials are indicated by *broken lines*. In normal recordings, the receptor potentials are masked by the action potential during this time and cannot be visualized directly.

of the axon. The action potential produces local currents that are necessary for the propagation of the action potential along the axon. Action potentials invading the bouton are converted into the calcium current, an amplitude-modulated signal, that initiates vesicle binding and subsequent release of transmitter from the bouton. Interaction between the transmitter and the postsynaptic membrane receptors generates amplitude-modulated postsynaptic potentials (PSPs) in the postsynaptic neuron. Finally, the PSPs are returned to action potentials, a frequency-modulated signal, at the initial segment in the postsynaptic cell.

One might well wonder why the nervous system uses this complex interaction between AM and FM signals to carry information. Since these modulation methods are different representations of the same information. Each method has

advantages in certain situations. Action potentials are necessary to transmit information over long distances. In some cases this distance can be several hundred centimeters. In large animals like whales and giraffes, the distances can be measured in meters! Local currents degrade with distance and over the course of a few micrometers diminish to the point of extinction. They cannot carry the information over the distances required without somehow being regenerated along the way. The action potential is the mechanism by which local currents are renewed by harnessing the potential energy of the membrane potential and creating new local currents. *The regenerative interplay between the action potential and the local current along the axon continuously restores the signal.* Because of this regenerative property, information can be transmitted over axons for indefinite distances with *no loss of signal quality.*

In contrast to the action potential, local currents can be combined easily and interact with one another. The speed and fluidity of local current interactions is essential to the function of the nervous system. This occurs predominantly in the soma and dendrites of neurons, where information is represented by the amplitude of PSPs (one form of local currents). The addition and subtraction of PSPs that occur in the neuron and the subsequent generation of the action potential at the initial segment is the mechanism by which computation (i.e., information transformation) is performed by the nervous system.

Representation

As individuals we are accustomed to perceiving several different sensory experiences. If our nervous system is intact, we can hear, see, taste, touch, and smell our environment. We also have sensations such as heat, cold, and **proprioception**

[L. *proprius,* one's own + *capio,* to take], the ability to sense the position of our limbs. The various forms of perception are called **modalities**. *Each modality is closely coupled with forms of energy that can be described in terms of physics and chemistry.* Furthermore, for each of these physical stimuli, there are specific receptors that are most sensitive to that form of energy. Under normal circumstances, a specific receptor is closely coupled to the energy form of the stimulus so that a receptor is affected by only one stimulus modality, which is called the **adequate stimulus**.

Receptors are the starting point of a chain of neurons that ascends the nervous system. Each chain carries only one sensory modality and is therefore *line labeled.* This means that the information in each chain of neurons is interpreted by central structures *according to the modality of the receptors connected to it in the periphery.*

TYPES OF RECEPTORS

There are five classes of receptors: photic, chemical, mechanical, thermal, and noxious (Table 4.1). Each class is closely coupled to a distinct form of energy and is most efficient in

Table 4.1. Classes of Sensory Receptors

Receptor Type	Stimulus	Perceptions
Photoreceptor	Photons	Light
Chemoreceptor	Specific molecules	Taste & smell
Thermoreceptors	Temperature	Heat & cold
Mechanoreceptors	Mechanical force	Touch, proprioception, sound, etc.
Nociceptors	Chemicals released by tissue damage	Sharp pain, dull pain, burning pain, etc.

transducing that form of energy into neuronal signals. **Photoreceptors** are exquisitely designed to respond to energy in the form of photons, perceived as vision. **Chemoreceptors** are cells that have evolved specific membrane proteins that recognize different steric classes of molecules, which are perceived as taste or smell. **Mechanoreceptors** respond to deformations that give rise to the sensations we generally call *touch;* **thermoreceptors** account for our perceptions of *temperature;* and **nociceptors** detect tissue damage that we perceive as *pain.* The latter three modalities are collectively called **somatosensory receptors** because they are located in the skin, muscles, joints, and viscera. Proprioceptors are a special class of mechanoreceptors located in muscles and joints that allow us to perceive the *position of our limbs* (see Chapter 6). Photoreception is discussed separately in the chapter on vision (see Chapter 11)

Chemoreceptors

Chemoreceptors are the transducers that mediate the gustatory and olfactory sensations. In general, a chemoreceptor molecule in the membrane binds specific molecules that, when bound, cause conductive channels through the membrane to open. Binding specificity is based on the steric relationship between the receptor molecule and the molecule to be bound. A number of different receptor molecules have been identified in both the gustatory and olfactory epithelia. It appears that each receptor is capable of evoking a separate sensation. In this way, a number of modalities of taste and smell can be established.

The chemoreceptors produce membrane hypopolarization by several mechanisms. For the simplest mechanisms, the receptor and the ionophore are the same molecule (Fig. 4.6*A*). For example, hydrogen ions can directly block certain K^+ channels, reducing the potassium current and hypopolarizing the cell. Other receptors act indirectly through G protein-coupled second-messenger systems (Fig. 4.6*B*) (see Chapter 3). Gustatory and olfactory systems do not play a major role in neurology, and therefore chemoreception will not be considered further in this text. The student with a special interest in this subject is referred to the references.

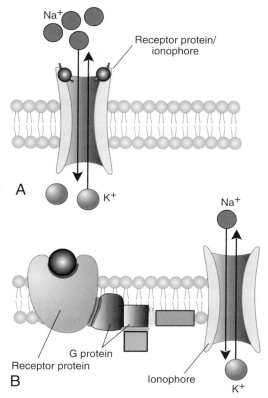

FIGURE 4.6.

There are two principal types of chemoreceptors, direct and indirect. The direct chemoreceptors combine the ionophore and the receptor mechanism in the same molecule (**A**) so that binding of the receptor directly affects the permeability of the ion channel. The indirect chemoreceptors (**B**) act through G-protein second-messenger systems.

Mechanoreceptors

To be sensitive to many forms of mechanical energy, the basic transducer, mechanically sensitive ionophores, is adapted by external structures to be most sensitive to a specific presentation of energy by the environment. For example, some mechanoreceptors are selectively sensitive to the indentation of the skin (in the case of the Merkel disks), others to the angular acceleration of the head (in the case of the hair cells in the vestibular ampullae). Their different sensitivities are derived not from differences in the receptor membrane but from the external structures associated with the receptor. Among the various types of receptors, the most elaborate secondary structures have evolved around the mechanoreceptors (Table 4.2).

Mechanoreceptors make the skin an exquisitely sensitive sensory organ. Hairless or **glabrous skin** [L. *glaber,* smooth] contains two types of specialized mechanoreceptors, Meissner corpuscles and Merkel disks. (Fig. 4.7). They are differentiated from one another by the elaborations that surround the mechanically sensitive receptor membrane. The location of these specialized mechanoreceptors, between the epidermis and the dermis, makes glabrous skin extremely sensitive to touch. Hairy skin contains special **peritrichal** [G. *peri,* around + G. *thrix,* hair] receptors at the base of the hair shafts. These are exquisitely sensitive to mechanical displacement of individual hairs. The dermis of both hairy and glabrous skin contains another set of touch-sensitive mechanoreceptors, pacinian and Ruffini corpuscles.

Touch receptors in the skin have differing spatial resolutions or **receptive fields**. In glabrous skin the receptive field of Meissner corpuscles and Merkel disks is very small, whereas the Ruffini and pacinian corpuscles, which lie deep in the dermis, have large receptive fields (Fig. 4.7). Furthermore, the somatosensory receptors are not evenly distributed over all parts of the body. In the fingertips, they are very close together. One can discriminate between two points that are only 1 or 2 mm apart. Two separate points on the back cannot be discriminated until they are about 40 to 50 mm apart.

Nociceptors

Another class of somatosensory receptors includes those that are activated by some form of tissue damage, but little is known about the mechanisms of nociceptive transduction. Nociceptors [L. *noceo,* to injure + L. *capio,* to take]

Table 4.2. Mechanoreceptor Subtypes

Modality	Location	Specialized Receptor
Touch intensity	Skin	Merkel disks
Touch velocity	Skin	Meissner corpuscle
Touch acceleration	Skin	Pacinian corpuscle
Hair movement	Hair shafts	Peritrichal receptors
Proprioception	Joints & muscles	Joint receptors[a]
Muscle dynamics	Muscle	Muscle spindle[a]
Muscle tension	Tendons	Golgi tendon organs[a]
Sound	Cochlea	Cochlear hair cells[b]
Head acceleration	Vestibular apparatus	Vestibular hair cells[c]

[a]See Chapter 6.
[b]See Chapter 10.
[c]See Chapter 9.

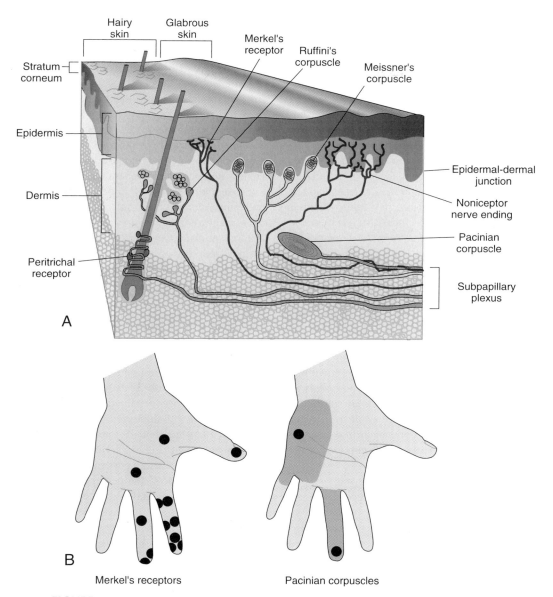

FIGURE 4.7.

   A, Skin contains a variety of mechanoreceptors. Most receptors in glabrous skin are at the epidermis-dermis junction. This location means that the slightest deformation of the skin is transmitted to the receptor. Peritrichal receptors at the base of hair shafts are also situated in a way that ensures exceptional sensitivity. The receptors located deep in the dermis are much less sensitive to light touch. The highly branched naked nerve endings infiltrate the interstitial spaces in the epidermis and are the principal nociceptors. **B,** The receptive fields of various mechanoreceptors are illustrated. The *black dot* indicates the point of greatest sensitivity. The size of the receptive field is highly correlated with the location of the receptor in the layers of the skin. Receptors located at the most superficial layers of the skin have the smallest receptive fields (*left*). Receptors located deep in the dermis have larger receptive fields (*right*).

appear to be simple naked axon terminals. Although the adequate stimulus for nociceptors is uncertain, their activation usually results in the *perception* of pain.

Nociceptors are traditionally classified as *mechanical, thermal, chemical,* and *polymodal*[1] based on the stimuli that excite them. Unlike benign thermal or mechanical stimuli, nociceptive thermal and mechanical stimuli must be of sufficient intensity to cause tissue damage. Under such conditions, K^+, a potent hypopolarizing agent, will be released into the interstitial space, where it will hypopolarize any axon it comes into contact with. Therefore it is questionable whether these nociceptors are truly specific to mechanical or thermal stimuli.

Nociceptors seem to be a class of chemoreceptors, but since they give rise to the perception of pain, they are not included within that class of specialized receptors. Two types of chemical signals other than free K^+ affect them. The first type directly affects the receptor membrane. The second sensitizes the receptor membrane by lowering its threshold to subsequent stimuli without directly causing it to initiate action potentials.

Among the first set of chemicals are substances released by damaged cells (see Table 4.3). When cells are disrupted, their contents diffuse throughout the surrounding tissue. Potassium ions, hista-

mine, and bradykinin are the most potent hypopolarizing products released by cell damage. No specialization of the axon membrane is required for it to be sensitive to the presence of K^+, since this ion will immediately hypopolarize any exposed neuron membrane. It is not known whether or not there are special receptors for histamine and bradykinin in the membrane of nociceptors.

Other chemicals, like prostaglandin E_2, sensitize nociceptors, lowering their threshold. Nociceptors themselves can release sensitizing agents. For example, substance P, a neuroactive peptide and putative neurotransmitter, can be released from nociceptors when they are invaded by action potentials antidromically. Once in the interstitial space, substance P causes mast cells to release histamine, a potent nociceptive stimulator (Fig. 4.8).

Thermoreceptors

Two different skin receptors have been discovered that respond to thermal stimuli. One set, the **heat receptors**, responds over a temperature range of approximately 32° to 45°C. There is a linear relationship between the firing rate of the axons associated with these heat receptors and temperature. Between about 45° and 50°C the firing rate drops to zero (Fig. 4.9). Below about 45°C, the perception of warmth correlates with temperature, but above 45°C the perception of warmth rapidly changes to pain. **Cold receptors** are sen-

---

[1]Nociceptors that respond to a combination of thermal, mechanical, and chemical stimuli are usually called polymodal nociceptors.

Table 4.3. Nociceptive Chemical Agents

Chemical Agent	Effect on Nociceptors
Potassium	Stimulates directly
Bradykinin	Stimulates directly
Histamine	Stimulates directly
Prostaglandins	Sensitizes receptor membrane
Substance P	Released by nociceptors

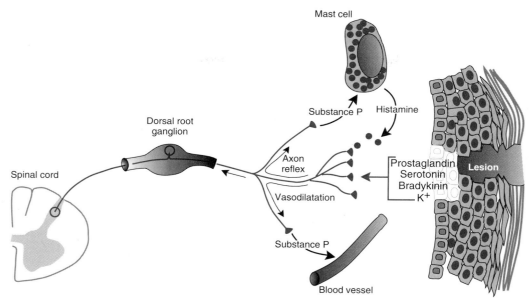

FIGURE 4.8.

Potassium ions from disrupted cells are potent nociceptive stimulants that directly affect the naked nerve endings. Bradykinin, histamine, and serotonin are also potent nociceptor stimulants. Axons terminating in nociceptors branch copiously in the epidermis. If one axonal branch is stimulated (central axon in figure), action potentials generated there are carried to the spinal cord. Action potentials will also invade other branches of the parent axon antidromically, a phenomenon known as an axon reflex. At the antidromically excited terminal, the neuroactive peptide substance P is liberated, causing mast cells to release histamine, which intensifies and spreads the nociceptive stimulation. Substance P also causes vasodilatation.

sitive to skin temperatures from about 24° to 34°C. Paradoxically, cold receptors also respond to temperatures between about 45° and 55°C.

SECONDARY PROPERTIES OF SOMATOSENSORY RECEPTORS

The neural representation of a sensory stimulus is an **analog** of that stimulus. One might suppose that the analog would correspond to the original stimulus in every detail. This is not possible. A physical stimulus has many descriptive properties. For example, a touch on the skin may be described in terms of intensity (pressure), location, duration, and modality (hair movement or skin touch). Furthermore, since each sensory modality has many properties, particularly properties associated with the dynamic aspects of the stimulus, different types of somatosensory receptors detect and encode the **dynamic** (veloc-

ity and acceleration), **intensity**, and **localizing** properties from mechanical stimuli. Therefore, most sensory receptors do not accurately encode all of the properties of a stimulus into a single signal. Being incomplete, *the analog representation within a labeled line of neurons is distorted with respect to the original signal.* Each of these modal properties is separately transmitted through the nervous system by separate sets of pathways.

Dynamics

When presented with a constant stimulus, the response of most receptors diminishes with time. This is called **adaptation**. Receptors may be classified according to how fast they adapt to a steady-state signal, being either rapidly adapting or slowly adapting (Fig. 4.10).

Slowly adapting receptors, like the Merkel

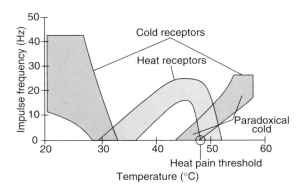

FIGURE 4.9.

Heat (*red*) and cold (*gray*) receptor stimulus response curves. Note that the neuronal response to heat rapidly diminishes to zero above 45°C. Cold receptors have a bimodal response, firing paradoxically above about 45°C.

disk, generate receptor potentials that are proportional to the amplitude of the signal (Fig. 4.10*A*). The amplitude of the receptor potential does not change very much for the duration of the signal. Within the brief physiological time course of most signals, the receptor potentials from slowly adapting receptors may be considered to be nonadapting.

Rapidly adapting receptors, like the Meissner corpuscle, respond primarily to the dynamic properties of the signal. They produce large receptor potentials at the beginning of a stimulus, the **on-response**. If the stimulus is maintained at a steady state, the receptor potential rapidly diminishes. When a constantly applied signal ceases, many rapidly adapting receptors generate a potential, signaling another transition in the stimulus, the **off-response** (Fig. 4.10*B*).

Adaptation is, generally speaking, not a property of the receptor membrane but a property provided by the elaborated structures attached to it. In the case of mechanoreceptors, these elaborations are simple encapsulations, such as that seen in the pacinian or Ruffini corpuscles.

The principal teleological reason for adaptation is to extract the dynamic information, *velocity* and *acceleration,* from the signal (Fig. 4.10*D* and *E*). For example, the slowly adapting Merkel disk is a somatosensory receptor that responds to the *steady-state* pressure of the stimulus without significant adaptation. It encodes *intensity* and *duration*. The rapidly adapting Meissner corpuscle, on the other hand, responds almost exclusively to

the *velocity* of the applied pressure. Finally, the pacinian corpuscle, a very rapidly adapting receptor, detects *acceleration*. All of these receptors have an axon membrane with mechanically sensitive ion channels. They are differentiated from one another by the external apparatus covering the naked axon that modifies the way the stimulus affects the receptor membrane (Fig. 4.10*D* and *E*).

Intensity

The physiologically significant range of intensity that most stimuli present to the nervous system covers six to eight orders of magnitude. Encoding such a wide range of intensity as neural signals cannot be accomplished directly because receptor potential amplitudes and action potential frequencies cannot accommodate this range. Action potentials have a dynamic range of about three orders of magnitude (FM); receptor potentials only about two (AM). To overcome this shortcoming, two mechanisms are important for encoding a wide dynamic intensity range.

First, different receptors within the same class (i.e., Merkel disks) may have different thresholds. If two receptors have the same dynamic range (e.g., two orders of magnitude) but the threshold of one is two orders of magnitude higher than the other, then together they can encode an intensity range of four orders of magnitude (Fig. 4.11). Given receptors of overlapping thresholds, the entire range of intensity information can be accommodated.

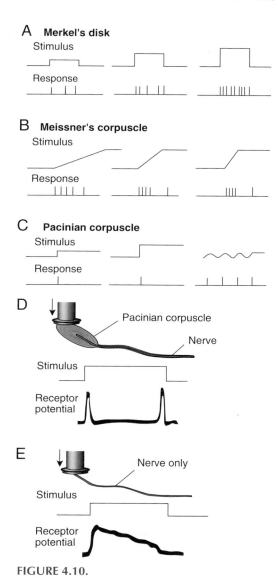

FIGURE 4.10.

Receptors can be classified according to their rate of adaptation. **A,** Slowly adapting receptors (e.g., a Merkel disk, shown here) respond in proportion to the amplitude of the stimulus and maintain the receptor potential during the entire stimulus. **B,** Rapidly adapting receptors respond primarily to changes in the intensity of the stimulus. The Meissner corpuscle shown here responds approximately in proportion to the velocity of the stimulus. **C,** The most rapidly adapting receptors respond approximately to the acceleration of the stimulus. The pacinian corpuscle, illustrated here, is a typical example. **D,** The pacinian corpuscle responds to both the onset and the termination of a stimulus. The very rapidly adapting receptor potential is shown in the lower tracing, with the

Second, The relationship between the stimulus intensity and the receptor potential is usually not linear. For most systems, this nonlinear relationship can be approximated by the following expression:

$$R \propto \log (S) \qquad (4.2)$$

where R is the receptor potential and S is the intensity of the stimulus. Receptors with nonlinear response characteristics, such as cold receptors, encode a greater range of intensity information at the expense of resolution.

Localization

Somatosensory information can be localized to specific anatomical sites. This is possible because the spatial relationship between the tactile receptors distributed over the body surface is maintained in the anatomical organization of the CNS neurons that receive information from those receptors. This creates a **somatotopic map** of the surface of the body in the structural arrangement of the CNS neurons. The map in the spinal cord is preserved at each subsequent synaptic level within the CNS. The mapping of the anatomical arrangement of the body in the structure of the nervous system is a fundamental feature of the nervous system and greatly affects its function. These functional attributes will be considered in many of the succeeding chapters.

Spatial discrimination within the maps is enhanced by synaptic interactions within the sensory nuclei as information is transmitted from one set of neurons to another. The most common form of synaptic interaction is a process known as **lateral inhibition.** Briefly, within the nuclei of the CNS, some neurons (transmission cells) send long axons that project to distant nuclei and short recurrent

stimulus shown above it. **E,** If the external connective tissue structures are removed from the pacinian corpuscle, the native response of the receptor membrane is revealed. The lower tracing is the receptor potential of the naked axon ending, showing relatively little adaptation. This shows that the mechanical characteristics of the capsule are important to the response characteristics of the receptor membrane.

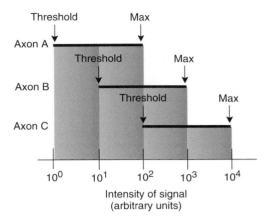

FIGURE 4.11.

A wide dynamic range of information can be carried by multiple information channels of narrow dynamic range if each channel carries a subset of the information. In this example, each axon is limited to two orders of magnitude. Axon A carries intensity information from 0 to 10^2 arbitrary units, axon B carries intensity from 10^1 to 10^3, and axon C carries intensity information from 10^2 to 10^4. Together the three axons are able to convey intensity information of four orders of magnitude even though each axon is limited to only two orders of magnitude.

collateral axons that terminate locally on inhibitory interneurons. The interneurons in turn terminate on adjacent transmission cells within the same nucleus (Fig. 4.12). Through the activity of the interneuron, each transmission cell is capable of inhibiting the activity of its neighbor. The most active neurons are able to diminish the activity of adjacent, less active neurons. Since the organization of the central tracts is mapped to the surface of the skin, the effect of lateral inhibition is to restrict the ability of sensations arising from the immediately surrounding regions of a punctate stimulus to reach higher levels in the CNS. By favoring the most active neurons, and suppressing less active neurons, boundary contrasts are enhanced.

Interpretation of Sensations

The adequate stimulus is the form of energy for which a receptor has the lowest threshold. The sensitivity of a receptor to a specific energy

modality may be increased both by *restricting the energy forms* that normally arrive at the receptor and by creating *specialized receptor mechanisms* for receiving that energy. In most cases, however, the specific sensitivity of receptors is not absolute. Receptors can generate neuronal signals even if they are stimulated by forms of energy for which they were not specifically adapted if the intensity is great enough. Photoreceptors, for example, can be stimulated mechanically by pressing on the eye. One's *perception* is light, not pressure, because the structures within the brain that receive the information *interpret* any signals received from the visual pathways as light. *The brain has no mechanism for distinguishing the absolute nature of the stimulus.* Similarly, under laboratory conditions, when isolated cold receptors are stimulated above 45°C, the subject reports the perception of cold, not warmth or pain. Just as in the case of pressing on the eye, the brain interprets the information based on its source as projected through the nervous system along a specific line-labeled chain of neurons. Were the temperature stimulus to be applied over a larger region of the skin so that nociceptors would also be stimulated, creating a more complex, information-rich signal, the brain would interpret the stimulus as pain.

The sensory modalities are all closely coupled with energy forms that can be quantified in terms of physics and chemistry because specific receptors, adapted to detect these forms of energy, have evolved. Each modality is presented to our consciousness as a **primary sensation**. We also have **derivative sensations**, sensations that are perceived, yet we have no specific receptors to detect them. The feeling of wetness is one example. We have no hydrologic receptors that sense the presence of water on our skin, yet we are aware of being wet. Itch, tickle, and pain are other examples.

These *derivative perceptions* that arise from neuronal activity represent certain combinations of primary sensations that are interpreted by our nervous system as unique. In contrast to the primary sensations that have separate tracts and nuclei within the CNS, there are no specific pathways reserved for derivative sensations. Although these derivative perceptions are not "real" in the sense that they do not correspond with objective

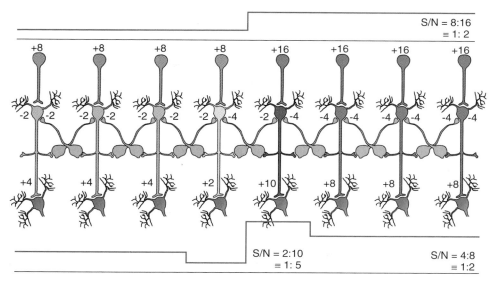

FIGURE 4.12.

The mechanism of lateral inhibition is schematically illustrated. A stimulus, having an arbitrary intensity of 8 units on one side of a boundary and 16 units on the other side is presented to a set of receptors (*top, dark red neurons*). The receptors synapse with a corresponding set of sensory neurons, providing the same arbitrary units of excitation to the sensory neurons. The axons of each sensory neuron provide recurrent collateral fibers that synapse with inhibitory interneurons (*light gray neurons*) that inhibit adjacent sensory neurons by 25% of the stimulus intensity. The final level of excitation is the simple sum of inhibition and excitation and is indicated at the boutons of the sensory neurons. The boundary ratio of the stimulus is 1:2. After passing through this simple neural network, the stimulus ratio is still 1:2 except at the boundary, where the ratio has increased to 1:5. Thus lateral inhibition incurs no overall loss of signal contrast over most of the stimulus field, and provides enhanced signal contrast at the boundary condition, making it more prominent. While greatly oversimplified, the principles illustrated by this neural network have been verified in numerous neuronal systems.

physical forms of energy, they are nonetheless "real" to the consciousness of our brains.

CASE HISTORY

Mr. S. M.

Mr. S. M. is a 21-year-old architecture student who was in good health until June 16, 1994. At that time he was repairing his motorcycle when a wrench slipped off from a nut, causing his right hand to smash into the handlebar. In great pain, he came to the student health center thinking he had broken his thumb.

I. MENTAL STATUS: Normal

II. CRANIAL NERVES: All normal

III. MOTOR SYSTEMS: All normal.

IV. SENSORY SYSTEMS: All normal except for a total anesthesia of the dorsal surface of the right thumb, sparing the palmar surface and the dorsal surface of the hand. Detailed examination revealed that there was no feeling to light touch or pin prick from the dorsal surface of the thumb (Fig. 4.13). Light pressure near the point of injury produced tactile sensations at the distal-most portion of the thumb, on the dorsal surface and near the thumbnail.

V. CEREBELLAR FUNCTIONS: Normal

VI. REFLEXES: All reflexes were present and normal.

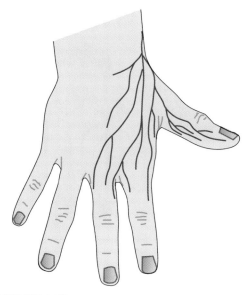

FIGURE 4.13.

The normal distribution of the cutaneous nerves to the dorsum of the hand. Note how crushing the lateral division of the superficial radial cutaneous nerve can cause the lack of sensation noted by Mr. S. M.

Comment: The handlebar struck Mr. S. M.'s thumb at the base of the metacarpal, near the insertion of the abductor pollicis longus. Superficial cutaneous branches of the radial nerve pass over this point before being distributed to the medial-dorsal surface of the thumb and hand. Although x-rays proved that no bones were fractured, apparently the bar crushed the lateral branch of the nerve, producing total anesthesia of the skin served. Light pressure at the point of injury could be expected to stimulate the crushed axons at the point of injury, sending action potentials to the CNS. The signals so received would be localized to the point where the receptors normally attached to those axons are located. This would account for the sensations perceived from the anesthetic areas of the thumb.

FURTHER APPLICATIONS

Peripheral nerves do not normally lie close to the surface; they lie deep to the surface, protected by the surrounding muscle and bone. There are specific locations, however, where peripheral nerves are relatively unprotected. Peripheral nerves, as in the present case, are commonly injured at these sites.

- Using your knowledge of gross anatomy, identify these sites and the nerves in jeopardy.
- What patterns of sensory and motor losses would you expect to observe following nerve injury at each of these locations?

The function of peripheral nerves can also be affected by simple compression that does not actually crush the nerve.

- What neuronal mechanisms would be affected by compression?
- How would that disrupt nerve function?

Compression syndromes frequently affect the carpal and ulnar tunnels.

- What is unusual about the anatomy of these regions that places the nerves passing through these tunnels particularly at risk?
- How can repetitive movements of the fingers and wrist exacerbate this situation?
- What patterns of sensory/motor dysfunction would you expect to observe with compression injuries at each of these locations?

SUMMARY

Physical stimuli are received by sensory receptors and converted into internal signals that create conscious perceptions. Transduction is the process by which the energy of the stimulus is converted into an energy form appropriate for the nervous system. Representation is the manner in which the neuronal energy is organized as an analog of the stimulus. Specific receptors exist that convert the energy of a stimulus into an amplitude-modulated (AM) receptor potential. The receptor potential is itself converted into a series of frequency-modulated (FM) action potentials.

Our sensations are divided into modalities that correspond to objective physical stimuli. The modalities have specific receptors and segregated neuronal pathways. There are five classes of receptors: photic, chemical, mechanical, thermal, and noxious. The latter three are collectively called somatosensory receptors. Somatosensory receptors usually adapt to constant stimuli, a property that makes them sensitive to velocity and acceleration. The dynamic range of receptor sensitivity is expanded by nonlinear transformation of the stimulus into the receptor potential. This transformation extends range at the expense of absolute selectivity. The physical organization of somatosensory receptors is maintained in the CNS, a process called mapping. Maps are maintained at all levels of the CNS.

The brain has no independent way of interpreting information arriving on the sensory pathways. Any form of energy that can stimulate a receptor will cause the brain to perceive a stimulus appropriate to the source. The sensory modalities are closely linked to objective physical forms of energy for which there are specific receptors. Subjective sensations are not associated with specific receptors. These sensations arise from neuronal activity that represents certain combinations of objective stimuli. While not real in the objective sense, these perceptions are real to the conscious awareness of the organism.

FOR FURTHER READING

Detwiler, P. Sensory Transduction. In: *Textbook of Physiology*, edited by Patton, H., Fuchs, A., Hille, B., Scher, A., and Steiner, R. Philadelphia: W. B. Saunders, 1989, pp. 98–129.

Dodd, J., and Castellucci, V. Smell and Taste: The Chemical Senses. In: *Principles of Neural Science*, edited by Kandel, E., Schwartz, J., and Jessell. T. New York: Elsevier, 1991, pp. 512–529.

Agur, A. *Grant's Atlas of Anatomy*, 9th edition. Baltimore: Williams & Wilkins, 1991.

Jessell, T., and Kelly, D. Pain and Analgesia. In: *Principles of Neural Science*, edited by Kandel, E., Schwartz, J., and Jessell, T. New York: Elsevier, 1991, pp. 385–399.

Kandel, E., and Jessell, T. Touch. In: *Principles of Neural Science*, edited by Kandel, E., Schwartz, J., and Jessell, T. New York: Elsevier, 1991, pp. 367–384.

Martin, J. H. Coding and Processing of Sensory Information. In: *Principles of Neural Science*, edited by Kandel, E., Schwartz, J., and Jessell, T. New York: Elsevier, 1991, pp. 329–340.

Martin, J. H., and Jessell, T. Modality Coding in the Somatic Sensory System. In: *Principles of Neural Science*, edited by Kandel, E., Schwartz, J., and Jessell, T. New York: Elsevier, 1991, pp. 341–352.

Partridge, L., and Partridge, L. D. *The Nervous System: Its Function and Its Interaction with the World*. Cambridge, MA: The MIT Press, 1993.

Phillips, J., and Fuchs, A. F. Somatic Sensation: Peripheral Aspects. In: *Textbook of Physiology*, edited by Patton, H. D., Fuchs, A. F., Hille, B., Scher, A. M., and Steiner, R. Philadelphia: W. B. Saunders, 1989a, pp. 298–313.

Phillips, J., and Fuchs, A. F. Gustation and Olfaction. In: *Textbook of Physiology*, edited by Patton, H. D., Fuchs, A. F., Hille, B., Scher, A. M., and Steiner, R. Philadelphia: W. B. Saunders, 1989b, pp. 475–502.

Shepherd, G. *Neurobiology*. New York: Oxford University Press, 1994.

Smith, C. U. M. *Elements of Molecular Neurobiology*. New York: John Wiley & Sons, 1989.

Szabo, R., and Steinbert, D. Nerve Entrapment Syndromes in the Wrist. *J. Am. Acad. Orthopaed. Surg.* 2:115–123, 1994.

5 The Somatosensory System

The brain receives information about the environment by means of its sensory systems; olfaction, taste, hearing, vision and the somatosensory system. This chapter considers the somatosensory system, a collection of receptors, tracts and nuclei that convey the sensations of light touch, proprioception, temperature and nociperception to consciousness. Somatosensory receptors are located in the skin, muscles, joints and viscera, a distribution that makes this system the largest and most varied of the sensory systems. While chiefly sensory, this system also plays a critical role in motor control by providing appropriate feedback to the motor system about joint position, muscle tension, the velocity of muscle contraction, and contact of the body with external surfaces. It is therefore appropriate to begin ones' examination of the function of the nervous system with this sensory system.

Anatomical Organization of Somatosensory Systems

The gross and microscopic features of the nervous system have been discussed previously (see Chapters 1 and 2). The intermediate structure, the organization of the tracts and nuclei, has been postponed until now. Before the sensory pathways can be presented, it will be necessary to describe the relationship of the prominent sensory tracts and nuclei at the various levels of the nervous system. The description that follows will focus on the sensory components within the CNS. The motor tracts and nuclei will be added later (see Chapters 6, 7, and 8). Finally, the cranial nerve nuclei will be described to complete the picture (see Chapter 9). Therefore, three trips through the neuro axis will be required to develop the full internal structure of the nervous system.

THE PERIPHERAL NERVES

The peripheral nerves consist of bundles of axons. There are two commonly recognized systems by which axons are described (Fig. 5.1). One system groups axons according to their diameter. They are labeled A, B, and C; the largest myelinated axons belong to group A, while the smallest unmyelinated axons belong to group C. The A group is further divided into four subgroups labeled by Greek letters α, β, γ, and δ. This nomenclature is used primarily, but not exclusively, to describe somatosensory afferent axons. The B group is used to designate the myelinated preganglionic axons of the autonomic nervous system (see Chapter 12). It is seldom used now.

The second system classifies certain sensory axons according to their origin, function, and conduction velocity. This system is used to describe afferent axons originating in muscles, tendons, and joints. They are designated by roman numerals in order of decreasing conduction velocity as, I, II, III, and IV, group IV being unmyelinated. Since diameter and conduction velocity are related, the two systems overlap. Unfortunately, since the two systems evolved independently, the overlap is not exact. Both nomenclature systems are in use and it is easy to become confused.

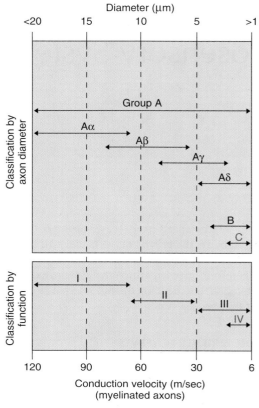

FIGURE 5.1.

The classification schemes of peripheral axons correlated with their diameter and conduction velocities. Axons indicated in *red* are unmyelinated.

THE SPINAL CORD

The **spinal cord** is organized into a central gray area surrounded by white matter. It is convenient to divide the gray matter of the spinal cord into a series of layers, or **laminae**, that extend its entire length, transcending the segmental divisions imposed upon it by the spinal nerves. The spinal cord gray matter was divided into ten laminae by Rexed, who used strictly cytoarchitectural methods to distinguish one lamina from another. Over the years it has been recognized that certain physiological properties correlate with the laminae. These observations have given Rexed's system a practical usefulness.

The laminae are labeled I through X (Fig. 5.2). They form nearly flat ribbons that lie parallel to the long axis of the spinal cord. The dorsal horn is composed of laminae I through VI. These

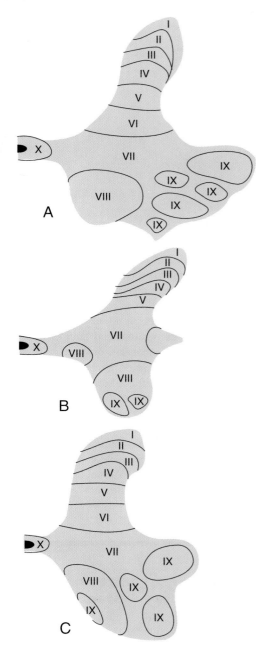

FIGURE 5.2.

The laminae of Rexed. **A,** Cervical; **B,** Thoracic; **C,** Lumbar.

laminae receive the primary afferent axons from the dorsal root and contain interneurons and projection neurons. Interneurons have short axons that form synapses within a few hundred micrometers of their soma. Projection neurons have long axons that transmit information to relatively distant portions of the CNS. The somatic motor neurons, the principal source of axons that leave the CNS, are located in lamina IX, while the autonomic motor neurons (see Chapter 12) are located in lamina VII. Interneurons associated with motor reflexes are located in lamina VIII. Lamina X includes the ependyma and a thin layer of gray matter surrounding the central canal.

The white matter of the spinal cord is separated into several regions by the pattern of the gray matter (Fig. 5.3). The area that lies dorsal and medial to the dorsal horns is called the **dorsal columns**. Below midthoracic levels, the **fasciculus gracilis** fills the entire dorsal column space. In the cervical and upper thoracic spinal cord, the dorsal columns are divided by the **dorsal intermediate sulcus** into the medial fasciculus gracilis and the lateral **fasciculus cuneatus** [L. *cuneo,* like a wedge].

The **lateral white columns** lie lateral to the gray matter and between the dorsal and ventral horns. They contain several motor tracts, which are described in the next chapter. The region medial and ventral to the ventral horns is occupied by the **ventral white columns**. The most important sensory tract in the ventral white columns is the **anterolateral system (ALS[1])**. **The ALS is composed of two principal groups of axons**, the spinothalamic tract and the **spinoreticular tract**.

A small collection of axons interconnecting the two halves of the spinal cord, the **ventral white commissure**, crosses the midline just ventral to the central canal.

THE BRAINSTEM

Descriptions of brainstem structures are based on their relationship to the fourth ventricle and cerebral aqueduct. The principal mass of neuronal tissue ventral to the ventricle is known as the **tegmentum** [L. *tego,* to cover], and one may speak of the medullary tegmentum, the pontine tegmentum, or the tegmentum of the midbrain. The **tectum** [L. *tego,* to cover, a roof] lies dorsal to the ventricle. In the midbrain, the tectum is composed of the corpora quadrigemini. In the

pons it is the cerebellum. There are no neuronal structures of the medullary tectum in the adult.

Occasionally one may speak of the walls of the brainstem. This is a little-used term and is useful only in referring to the **cerebellar peduncles**.

A large collection of axons lies ventral to the tegmentum, and this is commonly referred to as the **basis**. In the midbrain it consists of the **cerebral peduncles**. The basis is greatly enlarged in the pons and contains considerable gray matter, the **basal pontine nuclei**. The **corticospinal tracts** form the basis of the medulla. In the medulla, these tracts are also called the **medullary pyramids**.

Most of the brainstem tracts and nuclei are located in the tegmentum. Long ascending sensory and descending motor tracts traverse the entire brainstem. Shorter tracts connect brainstem structures with the cerebellum or establish intrinsic connections among cell groups within the brainstem. Most of the nuclei of the brainstem are associated with the cranial nerves or the cerebellum. With a few exceptions, the remaining cells in the tegmentum are considered part of the **reticular formation** [L. *reticulum,* a little net], an area of poorly organized groups of cells. The following brief description of brainstem structures emphasizes those associated with the somatosensory system.

The Medulla

The transition between the spinal cord and the brainstem occurs over a few millimeters and is marked by specific changes in the pattern and location of certain structures seen in the spinal cord (Fig. 5.4). The ventral horns become disorganized above the level of the decussation of the corticospinal tract and are no longer distinguishable. The area of the rapidly diminishing ventral horn becomes incorporated into the **medullary reticular formation**.

Rostral to the level of the obex, the dorsal horn enlarges and moves dorsolaterally, becoming separated from the remainder of the gray matter. It merges with a brainstem nucleus known as the **spinal nucleus of V[2]**. This nucleus is an important landmark that can be followed throughout

---

[1]Do not confuse the designation "ALS" as used here to signify a CNS tract with the same designation "ALS" used to designate the neurological disease amyotrophic lateral sclerosis.

[2] The roman numeral V refers to the trigeminal nerve, the fifth of the cranial nerves.

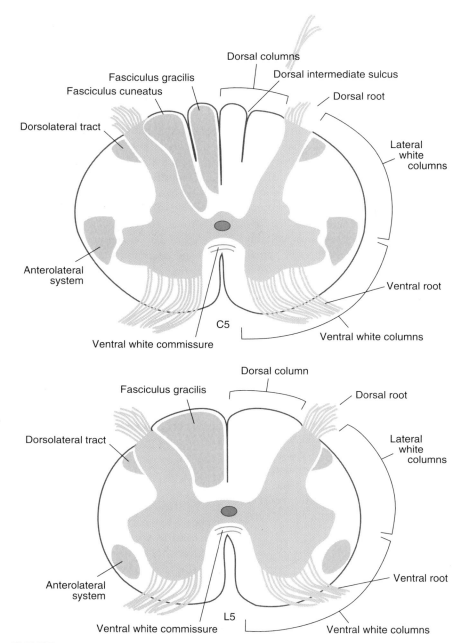

FIGURE 5.3.

Transverse sections of the spinal cord illustrating the various divisions of the white matter and the location of the principal somatosensory tracts. *Top,* cervical; *bottom,* lumbar.

the medulla and into the pons. Immediately lateral to the spinal nucleus of V lies the **spinal tract of V**. It contains primary afferent axons from the trigeminal nerve (CN V) that synapse in the spinal nucleus of V.

At the transition from the spinal cord into the medulla, the dorsal columns are gradually replaced by two pairs of nuclei, the **nucleus gracilis** and **nucleus cuneatus** (Fig. 5.4). The fasciculus gracilis, fasciculus cuneatus, and spinal tract of V all lie toward the surface of their respective nuclei. Together these three pairs of tracts and their associated nuclei occupy approximately the dorsal half of the caudal medulla. While the spinal nucleus and tract of V are constant features of the medulla, the dorsal column nuclei and their tracts stop approximately at the level of the obex.

The **medial lemnisci** [G. *lemniskos,* a woolen fillet] occupy the central space of the medulla. In cross section they appear as a pair of cigar-shaped structures that lie perpendicular to the long axis of the medulla, on either side of the midsagittal plane (Fig. 5.4). The medial lemnisci are another constant feature of the brainstem and are easily identified throughout its length.

The obex divides the caudal from the rostral medulla. Rostral to the obex, the fourth ventricle opens, and the shape of the medulla changes from round to rhomboid (Fig. 5.5). In the rostral

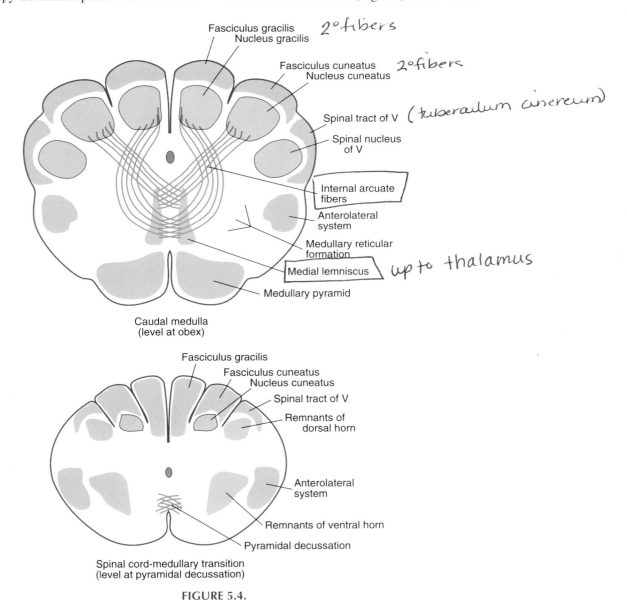

FIGURE 5.4.

Transverse sections of the caudal medulla.

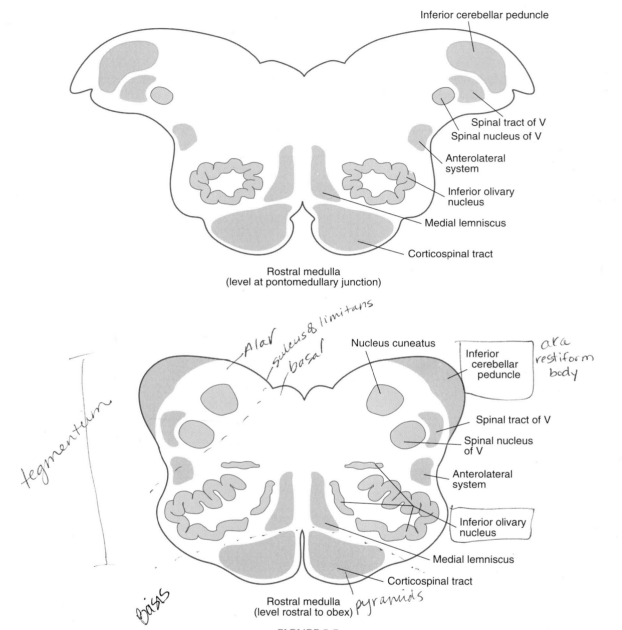

FIGURE 5.5.

Transverse sections of the rostral medulla.

medulla, the **inferior olivary nuclei**, which lie in a ventrolateral position on either side of the brainstem, become prominent. The spinal nuclei and tracts of V lie at the lateral margins of the medulla, with the ALS ventral to them and immediately dorsal to the inferior olives. The **inferior cerebellar peduncles** [L. *pes,* foot] (also

called the restiform body) occupy the dorsal lateral corners of the rostral medulla.

The Pons

The basis of the pons is the most conspicuous feature of the pons (Fig. 5.6). It contains a large

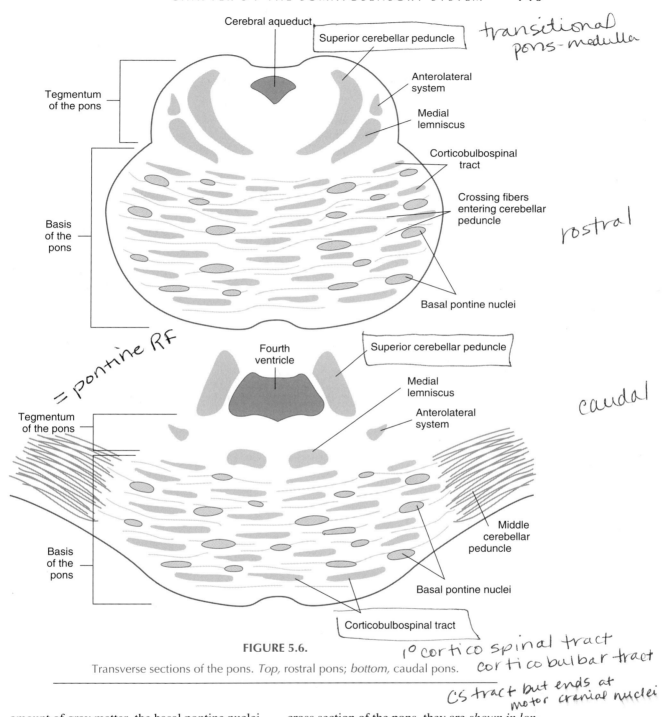

FIGURE 5.6.

Transverse sections of the pons. *Top,* rostral pons; *bottom,* caudal pons.

amount of gray matter, the basal pontine nuclei, and numerous axons. Axons from neurons in the basal pontine nuclei cross the midline and form the **middle cerebellar peduncles** (also called the brachium pontis). Since the axons of the middle cerebellar peduncles lie in the plane of cross section of the pons, they are *shown in longitudinal section* in Figure 5.6. The remaining axons are part of the corticobulbospinal tracts (see Chapter 6), which course the pons in a rostral-caudal direction. These axons are *shown in cross section* in the figure.

The area between the basis of the pons and the fourth ventricle is the tegmentum of the pons that contains the nuclei associated with the cranial nerves of the pons, the pontine reticular formation, and various other fiber tracts. At the caudal pons (Fig. 5.6), the medial lemnisci form a pair of oval structures on either side of the midline at the basis-tegmental junction. As one moves in a rostral direction, the medial lemnisci become flatter and move laterally until, at the most rostral end of the pons, they form a thin ribbon at the extreme lateral half of the basis-tegmental junction (Fig. 5.6). The ALS occupies a constant position at the ventrolateral margin of the pontine tegmentum. Throughout the length of the pons, the medial lemniscus becomes more laterally placed until at the rostral pons it joins the ALS. The **superior cerebellar peduncle** (also called the brachium conjunctivum) is a prominent structure of the rostral pons, forming part of the walls of the rostral pons and midbrain.

The Midbrain

As one moves from the pons into the midbrain, the superior cerebellar peduncles descend toward the center of the midbrain tegmentum

from their superolateral position, where they decussate [L. *decusso,* to make in the form of an X] (Fig. 5.7). The medial lemniscus and the ALS maintain their extreme lateral position in the midbrain tegmentum. The fourth ventricle narrows to form the **cerebral aqueduct**. Immediately surrounding the aqueduct is a prominent core of gray matter known as the **periaqueductal gray (PAG)**. An imaginary line passed horizontally through the aqueduct separates the midbrain tegmentum (ventral) from the tectum (dorsal), the latter consisting of the superior and inferior colliculi (corpora quadrigemini).

THE THALAMUS

As the midbrain blends into the thalamus [G. *thalamos,* a bedroom], the cerebral aqueduct lengthens in a dorsoventral direction, enlarging to become the third ventricle (Fig. 5.8). The third ventricle divides the thalamus at the midline, with the exception of the **interthalamic adhesion**, a small mass of gray matter that, in many individuals, connects a small part of the two halves of the thalamus.

Each half of the thalamus is an egg-shaped body that is divided into many separate nuclei. For convenience, these nuclei can be organized

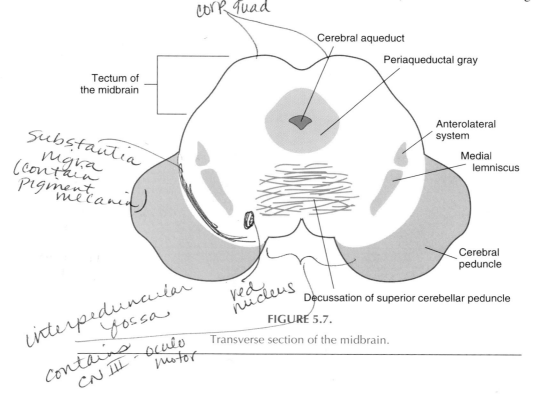

FIGURE 5.7.

Transverse section of the midbrain.

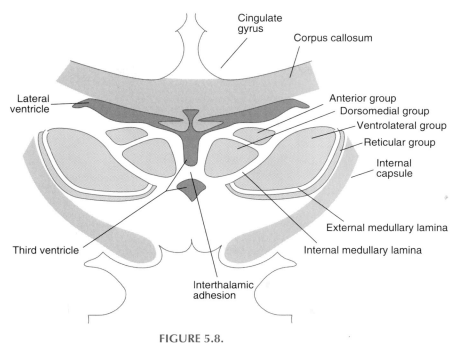

FIGURE 5.8.

Transverse section of the thalamus at the level of the ventrobasal complex.

into five groups, partly defined by a band of white matter, the **internal medullary lamina** (Fig. 5.9). At the anterior pole of the thalamus it splits, forming a pocket in which one finds the **anterior nuclear group**. The remainder of the thalamus is divided by the internal medullary lamina into a medial and a lateral division. Medial and dorsal to the internal lamina lies the **medial group**. Lateral and ventral to it is a collection of nuclei known as the **lateral group**. Within the internal medullary lamina one finds scattered cells and small nuclei. This constitutes the **intralaminar group**. An **external medullary lamina** surrounds the main body of the thalamus, and external to it lies a thin band of cells, the **reticular group**.

The thalamus is a complex structure serving many functions—sensory, motor, and cognitive. All sensory information, with the exception of olfaction, must pass through the thalamus before reaching the cerebral cortex. Similarly, motor command information must pass through the thalamus before a motor act can be initiated. Given this heterogeneity, the thalamus is discussed in several different parts of this book.

A common feature of all thalamic nuclei is the reciprocal connections they have with the ipsilateral cerebral cortex. Some nuclei are connected to very small parts of the cortex (**specific thalamic nuclei**); others have widespread connections (**nonspecific thalamic nuclei**).

The specific thalamic nuclei associated with somatosensory function are subdivisions of the lateral group (Fig. 5.9). This large group is subdivided into a ventral and a lateral division. Within the ventral division is the **ventrobasal complex**, a set of four nuclei: the **ventral posteromedial (VPM)**, the **ventral posterolateral (VPL)**, the **ventral posterosuperior (VPS)**, and the **ventral posteroinferior (VPI)**. The nonspecific nuclei associated with somatosensory function are from the intralaminar group. The intralaminar groups consist mostly of scattered cells within the white matter of the internal medullary lamina, but one collection is large enough to be named, the **centromedian (CM)**.

THE CEREBRAL CORTEX

The cerebral cortex is composed of a superficial gray layer with an underlying mass of white

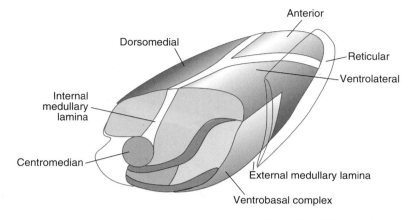

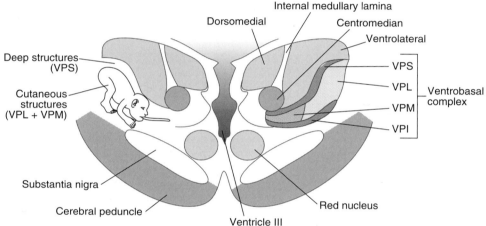

FIGURE 5.9.

Top, Schematic drawing of the right thalamus. Note how the internal medullary lamina divides the thalamus into three main groups of nuclei: the anterior, the dorsomedial, and the ventrolateral. Within the internal medullary lamina lie scattered cells and one large nucleus, the centromedian. These constitute the intralaminar group. External to the external medullary lamina lies the reticular group. *Bottom,* Drawing of a transverse section of the thalamus showing the location of the ventrobasal complex and its division into four nuclei. The figure to the *left* schematically represents the somatotopic arrangement of the VPL and VPM nuclei.

matter. The gray matter is divided into six layers that are designated by roman numerals in order from the surface. A detailed cytoarchitectural description of the six layers can be found in dedicated texts and will not be given here.

The principal cell type found in the cerebral cortex is the pyramidal cell, so named because of the triangular shape of its soma. An apical dendrite extends from the peak of the triangle and rises toward the surface of the cortex, crossing several cortical layers. Basal dendrites extend parallel to the surface from the base of the

soma. A single axon projects into the deep white matter of the cortex. The pyramidal cell is the only type to extend axons out of the cortical gray matter to remote parts of the nervous system. All other cell types in the cerebral cortex are interneurons in the sense that they only project to other cortical structures.

The pyramidal cells are classified into four groups according to the projection of their axon. **Short association fibers** project to adjacent gyri or gyri within the same lobe. **Long association fibers** interconnect lobes in the same hemisphere.

Callosal fibers cross the midline in either the anterior commissure or the corpus callosum. **Projection fibers** leave the cerebral cortex and innervate subcortical structures such as the basal ganglia, the thalamus, the brainstem, and the spinal cord. No pyramidal axons leave the CNS.

Projection fibers are the only pyramidal axons to leave the cerebral cortex. They collect as bundles in the subcortical white matter, forming the **corona radiata**. As they pass deeper into the core of the telencephalon, they become concentrated into a massive bundle of fibers called the **internal capsule**. The internal capsule passes through the caudate nucleus, then passes between the thalamus (dorsomedial) and the globus pallidus (ventrolateral) as it progresses through the diencephalon. In the midbrain the axons of the internal capsule become the cerebral peduncles. In addition to these descending fibers from pyramidal cells, axons from the basal ganglia and the thalamus join the internal capsule and ascend into the cerebral cortex.

The Somatosensory Pathways

There are two principal somatosensory pathways. One pathway, the **dorsal column system**, is rapidly conducting and highly localized. All the axons in the system are large-diameter myelinated fibers that are precisely ordered and represent a map of the body surface. The dorsal column system conveys *fine tactile, vibratory,* and *proprioceptive* sensations as well as *two-point discrimination* (the ability to distinguish the separation of two separate but simultaneous touches to the skin). The other pathway, the **anterolateral system** or **ALS**, is slowly conducting and diffuse. Spatial representation within this system is not highly localized. It conveys the *nociceptive and temperature* sensations.

In spite of these differences, these systems have much in common. Both convey information from the peripheral receptor to the cerebral cortex over a sequence of three neurons (Fig. 5.10). The first neuron, the **primary afferent axon**, has a specialized peripheral termination that is the sensory receptor. The soma of this neuron is located in the dorsal root ganglion. It makes synaptic connections in the ipsilateral gray matter of the spinal cord or medulla. The primary afferent axons entering the dorsal root segregate by size as they enter the spinal cord, with large myelinated axons (Aα, Aβ) forming the **medial division of the dorsal root**, while small myelinated (Aδ) and the unmyelinated (C) axons form the **lateral division**. The **secondary neuron** receives synapses from the primary afferent. Its axon *crosses the midline* and terminates in the thalamus. The soma of the **tertiary neuron** is located in the thalamus. Its axon passes through the internal capsule and terminates in the cerebral cortex, ipsilateral to its origin but contralateral to the origin of the stimulus. All sensory systems, with the single exception of olfaction, are based on this three-neuron pattern.

The three-neuron pattern, although quite useful, is an oversimplification. Each sensory system also expresses elaborations of this basic structure. Learning is simplified, however, if one first identifies the three-neuron core and later adds the elaborations.

THE DORSAL COLUMN PATHWAYS

The axons in the dorsal columns are the primary afferents of a rapidly conducting sensory system that carries the name of this fiber tract (Fig. 5.11). Fine *tactile sensations* and *two-point discrimination* arise from the encapsulated receptors in glabrous skin or the peritrichal receptors on the hair shafts, while vibratory sensations are usually ascribed to the pacinian corpuscles. Various types of receptors in the joints are responsible for signaling joint position (*proprioception*). All of these receptors convey information to the spinal cord over large myelinated Aα and Aβ axons.

The Primary Neuron

Primary afferent axons in the medial division of the dorsal root enter the dorsal column system. After entering the spinal cord, the primary afferent axons divide, sending a *short local branch* to the dorsal horn of the spinal cord, and another *long ascending branch* to the medulla.

The *short local branch* bifurcates close to the point of entry, one branch ascending and the

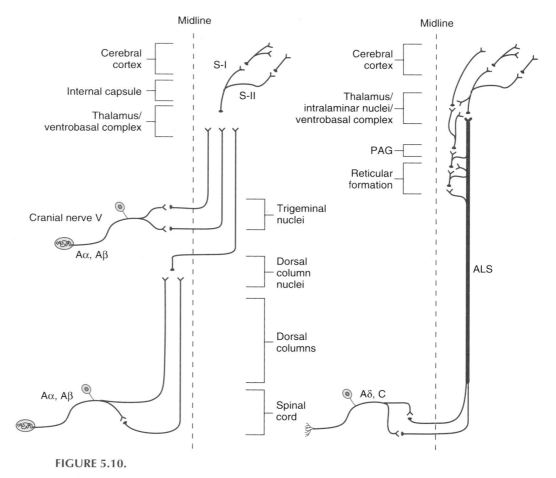

FIGURE 5.10.

Schematic representation of the somatosensory ascending pathways. Both the ALS and the dorsal column systems share the basic three-neuron structure, but note the two different places in the neuraxis where they cross the midline.

other descending one or two segments along the spinal cord in the dorsal columns. These short branches give off collaterals that penetrate the ipsilateral dorsal horn. Within the dorsal horn they synapse in laminae III, IV, and V (Fig. 5.12). The shape and location of the terminal fields of these axons differ somewhat depending on the sensory modality they convey. The terminations of some axons, such as those derived from hair follicle receptors, terminate in overlapping fields, forming a continuous projection field in the dorsal horn. Other axons, such as those from some slowly adapting receptors, terminate in discontinuous fields that form ball- or balloon-shaped patterns in the dorsal horn. In general, they all terminate in the intermediate laminae of

the dorsal horn, more or less continuously through the entire extent of the spinal cord.

The *long ascending branch* ascends the ipsilateral spinal cord in the dorsal columns and terminates in the medulla (Fig. 5.11). Long axons that enter the spinal cord below approximately the midthoracic level ascend in the fasciculus gracilis. Those that enter above the midthoracic level ascend in the fasciculus cuneatus. The axons of the dorsal columns terminate in the *ipsilateral* nucleus gracilis and nucleus cuneatus.

The primary afferent axons enter the dorsal column from the lateral aspect. As the tract ascends, more axons are added in the lateral position, increasing its size. In this manner, the axons in the dorsal columns become organized

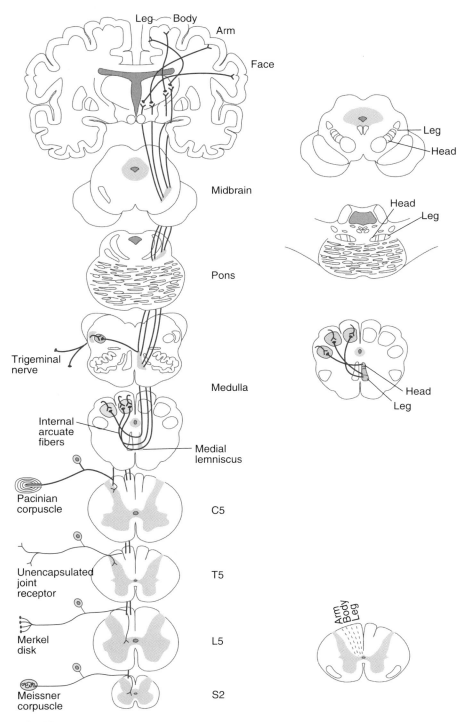

FIGURE 5.11.

The dorsal column system is illustrated here. On the *left* are the anatomical positions of the various components of the system at the principal levels of the nervous system. Small figures on the *right* depict the somatotopic arrangement of the fibers at the various levels.

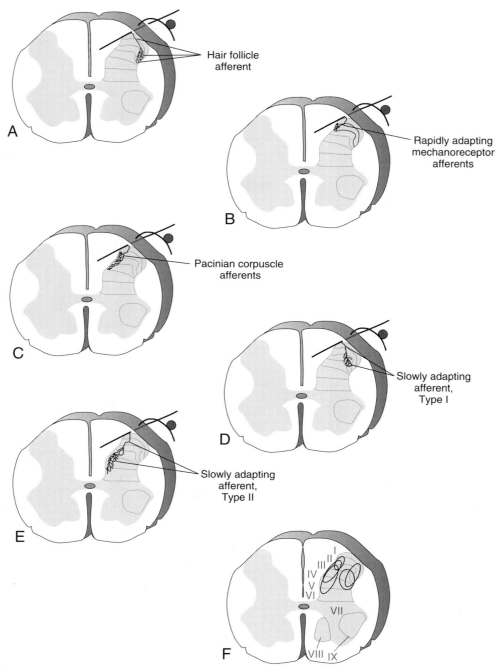

FIGURE 5.12.

The distribution of the short local branch of the large-diameter primary afferent axons in the dorsal horn is shown here in figures drawn from three-dimensional reconstructions of individually labeled and identified axons. Note that all the primary afferents have a common pattern whereby the afferent axon divides after entering the spinal cord and sends an ascending and a descending branch up and down the spinal cord. These branches send collaterals deep into the dorsal horn gray matter, where they further collateralize into a terminal arborization. The axons depicted here differ from one another according to their somatic origin. Axons from the leg are located most medial, while those from the arm are most lateral. This **somatotopic** arrangement is carefully preserved at all levels of the nervous system.

Somatosensory information from the face enters the CNS over primary afferent axons in cranial nerves V (trigeminal), VII (facial), IX (glossopharyngeal), and X (vagus)[3]. Just as in the spinal cord, the somatosensory primary afferent axons in the cranial nerves form two groups, a set composed of large-diameter Aα and Aβ axons, and a set composed of small-diameter Aδ and C fibers. The large-diameter axons enter the brainstem and bifurcate, sending a long descending branch into the medulla in the spinal tract of V (Fig. 5.13). These axons send collateral branches into the rostral third of the spinal nucleus of V, where they synapse. A short ascending branch terminates in the **chief sensory nucleus of V**.

The Secondary Neuron

The second-order neurons in the nucleus gracilis and nucleus cuneatus receive synaptic contact from the primary afferent axons of the fasciculus gracilis and fasciculus cuneatus, respectively. Axons arising from neurons within these nuclei leave their place of origin, course in a ventromedial direction, and cross the midline (Fig. 5.11). At that point they make an abrupt 90° turn and begin to ascend the brainstem. Where they arc across the medulla, these axons are known as the **internal arcuate fibers**. After they cross the midline they are known as the **medial lemniscus**. The internal arcuate fibers and the medial lemniscus both contain axons that

arise from the dorsal column nuclei and represent a continuous structure that simply carries different names to distinguish different locations. Second-order neurons from the spinal nucleus of V and the chief sensory nucleus of V send axons across the midline to join the medial lemniscus at its medial margin. The secondary axons arising in the trigeminal nuclei are separately named by some authors as the **trigeminal lemniscus**[4].

The axons in the medial lemniscus are arranged in a somatotopic pattern that is maintained at all levels of the brainstem (Fig. 5.11). Near its origin in the lower medulla, fibers from the nucleus cuneatus enter the dorsal portion of the tract, while axons from the nucleus gracilis form the ventral part. Since the medial lemniscus is a vertical structure in transverse sections of the medulla, the fibers are arranged in a head-up position. In the lower pons, the medial lemniscus begins to tilt horizontally such that the fibers from the head are dorsomedial, while those from the lower extremity are ventrolateral. In the midbrain, it is nearly horizontal and occupies an extreme lateral-ventral position in the tegmentum. The leg area occupies the most lateral portion of the medial lemniscus, while the head area occupies the most medial portion.

The medial lemniscus enters the thalamus and terminates in a precise somatotopic manner in the **ventrobasal complex** (Fig. 5.9). Signals originating from mechanoreceptors terminate primarily in the VPL and VPM nuclei; the axons of trigeminal origin enter the VPM nucleus, while those of spinal origin enter the VPL nucleus. The VPS nucleus receives strictly proprioceptive signals. The somatotopic arrangement fol-

---

[3] The cranial nerves VII, IX, and X innervate only a small area of skin near the external auditory meatus and the pinna. Their central connections follow those of the trigeminal nerve. Their contribution is detailed in the chapter on cranial nerves (see Chapter 9).

[4] There is some disagreement among various authorities concerning the exact location of the trigeminal lemniscus. I have chosen to describe only the traditional pathway because it seems to correlate best with clinical experience.

---

in the location in the dorsal horn where they terminate and in the pattern of their terminal fields. **A**, Hair follicle afferent; **B**, Rapidly adapting mechanoreceptor afferent; **C**, Pacinian corpuscle afferent; **D**, Slowly adapting afferent, type I; **E**, Slowly adapting afferent, type II; **F**, Summary of terminal fields in the dorsal horn, illustrating the extensive overlapping between groups.

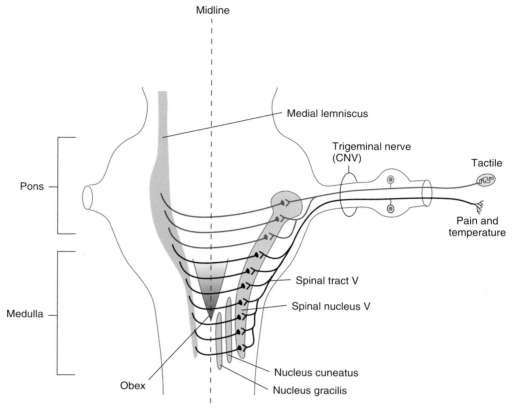

FIGURE 5.13.

A schematic drawing illustrating the relationship between the afferent axons of the trigeminal nerve and the principal sensory nucleus and the spinal tract and nucleus of V. Note that the nociceptive and thermal afferents terminate primarily in the caudal two-thirds of the spinal nucleus. Fine tactile and proprioceptive afferents terminate mostly in the principal nucleus and the rostral third of the spinal nucleus.

lows that of the medial lemniscus: head medial, leg lateral.

The Tertiary Neuron

Third-order axons arising from neurons within the ventrobasal complex project into the ipsilateral cerebral cortex through the posterior limb of the internal capsule. The axons terminate primarily in **Brodmann's areas 3a, 3b, 1, and 2** (see Fig. 1.7) of the **postcentral gyrus** (Fig. 5.14). The association between this area of the cerebral cortex and somatosensory information processing is so specific that this cortical area is frequently called **the primary somatosensory cortex** or **S-I**.

The somatotopic arrangement that is so characteristic of the dorsal columns, the medial lem-

niscus, and the ventrobasal complex of the thalamus is preserved in the cerebral cortex (Fig. 5.15). There, projections carrying information from the lower extremity terminate along the superomedial aspect of the postcentral gyrus. The representation of the upper extremity descends across the surface toward the lateral fissure. The face and internal surfaces of the mouth and throat are represented at the most inferolateral portion.

The surface area of the body is not represented proportionally in the somatotopic representation of the cerebral cortex. The upper extremity and thorax occupy about as much cortical surface area as the hand, and about one-third of the hand area is taken up by the thumb. The size of the face area is about the same as that of the hand, with about one-third of it being occupied by the

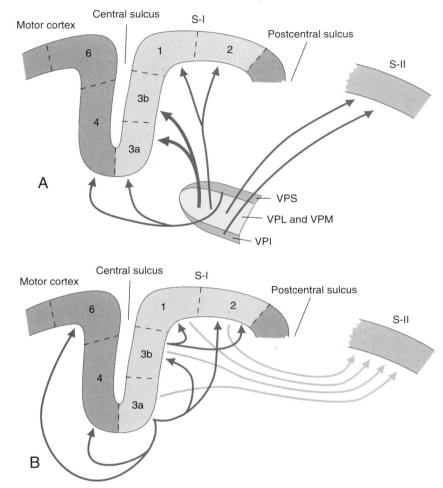

FIGURE 5.14.

A, This drawing illustrates the relationships between the ventrobasal complex of the thalamus and the somatosensory areas of the cerebral cortex. About 70% of the thalamic projections are from VPL and VPM to Brodmann areas 3a and 3b. **B**, Short association fibers interconnect the various regions in a series of cascades. All four of the S-I areas project to the secondary sensory area, S-II, in the parietal operculum.

lips. The map of somatotopic representation in the cerebral cortex corresponds quite closely with a map depicting the resolution of two-point discrimination; the areas with the finest discrimination have the largest cortical representation.

The primary sensory cortex is not functionally or anatomically homogeneous. The four Brodmann areas that constitute the primary sensory cortex do not receive thalamic projections equally. For example, about 70% of the thalamic axons from the VPL and VPM nuclei terminate in areas 3b and 3a, and the remaining axons are divided between areas 1 and 2. Furthermore, the thalamic projections into the cortex are divided by modality. Area 3a receives information primarily from muscle stretch receptors (see Chapter 6), while 3b receives input from cutaneous receptors. Information from rapidly adapting cutaneous receptors goes to area 1, while deep pressure receptors project to area 2. *Each of these modalities is separately mapped, creating parallel somatotopic cortical maps.*

There is an additional organizational pattern within the cerebral cortex formed by a vertical in-

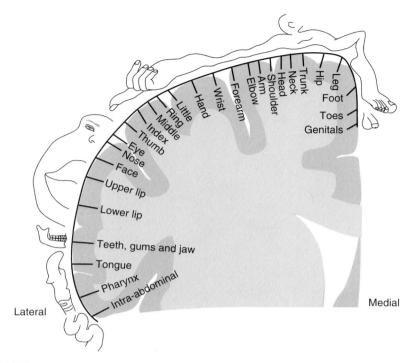

FIGURE 5.15.

A somatotopic map of the human cortical area 1 based on studies of awake, alert individuals undergoing cortical stimulation during brain surgery.

terrelationship among the neurons of the cortical gray matter. Cortical neurons are synaptically interconnected to form a functional unit called a **cortical column** (Fig. 5.16). Only about 300 to 600 cells form a column, and it is now recognized that *the cortical column is the principal computational unit of the cerebral cortex.* Each column has a distinct input-output relationship with the thalamus and with other cortical and subcortical structures. For example, within a single sensory column in area 3a, all the cells respond only to a stimulus of a single modality from a specific location. Adjacent to it, another column may serve a different modality from the same location on the skin. The concept of the cortical column is developed further in a later chapter (see Chapter 13).

Cortical areas are interconnected by cascading reciprocal connections (Fig. 5.14). For example, area 3a sends short association fibers to areas 3b, 2, 4, and 6 (areas 4 and 6 serve motor functions and will be discussed in detail; see Chapter 7). Area 3b projects to areas 1 and 2. All four primary somatosensory areas send projec-

tion axons to a small area of the operculum at the lateral sulcus that serves as a second, complete sensory cortex known as **S-II**. The S-II projection is also somatotopically arranged. Although S-II receives direct thalamic projections, chiefly from the VPI nucleus with a lesser contribution from the VPL and VPM nuclei, it cannot process sensory information independently of S-I.

The stimulus-response characteristics of cortical neurons change as one passes through the cascade from one cortical area to another. For example, neurons in areas 3a and 3b respond best to simple punctate stimuli appropriate to the receptive field of the cell. In areas 1 and 2, however, the cells require more complex stimuli in order to respond. Most of these cells are *motion sensitive,* that is, an object must be moving across the skin before the cell will respond. Other cells are *direction sensitive;* the cell preferentially responds to objects moving only in one direction, for example left to right across the hand but not right to left. *Orientation-specific* neurons prefer stimuli with a particular orientation but are indifferent to direc-

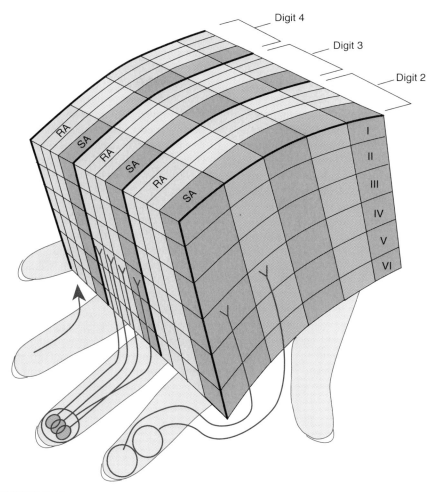

FIGURE 5.16.

Detail of area 3b shows how the cortex is divided by digit, and then each digit area is further subdivided into individual receptive fields. All the neurons in a vertical segment receive information from the same receptive field. This defines a cortical column.

tion. For example, they may respond to left-right or right-left stimulation, but not to proximal-distal or distal-proximal.

As the cortical areas become synaptically more removed from direct thalamic input, the response characteristics of the cortical neurons become more abstracted from the initial objective stimulus. *Motion and direction sensitivity are not properties of the stimulus that are encoded by receptors, they are an abstracted feature of the stimulus.* The ability to develop these abstract properties from a stimulus is unique to the cerebral cortex. No spinal, brainstem, or thalamic neurons have these properties The ability to process abstract information is essential to **stereognosis** (the ability to recognize objects by tactile shape alone), **graphesthesia** (the ability to recognize figure writing on the skin), and fine coordinated movements, especially of the fingers. The loss of these functions is particularly evident after cortical lesions to the postcentral gyrus.

THE POLYSYNAPTIC DORSAL COLUMN PATHWAY

There is a polysynaptic dorsal column pathway that *represents an elaboration of the three-neuron pattern*. The primary afferent fibers that contribute to the dorsal columns, as we have already described, send a short local branch into the ipsilateral dorsal horn, where they make synaptic contact with neurons in lamina III, IV, and V. Projection axons from lamina IV neurons ascend in the ipsilateral dorsal column to synapse in the nucleus cuneatus or nucleus gracilis, according to their point of origin (Fig. 5.10). From there, the signals they convey ascend to the thalamus with the previously described fibers of the medial lemniscus. This **polysynaptic dorsal column system** provides an alternate parallel pathway for tactile information. It also provides, by means of its synapses in the spinal cord, a mechanism by which other sensory modalities can modulate tactile sensations.

THE ALS PATHWAYS

The **anterolateral system** is named for the sensory tract located in the ventrolateral aspect of the spinal cord. This tract consists of secondary axons that originate in the spinal cord and terminate in the brainstem and medulla (Fig. 5.10). The system is not as homogeneous as the

dorsal column system. It is not discretely organized in the spinal cord, and it contains axons that synapse in the medulla and midbrain as well as the thalamus.

Nociceptive and thermoceptive signals originate from axons with free nerve endings and enter the spinal cord as small-diameter axons. This heterogeneous group of small primary afferent axons forms the lateral division of the dorsal root (Fig. 5.17). The myelinated axons belong to the Aδ group, whereas the unmyelinated axons are C fibers. These primary afferent axons bifurcate after entering the spinal cord. The branches ascend and descend one to three segments in a tract known as the **dorsolateral tract** (also known as Lissauer's tract). Axons within the dorsolateral tract send collateral branches that penetrate the dorsal horn gray matter, where they synapse.

The Aδ fibers make quite different synaptic connections in the spinal cord than do the C fibers. These differences are the basis for dividing the ALS into two groups, the spinothalamic and the spinoreticular tracts. Only the spinothalamic division follows the three-neuron pattern, so it will be described first.

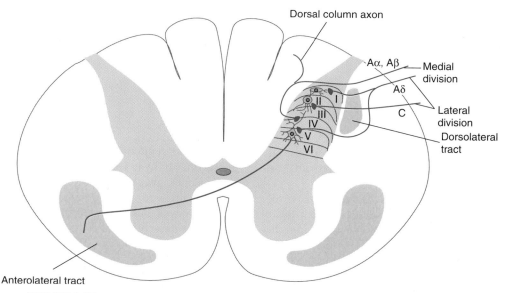

FIGURE 5.17.

The synaptic relationship between large-diameter afferents and small-diameter short local branches of primary afferents in the spinal cord.

The Spinothalamic Division

Primary afferent Aδ fibers are associated with sharp, brief, fairly well localized nociceptive sensations. Collateral fibers from these axons leave the dorsolateral tract and synapse with secondary neurons located in laminae I and V (Fig. 5.17). Neurons in both lamina I and lamina V send second-order axons to the contralateral thalamus. Because they have long axons that travel the entire length of the spinal cord and brainstem, they are called **projection neurons**. These axons cross the midline at approximately their level of origin, crossing the spinal cord in the ventral white commissure. On the contralateral side, they collect in the ventrolateral funiculus, where they make an abrupt 90° bend before ascending to the brainstem and thalamus as a part of the ALS (Fig. 5.18). A somatotopic organization of the ALS is established as new axons join this tract at the medial margin, creating a laminar structure, somatotopically organized.

The ALS maintains its ventrolateral position the entire length of the spinal cord and medulla, lying ventral to the spinal tract of V in the caudal medulla, and between the inferior olive and the spinal tract of V in the rostral medulla. It maintains this ventrolateral position through the pontine tegmentum. In the rostral pons, the medial lemniscus has moved far enough laterally to meet the ALS and, from that point onward, they travel together into the thalamus.

Spinothalamic axons in the ALS terminate in nuclei of the ventrobasal complex, in the VPL and VPM nuclei, and in the intralaminar group. Only the ALS axons that terminate in the VPL and VPM nuclei maintain a somatotopic relationship. Once at the thalamus, the dorsal column system and the spinothalamic division of the ALS share the same third-order neurons to the cerebral cortex.

Primary afferent axons from thermoreceptors for warm (C fibers) and cold (Aδ) sensations synapse in laminae I, II, and III. Thermal information must work through a series of synaptic connections in the dorsal horn until projection neurons in lamina V neurons are facilitated. These neurons send axons across the midline to join the ALS and project to the thalamus.

Primary afferent nociceptive and thermal pathways from the face are quite similar to those already described for fine touch and proprioception (Fig. 5.13). Small Aδ and C fibers enter the pons in the trigeminal nerve. These axons descend into the medulla in the spinal tract of V, where they terminate in the caudal two-thirds of the spinal nucleus of V. From this nucleus secondary neurons cross the midline and join the medial lemniscus, eventually to terminate in the ventral basal complex of the thalamus. From there tertiary neurons ascend to the cerebral cortex.

The Spinoreticular Division

The spinoreticular division of the ALS, as the name implies, ascends to the reticular formation from the spinal cord. This system does not follow the three-neuron pattern but represents an elaboration of the principal spinothalamic pathway.

Primary afferent C fibers are associated with long-lasting, dull aching or burning sensations. Collaterals of these axons leave the dorsolateral tract and synapse in laminae II and III of the dorsal horn (Fig. 5.17). These two laminae are filled with many small somas and are essentially devoid of myelinated axons. Therefore, in fresh cadaver material, they have a gelatin-like appearance, giving this area its traditional name, the **substantia gelatinosa**. The neurons in laminae II and III are all interneurons having short axons. They synapse within the spinal cord, forming a complex network of neuronal interconnections (**neuropil**) that has defied accurate anatomical description. Eventually, however, information from laminae II and III follows a polysynaptic course into lamina V, where it excites projection neurons that cross the midline and ascend with the ALS. These projection fibers terminate primarily in the reticular formation of the medulla and pons (Figs. 5.10 and 5.18). Neurons within the reticular formation form another polysynaptic ascending system throughout the length of the brainstem. Ultimately this polysynaptic system projects to the thalamus and terminates in a nonsomatotopic manner in the intralaminar nuclei. Neurons from the intralaminar nuclei project to wide areas of the cerebral cortex.

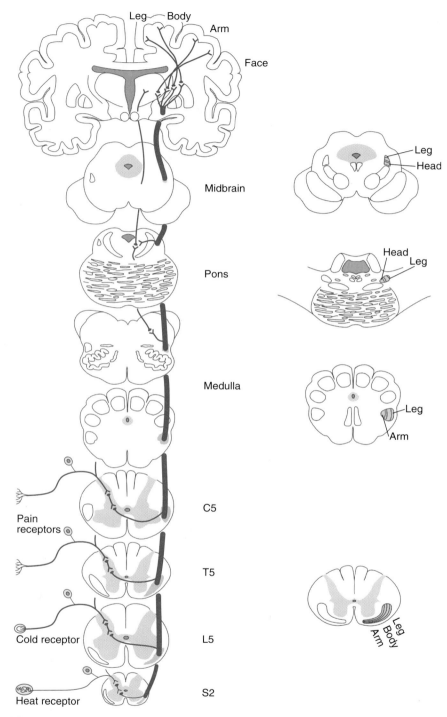

FIGURE 5.18.

The anterolateral system. On the *left* are shown the anatomical positions of the various components of the system at the principal levels of the nervous system. Small figures on the *right* depict the somatotopic arrangement of the fibers at the various levels.

Clinical Examination of Sensory Systems

Examination of sensory systems presents a special challenge to the physician because the sensory examination is the only part of the neurological evaluation that requires the full, honest participation of the patient. A successful sensory examination begins by communicating to the patient exactly what is expected. If the patient cannot understand and follow the instructions or attempts to lie about the results, the sensory examination may be invalid, but the physician has gained significant insight about the patient and his problem.

PATTERNS OF SENSORY LOSS

When one assesses somatosensory systems, one frequently must differentiate between lesions within the CNS, lesions to the spinal roots, and those affecting the peripheral nerves. Making these distinctions is facilitated by noting the *distribution* of sensory derangements projected on the body surface and the *modalities* affected. These distinctions are possible because of the way the various components of the somatosensory system are organized.

Peripheral Lesions

Complete lesions to spinal or peripheral nerves elicit sensory losses involving all sensory modalities and motor functions because all modalities are represented in the axons of these structures. Occasionally, pressure insults and a few other conditions will affect the largest axons first. In these cases, fine tactile, vibratory, and proprioceptive sensations may be compromised while leaving pain and temperature sensations relatively intact. These latter sensations will be aberrant, however, for their affect depends partly on signals transmitted by the large-diameter fibers (see discussion of pain below). Nearly everyone has experienced a nerve pressure palsy after having sat in an awkward position that caused pressure to be placed on a major peripheral nerve. In such cases one notices that the affected extremity is "asleep." Light touches to the skin cannot be perceived and proprioception is absent. Walking may be hazardous if the affected nerve is the sciatic, not due to paralysis, but because the lack of proprioceptive feedback makes controlling the limb impossible. Painful sensations are present, however, for tingling sensations are spontaneously present and pin pricks are readily felt, although they feel unnatural.

The distribution of sensory losses due to peripheral injuries is quite characteristic. The pattern is based on the segmental organization of the embryo that is established during development. This is reflected in the segmental organization of the adult spinal cord and is preserved in the distribution of sensory axons to the skin because *as the embryo expands and develops, the skin carries its segmentally derived innervation with it.* The area of skin supplied by the axons from a single dorsal root is known as a **dermatome**. The distribution of dermatome patterns over the embryonic limb buds is modified somewhat in the adult, but not as much as one might think. The essential segmental pattern remains in the adult; it is simply stretched to fit the extruded extremity (Fig. 5.19).

The development of the nerve plexuses complicates the innervation of the extremities. Each trunk, cord, division, and peripheral nerve contains varying sets of axons from several adjacent spinal roots (Fig. 5.20). In the adult extremities, *individual peripheral nerves contain axons that arise from several different spinal roots* and, conversely, *different peripheral nerves contain axons that arise from the same root.* The practical result is that, following *peripheral* nerve injuries, the pattern of sensory denervation observed on the skin is strikingly different from the pattern observed following *spinal* nerve injuries.

When a *spinal root is damaged, the sensory loss is revealed over the entire extent of the dermatome.* On the other hand, when the peripheral nerve is damaged, the subsequent sensory loss extends to the area of skin supplied a subset of axons from one or more adjacent spinal roots. Hence, only a part of the dermatome for each contributing root is affected. Therefore *the peripheral nerve pattern consists of fragments of several adjacent dermatomes.* It is very important for the physician to have a mental picture of

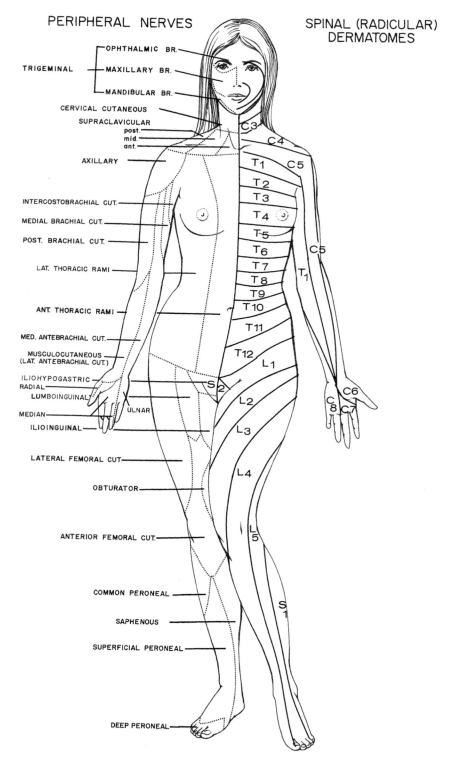

PERIPHERAL NERVES

SPINAL (RADICULAR) DERMATOMES

TRIGEMINAL
OPHTHALMIC BR.
MAXILLARY BR.
MANDIBULAR BR.

CERVICAL CUTANEOUS
SUPRACLAVICULAR
post.
mid.
ant.

AXILLARY

INTERCOSTOBRACHIAL CUT.
MEDIAL BRACHIAL CUT.
POST. BRACHIAL CUT.

LAT. THORACIC RAMI

ANT. THORACIC RAMI

MED. ANTEBRACHIAL CUT.
MUSCULOCUTANEOUS
(LAT. ANTEBRACHIAL CUT.)
ILIOHYPOGASTRIC
RADIAL
LUMBOINGUINAL
MEDIAN
ILIOINGUINAL

LATERAL FEMORAL CUT.

OBTURATOR

ANTERIOR FEMORAL CUT.

COMMON PERONEAL

SAPHENOUS

SUPERFICIAL PERONEAL

DEEP PERONEAL

C3
C4
T1 C5
T2
T3
T4
T5
T6 C5
T7 T1
T8
T9
T10
T11
T12
L1
S2
L2 C6
C8 C7
L3
L4
L5
S1

ULNAR

FIGURE 5.19.

The pattern of sensory derangement after irritation or destruction of individual spinal (dermatome pattern) and peripheral nerves.

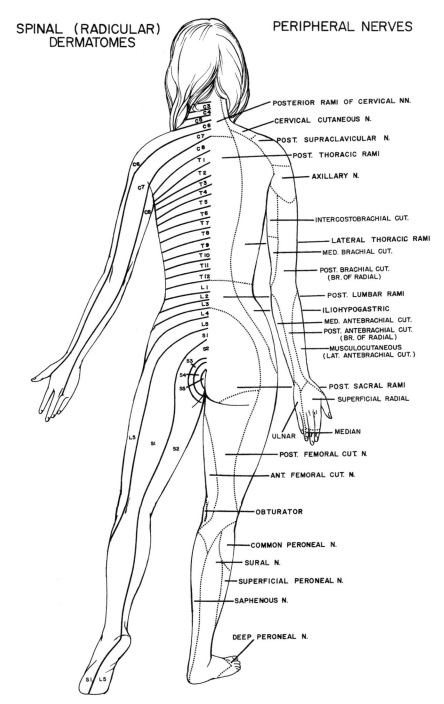

SPINAL (RADICULAR) DERMATOMES

PERIPHERAL NERVES

POSTERIOR RAMI OF CERVICAL NN.
CERVICAL CUTANEOUS N.
POST. SUPRACLAVICULAR N.
POST. THORACIC RAMI
AXILLARY N.
INTERCOSTOBRACHIAL CUT.
LATERAL THORACIC RAMI
MED. BRACHIAL CUT.
POST. BRACHIAL CUT.
(BR. OF RADIAL)
POST. LUMBAR RAMI
ILIOHYPOGASTRIC
MED. ANTEBRACHIAL CUT.
POST. ANTEBRACHIAL CUT.
(BR. OF RADIAL)
MUSCULOCUTANEOUS
(LAT. ANTEBRACHIAL CUT.)
POST. SACRAL RAMI
SUPERFICIAL RADIAL
MEDIAN
ULNAR
POST. FEMORAL CUT. N.
ANT. FEMORAL CUT. N.
OBTURATOR
COMMON PERONEAL N.
SURAL N.
SUPERFICIAL PERONEAL N.
SAPHENOUS N.
DEEP PERONEAL N.

FIGURE 5.19.—continued

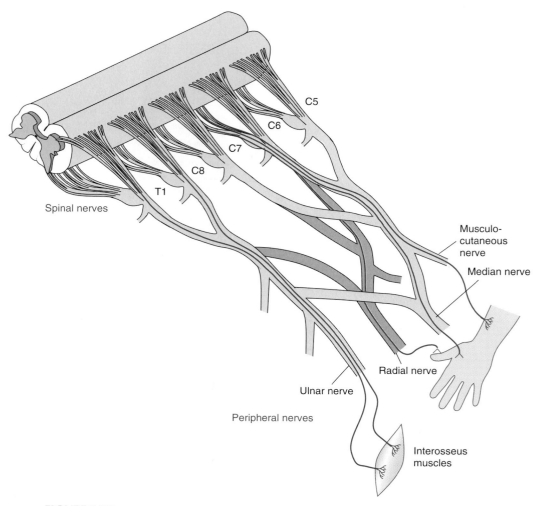

FIGURE 5.20.

A segment of the brachial plexus is illustrated to show how a single spinal nerve may contribute to several peripheral nerves and how a single peripheral nerve may receive axons from more than one spinal nerve.

the dermatomes and peripheral nerve patterns so that sensory losses can be tested in a methodical and meaningful way. The most diagnostic patterns are to be found on the hands and feet. The student is advised to memorize them (Table 5.1).

Peripheral neuropathy is a generalized loss of sensory and motor functions of peripheral nerves. Because peripheral neuropathy is so often caused by toxins or metabolic disturbances, the larger, most metabolically active neurons are usually affected first. These are also neurons with the longest axons; therefore, symptoms usually begin at the distal portion of the extrem-

ities. This produces a "stocking-and-glove" pattern of sensory loss. This stocking-and-glove pattern is distinct from the dermatome and peripheral nerve patterns and is therefore diagnostically important.

Central Lesions

Central lesions are often characterized by dissociation of functions. Motor functions can be affected, with the sparing of sensory functions from the same area of the body and vice versa. Pure sensory dissociations can also occur.

Table 5.1. Spinal Nerve Innervation of Surface Landmarks

Root	Diagnostic Pattern
C6	Thumb
C7	Index and middle fingers
C8	Ring and little fingers
T4	Nipples
T10	Umbilicus
L4	Patella and large toe
L5	Middle three toes and sole of foot
S1	Little toe and lateral aspect of foot

This means that only some of the sensory modalities may be lost over an affected part of the body where other modalities are preserved. This is possible because the dorsal column system crosses the midline in the brainstem, whereas the ALS crosses at the spinal cord. For example, hemisection of the spinal cord (**Brown-Séquard** syndrome) will produce sensory losses below the level of the lesion. Fine tactile, proprioception, and vibratory sensations will be lost from the side of the body *ipsilateral* to the lesion because the dorsal columns carry primary afferent axons that have not yet crossed the midline. Nociperception and temperature sensations will be lost on the contralateral side because the ALS contains secondary axons that have already crossed the midline. Lesions to the brainstem or cerebral cortex that involve sensory structures

will produce a deficit in all somatosensory modalities on the entire side of the body *contralateral* to the lesion. If the lesion is in the brainstem, there is usually involvement of the cranial nerves also (see Chapter 9). Memorizing these basic patterns of sensory loss is essential if one is to make neurological diagnoses (Fig. 5.21).

Many other spinal cord lesions produce characteristic patterns of combined sensory and motor losses that are quite diagnostic once the underlying anatomy is understood. One of these lesions is caused by a spinal syrinx and is reported in the case of Mrs. J. R. below. A further description of other classic spinal cord lesions will be deferred until after the motor systems have been discussed.

PERFORMING THE SOMATOSENSORY EXAMINATION

There are three objectives of the somatosensory examination: to determine if there is a sensory loss, to establish its pattern, and to identify the modalities involved. These three elements are important for localizing the lesion. Peripheral lesions are characterized by the *pattern* of sensory loss affecting all modalities. Spinal nerve injuries produce patterns of loss that follow dermatomes, whereas nerve lesions produce spotty, patchy patterns that are appropriate to the peripheral nerve. Central lesions frequently produce somatosensory dissociations or loss of all modalities over half of the body.

The dorsal column system can be tested by lightly touching the skin with a fine wisp of cotton. This is equally effective on glabrous as well as hairy skin. In the latter case, one should be careful to touch only the hairs. The normal patient will have no difficulty perceiving even the slightest touch of the cotton, the only exception being on heavily calloused hands or feet. As another test of the dorsal column system, one can firmly place the base of a 128-Hz tuning fork on a bony protrusion, such as the fibular or tibial malleolus, the patella, the styloid process of the radius, or the olecranon of the ulna. The normal patient will have no difficulty perceiving the vibration of the tuning fork.

Finally, one can test the proprioceptive component of the dorsal column system by holding the patient's finger or toe and ever so slightly

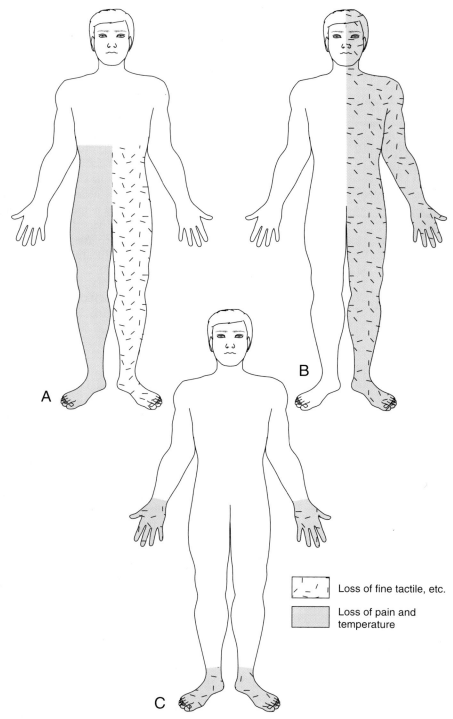

FIGURE 5.21.

The patterns of sensory dissociation are often important for assessing the location of a lesion. **A**, The "harlequin" pattern characteristic of a spinal hemisection. **B**, The division of the body along the midsagittal plane is characteristic of lesions to the brainstem or cerebral cortex. **C**, The "stocking-glove" pattern is characteristic of peripheral neuropathies.

bending one of the joints. The normal patient will be able to perceive the slightest motion. A patient with a severe lesion may not be able to sense any joint movement, no matter how gross. The **Romberg test** is a valuable test of proprioception from the lower extremities. The patient is asked to stand with his feet together. If he is steady and does not fall, he is directed to close his eyes. If the proprioceptive sense is lost, the patient will fall (be sure to be prepared to catch him!). In such cases, the patient can stand with his eyes open because the visual sense allows him to detect and compensate for the slight unsteadiness we all have when standing on a narrow base. The patient will fall if deprived of vision when the sense of joint position is missing from the lower extremities.

The ALS can be independently tested by sharply poking the skin with a pin.[5] If there is damage to the ALS, the patient may perceive the presence of the pin, but it will be dull and not annoying. The normal patient will perceive the pin as a sharp prick that is painful. Since an individual's perceptions of sharpness and pain vary, it is best to compare one side against the other. For example, prick the arm at the C8 dermatome on the left and then test the same dermatome on the right. Ask the patient if the two pin pricks seem about the same. Alternatively, one can circumscribe the extremity, crossing dermatomes. The patient with a lesion will report a definite demarcation between sharp and dull sensations. The ALS can be further tested by placing a cold instrument (e.g., the side of the nonvibrating tuning fork used above) on the skin at various locations and asking the patient if it feels warm, cold, or neutral. Normal patients will report that the instrument feels cold.

The somatosensory examination rarely reveals absolute anesthesia. This is due to the considerable overlap of the terminal fields of the primary afferents in the dorsal horn of the spinal cord. What a patient experiences is a *change in the perception* of the stimulus as it passes from a normally innervated area to one that has been denervated. This is especially true if the lesion involves a spinal root. Therefore is it important to the success of the sensory examination to ask the patient to make comparisons between sensations presented at two areas, either by stimulating the two extremities in turn or by crossing dermatomes on a single extremity.

Pain

Pain is not a sensation, it is a perception. The mechanisms by which nociceptive sensations are interpreted as emotionally disturbing painful perceptions is not known. Pain is, nevertheless, a real phenomenon with both personal and social repercussions. Millions of people suffer disabling chronic pain for which modern medicine can offer little treatment.

The role of the cerebral cortex in the perception of nociceptive stimuli is unclear. Although sharp pain is well localized, dull burning pain is not. This is not unexpected, since sharp pain is conveyed directly to the specific thalamic nuclei (VPL and VPM) by spinothalamic pathways. Dull, burning pain is conveyed by polysynaptic pathways through the reticular formation to the intralaminar, nonspecific thalamic nuclei that project to widespread areas of the cerebral cortex. The perception and localization of nociperception appear to be facilitated by the synergistic participation of both the dorsal column system and the spinothalamic division of the ALS. The dorsal column system is necessary for localizing pain. Patients with dorsal column lesions are unable to localize painful stimuli very well, saying something like "I hurt on the left somewhere."

Pain is an abstraction of nociperception. Pain and nociperception can be separated. For example, patients suffering pain and given morphine will frequently be able to describe the nociceptive sensations, yet also report that the *affect* of pain is absent. Patients with certain thalamic lesions have the opposite experience; they may have total numbness on the contralateral side of the body, yet report excruciating pain "from somewhere." In spite of the clear and obvious dissociation between nociperception and pain, no cortical representation of pain has been discovered. Pain remains a medical mystery. In the following section we will consider some of the clinically

---

[5]One frequently draws blood during this examination, so it is important to use a new, clean pin for each patient. In the hospital, one can always identify neurologists because of the number of safety pins attached to their bags.

relevant aspects of pain and attempt to develop a basis for understanding this phenomenon.

PHANTOM PAIN

As a result of certain types of lesions to the nervous system, individuals can perceive pain that is not associated with injury to the body. Because of the discrepancy between obvious injury and the localization of the sensations, this is commonly called **phantom pain**. Lesions that cause phantom pain can be either peripheral or central. **Peripheral lesions** usually involve both small- and large-diameter primary afferent axons. Amputation of a limb, for example, frequently causes the patient to feel painful sensations localized to the missing extremity. This same type of pain can accompany entrapment lesions, wherein whole peripheral nerves become irritated and inflamed due to constant pressure or intermittent insult. For example, the median nerve is frequently injured at the flexor retinaculum (**carpal tunnel syndrome**), an injury that often results in pain that involves not only the hand, but the entire arm. The extent of the pain goes far beyond the territory of the nerves injured at the carpal tunnel. Central lesions that produce phantom pain are usually found in the ventrobasal complex of the thalamus after vascular insult such as a stroke (**thalamic pain syndrome**). With either central or peripheral lesions, patients usually describe the affect of the pain as unlike, and more excruciating than, anything previously experienced.

REFERRED PAIN

Sensations arising from the viscera are poorly localized and generally only painful. Visceral afferents enter the spinal cord over the dorsal root and synapse in the dorsal horn. In the CNS, however, there are no specific pathways devoted to visceral sensations, and the somatotopic maps of the body carry no separate representation of the viscera. Probably, after entering the spinal cord, visceral pain afferents ultimately excite the same transmission pathways that somatic pain afferents use. Therefore, the *visceral and somatic sensations share the same transmission tracts*. Since the brain normally interprets these signals to be of somatic origin, visceral pain is mistakenly perceived to be from the body surface. This phenomenon is called **referred pain**.

The viscera, like the skin, carry their embryonic innervation with them during development. Many of the viscera migrate great distances from their embryonic origin, and therefore pain referred from these viscera may be perceived from locations remote from the actual site of the stimulus. For example, blood from abdominal bleeding can irritate the diaphragm, causing pain. The pain will not be localized to the abdomen; it will be referred to the scapular area because the diaphragm originates from cervical myotomes. Table 5.2 lists some important organs and the dermatome to which pain from each organ is referred.

SURGICAL INTERVENTION

The ALS is anatomically important for pain perception. Stimulating the tract causes severe pain that is localized to the contralateral side of the body. This fact suggests that intractable pain might be alleviated by cutting the dorsal roots (**rhizotomy**) or the ALS (**tractotomy**) in the spinal cord. After such operations, the patient usually experiences prompt and profound relief from his pain. Unfortunately, the relief is usually transitory. The pain almost always returns, usually in a form more intense and unmanageable than before surgery. Recent experimental

Table 5.2. Principal Dermatomes for Referred Pain and Visceral Origin

Somatic Projection	Visceral Organ
C3–C4	Diaphragm
T1–T8	Heart
T10	Appendix
T10–T12	Testes & prostate
T10–T12	Ovaries & uterus

evidence from animal studies has shown that deafferentation causes some neurons in lamina V to become tonically active. Since this is the origin of many ALS projection neurons, the inference is that these abnormally active neurons are the source of the pain signals. With time these neurons may adapt and the abnormal spinal activity will subside, only to be replaced by abnormal spontaneous activity in the thalamus. This type of abnormal spontaneous activity probably accounts for the return of pain after tractotomy.

Therapeutic lesions to the thalamus have also proved to be of questionable value. Ablation of the ventrobasal complex usually decreases sensations from the contralateral side of the body. However, the affect of pain remains, leaving the patient more miserable than before. Lesions to the intralaminar group are said to relieve pain without altering sensation, but the results have often been unpredictable.

Therapeutic lesions to the cerebral cortex designed to alleviate intractable pain provide a special challenge to the surgeon because *there is no cortical representation for pain*. Direct stimulation of the cerebral cortex results in a variety of sensations, including numbness, tingling, pressure, and visual or auditory perceptions, but cortical stimulation does not elicit pain. This remarkable fact allows neurosurgeons to operate on the brain of conscious individuals. It also means that cortical ablation cannot be used to eliminate pain. Various cortical ablations have been directed at nonsensory parts of the cerebral cortex in an effort to eliminate the affect of pain. These operations are extremely controversial and have been difficult to evaluate dispassionately. The rationale for cortical ablations to relieve pain will be discussed later (see Chapter 13). Because consistent beneficial results have been difficult to document, surgical approaches to pain therapy have largely been abandoned.

PAIN-MODULATING SYSTEMS

Since the dawn of human history, extracts from the opium poppy have been used to alleviate pain. **Morphine**, the modern purified drug extracted from opium, is still the best analgesic. The effectiveness of morphine implies that specific regulatory pathways and specific morphine receptors must exist. The recent discovery of en-

dogenous opioid receptors in the brain stimulated an intensive search for these pathways, a search that has borne some fruit.

Since the initial discovery of morphine receptors, three families of pain-modulating proteins, **endogenous opioids**, have been discovered; the **endorphins**, the **enkephalins**, and the **dynorphins**. Each family is derived from a separate gene that encodes precursor proteins. Although there are chemical differences in the structure of the three families, more remarkable is the fact that all three families produce active peptides that share the common amino acid sequence tryptophan, glycine, glycine, and phenylalanine (*Try-Gly-Gly-Phe*).

Three separate opioid receptors have also been discovered; **mu**, **delta**, and **kappa**. The endorphins bind best to the mu receptor, which is the most widely distributed of the three. The enkephalins bind best to the delta receptor, while the kappa receptor seems to be specific for the dynorphins. All three receptors bind morphine, and there is considerable cross-reactivity, especially between endorphins and enkephalins with the mu and kappa receptors.

Periaqueductal Gray Matter

Descending pathways are capable of modulating pain perception, and they are closely linked with endogenous opioid systems. The most prominent of these pathways originates in the periaqueductal gray (PAG) matter and descends into the medulla, where it synapses. The medullary neurons project into the spinal cord, where they synapse in the dorsal horn and inhibit nociceptive pathways. A number of nuclei and neurotransmitters are associated with this pathway.

PAG neurons send axons into the **nucleus raphe magnus (NRM)** and the **nucleus reticularis paragigantocellularis (NRPG)** in the medullary reticular formation (Fig. 5.22), where they make excitatory synaptic contacts. Neurons from the NRM and NRPG send axons into the spinal cord. These reticulospinal axons travel in the dorsal aspect of the lateral white column and synapse in the dorsal horn (probably laminae II and III).

Most of the neurons from the NRM secrete serotonin and excite interneurons in the spinal cord. These interneurons appear to presynapti-

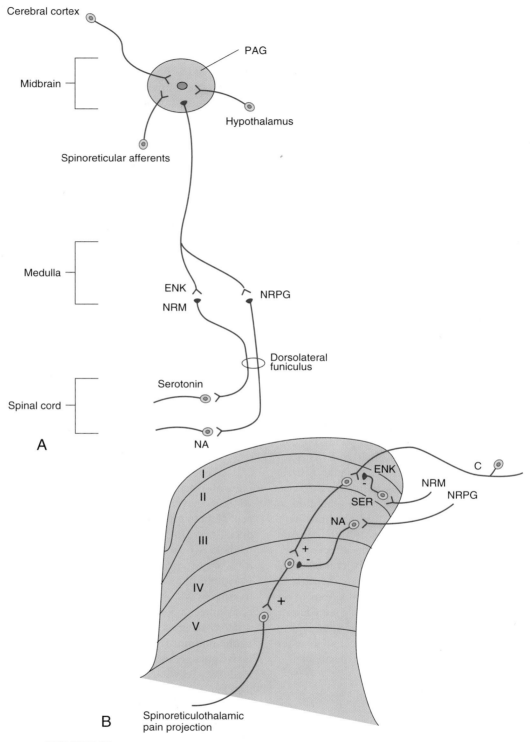

FIGURE 5.22.

   A, Neurons in the PAG receive axons from diverse areas of the nervous system, including the hypothalamus and the spinal cord. Axons from the PAG project to the nucleus raphe magnus (*NRM*) and the nucleus reticularis paragigantocellularis (*NRPG*) in the medulla. PAG neurons apparently secrete enkephalin (*ENK*) as either a neurotransmitter

cally inhibit primary afferent C fibers by secreting enkephalin. Neurons from the NRPG also descend in the dorsal lateral white column, but they secrete noradrenalin on interneurons of the dorsal horn. Activation of these interneurons inhibits the thalamic projection neurons of lamina V through a nonopioid polysynaptic pathway. The details of this circuit are not known; however, when activated, this descending control system exerts a strong analgesic effect. This knowledge has been used successfully to treat patients suffering from chronic pain by exciting PAG neurons with indwelling electrodes.

Neurons in the periaqueductal gray matter of the mesencephalon can be activated by many parts of the brain (Fig. 5.22). It receives ascending sensory projections from the spinoreticular division of the ALS. It also receives descending projections from the hypothalamus and many parts of the cerebral cortex. PAG neurons have mu receptors that can be activated by circulating opioids or opioids secreted by axons synapsing in the PAG. The thalamic and cortical projections into the PAG provide a potential explanation for the suppression of pain during periods of high stress and anxiety, a phenomenon that has frequently been reported by soldiers wounded in battle and civilian accident victims.

Gate-Control Mechanisms

In addition to the descending pain-modulating system, pain pathways can be modulated by spinal afferents. The **gate control** hypothesis suggests that, in addition to the connections already described, large-diameter primary afferent axons (Aα and Aβ) make excitatory synaptic contact with interneurons in the dorsal horn (Fig. 5.23). These interneurons are inhibited by pathways originating from the small-diameter afferents (Aδ and C). When activated, the interneurons suppress primary afferent C fibers by presynaptic inhibi-

tion. Thus the hypothesized interneuron acts as a gate, controlling the transmission of pain stimuli conveyed by C fibers to higher centers.

One prediction of this hypothesis is that sufficient stimulation of the *large-diameter afferents could block transmission of nociceptive input from the small-diameter axons* by the action of the interneuron. Although the detailed synaptic interrelationships suggested by the gate control hypothesis have yet to be proved, its essential correctness has been demonstrated clinically. Electrodes placed on the dorsal columns have been used to antidromically (backward conduction) stimulate the large-diameter primary afferent axons and successfully reduce or eliminate previously intractable pain. **Transcutaneous electrical neural stimulation (TENS)** is a less invasive procedure that is also based on gate control theory. With this type of stimulator, electrodes are placed on the skin over a peripheral nerve. Large-diameter axons can be preferentially activated by transcutaneous stimulation because they have the lowest threshold to electrical stimulation. TENS units have been quite successful in the treatment of chronic pain.

CASE STUDIES

Ms. J. R.

HISTORY: Ms. J. R. is a 67-year-old previously healthy black female who has come to the emergency room complaining of swelling and decreased mobility of her right arm. She states that five days ago she noticed a burning sensation under her right axilla and weakness in her right arm, which has persisted to the present time. She has no pain at the present time.

In the past she has had bilateral ear surgery (diagnosis not stated) and bilateral carpal tunnel surgeries. She began using a cane about a year ago because she developed a limp at that time.

PHYSICAL EXAMINATION, March 31, 1994: Ms. J. R.'s temperature was 95.2°F, heart rate

---

or a neuromodulator. Neurons from the NRPG secrete noradrenalin (NA); those from the NRM secrete serotonin. Both systems project to the spinal cord. **B,** The spinal terminations of the descending fibers from the NRPG and the NRM. Both systems inhibit nociceptive pathways by activating interneurons in the dorsal horn that inhibit the spino-reticulothalamic pain projection pathway.

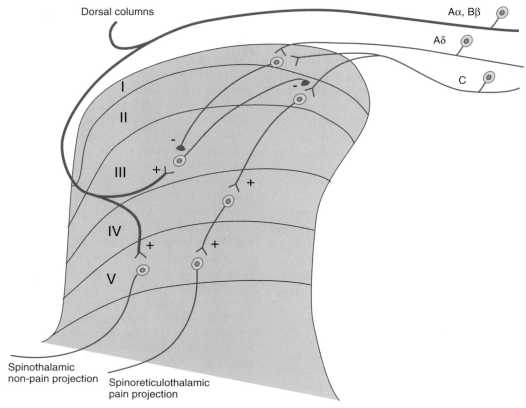

Dorsal columns

Aα, Bβ

Aδ

C

I

II

III

IV

V

Spinothalamic
non-pain projection

Spinoreticulothalamic
pain projection

FIGURE 5.23.

A schematic illustration showing a possible synaptic implementation of the gate control hypothesis in the spinal cord. The basic principle of this hypothesis is the prediction that large-diameter afferents, conveying non-noxious information, can block the transmission of nociceptive information from small-diameter afferents. Note that there is a balance between the small-diameter axons (Aδ and C fibers) that disinhibit the C fiber input and the large-diameter axons (Aα and Aβ) that inhibit it.

80, respirations 20, blood pressure 152/82. She was in good spirits and showed no signs of discomfort or distress. A burn scar was noted on her left shoulder, which she stated was the result of using a heating pad ten years ago. She no longer uses heating pads.

NEUROLOGICAL EXAMINATION, March 31, 1994

I.   MENTAL STATUS: Normal.

II. CRANIAL NERVES: All normal.

III. MOTOR SYSTEMS:[6]

A. The strength in her legs was equal bilat-

erally. Weakness and lack of mobility in the right upper extremity was noted. The left hand grip was weaker than the right.

B. Wasting of the interosseous muscles of the left hand was noted.

C. She walked with a slight limp, favoring the right side. She was steady on her feet. Deep knee bends and hopping on one foot were deferred.

IV. REFLEXES:

A. MSRs were absent from the biceps, triceps, and brachioradialis muscles. MSRs were increased from the quadriceps and gastrocnemius muscles.

B. Babinski sign was present bilaterally.

[6]Defer consideration of motor systems and reflexes until after reading about the motor systems (see Chapters 6, 7, and 8).

V. CEREBELLUM: Normal.

VI. SENSORY SYSTEMS:

   A. There was a bilateral decreased perception of pin prick and temperature from the base of her neck to 2 cm above the umbilicus.

   B. Perception of wisps of cotton stroked across her skin was equal in all four extremities.

   C. She was able to stand unassisted with her eyes open and closed (Romberg test).

LABORATORY STUDIES:

   1. X-rays revealed a displaced fracture of the distal head of the right humerus and severe degeneration of the left elbow and right shoulder joints (neurotrophic joint).

   2. MRI of the brain and spinal cord revealed an Arnold-Chiari malformation and syringomyelia of the spinal cord from the cervicomedullary junction to the midthoracic level (Fig. 5.24).

*Comment: **Syringomyelia** [G. syrinx, tube + myelos, marrow], **as the name implies, is a cavitation of the spinal cord. This rare disorder is of unknown etiology. It appears that in most cases, the central canal of the spinal cord enlarges and becomes filled with fluid. The cavity usually, but not invariably, communicates with the CSF compartment. The deformity frequently begins in the cervical spinal cord and gradually expands. It may extend into the thoracic cord, as in the present case, or into the medulla, in which case the condition is known as syringobulbia. In the present case, the cavity spontaneously decompressed, which accounts for the distorted shape of the spinal cord in the MRI. Spontaneous decompression is extremely unusual.***

   The cavity first appears near the center of the cord, where its expansion destroys the secondary fibers of the ALS where they cross in the ventral white commissure. Consequently the initial symptom is the bilateral loss of pain and temperature perception with the preservation of proprioception and light touch and vibratory sensations over the upper extremities. The pattern will follow several dermatomes, depending on the rostral-caudal extent of the cavity.

   As the circumference of the cavity expands, other structures become compromised, including

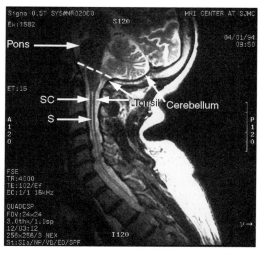

FIGURE 5.24.

A T1-weighted MRI in the midsagittal plane from Ms. J. R. revealing a large syrinx that extends from the cervicomedullary junction to midthoracic levels. The full extent of the syrinx cannot be seen in this image because it passes out of the plane of section in the upper thoracic region. In this T1-weighted MRI, the CSF is light, while neural tissue is dark. The spinal cord, delineated by a pair of *arrows (SC),* is very narrow, much smaller than normal, leaving a very large subarachnoid space in the cervical region of the spinal cord. This suggests that the syrinx was at one time much larger and spontaneously ruptured into the subarachnoid space, causing the spinal cord to collapse. Ms. J. R. also exhibits a very mild Arnold-Chari malformation, which is an extension of the cerebellar tonsils into the opening of the foramen magnum. Here the cerebellar tonsils are pointed out extending below the base of the skull (*dashed line*). This is an incidental finding in this case.

the ventral horns and the lateral columns. This gives rise to lower motor symptoms in the upper extremity and upper motor symptoms in the lower extremity (see Chapters 6 and 7). Involvement of the dorsal columns leads to loss of proprioception and fine tactile and vibratory sensations—losses that may involve all four extremities, depending on extent of the expansion. Since the expansion may not be symmetrical, the involvement of these structures may be quite capricious.

   Syringomyelia, as demonstrated in the present case, is commonly associated with the Arnold-Chiari malformation and with neurotrophic joint disorders. The former disorder is characterized by the protrusion of the cerebellar tonsils into the foramen magnum. Neurotrophic joint disorder is a

trophic enlargement and malformation of the joint associated with certain neurological diseases, particularly syringomyelia and neurosyphilis. The interested student is referred to the references.

FURTHER APPLICATIONS

The loss of pain and temperature perception following lesions to the ALS is dramatically illustrated by Ms. J. R. who, in spite of a fractured humerus, was in no particular discomfort or pain.

- Identify the level of involvement of the spinal cord from the pattern of anesthesia to pin prick. How closely does this agree with the MRI evaluation? Make a sketch, illustrating the pattern of sensory losses.

- How does the preservation of fine tactile sensations limit one's hypotheses about the scope of the lesion?

- What is the significance of the Romberg test? Do the results of this test help one to determine the extent of the lesion? How?

- How does the fact that Ms. J. R. burned herself on a heating pad affect your thinking about this case? How does this relate to determining the time course of the disease and the veracity of the patient?

- It is common for patients to experience pain from the affected area when the symptoms of syringomyelia are first noticed. How do you explain this? The chronic pain subsides with time, leaving the area anesthetic. Can you explain this?

Mr. H. D.

Mr. H. D. is a 40-year-old white male who 20 years ago suffered an injury to his right brachial plexus following a bicycling accident. He experienced immediate total motor and sensory loss to his entire right arm. Since the accident he has experienced severe burning pain from the right arm continuously for nearly 20 years. Four years after his accident, his right arm was amputated, a procedure that provided no relief from his pain. Ten years ago, electrodes were implanted on the dorsal columns between C2 and C5. Dorsal column stimulation was maintained for 6 months. No relief from his pain was achieved.

Four years ago, Mr. H. D. submitted to another surgical procedure designed to ablate the right dorsal horn between C5 and C7. At the time of the operation, a cervical laminectomy (removal of the dorsal laminae of the vertebrae) from T1 through C5 revealed complete avulsion of both the dorsal and ventral roots. After identifying the former dorsal root entry zone, the surgeon inserted an electrocautery needle about 3 to 4 mm into the right spinal cord at an angle of about 20°. After passing a brief electrical current sufficient to cauterize a sphere of tissue about 1 to 2 mm in diameter, the needle was removed. This process was repeated 15 times between C5 and C7.

Immediately on recovery, Mr. H. D. had, for the first time in 20 years, significant (subjectively 50% reduction) relief from the pain localized to his right arm. In addition he had some slight weakness in his right leg. There was normal response to pin prick from all parts of his body, but he had a decreased appreciation of light touch (cotton wisp) and vibration from his right leg. The vibration stimuli he sometimes reported as a burning sensation. These difficulties cleared in 2 weeks.

Comment: Dorsal root avulsion (ripping the roots out of the cord) frequently results in intractable phantom pain because, as most authorities believe, normal afferent stimulation continuously suppresses thalamic projection cells in lamina V. The removal of afferent stimuli, especially from the large-diameter axons, removes this inhibition and the projection neurons become spontaneously active. This activity is interpreted by higher centers as pain distributed over the phantom limb. Avulsion is more likely to produce phantom pain, according to this theory, than simple amputation because the primary afferent neuron is totally destroyed, thus releasing central mechanisms from tonic inhibition. Simple amputation leaves the dorsal root and the dorsal root ganglion intact. It is possible in such cases for the cell body in the dorsal root ganglion to survive, and, if it does, the central connection of the primary afferent may be maintained. Consequently its inhibitory function may be maintained.

FURTHER APPLICATIONS

- Explain why amputation of the arm could not alleviate chronic phantom pain in this case.

- Explain why one could have predicted that dorsal column stimulation would be ineffective in this case. Would TENS be more likely to be successful? Explain.

- What is the rationale for coagulating the dorsal horn to alleviate chronic phantom pain? How

does this procedure differ from ALS tractotomy?

- Would you expect the procedure described
here to provide more permanent relief from
chronic phantom pain than ALS tractotomy?
Explain.

Neurosurgery produces edema in the structures surrounding damaged neural tissue. The
edema dissipates with time.

- Make a sketch of the dorsal horn and fill in the
area of the surgical lesion. Label the structures
adjacent to the lesion and try to predict the possible complications of this type of surgery if the
lesion is not placed accurately.

- Explain the postoperative sensory findings in
this case. Which findings are the direct result
of the coagulation of neural tissue and which
are secondary to edema?

SUMMARY

There are two CNS systems that carry somatosensory information from the periphery to
the cerebral cortex: the dorsal column system and
the anterolateral system (ALS). Both systems have
a basic three-neuron structure: (a) the primary afferent enters the CNS over a dorsal root or cranial
nerve and synapses ipsilateral to its point of entry;
(b) the secondary neuron receives synaptic input
from the primary afferent, crosses the midline, and
terminates in the thalamus; (c) the third-order neuron leaves the thalamus, traverses the internal
capsule, and terminates in a primary sensory cortex, ipsilateral to its thalamic origin but contralateral to the source of the sensory information it carries. Both of these systems are described in detail
in the main text; only the principal pathways are
summarized below.

Dorsal column system: Primary axons of this
system convey fine tactile, proprioception, and vibratory sensations. They are large-diameter Aα or
Aβ fibers that ascend in either the fasciculus gracilis or the fasciculus cuneatus. On reaching the
medulla, they synapse in the nucleus gracilis or
nucleus cuneatus, depending on their origin. Secondary neurons from these nuclei send axons that
cross the midline as internal arcuate fibers and ascend the brainstem as the medial lemniscus. They
terminate primarily in the thalamic nuclei (VPL
and VPS). Axons carrying fine tactile and proprioceptive information from the face enter the CNS
over the trigeminal nerve and synapse in the chief
sensory nucleus of V and the rostral two-thirds of

the spinal nucleus of V. Secondary axons from
these nuclei cross the midline, join the medial
lemniscus, and terminate in the VPM and VPS nuclei. From the thalamus, third-order axons pass
through the posterior limb of the internal capsule
to terminate in Brodmann's areas 3a, 3b, 1, and 2,
also known as the primary somatosensory cortex
or S-I.

Anterolateral system: Primary axons of this system convey nociceptive and thermal sensations.
They are small-diameter Aδ and C fibers that enter the spinal cord and synapse in the dorsal horn.
Secondary neurons arise from several areas of the
dorsal horn, cross the midline in the ventral white
commissure, and ascend to the thalamus as the
ALS. They synapse in the VPL, VPI, and intralaminar nuclei. From the face, similar small-diameter
primary afferents enter the CNS over the trigeminal nerve and synapse in the caudal two-thirds of
the spinal nucleus of V. Secondary axons cross the
midline and join the medial lemniscus and synapse
in the thalamic VPM and VPI nuclei and the intralaminar nuclei. From the thalamus, nociceptive
and thermal information reaches S-I via third-order neurons that traverse the internal capsule.

A precise somatotopic relationship is maintained at all levels for both of these systems. In the
cerebral cortex, multiple somatotopic maps exist,
organized by modality, pain being conspicuously
absent. These separately mapped areas are connected with one another in a series of cascades by
short cortical association axons. Cortical neurons
respond to more and more specific stimuli as one
follows these cascades away from the principal
thalamic synapses in the cortex. The more remote
cortical neurons respond best to stimuli that are
abstracted from the primary modality.

The manner in which the nervous system ab-
stracts the perception of pain from nociceptive
and tactile information is not known. Pain can be
present, even in the absence of objective nociceptive stimuli (phantom pain), possibly because
transmission neurons in the spinal cord or thalamus become spontaneously active in the absence
of normal peripheral stimuli. Surgical ablations at
the level of the spinal cord, thalamus, and cerebral cortex have not proved to be of long-term
benefit to the patient.

Pain can be modulated by descending systems,
the most prominent originating in the PAG and
synapsing in the medullary reticular formation
(NRM and NRPG). These reticular nuclei send axons into the spinal cord in the dorsolateral funiculus, where they can suppress thalamic transmission neurons through both opioid and non-opioid

polysynaptic pathways. The gate control theory offers a hypothetical model to explain how activity on large-diameter afferents is able to suppress or moderate painful stimuli.

Dermatomes represent the pattern of distribution of spinal nerves on the skin. Peripheral nerves, being composed of some of the axons from one or more spinal nerves, project patterns that are fragments of several adjacent dermatomes. Recognizing the patterns of sensory losses following lesions to spinal and peripheral nerves is essential to neurology. Understanding the dermatomes also helps to explain referred pain, the localization of painful visceral stimuli to the body surface. Visceral pain is localized to the dermatomes that correspond with the embryonic origin of the organ, not to its adult position in the body.

FOR FURTHER READING

Adams, R., and Victor, M. *Principles of Neurology.* New York: McGraw-Hill, 1993.

Barr, M., and Kiernan, J. *The Human Nervous System.* Philadelphia: J. B. Lippincott, 1993.

Brodal, A. *Neurological Anatomy In Relation to Clinical Medicine.* New York: Oxford University Press, 1981.

Brown, A. G. *Organization in the Spinal Cord: The Anatomy and Physiology of Identified Neurons.* Berlin: Springer-Verlag, 1981.

Carpenter, M. B. *Core Text of Neuroanatomy.* 4th edition. Baltimore: Williams & Wilkins, 1991.

Greer, M. Arnold-Chiari Malformation. In: *Merritt's Textbook of Neurology,* Ch. 27, edited by Rowland, L. P. Philadelphia: Lea & Febiger, 1989, pp. 480–483.

Haines, D. *Neuroanatomy: An Atlas of Structures, Sections, and Systems.* Baltimore: Urban & Schwarzenberg, 1991.

Jessell, T., and Kelly, D. Pain and Analgesia. In: *Principles of Neural Science,* Ch. 27, edited by Kandel, E., Schwartz, J., and Jessell, T. New York: Elsevier, 1991, pp. 385–399.

Mancall, E. Syringomyelia. In: *Merritt's Textbook of Neurology,* Ch. 124, edited by Rowland, L. P. Philadelphia: Lea & Febiger, 1989, pp. 687–691.

Martin, J. H., and Jessell, T. Anatomy of the Somatic Sensory System. In: *Principles of Neural Science,* Ch. 25, edited by Kandel, E., Schwartz, J., and Jessell, T. New York: Elsevier, 1991, pp. 353–366.

Rowland, L. *Merritt's Textbook of Neurology.* Philadelphia: Lea & Febiger, 1989.

Sacks, O. Phantoms. In: *The Man Who Mistook His Wife for a Hat,* Ch. 6. New York: Harper & Row, 1987, pp. 66–70.

6 Spinal Mechanisms of Motor Control

Locomotion is one of the most important attributes of living things because it liberates an organism from some of the restraints imposed on it by its immediate environment. The constant improvement in motor skills is one of the hallmarks of evolution, and the development of the nervous system accompanies increased motor performance. In mammals, more than half of the nervous system is directly devoted to motor performance. Consequently, damage to the nervous system is almost always accompanied by a degradation of motor skills. Therefore, assessment of the motor system is central to the examination of the nervous system by the physician.

Motor control mechanisms originated as very simple reflexes that regulate muscles acting at one joint. As animals became more complex, mechanisms evolved to coordinate muscles acting across multiple joints within an extremity and between extremities. Eventually an elaborate command center evolved to detach the initiation of motor acts from reflexes triggered by environmental cues.

This evolutionary history is reflected in the anatomical and physiological organization of the mammalian nervous system. The spinal cord and brainstem contain the circuitry that implements reflex coordination around joints and between extremities. The basal ganglia and cerebral cortex are the major command centers, while the cerebellum refines all motor acts by coordinating the timing among the individual motor groups. This chapter presents the fundamental regulatory mechanisms that are found in the spinal cord—mechanisms that affect the motor neurons directly. Subsequent chapters will present higher-order systems that are more abstractly involved in initiating and regulating motor activity.

The Muscle Sensory Receptors

To move an extremity accurately from one place to another, the motor control system must first have information about the current position and orientation of that extremity and the length of the muscles that control it. As a motor act progresses, it is also necessary to have continuous information about the state of the muscles, the position of the extremity, and the velocity with which its relationship to the target is changing. Much of this information is made available to the nervous system by specialized receptors that are located in the muscle, its tendons, and the joints. There are two types of muscle sensory receptors, muscle spindles and Golgi tendon organs. These have been studied extensively, and consequently their role in the regulation of motor activities is reasonably well understood. The joint receptors have not been so extensively studied; consequently there is little information concerning their physiological role in motor control.

THE MUSCLE SPINDLE

The principal specialized sensory receptor in muscle is a complex encapsulated receptor called the **muscle spindle apparatus**. It contains both sensory and motor elements. The capsule is shaped like the spindle of an old spinning wheel, which gives it its name. The spindle apparatus is small, consisting of two to 14 specialized muscle fibers called **intrafusal muscle fibers** [L. *fusus,* a spindle] (meaning inside the spindle apparatus) that are enclosed within a connective tissue capsule (Fig. 6.1). Surrounding the muscle spindle apparatus are the main contractile muscle fibers, the **extrafusal muscle fibers**. Every striated muscle contains large numbers of muscle spindles.

There are two kinds of intrafusal fibers, named after their histological appearance. One type, called the **nuclear bag**, is composed of a modified multinucleated muscle fiber whose central,

sensory portion forms a noncontractile, bag-like structure. There are two types of bag fibers, **dynamic** and **static**, that differ from each other in their response to muscle stretch. The second intrafusal fiber type is called the **nuclear chain** because its central portion contains a number of nuclei arranged in a chain-like series along the length of the fiber. The distal portion of both types of intrafusal fibers is striated muscle that retains its contractile properties. *Since the contractile portion of the muscle spindle is small, contraction of the intrafusal muscle fibers cannot cause muscle movement.* A typical muscle

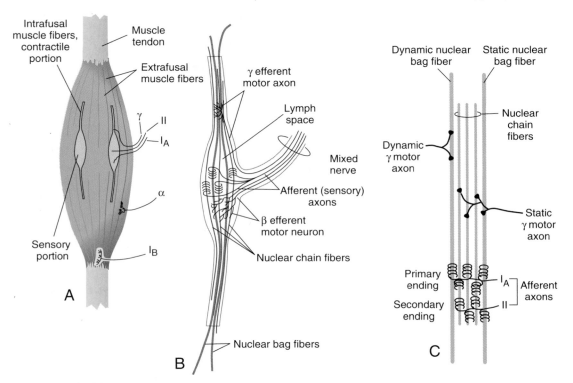

FIGURE 6.1.

A, The muscle spindle apparatus is arranged in parallel with the main contractile elements (extrafusal fibers) of the muscle. The Golgi tendon organs are arranged in series. This arrangement is shown here. The sensory elements are exaggerated in size for clarity. **B**, Drawing of a cat muscle spindle. The receptor is innervated by a mixed nerve, one containing both sensory and motor fibers. Note how the afferent (sensory) axons branch before terminating. They innervate more than one sensory fiber. The efferent (motor) axons are of the γ or β type. **C**, This schematic drawing of a muscle spindle apparatus illustrates the typical arrangement of afferent and efferent axons. The I$_A$ axon innervates both the dynamic and static bag fibers as well as the chain fibers. The group II axon does not innervate the dynamic bag fibers. Dynamic γ-efferent axons innervate only the contractile portion of dynamic bag fibers. Static γ-efferent axons innervate the contractile portion of all the intrafusal fibers except the dynamic bag fibers. The exact innervation of the β fibers (not shown) is not known.

spindle apparatus contains two bag fibers, one of each type, and about five chain fibers.

The spindle apparatus is innervated by two kinds of afferent axons (see Table 6.1) that make contact with the central, receptor portion of the intrafusal fibers (Fig. 6.1). **Group I_A** myelinated axons (12 to 20 μm) innervate the sensory portion of bag and chain fibers. They spiral around the central part of the intrafusal fibers before leaving the capsule. A single I_A axon will have branches that arise from all of the intrafusal fibers of a muscle spindle. The terminals of the I_A axons are called the **primary** or **annulospiral** receptors. Smaller, **group II** myelinated axons

(4 to 12 μm) innervate static bag fibers and all of the chain fibers. None innervate dynamic bag fibers. The terminals of group II axons are called **secondary** or **flower spray** receptors.

The efferent innervation of a muscle arises from three sets of motor neurons (see Table 6.2). The first set, known as **γ motor neurons**, innervates only the intrafusal muscle fibers and for this reason are often called **fusimotor neurons**. The γ motor neurons have small (3 to 8 μm) myelinated axons that only innervate the contractile portion of intrafusal muscle fibers. These fusimotor neurons can be further categorized on the basis of their function into **dynamic** and **sta-**

Table 6.1. Classification of Muscle Afferents

Receptor	Axon	Size (μm)	Conduction Velocity (m/sec)
Muscle spindle; primary	I_A	12–20	72–120
Muscle spindle; secondary	II	4–12	24–72
GTO	I_B	12–18	72–108

Table 6.2. Classification of Muscle Efferents

Motor Neuron	Target	Size (μm)	Conduction Velocity (m/sec)
α Motor neuron	Extrafusal muscle	8–13	44–78
γ Motor neuron; dynamic	Intrafusal muscle; dynamic bag, static bag, and flower spray	3–8	18–48
γ Motor neuron; static	Intrafusal muscle; static bag and flower spray	3–8	18–48

tic fibers. Dynamic γ motor axons innervate only the dynamic bag intrafusal fiber, while the static γ motor axons innervate the static bag and all the chain fibers, but they do not innervate the dynamic bag fibers. *Contraction of the intrafusal muscle fibers can only place tension on the central, sensory portion of the intrafusal muscle fiber, it cannot cause movement.*

The second set of efferent axons, the **α motor neurons**, have large myelinated axons (8 to 13 μm) that innervate only extrafusal muscle fibers. They are able to initiate movement through extrafusal muscle contraction. The third set of efferent axons are a hybrid class of axons, the **β motor neurons**, that innervate both the intrafusal and the extrafusal muscles.[1]

The main mass of the muscle consists of contractile muscle fibers known as the extrafusal muscle fibers. The muscle spindles lie within the extrafusal muscle mass. The intrafusal muscle fibers have tendons that merge with the tendons or fascia of the adjacent main contractile elements of the muscle (Fig. 6.1). This creates a physical arrangement whereby the entire spindle apparatus lies *in parallel* with the extrafusal muscle fibers. In contrast, the receptors of the nuclear bag and chain are *in series* with the intrafusal muscle fibers.

There are two ways of stimulating the muscle spindle receptors: first by lengthening the whole muscle, and second by stimulating the contractile portion of the intrafusal muscle. Appreciating this arrangement is critical to understanding how this complex sensory receptor works.

The sensory portion of the intrafusal muscle fibers is sensitive to tension. Stretching the receptor hypopolarizes it and initiates action potentials in the group I_A and II axons. Tension can be applied to the receptor by *lengthening* the intrafusal muscle, because the intrafusal muscle and the receptor lie in series. Much like stretching a rubber band, lengthening the intrafusal fiber will cause tension to increase throughout its length, including the central, sensory portion. Physiologically, the intrafusal muscle, with its receptor, is lengthened by *stretching the entire muscle (intrafusal*

and extrafusal), since the extrafusal and intrafusal fibers are arranged in parallel.

Tension can also be applied to the sensory portion of the muscle spindle apparatus by *stimulating the contractile portion of the intrafusal muscle.* Since the intrafusal muscle is not strong enough to cause movement of the entire muscle mass, the tension of intrafusal muscle contraction cannot be relieved by joint movement. The tension will be directly transferred to the receptor, hypopolarizing it.

THE GOLGI TENDON ORGANS

Contrary to its name, the **Golgi tendon organ** (GTO) is rarely found in tendons. Rather it is more consistently located within the aponeurotic sheaths of muscle attachments. Golgi tendon organs are encapsulated sensory receptors consisting of collagen fibrils intertwined with the naked nerve ending from large-diameter **Group I_B** (12 to 18 μm) myelinated axons (Fig. 6.2). Tension applied to the collagen fibrils apparently squeezes and distorts the bare axon membrane, hypopolarizing it and triggering action potentials on the I_B axon.

The Golgi tendon organs are arranged *in series* with the extrafusal muscle fibers, an arrangement that ensures that *tension* on the whole muscle will be transferred to the GTO. About ten extrafusal muscle fibers are attached to a single GTO capsule. The arrangement ensures that tension from only a few motor units will be applied to an individual GTO. The converse is also true; individual muscle fibers of a single motor unit are distributed among several GTO receptors. The CNS connections that take advantage of this arrangement will be described shortly.

Physiological Responses of Muscle Receptors

Muscle sensory receptors are able to sense the steady-state length of the muscle as well as dynamic changes in muscle length, tension, and the velocity with which muscle length changes. The static (steady-state length) and dynamic (velocity) properties of the muscle receptors can be studied in isolated nerve-muscle preparations in

---

[1]While all vertebrates have muscle spindles, only in mammals is the efferent innervation of the intrafusal and extrafusal muscles separated. In nonmammals, all motor neurons are of the β class. Within mammals, β innervation is retained, but it becomes less dominant in primates, including humans.

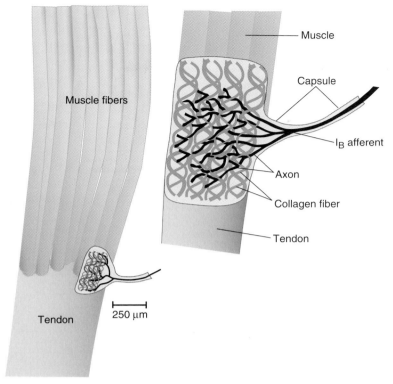

FIGURE 6.2.

The Golgi tendon organ is a specialized encapsulated ending. The collagen strands of the aponeurosis wrap around the bare ending of the I_B axon. Tension on the muscle is transferred to the collagen strands that squeeze the bare axon, hypopolarizing it.

which all connections of the muscle nerves with the CNS have been severed. Under these isolated conditions, stretching (lengthening) the entire muscle mass or stimulating either the α or γ motor neurons reveals important differences in the response properties of the primary and secondary receptors (Fig. 6.3A). The isolated response properties of muscle sensory receptors will be discussed briefly so that one can appreciate how the integrated system participates in motor control.

MUSCLE STRETCH

When a muscle at one length is stretched and held steadily at a different length (linear stretch), both the group I_A (primary receptor) and group II (secondary receptor) axons will fire at an increased rate that is proportional to the new length of the muscle. This is a static property of the receptors because their response remains more or less constant as long as the muscle length is held constant. The group II axon that arises from static bag and chain fibers displays only static responses. The group I_A axon, however, in addition to its static properties, also displays dynamic properties since it responds with a burst of activity during the stretch itself (Fig. 6.3B).

The dynamic properties of the primary receptor are best revealed during small-amplitude sinusoidal stretches of the muscle (Fig. 6.4). During sinusoidal stretching, one can observe that the primary axon fires only when the muscle is lengthened, and it ceases to fire when the muscle is allowed to shorten by passive elastic relaxation. During sinusoidal stretching there is no steady state condition. Consequently the secondary axon does not appreciably change its rate of firing as long as the amplitude of the sinusoid is small.

Under the experimental conditions just described, the increase in overall tension on the muscle is very slight and the change in tension is distributed over all of the individual muscle fas-

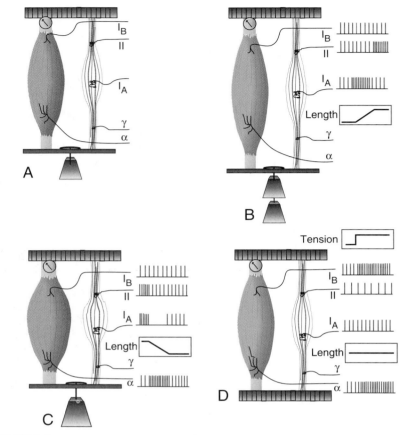

FIGURE 6.3.

　A, A typical muscle is represented here schematically, with the muscle spindle apparatus separated from the extrafusal muscle to emphasize their parallel relationship. The weight symbol is used to indicate a constant steady resting load on the muscle. The thick horizontal bars symbolize a fixed, reference platform. Action potentials from the various axons are indicated to the right. Muscle length and tension are also indicated where appropriate. **B**, Linear stretch. The extra weight symbol indicates that the muscle has been stretched from its initial state shown in **A**, and is under slightly greater tension. **C**, Isotonic contraction. Stimulation of the α motor neuron causes the muscle to contract. Note particularly that the tension on the muscle is the same as in **A**. **D**, Isometric contraction. The weight at the bottom is replaced with a fixed platform so that the muscle cannot shorten during α motor neuron stimulation. See text for details concerning the responses of the various receptors.

cicles. Therefore, so little tension is transferred to the Golgi tendon organs that they are not stimulated enough to cause the I_B axons to increase their firing rate.

ISOTONIC CONTRACTION

　With an isolated muscle preparation, one can selectively stimulate only the α motor neurons to generate an **isotonic** contraction (constant load) of the extrafusal muscle, leaving the γ motor neurons and their associated intrafusal muscle fibers unstimulated. Under this artificial condition, when the whole muscle mass shortens, the muscle spindle apparatus will passively shorten due to its intrinsic elastic properties (Fig. 6.3C). As the intrafusal fibers in the spindle shorten, the firing rates of both the group I_A and II axons

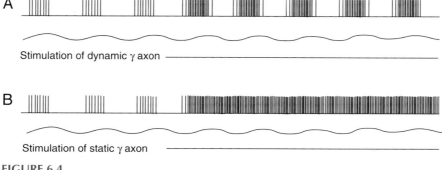

FIGURE 6.4.

The effects of stimulation the dynamic (**A**) and the static (**B**) γ axons is shown in this recording from a I_A axon, from an actual experiment. The muscle was stretched with a sinusoidal pattern (1 mm peak to peak). The horizontal bars indicate stimulation of the γ-efferent axons. Note that stimulating the dynamic γ axon (**A**) increases the rate of I_A firing only during the stretching phase of the sinusoid. In sharp contrast to this, stimulation of the static γ-efferent axon (**B**) increases the overall firing rate of the I_A axon. Even though the axon now fires during the passive, shortening phase of the stimulus, the rate of firing during the lengthening phase is not as high as when the dynamic γ axon was stimulated.

are *reduced* because tension in the central, receptor portion of the intrafusal fibers is relieved.

During an isotonic contraction, the overall tension on the extrafusal muscle is unchanged because the load remains constant. Therefore, during the contraction the Golgi tendon organs are not subjected to any changes in tension, so the firing rates observed on the I_B axons remain constant.

ISOMETRIC CONTRACTION

If an enthusiastic weight lifter attempts to lift a load that he cannot move, tension will increase in the muscles during the contraction, but the muscles will not shorten. This is an **isometric** contraction. The length of the intrafusal muscle fibers and the tension on their receptors will remain unchanged. Consequently, there will be no change in the firing rate observed on the I_A and II axons (Fig. 6.3*D*). The Golgi tendon organs, on the other hand, will have all of the tension transferred to them since they are in series with the extrafusal muscle fibers. If the tension is intense enough, one will be able to observe an increase in the firing rate of the I_B axon.

FUSIMOTOR STIMULATION

The central nervous system is able to adjust the sensitivity of the primary and secondary receptors. This is accomplished by stimulating γ motor axons, causing the intrafusal muscles to contract. The dynamic and static gamma systems are independently regulated.

Stimulation of γ motor neurons causes the contractile elements of the intrafusal muscle fibers to shorten. Since these tiny muscles are too weak to cause the whole muscle mass to contract, *gamma stimulation, in effect, is an isometric contraction of the intrafusal muscle.* Tension rapidly rises on the central, sensory element of the fiber, hypopolarizing it and triggering action potentials on the group I_A and II sensory axons.

Stimulation of the dynamic γ fibers enhances the response observed on the I_A axon only during muscle lengthening. The static properties of the I_A and II axons are unchanged. This is best observed during low-amplitude sinusoidal stretching of an isolated muscle (Fig. 6.4). In contrast to this, stimulation of the static γ motor neurons enhances the response observed on the group II axons, in effect lowering their threshold. The overall firing rates of the axons are increased. In other words, stimulating static γ motor neurons is like turning up the volume of a phonograph; all of the static and dynamic responses are still present, only at a greater intensity.

The Reflex Pathways

A number of local circuits within the spinal cord regulate the coordinated action of a muscle when the muscle is connected to the spinal cord through the dorsal and ventral roots. The simplest of these connections are known as **reflexes**. A reflex has five components: (*a*) a sensory receptor, (*b*) an afferent path to the CNS, (*c*) one or more synapses within the CNS[2], (*d*) an efferent path, and finally (*e*) an effector (Fig. 6.5). There are several clinically relevant reflexes involving muscle sensors. Their salient features will be presented here, although their full clinical significance cannot be fully developed until more of the motor system has been presented.

---

[2]To avoid labeling all CNS activity as a reflex, one usually limits the number of CNS synapses to involve not more than two or three neurons. This restriction is not universally accepted, however, and some authors consider complex behavior, involving perhaps thousands of synaptic links, a "reflex."

THE MUSCLE STRETCH REFLEX

As the name implies, the **muscle stretch reflex (MSR)** is invoked by rapidly stretching a muscle. This stimulates, primarily, the dynamic nuclear bag fibers in the muscle spindle, causing an increase in the firing rate of the I_A and group II axons. I_A afferent axons enter the spinal cord over the dorsal root and synapse with three groups of neurons: motor neurons in lamina IX, interneurons in lamina VII, and interneurons in laminae V and VI. Group II afferents synapse with two groups of spinal neurons, motor neurons in lamina IX and interneurons in laminae IV through VI (Table 6.3).

Group I Afferents

The first group of neurons with which afferents from primary muscle spindle receptors (I_A) synapse is the α motor neurons in lamina IX of the ventral horn (Fig. 6.6*A*). A single I_A axon forms synapses with about 80% of the motor neurons innervating the **homonymous** [G. *homonymous,* same name] muscle (the same muscle) from which it originated. These synapses produce very

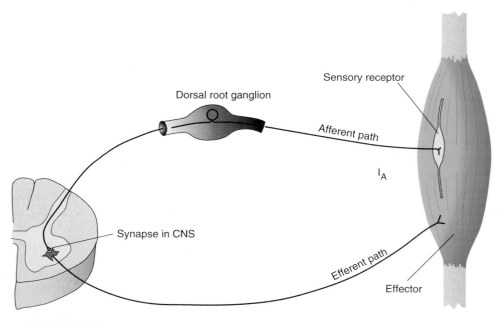

FIGURE 6.5.

The five elements of a reflex are illustrated here using the components of the muscle stretch reflex (MSR).

Table 6.3. Termination of Muscle and Cutaneous Axons in the Spinal Cord

Afferent Axon	Laminae of Termination	Target Neuron Pool
MUSCLE AFFERENTS		
I_A	IX	Motor neurons
	VII	Inhibitory interneurons
	V–VI	Mixed interneuron pools
II	IX	Motor neurons
	IV–VI	Mixed interneuron pools
I_B	V–VII	Inhibitory interneurons
CUTANEOUS AFFERENTS		
FRAs (II–IV)	III–VII	Mixed interneuron pools

large (about 100 μV) excitatory postsynaptic potentials (EPSPs) that strongly facilitate the neuron.

The I_A axon also synapses on α motor neurons that innervate **heteronymous** [G. *heteronymous,* different name] **synergistic** muscles (muscles acting on the same joint with a similar action). They facilitate about 60% of the synergistic motor neurons and produce somewhat weaker EPSPs (about 70 μV) than those produced on homonymous motor neurons.

The α motor neurons, if facilitated above threshold, fire, sending action potentials back to the homonymous and synergistic muscles, causing them to contract. Thus *muscle stretch causes a reflex muscle contraction to oppose the stretch.* The connections that constitute this reflex pathway involve a single synapse, i.e., the I_A afferent on the α motor neuron, and for this reason the muscle stretch reflex may also be called the **monosynaptic reflex.**

Primary muscle spindle afferents form synapses with a second group of neurons. These are excitatory synapses on interneurons in lamina VII that in turn make inhibitory synapses on α motor neurons. These connections are also quite specific. The α motor neurons that are inhibited by this pathway all serve **homonymous antagonist** muscles—muscles having opposite action on the same joint from which the initial stretch originated. By hyperpolarizing antagonist α motor neurons, the number of action potentials reaching the muscle is decreased, lessening the tension on the antagonist muscle.

The third group of neurons with which I_A axons synapse in the spinal cord are found in laminae V and VI of the dorsal horn. In these laminae I_A afferent axons produce EPSPs on a different set of interneurons that are involved in a wide variety of motor functions. These interneurons receive additional sensory input from other sources that will be discussed below. They also receive motor commands from supraspinal sources that will be discussed in Chapter 7. The interneurons of laminae V and VI are a heterogeneous group. Some synapse on other interneurons, others synapse on motor neurons. Not all of the connec-

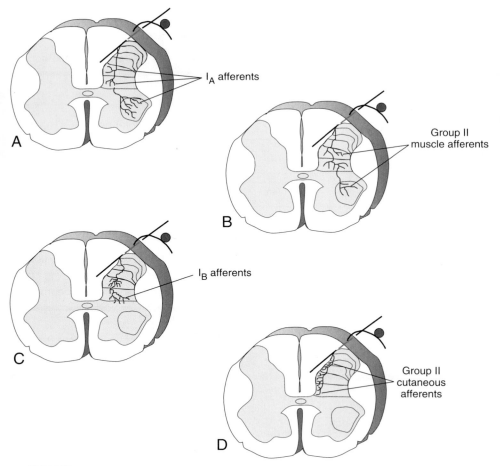

FIGURE 6.6.

Terminal fields of principal sensory neurons in the spinal cord. Although only one slice through the spinal cord is shown here for each axon group, the distribution of terminal fields extends a considerable distance along the spinal cord (see Chapter 5).

tions of these interneurons are known. Because of the diversity of inputs, the interneurons of laminae V and VI form the major motor integrating center of the spinal cord.

Group II Afferents

Group II afferents from secondary muscle spindle receptors also synapse in the spinal cord and influence α motor neuron activity (Fig. 6.6B). Since the secondary receptors are poorly stimulated by rapid stretch, their influence is minimal during the muscle stretch reflex, but their static characteristics help to determine the overall level of excitability of the motor neuron pool.

Group II axons terminate primarily in two areas in the spinal cord. Like the group I_A axons, group II axons synapse monosynaptically on α motor neurons in lamina IX. These synapses are less potent than those from I_A afferents, producing EPSPs with an amplitude of about 24 μV. Their distribution is also less pervasive; they synapse on only about 50% of the homonymous and 20% of the synergist motor neurons.

Group II axons also terminate in laminae IV through VI and, like the group I_A axons, excite interneurons. There are two parallel sets of interneurons; one set excites flexor motor neurons and inhibits extensors, while the other set inhibits flexors and excites extensors. These two parallel pathways are under supraspinal regulation such that the effect of group II activity can facilitate either flexors or extensors, depending on the nature of the supraspinal signals. That is, descending systems can facilitate one of the parallel interneuronal pathways in preference to the other and thereby regulate the effect of group II excitation. Under normal physiological conditions, the system is biased in favor of flexor ex-

citation by group II afferents. But supraspinal activity can reverse the effect (Fig. 6.7).

Group I_B Afferents

Primary afferent axons from Golgi tendon organs, group I_B, terminate in the spinal cord in laminae V through VII (Fig. 6.6C). There they make excitatory synapses on interneurons that subsequently inhibit homonymous and syner-

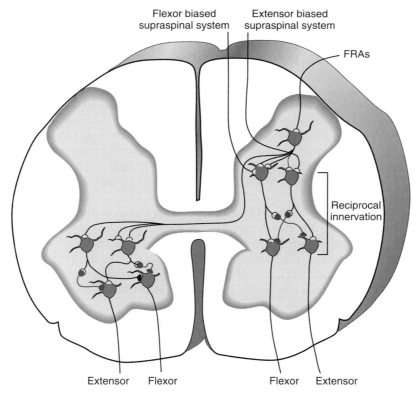

FIGURE 6.7.

Reciprocal innervation is implemented by the anatomical arrangement of interneurons in the spinal cord. Depicted here are sets of interneurons that, when activated, produce opposite effects on motor neuron pools innervating antagonist muscle groups. Flexion reflex afferents (FRAs) initiate complex polysynaptic reflexes in the spinal cord. The simplest connections shown here illustrate the basic pattern. FRAs facilitate ipsilateral flexor motor neuron pools and inhibit extensor motor neuron pools. They simultaneously produce the opposite effect on the contralateral side. The effects of FRA stimulation are widespread, involving synergistic muscle groups in an entire limb and their contralateral counterpart. The normal flexion response seen after FRA stimulation can be reversed by supraspinal systems. This has led to the hypothesis that FRA stimulation facilitates parallel pathways in the spinal cord that can potentially facilitate either flexor or extensor reflexes. This is illustrated schematically here by the use of an interneuron. The extra synapses on one set of interneurons simply indicates that the system is normally biased in favor of the flexor response.

gistic α motor neurons while facilitating antagonists. Like the I_A axon, the I_B exerts its primary influence on motor neurons that affect movement around the same joint. However, the I_B also exerts its influence over a wider area, affecting, to some degree, most of the motor neurons innervating muscles in the entire limb.

Under ordinary physiological conditions, the Golgi tendon organs play an important role in adjusting tension among the various motor units within a muscle. This is accomplished by inhibiting motor units that carry a disproportionate amount of the load. Only a limited number of muscle fibers, innervated by separate motor units, are attached to an individual GTO capsule, and each GTO measures the tension developed only by those motor units. The I_B afferent axon originating in a GTO primarily inhibits the α motor neurons that innervate the same motor units attached to the GTO capsule. When contraction of a muscle is initiated by a generalized facilitation of the motor neuron pool innervating that muscle, certain motor units within that pool will inevitably carry more of the tension than others. However, since the I_B inhibition will most affect the motor units that are under the most tension, those motor units will relax slightly. Therefore, the primary physiologic role of I_B inhibition is to regulate α motor neuron excitability to evenly distribute tension among motor units.

THE CLASP KNIFE REFLEX

The **clasp knife reflex** is evoked by attempting to stretch a muscle during an isometric contraction. The attempted stretch rapidly and dramatically increases tension in the muscle, eliciting a massive inhibition of the homonymous and synergistic α motor neurons. The initiation of the inhibition is triggered suddenly, at high tension, causing the limb to rapidly collapse, much like closing the blade on a jack-knife, hence the name "clasp knife" reflex. Power weight lifters sometimes experience this reflex if they try to lift too heavy a load. During the lift, muscle tension increases rapidly. If the tension is too great, partway through the lift the muscles will suddenly lose power for what appears to be no reason and the bewildered weight lifter will drop the load.

The spinal pathways responsible for this reflex are not known. One hypothesis suggests that the Golgi tendon organs sense the tension on the muscle and inhibit the α motor neurons via the I_B inhibitory interneuron. The principal difficulty with this hypothesis is that Golgi tendon organs characteristically respond to low tension on individual motor units. They do not have the high threshold required to produce the threshold effect characteristic of the clasp knife reflex. The high-threshold trigger could be a function of spinal interneurons, but such interneurons have not been identified. Another hypothesis suggests that the group II afferents, through inhibitory interneuron pathways, are responsible for the reflex. It is further suggested that group III afferents from fascia and joint receptors may participate. Again, this hypothesis does not explain the threshold phenomenon.

RECURRENT INHIBITION

The axons of α motor neurons not only innervate muscle, they also send recurrent collateral branches into the spinal gray matter, where they synapse on a special class of interneurons called **Renshaw cells** (Fig. 6.8). Renshaw cells are inhibitory interneurons. They synapse on

FIGURE 6.8.

Renshaw cells are inhibitory interneurons that are excited primarily by recurrent collaterals from α motor neurons. They reciprocally innervate pools of α motor neurons, disfacilitating the antagonist pools. Renshaw cells are also under supraspinal regulation. Recurrent inhibition (Renshaw inhibition) is a mechanism that reduces the size of a facilitated neuron pool and ensures that only the most facilitated cells will remain active. This mechanism focuses motor control in exactly the same manner that lateral inhibition focuses the effects of sensory stimuli. In the schematic example given here, the inhibitory levels are set to −1 for each synapse. The excitatory levels are as indicated.

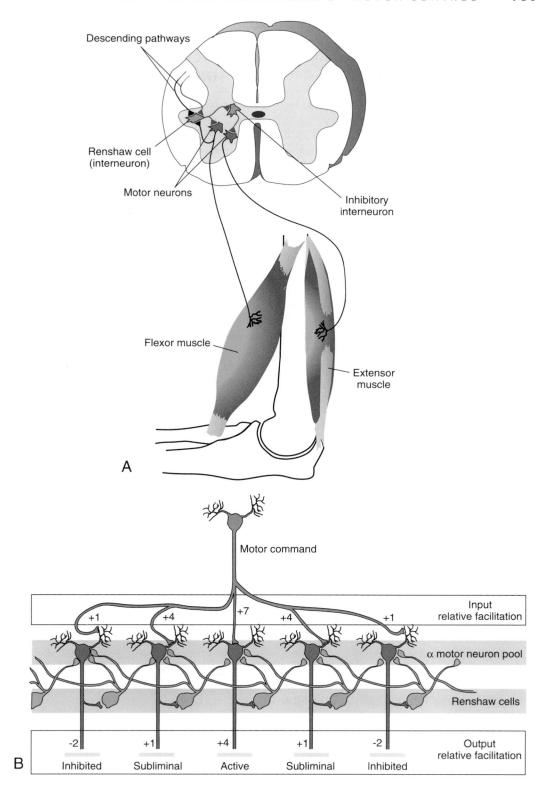

homonymous and synergistic α motor neurons and on interneurons that are inhibitory to antagonist α motor neurons. Thus Renshaw cells inhibit homonymous and synergistic motor neurons and *disinhibit* antagonists. Inhibition of motor neurons by Renshaw cells is called **recurrent inhibition**.

Recurrent inhibition is best understood as a type of lateral inhibition (see Chapter 5). Just as lateral inhibition serves to focus sensory activity and to increase the contrast boundaries between adjacent stimuli, recurrent inhibition via the Renshaw cell focuses motor activity to the most facilitated motor neurons. In this way, broad facilitatory and inhibitory patterns that are established by interneurons in laminae IV through VI on α motor neuron pools are refined by the activity of the motor neurons themselves, acting through the Renshaw cell. Motor neurons marginally excited above threshold will be rapidly inhibited to just below threshold. Neurons that are facilitated to a point just below threshold are said to be in the **subliminal fringe** because they can be rapidly brought to threshold by minimal facilitation. They are very much like a cocked gun with a hair trigger.

THE FRA REFLEX

Motor neurons are influenced by afferent axons that arise from cutaneous sensors as well as from the muscle sensors discussed above. The cutaneous afferents consist of axons belonging to axon groups II, III, and IV (Fig. 6.6*D*). They carry sensations from nociceptors, tactile receptors, and joint receptors. Group II cutaneous afferents[3] terminate in laminae III through VII, where they make excitatory synapses on interneurons. Normally, stimulation of group II, III, and IV cutaneous afferents, acting through interneurons, facilitates flexor motor neurons innervating the ipsilateral extremity and inhibits the extensor antagonists (Fig. 6.7). For this reason, they are commonly referred to as **flexion reflex afferents** or **FRAs**. The effects of FRA stimulation are distributed over the entire limb.

The basic flexion reflex pattern can be reversed by the influence of supraspinal systems. This is similar to the ability of supraspinal systems to reverse the normal flexor-biased action of group II muscle afferents. (Fig. 6.7). Therefore, it is hypothesized that FRA and group II muscle afferents share the same intraneuronal pathways in the spinal cord.

In the intact animal (or human), stimulation of FRAs produces a generalized flexion of the entire limb, even though only a small area of skin may have been stimulated. For example, stepping on the sharp point of a thumbtack with bare feet causes one to reflexively withdraw the foot by flexing all the joints in the leg. The interneurons that mediate this FRA reflex also send collateral axons across the midline of the spinal cord, where they facilitate additional interneurons that ultimately facilitate extensor motor neurons and inhibit flexors (Fig. 6.7). This makes sense if one considers our hapless barefoot wanderer. When the leg is reflexively withdrawn to remove the foot from the tack, the weight is transferred to the contralateral leg. A reflex facilitation of extensors contralateral to the noxious stimulus will obviously be desirable to avoid having that leg collapse from the additional weight.

RECIPROCAL INNERVATION

All of the afferent pathways described above involve, at least in part, interneurons that provide a means by which one pool of motor neurons is facilitated while its antagonist pool is inhibited (Fig. 6.7). This anatomical arrangement is known as **reciprocal innervation**. Its utility is obvious, since it is desirable to relax the antagonist of a muscle that is contracting in order to facilitate smooth joint movement. This inhibition is automatically accomplished by the connections of interneurons within the spinal cord. This minimizes the detail of central motor commands.

Spinal Integration of Motor Commands

Except for local spinal reflexes, all motor commands are initiated in supraspinal structures

---

[3]Due to the technical difficulties of marking individual axons, the terminal fields of single group III and IV cutaneous afferents have not been mapped within the spinal cord. One might hypothesize that these axons would terminate in roughly the same laminae as the group II cutaneous afferents.

and sent to the spinal cord over the descending motor tracts. These descending control systems exert their influence indirectly through interneurons and, to a lesser degree, directly on motor neurons. The direct activation of α motor neurons has certain advantages with regard to speed and precision of movement (see Chapter 7). Reflecting this, direct activation of α motor neurons has increased with phylogeny. However, most motor control, even in the most evolutionarily derived mammals, is exerted through interneurons, a process that takes advantage of the intrinsic coordinating circuitry in the spinal cord. This approach relieves higher centers from having to compute all the details of a motor act.

Group II muscle afferents facilitate two parallel interneuronal paths, one that facilitates flexors and the other that facilitates extensors (Fig. 6.7). This apparent irrational anatomical arrangement can be understood if one hypothesizes that neither path can be effective unless there is continuous, subliminal facilitation from supraspinal systems. By facilitating one of the two antagonistic parallel pathways in the spinal cord, higher descending control systems can regulate the balance between flexor and extensor muscle groups with simple generic commands.

COACTIVATION OF α AND γ MOTOR NEURONS

The α motor neuron innervates only the extrafusal muscle. When the muscle contracts, the muscle spindle apparatus within it will be unloaded (i.e., there will be a release of tension on the spindle), and consequently fewer action potentials will be returned to the spinal cord over the I_A and II axons from the muscle spindle (Fig. 6.9A). This is a negative feedback system. It works very well to maintain a constant muscle length under varying loads, but it interferes with the execution of commands intended to alter muscle length.

This problem can be overcome if α and γ motor neurons are facilitated simultaneously by command signals. In this way, the intrafusal muscle will shorten at the same time that the extrafusal muscle contracts, and a steady tension on the muscle spindle receptors will be maintained (Fig. 6.9B). I_A facilitation of the motor neurons will not decline during active muscle

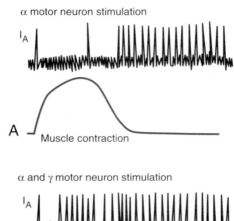

α motor neuron stimulation

I_A

A   Muscle contraction

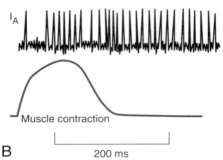

α and γ motor neuron stimulation

I_A

Muscle contraction

B   200 ms

FIGURE 6.9.

A, Muscle spindle afferent axons reduce or even cease firing during muscle contraction if α motor neurons are stimulated alone. **B,** If the α and γ motor neurons are stimulated concurrently, the muscle spindle afferent continues to fire during muscle contraction.

contraction. In the intact animal this is accomplished so precisely that during voluntary movements, I_A activity remains approximately constant as the muscle contracts.

The problem can also be solved by using the gamma system to establish a new baseline muscle length and allowing the monosynaptic reflex circuitry to regulate the extrafusal muscle tension. For example, stimulating γ motor neurons alone produces an increase in tension on the intrafusal muscles, a tension that is too weak to effect joint movement (Fig. 6.10A). But the increased intrafusal tension does increase the tension on both the primary and secondary receptors, causing an increase in the frequency of action potentials on the I_A and II axons (Fig. 6.10B). This in turn exerts a powerful facilitation on the homonymous α motor neuron pool that does affect muscle contraction and joint movement (Fig. 6.10C). As the muscle contracts under this reflex facilitation, the

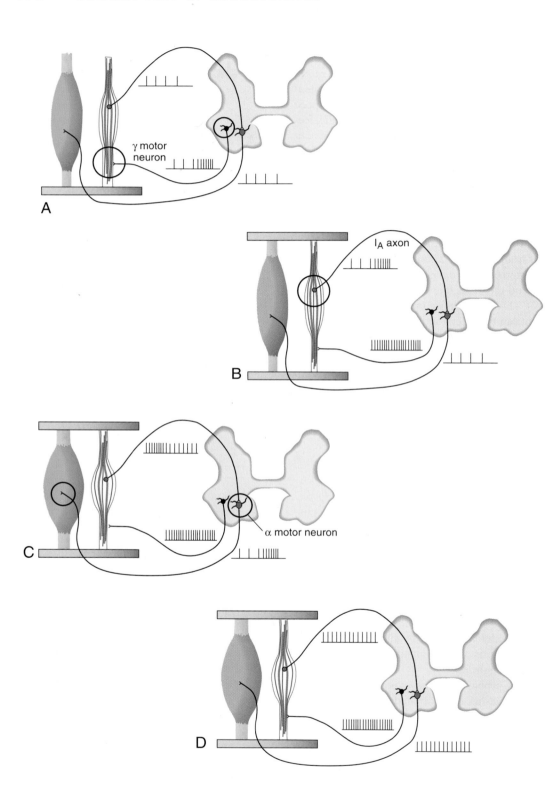

muscle spindle becomes unloaded, reducing the I_A and II reflex facilitation until movement ceases (Fig. 6.10D). The muscle comes to rest, but at a new equilibrium length. In this way, *gamma activation determines the muscle length around which further motor regulation is established.*

Independent gamma activation is also necessary because the muscle spindles have a limited effective range—they cannot respond over the entire range of muscle length. To be maximally effective, the tension on the spindle must be maintained within a narrow range so that increases or decreases in muscle length will cause appropriate changes in receptor activity. Activity on the γ motor neuron adjusts the baseline tension on the muscle spindle receptor, matching it to the desired resting length of the muscle.

RECRUITMENT OF MOTOR NEURONS BY SIZE

One of the more important recent discoveries is the observation that motor neurons are activated in a specific order during voluntary contraction of a muscle, and that they are deactivated in the opposite order as the contraction is voluntarily released (Fig. 6.11).

The significance of this observation lies in the fact that there are two types of extrafusal muscle fibers. One type has a relatively slow rate of contraction and is fatigue resistant. The other type has a fast rate of contraction, but it fatigues rapidly. The **slow, fatigue-resistant fibers** develop tension slowly as action potential frequency increases. They have a broad dynamic range, being able to increase their tension 10 or 20 times. The **fast, fatigable fibers** develop tension rapidly. This is accomplished at the expense of dynamic range, for the fast, fatigable

fibers can produce only about a twofold increase in tension.

There is a precise relationship between α motor neuron size and muscle fiber type; slow, fatigue-resistant fibers are innervated by small α motor neurons, while the fast, fatigable muscle fibers are innervated by large α motor neurons. This relationship is important because neuron size determines the order in which motor units are recruited.

As supraspinal commands facilitate a motor neuron pool, small α motor neurons are, in general, brought to threshold before the larger α motor neurons. This is due to the fact that individual EPSPs generate a larger hypopolarization in small neurons than in large ones because the amplitude of an EPSP at the trigger zone of a

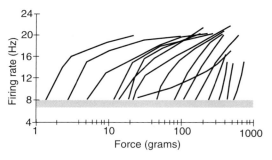

FIGURE 6.11.

The recruitment of motor units during voluntary contraction of the extensor digitorum communis in the human. Each line represents the rate of modulation of a single motor unit. Note that the slopes of the lines corresponding to the first units to be activated as the force of contraction is increased (slow, fatigue-resistant motor units) are much lower than those for the last units to be activated (fast, fatigable motor units).

FIGURE 6.10.

This cartoon illustrates the consequences of stimulating γ motor neurons without simultaneously stimulating α motor neurons. The bars near the axons represent action potential on that axon. **A,** Activity on the γ motor neuron causes the contractile portion of the intrafusal muscle to contract. **B,** Contraction of the intrafusal muscle hypopolarizes the sensory portion of the fiber, initiating action potential on the I_A axon. **C,** Activity on the I_A axon causes the α motor neuron to fire and consequently the extrafusal muscle to contract, unloading the intrafusal fiber. **D,** The system reaches a new equilibrium at a shorter muscle length, but the tension on the intrafusal fiber is approximately the same in **D** as it was in **A**.

neuron is related primarily to the total membrane resistance of the cell. Small neurons have a higher total membrane resistance than large neurons because they have less membrane surface area. Consequently, all other variables being equal, the inward current flow initiated by an excitatory synapse will generate a larger EPSP in a small neuron than in a large one.[4] Therefore, if a given level of facilitation is distributed among all the motor neurons in a pool, the smallest neurons will be brought to threshold first. This means that the steady, baseline muscle activity is borne by the least fatigable type of muscle fiber, leaving the fastest fibers in reserve for quick, short-lasting movements that are superimposed on the steady-state background tension.

DEVELOPMENT OF MUSCLE TENSION

The modulation of individual α motor neuron firing rates regulates tension produced by individual motor units. Under physiological conditions, human α motor neurons fire between about 8 and 25 action potentials per second. At the lowest rate of discharge, the motor unit produces about 10 to 25% of its maximum tension. As the firing rate of the α motor neurons increases, muscle tension also increases.

Tension in a whole muscle is also regulated by the recruitment of more and more motor units. The smooth application of tension during voluntary contraction is brought about both by increasing the force of contraction of individual motor units (modulated by α motor neuron firing rate) and the recruitment of additional motor units (Fig. 6.11). The intrinsic organization of the spinal cord and the relationship between the motor neurons and the muscle type ensures that the slow, fatigue-resistant motor units will be recruited first and the fast, fatigable motor units will be activated last. Therefore, *descending motor commands need not be encoded for muscle fiber type*. Only a general level of facilitation of the whole motor neuron pool needs to be incorporated in the motor command.

---

[4]This is a simple consequence of Ohm's law ($E = IR$). If the synaptic current (I) is constant, then a larger voltage (E) will be generated over a larger resistance (R). Of course, this is an oversimplification. One must also consider the electrotonic distance between the trigger zone and the synapse and a number of other factors, but these factors only modify the basic principle, which is that small neurons are more easily excitable than large ones.

Injury and Regeneration of the Peripheral Nervous System

To understand how testing the muscle stretch reflex is useful in making neurological diagnoses, one first needs to have a fundamental understanding of how injury to the peripheral nervous system can affect the MSR and muscle. The peripheral nerve may be injured in several different ways. It can be cut cleanly in two; it can be constricted or crushed without severing the whole nerve; or it can be compromised by metabolic, immunologic, or vascular disease. The neuronal response to injury in each of these cases is the same. The final result, however, differs, and understanding these differences is important in determining the proper treatment of nerve lesions and evaluating the prognosis.

DEGENERATION

When injured, neurons undergo a specific set of degenerative changes. These changes are of two types, those that occur proximal to the lesion and those that are distal to it.

Orthograde Degeneration

When an axon is severed, the distal portion is removed from its source of metabolic nourishment. Without a continuous supply of trophic substances from the soma, the isolated axon soon dies. The death of the axon releases from cisterns and sequestering molecules intra-axonal calcium, which when released activates several enzymes that facilitate the disintegration of the axon. This process is called **orthograde** or **wallerian degeneration**.

With the death of the axon, secondary changes occur. Collagen and fibronectin production is stimulated in fibroblasts, causing the endoneurium and perineurium to become hypertrophied. Myelin cannot be maintained by the Schwann cell in the absence of an intact axon. It becomes disorganized, disintegrates, and is phagocytized by Schwann cells and macrophages. Furthermore, the Schwann cells undergo rapid proliferation and organize themselves into columns of cells (**bands of Büngner**) that take up the space vacated by the degenerated axon. These bands of Büngner, also known as **Schwann tubes**, are critical in the regenerative process (see below). If the axon does not regenerate, the Schwann cells will eventually be unable to maintain themselves. In about one year the Schwann tubes will be replaced with connective tissue and the Schwann cells by fibroblasts.

Retrograde Degeneration

The neuronal cell body also undergoes characteristic changes when its axon is severed. The signal precipitating these changes is apparently the loss of materials returned from the axon terminals to the soma by retrograde (from the periphery to the soma) axoplasmic flow. Instead the soma is exposed to materials returned from the site of the lesion. These materials may be enzymes released by Schwann cells, or damaged axons. Although the precise nature of the signal is not known, the effect is apparent within hours. The soma becomes swollen and the nucleus moves toward the periphery of the cell. The most conspicuous events are the reorganization of the rough endoplasmic reticulum and the mobilization of the ribosomes. With the disorganization of the rough endoplasmic reticulum, the cell becomes less basophilic. Its appearance is pale by light microscopy; hence, the term **chromatolysis** [G. *chroma,* color + G. *lysis,* dissolution] was used by the original light microscopists, a description that survives to this day.

Effects of Muscle Denervation

Denervated muscle also undergoes conspicuous changes. These changes occur in three stages, signaled by fasciculations, fibrillations, and atrophy.

As the distal segment of the axon degenerates, injury potentials, caused by hypopolarization at the sites of membrane breakdown, cause spontaneous action potentials to travel along the intact, distal portions of the axon and stimulate the muscle. As long as the distal axon is intact, these action potentials are delivered to a complete motor unit and elicit a coordinated twitch. These contractions are visible to the casual observer and are known as **fasciculations** [L. *fascis,* bundle] (Fig. 6.12*A*). *Fasciculations are among the earliest signs of muscle denervation.*

As the axon degenerates further, the terminal branches become isolated. The injury potentials are still generated and still cause muscular contractions. However, since the motor units are no longer joined by a functioning axon, the muscle cells no longer contract as a unit. The contractions are small and are not visible on the body surface (Fig. 6.12*B*). The **fibrillations**, as they are called, may be detected electrically by placing an electrode in the muscle and observing the muscle action potentials on an oscilloscope. This type of recording is the **electromyogram** or **EMG**, a laboratory study that is very useful in documenting the course of muscle denervation.

After the distal axon has completely degenerated, the muscle becomes electrically silent, and degenerative changes in the muscle become apparent. Deprived of trophic substances delivered to the muscle by the neuron, the muscle undergoes **denervation atrophy**. This atrophy of denervation is not to be confused with the **atrophy of disuse** that occurs during immobilization. Denervation atrophy is much more profound and is a very useful clinical sign.

As a prelude to reinnervation, the denervated muscle develops a large number of acetylcholine receptor sites that serve as targets for regenerating axons. The denervated muscle can maintain these isolated receptors for about 2 years, after which, if there is no reinnervation, they are lost. The presence of the un-innervated receptor sites also makes the muscle hypersensitive to circulating acetylcholine, a condition that contributes to fibrillations.

REGENERATION

Injured neurons, after a period of nuclear reorganization, attempt to regenerate their original structure in order to restore function. Since the embryonic conditions under which the neuron originally developed no longer exist, regeneration is not always successful. The success or failure depends on many factors, most of which are not currently understood. A few factors will be discussed below.

Peripheral Nerve Regeneration

Injured neurons increase their production of proteins several fold. The injured neuron is not

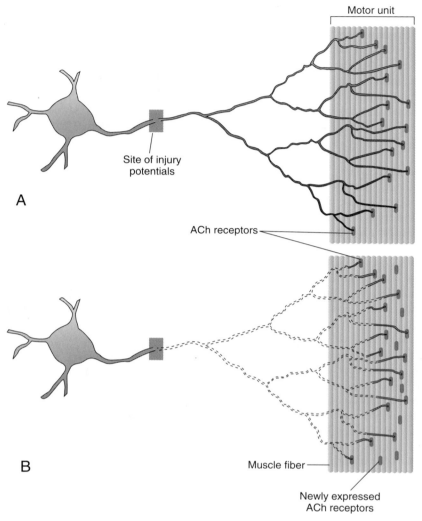

FIGURE 6.12.

Fasciculations and fibrillations are caused by injury potentials generated at the site of injury to a motor neuron axon. **A,** If the injury is proximal to the distal branching of the motor neuron axon, injury potentials will cause all the elements of the motor unit to contract simultaneously, producing a coordinated twitch (fibrillation) that is visible on the surface of the body. **B,** As the distal axon degenerates, the distal branches become separated. Each has its own site where injury potentials are generated. The individual muscle fibers no longer contract as a unit, so the twitches (fibrillations) are uncoordinated among the individual muscle fibers and are not visible on the surface. In addition, as a consequence of denervation the muscle fibers express numerous ACh receptors that make the muscle fibers very sensitive to circulating ACh.

simply producing more protein, but the *types* of proteins produced change. The intact neuron produces materials that are necessary for synaptic transmission and general cell maintenance. The injured neuron produces materials necessary for the reconstruction of the distal axon and

all of its terminals. "Growth-associated proteins" have been identified whose transcription is triggered by axonal injury. They are not produced by intact neurons.

To accommodate this increase in protein synthesis the *orthograde* (from the soma to the periphery) axoplasmic transport mechanisms are stimulated. The rate of slow transport is increased as much as twofold. Although the rate of fast axoplasmic transport is not altered, the amount of material transported increases as much as 250%.

About two days after peripheral nerve injury, the proximal stump of the injured axon develops a specialized, bulbous ending, the **growth cone**. Emanating from the growth cone are many (up to 50) tendrils or sprouts that elongate and wave about in the interstitial space. If one of these tendrils contacts a favorable surface, its movement diminishes and the remaining sprouts contract back into the body of the growth cone. The bulbous segment is then expressed to the tip of the remaining sprout, a process that looks very much like a snake swallowing a toad. After the growth cone has established itself at the tip of the terminal sprout, new tendrils appear and the process repeats.

Apparently this regenerative process is quite dependent on the favorable interaction between the hydra-like tendrils and the environment in which they find themselves. The Schwann cells are a critical part of this environment. **Laminin** and **fibronectin**, two glycoproteins, are found in the **basal lamina** of the Schwann cell. Both seem to stimulate the formation of growth cones and sprouts. Laminin, in particular, is a very favorable substrate for neurite attachment. Since a Büngner band consists of a string of Schwann cells and basal lamina that occupy the space of a degenerated axon, a growing neurite that enters one of these bands can grow along the path of the degenerated axon.

Reinnervation of Muscle

When a regenerating motor neuron reaches a muscle, it is attracted to the acetylcholine receptors that have blossomed on its surface. Although the process is not completely understood, apparently the acetylcholine receptor sites secrete a chemotactic substance that attracts the regenerating axon. At some of these receptor sites, a new myoneural junction is established. As myoneural junctions develop on the muscle, the unoccupied receptor sites are reabsorbed. The reinnervated muscle, being resupplied with trophic substances from the axon, increases in size; the atrophy becomes much less prominent. However, since reinnervation of the muscle is rarely complete, the muscle never regains its normal size.

Regeneration of Sensory Receptors

Sensory receptors can also be regenerated, although the process is not well understood. Pain is usually the first sensation to recover. The simplicity of the nociceptor terminals (free nerve endings) probably allows them to become functional at an early stage of regeneration. Other, more specialized touch receptors can be renewed, but the full complement of sensory receptors is not restored, so natural sensory perceptions from a denervated area are never completely reestablished.

Practical Considerations

From this description it is obvious that successful axonal regeneration will be much more likely if the peripheral nerve is simply crushed and not completely severed (Fig. 6.13A). Following crush injuries, the regenerating axon is almost always directed into the Büngner band that was derived from the distal portion of itself. The axon is usually able to regenerate, and good function can be restored (Fig. 6.13B).

If the nerve is completely severed there are several new problems. For example, there may be a gap between the proximal and distal ends of the nerve. Growing axons can bridge gaps of several millimeters, but they may enter Schwann tubes that lead to inappropriate terminating sites (Fig. 6.13C). In such cases a sensory axon may send a growth cone into a tube that terminates on a muscle body and vice versa. Or motor axons may regenerate to muscles that are different from the ones they originally innervated. If the gap is very large the neurites may not be able to cross it and find their way into a Schwann tube (Fig. 6.13C). Sensory neurons that wander beyond the protective field of the Schwann cells may coil into a whorl of axons and connective tissue known as a **plexiform neuroma**. These neuromas can be the source of intense pain. More frequently, however, lost axons simply stop growing and regeneration ceases. Eventually, unconnected axons degenerate completely, and ultimately the entire neuron dies.

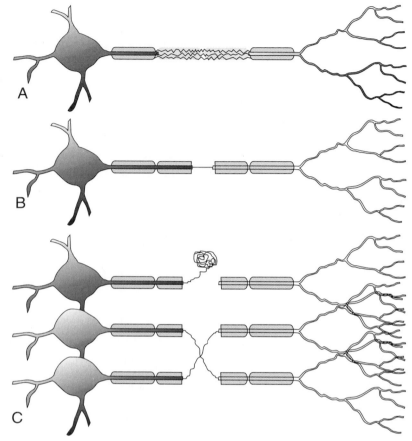

FIGURE 6.13.

Successful regeneration depends on a number of factors, one of which is the severity of the injury. **A,** If the peripheral nerve is crushed, leaving the connective tissue intact, growth cones from the proximal segments of injured axons have little difficulty entering the Schwann tubes. Restoration of function is usually complete. **B,** If the nerve is severed, but the stumps can be anastomosed carefully, growth cones have a good chance of entering appropriate Schwann tubes. Restoration of function is possible, but rarely complete. **C,** If there is a large gap between the stumps of injured neurons, growth cones may not enter Schwann tubes. In such cases, a plexiform neuroma is formed. In other cases, motor axons may enter Schwann tubes that are derived from sensory nerves and vice versa. Little functional recovery occurs in such cases.

Microsurgical techniques allow one to suture severed peripheral nerves together, fascicle by fascicle, a process that is purported to improve functional recovery. This tedious process is not always possible. If a large gap exists, the surgeon may need to insert a nerve graft between the stumps; with this procedure, fascicular realignment is impossible. Yet significant functional recovery is frequently attainable in such cases. Recent evidence suggests that it may not be as necessary to precisely reanastomose fascicles as was once believed. Under laboratory conditions, severed nerves that are re-apposed with a slight gap between the nerve ends regenerate with extraordinary precision. It appears that the regenerating axons have a remarkable ability to select an appropriate Büngner band that will lead them to an appropriate target. Although a chemotactic signal is suspected, the process is not understood.

Clinical Evaluation of Motor Function

Localizing a nervous system lesion is the essential first step in forming a correct neurological diagnosis. Many neurologists approach the problem of localization by mentally dividing the nervous system into several functionally distinct regions. Usually the patient's history, signs, and symptoms allow one to exclude the lesion from one or more of these regions. Further analysis then allows one to refine the location further.

Two of the functionally distinct regions of the nervous system are known as the "**upper motor neuron**" (**UMN**) and the "**lower motor neuron**" (**LMN**). This nomenclature is misleading, for neither of these "neurons" exists as a neuron. Each is a set of tracts and nuclei that form a clinically recognizable functional system. *The concept of the nervous system consisting of an upper and a lower motor neuron is a generalization about the anatomical organization of the nervous system.* Understanding the principles that define this concept is essential to making neurologic diagnoses.

The ability to distinguish between lesions to the LMN and the UMN requires an understanding of how the physician uses the MSR to test the "lower motor neuron" and to evaluate neural dysfunction. By definition, *the "lower motor neuron" consists of those parts of the nervous system that, when injured, will cause a decrease in the strength of the muscle stretch reflex.*

NEUROLOGY OF THE MSR

Testing the muscle stretch reflex is fundamental to the neurological examination because *it allows one to separate upper from lower motor neuron injuries.* The physician uses a reflex hammer to sharply strike a tendon, e.g., the patellar tendon. This places a rapid stretch on the muscle that, as we have just seen, *stimulates primarily the dynamic bag receptors,* causing a reflex contraction of the homonymous muscle. This reflex is commonly called the "deep tendon reflex" or "DTR" because of the emphasis on striking the

tendon with the reflex hammer[5]. This unfortunate term is so ingrained in the medical parlance that it will probably never be replaced. The student is cautioned, however, to avoid associating the muscle stretch reflex with the Golgi tendon organ receptor. To avoid that confusion, this text will use the term MSR to designate this reflex, a usage that is gaining popularity among neurologists.

The responsiveness of the MSR can be altered by trauma or disease. Only three findings can be observed: the reflex may be **hypoactive**, it may be **normal**, or it may be **hyperactive**. If the muscle, in response to the reflex hammer, does not contract with a quick strong jerk, the reflex is said to be hypoactive or reduced. It may be reduced so much that it is absent. The reaction of the muscle may be faster and stronger than normal, in which case the MSR is said to be hyperactive, elevated, or brisk (see Chapter 7). Many physicians use the nomenclature suggested in Table 6.4 to describe the MSR response. The determination of whether an MSR is elevated or reduced or normal is not always obvious. Learning to make such judgments requires experience.

DECREASED MSR

Injury to any part of the peripheral nervous system including the muscle will cause the MSR to be reduced or eliminated. Consider for a moment all the elements of the nervous system that must perform correctly if this simple reflex is to function properly. The muscle sensory receptors (spindle apparatus) must be intact along with their primary afferent neurons (I_A). The axons of these neurons lie in the peripheral nerve and the dorsal roots. Thus any injury of the muscle nerve or the spinal roots that supply it will affect the reflex. The spinal cord, including the motor neurons, must be intact at the level where the root enters. Then the ventral root and peripheral nerve must be considered, and finally, the myoneural junction and the muscle itself. *If any of these elements is damaged or destroyed, the MSR will either be less responsive than normal or in extreme cases be absent altogether.* The careful physician must consider all elements of

---

[5]They are also often referred to as "muscle jerks" because the muscle responds to the reflex hammer with a quick jerk.

Table 6.4. Clinical Notation Commonly Used to Describe MSR Responses

Description	Interpretation
+ +	Very brisk, definitely abnormal
+	Brisk, probably abnormal
Normal	Within normal range
–	Weak, probably abnormal
– –	Very weak, definitely abnormal
Absent	Not perceptible

the reflex path in forming conclusions about the motor system.

To illustrate, **peripheral neuropathies**, a general class of diseases that affects the peripheral nerves (see Chapter 5), often are associated with diminished reflexes because the disease process interferes with axonal transmission. A **compression injury** at the union of the dorsal and ventral root caused by a herniated intervertebral disk is a very common cause of decreased reflexes in the extremities. **Tabes dorsalis**, a condition resulting from long-term infection with syphilis, leads to the destruction of the large-diameter axons in the dorsal root, including the I_A axon. This results in extreme paroxysmal pain and, due to the loss of the I_A axons, decreased muscle stretch reflexes. **Poliomyelitis**, a viral disease that frequently kills motor neurons in the spinal cord and brainstem, leads to paralysis and decreased or absent reflexes. **Myasthenia gravis**, a disease process that interferes with the chemical communication between the motor endplate and the muscle, is manifested clinically by increasing fatigue and progressively diminished reflexes with exercise. Finally, the **muscular dystrophies** lead to decreased or absent reflexes due to primary muscle disease that leads to eventual failure of the muscle.

This diverse group of neurological diseases have in common two elements: *they all cause decreased or absent myotatic reflexes,* and they *all affect the peripheral nervous system or muscle.* This association between decreased MSRs and the aforementioned parts of the nervous system led to the clinical concept of the "lower motor neuron." The concept is extraordinarily useful. With simple observations and a reflex hammer, the physician can quickly determine whether the neurological problem affects the central or the peripheral nervous system, and localize it to a relatively small region. *Correctly using and interpreting the muscle stretch reflex is absolutely critical to the neurological examination.*

AUXILIARY SIGNS

A reduction in the MSR is the only sign necessary to localize a lesion to the LMN. Under some circumstances it may be difficult to be certain that a reflex is depressed. Therefore ancillary signs are extremely useful in corroborating one's interpretation of the MSR and determining the presence of lower motor neuron disease. Most prominent among these ancillary signs is the presence of fasciculations. Since damage to the α motor neuron will denervate the muscle, the presence of fasciculations is a useful indirect sign of nerve damage. At a later stage in the course of the disease, observing pronounced muscle atrophy serves the same purpose.

CASE STUDIES

Mr. E. A.

HISTORY: Mr. E. A. is a 55-year-old white male, employed as a tool and dye maker for a large corporation. He was first seen by his physician on April 11, 1990, complaining of numbness in his feet. This numbness was of several months' duration. At that time, Mr. E. A. was diagnosed as an alcoholic. He refused all treatment at that time. He was next seen on July 7, 1990, complaining of generalized weakness and numbness in his lower extremities. At this time he said his symptoms had gotten worse since his last visit.

NEUROLOGICAL EXAMINATION, July 7, 1990:

I.   MENTAL STATUS: Normal.

II. CRANIAL NERVES: Normal

III. MOTOR SYSTEMS: His gait was wide-based and ataxic. His upper-extremity strength was bilaterally symmetrical and seemed normal. There was no drift as he held his arms out in front of himself. Lower-extremity strength was symmetrical but weak to direct strength testing and confirmed by his inability to walk on his toes or heels.

IV. REFLEXES: Both ankle MSRs (gastrocnemius-soleus muscle) were depressed (—). Patellar (quadriceps muscle) and upper-extremity MSR reflexes were normal. Plantar reflexes were flexor (normal, see Chapter 7).

V.   SENSORY SYSTEMS: Perception of pin prick was bilaterally dull and vibration sense absent from his knees down to and including his feet. Mr. E. A. could not determine the position of his toes or ankles when they were passively positioned by the examiner. Sensations from the upper extremities were symmetrical and perceived as normal.

Comment: The first steps in analyzing a neurological case is to attempt to determine (a) the time course of the illness and (b) the distribution of the signs and symptoms. Identifying the time course helps one to establish the etiology of the disease, while the distribution of symptoms helps to establish the location of the lesion. Both of these factors are important to the present case.

From the history we know that Mr. E. A.'s problems are localized to his lower extremities. The impression gained from the history is confirmed during the neurological examination. The decreased MSR from the gastrocnemius-soleus muscles establish the peripheral nature of the disease process, the involvement of "the lower motor neuron," and the localization to both lower extremities.

The loss of pin prick and vibration sensations from the lower extremities involves the entire distal portion of the extremity. This "stocking-glove" pattern is characteristic of diffuse peripheral nerve lesions and is in sharp contrast to the dermatome pattern associated with lesions of the roots or the patchy patterns associated with localized damage to a peripheral nerve (see Chapter 5). Mr. E. A. has difficulty walking in part because of weakness but also because the sensory positional information from the skin and joints of his lower extremities is not being received in the spinal cord. The examination of joint position sense clarifies this. The importance of sensory feedback in motor control will be discussed further in the next chapter.

The history also reveals that Mr. E. A. is suffering from a long-term (several months), slowly progressing problem. It originally manifested only sensory symptoms and has progressed to involve both sensory and motor functions. Slowly progressive problems are not likely to be caused by strokes (intracranial thrombosis or hemorrhage) or traumatic injury. Expanding intracranial masses, metabolic, inflammatory, or immunological diseases and toxins are more likely explanations.

Mr. E. A. presents with a typical picture of peripheral neuropathy. Peripheral neuropathy, also known as polyneuropathy, describes a clinical syndrome that includes both motor and sensory losses caused by diffuse lesions to peripheral nerves. The affected areas cannot be associated with the territory of a single nerve or spinal root. Frequently the pattern of loss is bilaterally symmetrical and has a "stocking-glove" distribution.

The lesions causing these losses are numerous and may be inherited or acquired. Diabetes mellitus is a very common cause of peripheral neuropathy. In the later stages of this disease, the microcirculation is disrupted. Many organs suffer vascular damage, including the peripheral nerves. Peripheral neuropathies are also commonly associated with toxins. These include alcohol, heavy metals, metabolic toxins (associated with diseases such as diphtheria, or malignancies), and prescription drugs. Nutritional and vitamin deficiencies, particularly involving the B group, are also associated with peripheral neuropathy. A very large number of peripheral neuropathies are simply idiopathic [G. idios, individual. + G. pathos, suffering], i.e., a disease of unknown cause.

Peripheral neuropathy is a common sequela to chronic alcohol abuse. The primary cause of the neuropathy is not known. Nutritional deficiency, particularly thiamine deficiency, is a contributing factor, especially for someone with Mr. E. A.'s history of alcohol abuse.

Mr. E. A. was again encouraged to seek treatment for his alcoholism. He again refused. He was prescribed vitamin supplements and advised to improve his nutritional status.

FURTHER APPLICATIONS

- Explain the decreased MSR in cases of peripheral neuropathy. How is your explanation consistent with the loss of position sensation from the toes? Does your explanation also explain the ataxia?

- Explain the causal relationship between diabetes mellitus and peripheral neuropathy.

- Consider possible mechanisms that could cause the characteristic "stocking and glove" distribution of the sensory losses in peripheral neuropathy.

Ms. N. B.

HISTORY: Ms. N.B. is a 63-year-old white female who went to her family physician complaining of pain in her back and extending down the side and back of her right leg. This pain began in April of 1990 and had been getting progressively worse. The pain was very severe and after a week of complete bed rest remained undiminished. She was admitted to the hospital on May 20, 1990, and after another week of complete bed rest, this time with pelvic traction, her condition remained unchanged.

NEUROLOGICAL EXAMINATION, May 20, 1990:

I.   MENTAL STATUS: Normal

II. CRANIAL NERVES: Normal

III. MOTOR SYSTEMS: She walked with a slight limp, favoring her right leg. She could walk on her heels, but had some difficulty with toe walking, being unable to maintain the extension of her right foot. No other weakness could be demonstrated in any of Ms. N.B.'s extremities. There were no fasciculation and no apparent atrophy.

IV. REFLEXES: All MSRs were normal and symmetrical except that from the right Achilles tendon, which was absent. Plantar reflexes were bilaterally flexor (normal).

V. SENSORY SYSTEMS: Ms. N.B. was in great pain when she sat in a chair or stood. She was most comfortable while lying in bed with her knees flexed, although she was not pain free even then. Passive straight leg raises were limited by pain on the right to 30°. The left leg could be raised nearly pain free to 90°. There was dullness to pin prick over the lateral border of her right lower extremity and foot.

Comment: Lower back pain is a very common complaint that is often not successfully resolved. Treatment for Ms. N. B. is likely to be successful because she has very objective, localized findings.

From the history one learns that Ms. N. B.'s pain is not associated with any notable injury. The pain is also highly localized to her right back and leg and cannot be relieved with enforced bed rest. Also one should note that the pain got progressively worse, reached a plateau, and has remained undiminished since.

Most notable among the objective findings of the neurological examination are weakness and an absent MSR from the right gastrocnemius-soleus muscle. This finding establishes the peripheral nature of the problem. The fact that there is no other weakness and all the other reflexes are normal makes the absent MSR more significant.

The dull pin prick sensation from the lateral aspect of the right foot corresponds to the territory of the S1 dermatome (see Chapter 5). Particularly noteworthy is the fact that the gastrocnemius-soleus muscle receives its primary innervation from the S1 spinal root (see Chapter 5). Therefore, two independent lines of evidence, one sensory and one motor, help to localize the problem to a single spinal root.

Weakness, pain, and loss of somatosensory perceptions—all attributable to one or two adjacent spinal nerves—are the cardinal signs associated with the spinal disk syndrome. *This syndrome is caused by a displaced intervertebral disk that places pressure on one or two spinal nerves, close to their egress from the spinal column. Therefore, symptoms are of the LMN type, involve both sensory and motor signs, and are highly localized. The present case is typical in these regards.*

Straight leg raises (placing the patient on his back, raising each fully extended leg independently, and noting the angle of the leg to the bed) are not diagnostic in these cases, but are traditionally performed in an attempt to quantify the severity of the symp-

toms. *Raising the leg stretches the sciatic nerve and can increase the pressure on the entrapped root. In many of the cases, however, raising the leg contralateral to the injured root causes increased pain, while raising the ipsilateral leg does not.*

At this point one would entertain the diagnosis of right S1 spinal nerve entrapment at the L5-S1 vertebral interspace caused by a protruding intervertebral disk. A disk in this position could pressure either the L5 or S1 root. One presumes the S1 root is involved because of the dermatome data. The S1 root also innervates the gastrocnemius muscle, and pressure on the nerve root would explain the loss of Achilles reflex. One would seek to verify the diagnosis by using ancillary studies.

In this case, an MRI study of the spine was ordered that subsequently revealed a displaced intervertebral disk (Fig. 6.14).

Ms. N. B. was operated on the following day. The intervertebral disk was removed from between the L5 and S1 vertebrae without incident. Ms. N. B. was placed on a supervised program of physical therapy. Six months after her surgery she was fully recovered, had no pain, and had resumed her normal lifestyle.

FURTHER APPLICATIONS

• Suppose the L5 spinal root were affected instead of the S1 root. What clinical picture would you expect to see?

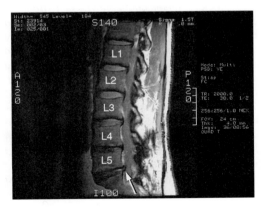

FIGURE 6.14.

An MRI of a lateral view of the lumbar spine showing the intervertebral disks between the bodies of the lumbar vertebrae labeled L1 to L5. In the case of Ms. N. B., the intervertebral disk at the L5 and S1 interspace (*arrow*) is displaced into the spinal canal, compressing the S1 spinal root.

• In this case, the S1 root was affected by the intervertebral disk located at the L5-S1 interspace where the L5 root leaves the spinal canal. Can you explain this apparent discrepancy?

• Most intervertebral disks rupture to one side, but occasionally one will protrude at the midline. Other than the obvious bilateral effects, what other consequences might you expect from a midline displacement that you would not expect to see from a lateral displacement?

• In most cases, the displacement of an intervertebral disk occurs in the lumbar region. However, displaced cervical disks are not uncommon. What signs would one expect to observe if the disk lying in the C6-C7 interspace were to become displaced and place pressure on the left spinal nerve?

Ms. S. B.

HISTORY: Ms. S. B. is a 25-year-old housewife. On October 2, 1978, she noticed severe pain in her right buttock which progressed and involved her whole right leg by the 4th of October. By the 5th, all of her muscles ached, especially her back, but over the next five days, the muscle pain largely abated.

On the morning of October 6th, she was able to walk, but her left calf felt tight. She went to bed for a couple of hours, but when she got up, her left leg would not support her. The weakness progressed rapidly for the next two or three days and involved both legs. She was admitted to the hospital on October 10th.

NEUROLOGICAL EXAMINATION:

I.   MENTAL STATUS: Normal

II. CRANIAL NERVES: Normal

III. MOTOR SYSTEMS: There was a severe wide spread and global flaccid weakness in both lower limbs. On the right, she could not move her entire lower extremity except for the toes, and these just barely. On the left, she could flex her hip slightly, raising her knee two inches off the bed against gravity. She had slight movement of her left ankle and good movement of her left toes.

IV. REFLEXES: All MSRs were absent from the lower limbs except for the left ankle (—). The left and right plantar reflexes were absent. The lower trunk was weak with a positive **Beevor's sign.**[6]

V. CEREBELLUM: Normal function of the upper extremities. The lower extremities were not testable.

VI. SENSORY SYSTEMS: Sensory systems were normal except for some superficial tenderness on the front of her thighs and upper arms. Neck flexion caused pain down her back and in her hips (**Lhermitte's sign**).

Comment: Throughout the hospitalization, her neurological state remained the same except that she lost the signs of meningeal irritation. Two lumbar punctures were performed, one on admission and the other 14 days later.

The distribution of symptoms is restricted to the lower extremities, consisting of LMN weakness (frank paralysis in this case) and an intact sensorium. The weakness is not global. Not only are the upper extremities spared, but the paralysis is not complete, even in the affected extremities, where some muscle groups are affected more than others. The time course is characterized by a prodromal illness, intense pain that subsides to be replaced with weakness or paralysis. Examination reveals that the pain can be attributed to meningeal irritation.

The pattern of (a) meningeal irritation and (b) extreme muscle pain followed by a (c) rapidly developing paralysis of the lower motor neuron type with (d) no sensory involvement other than localized pain is typical of poliomyelitis. This follows from the fact that polio is a viral infection of the CNS (meningeal signs) that kills only α motor neurons (LMN paralysis with no sensory loss). While destruction of motor neurons occurs most frequently in the lumbar spinal cord, any motor neuron can be affected. The attack on motor neurons is random and the subsequent weakness may be randomly distributed among muscle groups, even within the same extremity.

Ms. S. B. had never been immunized against polio. When she was a child, the polio vaccines were new and her parents refused to have her inoculated. On September 14, 1978, her first child received his first trivalent oral polio vaccine. Eighteen days later, Ms. S. B. developed her first signs

of polio infection. In this case, the neurologist who examined Ms. S. B. sent rectal swabs to the Centers for Disease Control in Atlanta. From these swabs, they were able to isolate, culture, and then positively identify type 2 polio virus of the vaccine type. Similar isolation and identification was made from stool samples from Ms. S. B.'s baby. Subsequent serum samples from Ms. S. B. showed very high titers for type 2 polio virus.

In Ms. S. B.'s case, there is unequivocal evidence that she was shedding live vaccine polio virus, and there is little doubt that this infection was the cause of her paralysis. Since she was not inoculated by her physician, she was undoubtedly infected by her baby subsequent to his vaccination by his pediatrician.

It is well known that the live virus oral polio vaccine can cause paralytic polio in some people. Ms. S. B. is one of those people. Examining the population as a whole, the risk is low: about one case in 6 million doses in the United States. Such statistics have led to recommendations that all children be immunized with oral polio vaccine (OPV). The rare case of paralysis is considered to be a statistically acceptable risk. This statistic is based on intentionally immunized children. It ignores accidentally infected adults.

It has been observed for over 20 years that nonimmunized adults over the age of about 18 are much more susceptible to vaccine-induced paralytic polio than are children. The reasons for this are not known. More recently, it has become apparent that immunocompromised adults and children are also at risk for developing paralytic polio from the oral vaccine. These people may be compromised due to immunosuppressive chemotherapy or HIV infection. These observations are important, for in a typical family environment all at-risk adults and children who care for an immunized infant will almost certainly become contaminated with OPV. Therefore, the inescapable conclusion is that nonimmunized adults and immunocompromised individuals are at particular risk for contracting paralytic polio if an infant in that household is vaccinated with OPV.

Certainly children should be vaccinated against this terrible disease. However, before an infant is given oral polio vaccine the physician must take a complete family history in an effort to identify immunodeficient and unvaccinated adults in the household. Nonimmunized adults can be safely protected by using the killed virus polio vaccine. It carries no risk for paralysis. Once protected in this way, they and the infant can subsequently be safely vaccinated with the live virus vaccine. If immunocom-

---

[6]The patient lies on his back and is asked to contract his abdominal muscles. If the lower part of the rectus abdominis is weak, the umbilicus will move toward the head (positive sign). If all of the muscles are equally strong, the umbilicus will remain stationary.

promised individuals are present in the household, the infant may have to be vaccinated with the IVP only.

FURTHER APPLICATIONS

- Explain the presence of Beevor's sign and its significance in the case of Ms. S. B.

- Explain the fact that some muscle groups in the same extremity are affected, while others are spared.

- The World Health Organization has predicted that, because of the widespread use of live polio vaccine, polio will be eradicated as a human disease by the end of the century. If that goal is achieved, will unvaccinated people like Ms. S. B. still be at risk? Explain.

- All three of the preceding cases present with decreased MSRs, yet each case results from entirely different pathology. Carefully explain the decreased MSRs in each case, paying particular attention to the different loci of the lesions, yet the similarity in the observed results.

- Reconsider these cases, paying particular attention to the distribution of the signs and symptoms. How does the distribution of weakness and sensory losses help one to establish the site of injury to the nervous system?

SUMMARY

Striated muscle is endowed with a number of sensory receptors that convey critical information about muscle dynamics to the CNS. The muscle spindle apparatus is a sensory receptor that furnishes information to the central nervous system about muscle length and the velocity of length changes. The muscle spindle apparatus consists of specialized intrafusal muscle fibers enclosed in a capsule. There are two anatomically distinct intrafusal fibers, the nuclear bag and the nuclear chain. All of the chain fibers and some of the bag fibers are static receptors, responding primarily to muscle length. A separate type of bag fibers are dynamic, responding primarily to the velocity of length changes. The Golgi tendon organ is a sensory receptor that furnishes information about muscle tension.

There is an afferent and an efferent innervation of muscle. Two types of afferent fibers originate in the muscle spindle apparatus, designated as group I_A, and group II axons. I_A axons originate in all bag and chain fibers and have both dynamic and static properties. Group II axons originate in static bag and chain fibers and have primarily static properties. The group I_B axons originate in the Golgi tendon organs.

The efferent innervation of muscle falls into three classes: α, β, and γ motor neurons. The α motor neurons innervate the main muscle mass, the extrafusal muscle, that provides the power necessary to effect movement. The α motor neurons innervate the intrafusal muscle, the contractile portion of the muscle spindle apparatus. Although these muscles are small and cannot cause movement, they do regulate the sensitivity of the muscle spindle receptors. The β motor neurons innervate both intrafusal and extrafusal muscle fibers. Their physiological and clinical significance is not understood at this time.

The muscle sensory receptors make specific connections in the spinal cord that affect motor neuron activity. The I_A and II axons make excitatory monosynaptic synapses on homonymous α motor neurons. The I_B axon synapses on interneurons that in turn make inhibitory synapses on homonymous α motor neurons. All of these axons also make numerous synapses on interneurons. The interneurons are part of a synaptic network that coordinates synergistic neurons and provides for reciprocal innervation of antagonistic motor neuron pools.

The muscle stretch reflex (MSR) is an important reflex that results from a rapid stretching of a muscle. The dynamic bag fibers are stimulated almost exclusively. The signal enters the spinal cord over the I_A axon that makes excitatory synapses directly on α motor neurons. The motor neuron returns to the homonymous muscle, causing it to contract. The clasp knife reflex is not fully understood, but probably originates in Golgi tendon organs. The signal enters the spinal cord over the I_B axon, where it inhibits the homonymous α motor neurons through the action of an interneuron. The FRA reflex originates from cutaneous receptors. The signal enters the spinal cord over group II, III, and IV afferent axons that make polysynaptic connections. Stimulation of FRAs typically elicits an ipsilateral flexion and a contralateral extension of the extremities.

Most supraspinal motor commands indirectly affect motor neurons, passing first through interneurons. This strategy allows local spinal circuits to automatically provide for reciprocal innervation and synergistic coordination. Most motor commands also coactivate both α and γ motor neurons so that the muscle spindle does not become unloaded during muscle contraction. Activation of γ

motor neurons alone causes a reflex contraction of the extrafusal muscle that establishes a new resting length for the muscle.

During voluntary contraction, tension is smoothly applied to a muscle. This is accomplished by modulating the frequency of action potentials on the α motor neurons as well as by recruiting more motor neurons. Motor neurons are recruited in a specific order, from the smallest to the largest. Small motor neurons innervate slow, fatigue-resistant muscles, while the large motor neurons innervate fast, fatigable muscle.

If an axons is severed, the distal portion undergoes orthograde and retrograde degenerative changes. The neuron soma becomes chromolytic. The Schwann cells form bands of Büngner (Schwann tubes) along the path of the degenerated distal axon. The distal end of a severed axon forms a growth cone that extends itself in an effort to regenerate the axon. Growth cones are able to follow Schwann tubes and under favorable conditions can form a new functional axon. Denervated muscle undergoes characteristic changes, first developing fasciculations that are followed by fibrillations. Eventually denervated muscle will atrophy. If the muscle is re-innervated by a motor axon, muscle function can be restored.

The "lower motor neuron" (LMN) is a clinical concept. The LMN consists of all of the peripheral components of the muscle stretch reflex (MSR), including the muscle. Injury or disease can affect the MSR, causing it to be either hyperactive or hypoactive. Lower motor neuron pathology is characterized by hypoactive or absent muscle stretch reflexes. Diminished MSRs may also be accompanied by auxiliary signs such as fasciculation, fibrillations, and atrophy.

FOR FURTHER READING

Adams, R., and Victor, M. *Principles of Neurology*. New York: McGraw-Hill, 1993.

Binder, M. Properties of Motor Units. In *Textbook of Physiology*, Ch. 23, edited by Patton, H. D., Fuchs, A. F., Hille, B., Scher, A. M., and Steiner, R. Philadelphia: W. B. Saunders, 1989a, pp. 510–521.

Binder, M. Peripheral Motor Control: Spinal Reflex Actions of Muscle, Joint, and Cutaneous Receptors. In *Textbook of Physiology*, Ch. 24, edited by Patton, H. D., Fuchs, A. F., Hille, B., Scher, A. M., and Steiner, R. Philadelphia: W. B. Saunders, 1989b, pp. 522–548.

Binder, M. Functional Organization of the Motoneuron Pool. In *Textbook of Physiology*, Ch. 25, edited by Patton, H. D., Fuchs, A. F., Hille, B., Scher, A. M., and Steiner, R. Philadelphia: W. B. Saunders, 1989c, pp. 549–562.

Brodal, A. *Neurological Anatomy In Relation to Clinical Medicine*. New York: Oxford University Press, 1981.

Brooks, V. *The Neural Basis of Motor Control*. New York: Oxford University Press, 1986.

Brown, A. G. *Organization in the Spinal Cord: The Anatomy and Physiology of Identified Neurons*. Berlin: Springer-Verlag, 1981.

Centers for Disease Control. Recommendation of the Immunization Practices Advisory Committee: Poliomyelitis Prevention. *MMWR* 31:22, 1982.

Fuchs, A. F., Anderson, M., Binder, M., and Fetz, E. The Neural Control of Movement. In *Textbook of Physiology*, Ch. 22, edited by Patton, H. D., Fuchs, A. F., Hille, B., Scher, A. M., and Steiner, R. Philadelphia: W. B. Saunders, 1989, pp. 503–509.

Fulginiti, V. The Problems of Poliovirus Immunizations. *Hosp Pract* 15:61–71, 1980.

Lundborg, G. *Nerve Injury and Repair*. Philadelphia: Churchill Livingstone, 1988.

Partridge, L., and Partridge, L. D. *The Nervous System: Its Function and Its Interaction with the World*. Cambridge, MA: MIT Press, 1993.

Reier, P., Bunge, R., and Seil, F. *Current Issues in Neural Regeneration Research*. New .York: Alan R. Liss, 1988.

Rowland, L. *Merritt's Textbook of Neurology*. Philadelphia: Lea & Febiger, 1989.

Salk, D. Eradication of Poliomyelitis in the United States: I. Live Virus Vaccine-Associated and Wild Poliovirus Disease. *Rev Infect Dis* 2:228–242, 1980.

7 Descending Motor Systems

The motor circuits in the spinal cord are regulated and controlled by motor systems in the brainstem and cerebral cortex. These regulatory systems are composed of several separate collections of neurons, each of which send long tracts of axons into the brainstem and spinal cord, where they terminate. Consequently, they are called the descending motor systems. Like the ascending sensory systems (see Chapter 5), the descending motor systems are a parallel set of pathways that have overlapping as well as complementary functions.

Four areas of the cerebral cortex are primarily concerned with generating motor commands. While these areas are interconnected, each is concerned with different aspects of motor control. These areas are, in turn, closely associated both functionally and anatomically with the basal ganglia and thalamus (see Chapter 8). Together, the four cortical areas and the subcortical motor nuclei of the thalamus and basal ganglia produce the motor commands that descend to the brainstem and spinal motor areas. Motor commands reach the brainstem and spinal cord in a form that does not ensure precise motor performance. Servocontrol mechanisms in the spinal cord and cerebellum (see Chapter 8) fine tune the motor commands, making supple, graceful, coordinated movement possible.

The Descending Motor Pathways

The descending motor pathways can be organized into four sets of tracts, a division based on the origin of the cell bodies. They are the **corticobulbospinal tract**, the **rubrospinal tract**, the **reticulospinal tracts**, and the **vestibulospinal tracts**. Each tract occupies a specific location in the brainstem and spinal cord.

Motor neurons in the spinal cord are not randomly scattered in lamina IX, but are precisely organized in a somatotopic arrangement that reflects the distribution of muscles in the extremities (Fig. 7.1). If one were to draw an imaginary line bisecting the ventral horn, in general, motor neurons located dorsal to that line would innervate flexor muscles and those ventral to it, extensors. Furthermore, the motor neurons that lie most medial would supply proximal muscles, while those lying most lateral, the distal muscles. This arrangement of motor neurons places the descending motor tracts in the lateral white column, such as the corticospinal tract, in a favorable position to preferentially innervate flexor motor neurons. The descending motor tracts that are in the ventromedial white column, such as the vestibulospinal tracts, are in a favorable position to preferentially innervate extensor motor neurons.

THE CORTICOBULBOSPINAL TRACT

The **corticobulbospinal (CBS) tract**[1] originates in the cerebral cortex and projects onto the

---

[1]Most physicians will refer to the CBS tract as the "pyramidal tract." Strictly speaking, the pyramidal tract refers only to the axons that are in the medullary pyramids. These axons represent only a small subset of the corticobulbospinal tract. The use of "pyramidal tract" as a synonym for the CBS tract is not only inaccurate, but has led to the even more unfortunate terms "pyramidal syndrome," "extrapyramidal system," and "extrapyramidal syndrome." A number of authors have advocated that these "pyramidal" terms be abandoned in favor of terms that express more accurately the anatomical and clinical situation.

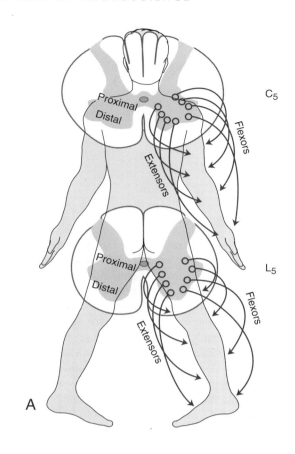

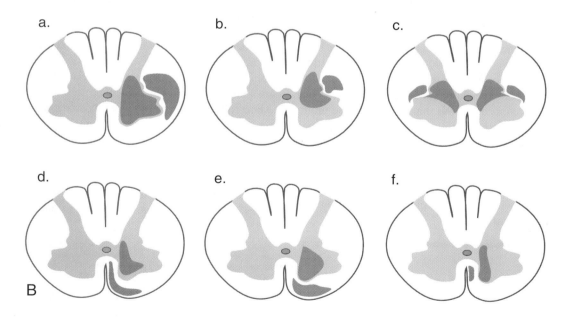

brainstem and spinal cord motor systems. This tract is frequently divided into the **corticobulbar tract** (those fibers that terminate in the brainstem) and the **corticospinal tract** (those fibers that terminate in the spinal cord). Taken together, the corticobulbar and corticospinal tracts represent the entire *voluntary cortical drive* to brainstem and spinal motor systems. As such, they have much in common, and it is logical to consider them as a single system under a single name.

Origin and Course of the CBS Tract. The CBS tract originates in three large areas of the cerebral cortex (Fig. 7.2). Brodmann's **area 4** in the frontal lobe is the most specific cortical motor area and is usually designated the **primary motor cortex** or **M-I**. About 31% of CBS axons originate in M-I. Another 29% originate in Brodmann's **area 6**, a motor region just anterior to M-I. Area 6 is divided into a lateral part, usually called the **premotor area (PMA)** and a medial part, usually called the **supplementary motor area (SMA)**. The remaining 40% of the CBS tract arises from the parietal lobe, specifically **areas 3, 1, and 2 (S-I)** (31%), and **areas 5, 7, 39, and 40** (9%), the latter known simply as the **posterior parietal cortex**.[2] The axons from these cortical areas destined to become the CBS tract leave the cerebral cortex and collect together

in the posterior portion of the posterior limb of the internal capsule. At the level of the mesencephalon, the axons occupy a small region in the middle portion of the cerebral peduncle.[3] This tract becomes disorganized in the pons, but becomes easily identifiable in the medulla as the medullary pyramid. In the spinal cord the CBS tract lies in the lateral white column.

The primary motor area of the cerebral cortex, M-I, is somatotopically organized in a way that is essentially identical to the organization of the primary sensory cortex (Fig. 7.3). Other motor areas, the SMA and the PMA, are somatotopically organized as well. One should note that the motor areas are supplied by two different cerebral arteries (see Chapter 1). The leg area of M-I and most of the SMA is supplied by the anterior cerebral artery. The trunk, hand, and face area of M-I and the entire PMA are supplied by the middle cerebral artery.

The columnar organization of the cerebral cortex found in S-I is also found in M-I, with several columns serving slightly different motor functions for the same general somatotopically defined area. Some motor columns are organized in such a way that they initiate activation of individual muscles. Other columns, when

---

[2]The frontal eye fields, Brodmann's area 8, are included in the CBS system by a few authors. Since the initiation and regulation of eye movements is considerably different from all other voluntary motor activity, is seems appropriate to keep area 8 separate from the CBS tract and postpone its consideration.

[3]Many authors continue to insist that the CBS tract occupies the middle two-thirds of the cerebral peduncle. This is clearly not possible. There are approximately 1 million corticospinal axons. There are no good studies giving us the number of corticobulbar axons, but they are probably no greater in number than the corticospinal fibers. Therefore, at most, the CBS tract at the level of the cerebral peduncle consists of 2 million axons. Since each cerebral peduncle contains no less than 20 million axons, the CBS tract can hardly occupy more than 10% of its area.

FIGURE 7.1.

A, The motor neurons in the spinal cord are distributed systematically within the gray matter of the ventral horn. Motor neurons innervating proximal muscles are located medially, while those innervating distal muscles are located laterally. Flexors are located along the dorsolateral margin, and extensors are located along the ventromedial margin. **B**, These diagrams illustrate the principal patterns of the descending motor pathways. The location of the descending axons is shown in the white matter. The area where these axons terminate in the spinal cord is shown in the gray area. These tracts can be organized according to their location in the spinal cord, either in the lateral white columns or the ventromedial white columns. Accordingly, the former chiefly innervate α neurons affecting flexor muscles, while the latter affect extensors. LATERAL WHITE COLUMNS: *a*, Lateral corticospinal tract; *b*, Rubrospinal tract; *c*, Medullary reticulospinal tract. VENTROMEDIAL WHITE COLUMNS: *d*, Pontine reticulospinal tract; *e*, Lateral vestibulospinal tracts; *f*, Medial vestibulospinal tracts.

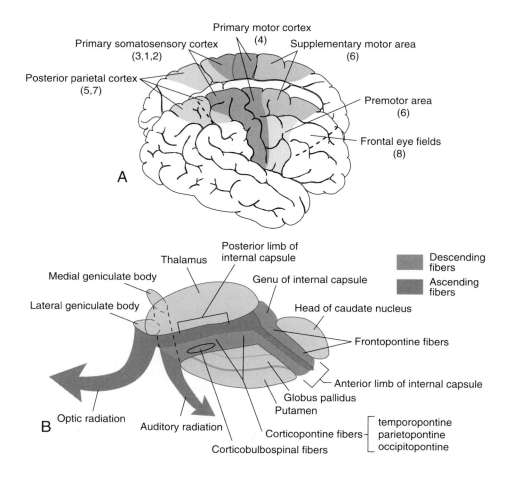

Primary motor cortex
(4)

Primary somatosensory cortex
(3,1,2)

Supplementary motor area
(6)

Posterior parietal cortex
(5,7)

Premotor area
(6)

Frontal eye fields
(8)

A

Posterior limb of
internal capsule

Thalamus

Medial geniculate body

Genu of internal capsule

Lateral geniculate body

Head of caudate nucleus

Descending
fibers

Ascending
fibers

Frontopontine fibers

Anterior limb of internal capsule

Optic radiation

Auditory radiation

Globus pallidus
Putamen

temporopontine
parietopontine
occipitopontine

Corticopontine fibers

Corticobulbospinal fibers

B

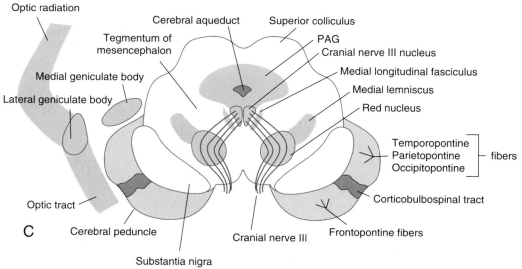

Optic radiation

Cerebral aqueduct

Superior colliculus

Tegmentum of
mesencephalon

PAG
Cranial nerve III nucleus

Medial geniculate body

Medial longitudinal fasciculus

Lateral geniculate body

Medial lemniscus

Red nucleus

Temporopontine
Parietopontine
Occipitopontine

fibers

Corticobulbospinal tract

Optic tract

Frontopontine fibers

C

Cerebral peduncle

Cranial nerve III

Substantia nigra

stimulated, cause contractions of several synergistic muscles simultaneously—contractions that produce movement in a preferred direction. Thus, just as in the sensory cortex, in the motor cortex there are numerous parallel systems of cortical columns, each serving related, but slightly different, functions.

Terminations of the CBS Tract. Axons of the CBS tract terminate in the brainstem and spinal cord. The corticobulbar division synapses in the **red nucleus**, motor areas of the **reticular formation** (see below), as well as in the various **motor nuclei of the cranial nerves**[4] (Fig. 7.4). The corticobulbar tract synapses ipsilateral or contralateral to the cells of origin, depending on the target. Corticobulbar axons destined for the red nucleus terminate exclusively on the ipsilateral side, but axons synapsing in the reticular formation project bilaterally. The terminations of corticobulbar axons in the cranial nerve motor nuclei vary widely; some nuclei receive strictly contralateral innervation, while others receive bilateral innervation. These differences related to the cranial nerves will be discussed as each nerve is presented (see Chapter 9).

In contrast to the corticobulbar tract, the vast majority of axons in the corticospinal tract cross the midline before terminating. The majority cross at the pyramidal decussation, becoming the **lateral corticospinal tract** in the spinal cord. A variable number of the corticospinal axons, probably never more than 10% of the total tract, continue uncrossed in the spinal cord as the **ventral corticospinal tract** (Fig. 7.4). Most of these axons cross the midline at the level of their termination.

Corticospinal fibers terminate in nearly the entire gray matter of the spinal cord, specifically laminae IV to IX. Corticospinal axons originating in the parietal lobe terminate in the more dorsal regions of the spinal gray matter (laminae IV to VI), while the terminations of axons from the frontal lobe are more broadly distributed (laminae V to IX) (Fig. 7.1). It is clear from the location of corticospinal terminations that this tract must synapse primarily on interneurons. Consequently, excitation of α and γ motor neurons by the corticospinal tract is indirect, effected through polysynaptic pathways. There are a few direct, monosynaptic connections by corticospinal axons on α motor neurons, particularly in primates and humans. However, even in the human, where monosynaptic connections are most numerous, they represent no more than a few percent of the total contacts the corticospinal tract makes in the spinal cord.

In the spinal cord, the corticospinal tract courses in the lateral white column. From this position, one would expect that this tract would facilitate primarily flexor motor neurons; while this is generally true, the flexor dominance pertains mostly to the proximal muscles. Both flexor and extensor motor neurons supplying the distal muscles are facilitated by the corticospinal tract. Furthermore, the corticospinal neurons innervating distal muscles are more likely to synapse with motor neurons monosynaptically. Therefore, there is a definite functional bias in the corticospinal system toward fine control of the distal extremities and coarser regulation of the proximal flexors.

THE RUBROSPINAL TRACT

The red nucleus is a nearly spherical body located in the mesencephalon. In fresh cadaver sec-

---

[4]Some authors refer to corticorubro or corticomedullary tracts when discussing those specific connections.

FIGURE 7.2.

A, The principal motor areas of the human cerebral cortex. The numbers refer to the Brodmann areas. The supplementary motor area (SMA) and the premotor area (PMA) are both located in area 6. **B**, A horizontal section through the brain illustrating the relationship between the basal ganglia, the thalamus, and the internal capsule. Within the internal capsule, the locations of the corticobulbar and corticospinal axons are shown and their relationship to other ascending and descending pathways. **C**, The location of the corticobulbospinal fibers in the cerebral peduncle at the level of the red nucleus.

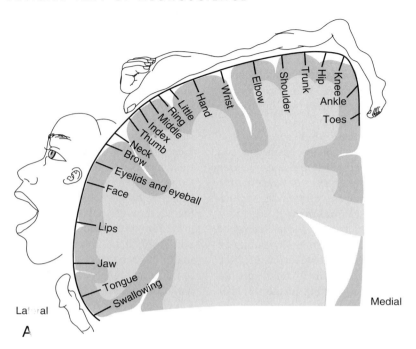

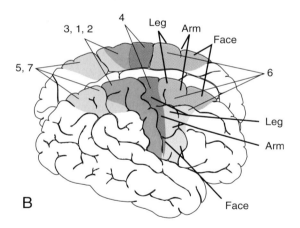

FIGURE 7.3.

  A, The somatotopic organization of the human M-I, shown here, closely mirrors the organization of S-I (see Fig. 5.15). **B**, The principal areas of the cerebral cortex that contribute to the corticobulbospinal tract. The somatotopic arrangements of M-1 (*dark red*), the SMA (*medium red,* area 6) and the PMA (*light red,* area 6) are shown.

tions the nucleus has a salmon color due to the high iron content of its cells. Once thought to be vestigial in the human, the rubrospinal tract is now known to be an important descending motor system in all vertebrates, including humans. Axons originating in neurons of the red nucleus form the **rubrospinal tract** (Fig. 7.5). Rubrospinal ax-

ons cross the midline immediately after leaving the nucleus, pass through the brainstem, and merge with the corticospinal tract in the cervical spinal cord. Within the spinal cord the rubrospinal tract and the corticospinal tract are intermingled. Rubrospinal axons terminate in approximately the same laminae as corticospinal axons (Fig.

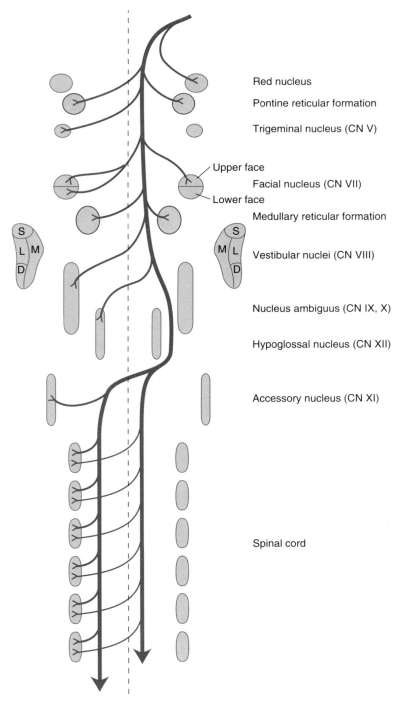

Red nucleus

Pontine reticular formation

Trigeminal nucleus (CN V)

Upper face

Facial nucleus (CN VII)

Lower face

Medullary reticular formation

Vestibular nuclei (CN VIII)

Nucleus ambiguus (CN IX, X)

Hypoglossal nucleus (CN XII)

Accessory nucleus (CN XI)

Spinal cord

FIGURE 7.4.

Schematic diagram depicting the course and principal termination sites of the corticobulbospinal tract.

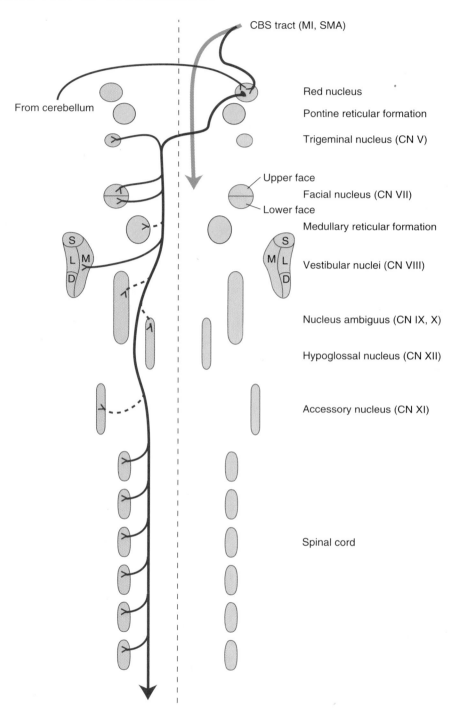

FIGURE 7.5.

A schematic illustration of the course and principal sites of termination of the rubrospinal tract.

7.1). Efferents from the red nucleus also terminate in various nuclei of the brainstem, including the facial nucleus, the trigeminal nucleus, the vestibular nuclei, and the dorsal column nuclei.

The red nucleus receives most of its afferent connections from the cerebellum. This pathway will be discussed in Chapter 8. Of interest here are the substantial numbers of afferents it receives from the cerebral cortex. Monosynaptic corticorubric afferent connections have been mapped in several species, and a somatotopic arrangement between the cerebral cortex and the red nucleus has been established. For example, pyramidal cells in the leg area of the motor cortex synapse on neurons in the leg area of the red nucleus that in turn synapse on neurons in the lumbar spinal cord. Cortical afferents to the red nucleus arise primarily from the ipsilateral primary motor area (M-I), but there is a significant contribution from the ipsilateral supplementary motor area (SMA) also. Most of the cortical axons synapsing in the red nucleus appear to terminate there and are true corticorubric axons. However, at least some corticospinal axons send collateral fibers into the red nucleus, where they synapse.

THE RETICULOSPINAL TRACTS

The reticular formation consists of the brainstem gray matter that is not well organized into obvious nuclei. Recently, areas of the reticular formation have been associated with specific functions, and neurons within these areas have been described and recognized as nuclei. Few named nuclei in the reticular formation are as discrete and prominent as the more typical nuclei associated with the cranial nerves. But this does not diminish their importance. The motor nuclei of the reticular formation are discussed here. Other reticular nuclei are presented in subsequent sections.

The **medullary reticulospinal tract** arises from the **nucleus reticularis gigantocellularis** in the medulla (Fig. 7.6). It descends *bilaterally* to all spinal levels, intermingled in the spinal cord with the rubrospinal and the corticospinal tracts in the lateral white columns (Fig. 7.1). Terminating primarily in laminae VII and VIII, it synapses on interneurons that facilitate flexor α and γ motor neurons. The **pontine reticulospinal tract** arises from two pairs of nuclei,

the more rostral **nucleus reticularis pontis oralis** and the **nucleus reticularis pontis caudalis**. Axons arising from these nuclei descend *ipsilaterally* to all spinal levels, coursing in the ventromedial funiculus. They also terminate primarily in laminae VII and VIII, facilitating, through interneurons, extensor α and γ motor neurons.

The reticular motor neurons that give rise to the reticulospinal tracts receive afferents from the spinal cord, the cerebellum, and the cerebral cortex (Fig. 7.6). Some of the ascending spinal connections of the **ALS (spinoreticular tract)** provide sensory input into the reticular motor nuclei, although the exact sensory modalities represented have not been worked out. Cerebellar afferents into the reticular formation will be discussed in Chapter 8. Cortical afferents into the reticular motor nuclei arise from M-I and the PMA. The cortical projections are distributed *bilaterally* in the reticular formation, and many monosynaptically excite the reticulospinal neurons that project into the spinal cord.

THE VESTIBULOSPINAL TRACTS

The vestibular complex consists of four nuclei—the **superior**, the **lateral**, the **medial** and the **inferior vestibular nuclei**—all of which receive afferent connections from the vestibular nerve (part of cranial nerve VIII). Several long fiber tracts originate in the vestibular complex. Only two of them, the **medial vestibulospinal tract** and the **lateral vestibulospinal tract**, are motor tracts that descend into the spinal cord (Fig. 7.7).

The **lateral vestibulospinal tract** originates in the **lateral vestibular nucleus**. It descends to all levels of the spinal cord, coursing in the *ipsilateral* ventral medial funiculus (Fig. 7.1). Its axons terminate in laminae VII and VIII. A few are known to synapse on α motor neurons in lamina IX, but this is clearly the minority. As one might expect, given the ventral and medial location of this tract, it primarily facilitates extensor α and γ motor neurons.

The **medial vestibulospinal tract**[5] originates in the **medial vestibular nucleus**. It descends to cervical and high thoracic levels only; there are

---

[5]Formerly called the medial longitudinal fasciculus (MLF) of the spinal cord, this tract is now called the medial vestibulospinal tract by most authors.

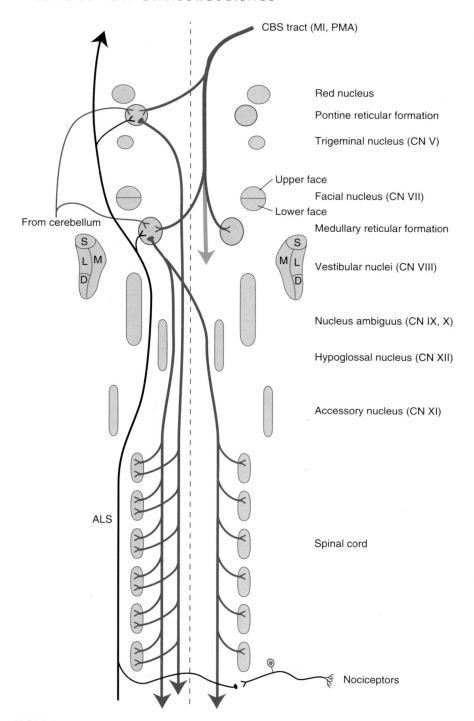

FIGURE 7.6.

This schematic illustrates the course and principal connections of the reticulospinal tracts.

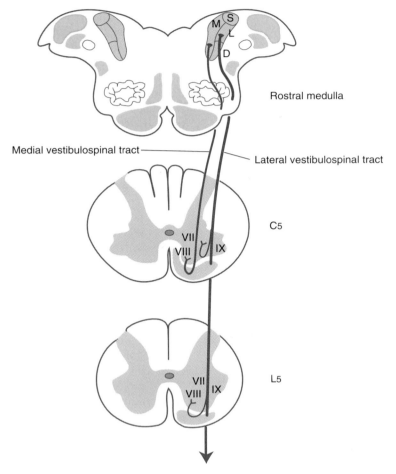

Medial vestibulospinal tract

Lateral vestibulospinal tract

Rostral medulla

C5

L5

FIGURE 7.7.

This diagram illustrates the major features of the vestibulospinal tracts. Note that the medial vestibulospinal tract descends only to cervical levels, while the lateral tract innervates the entire spinal cord.

no projections to the lumbar spinal cord (Fig. 7.1). Like the lateral vestibulospinal tract, it synapses on interneurons in the ipsilateral laminae VII and VIII, and it makes a few monosynaptic connections with α motor neurons in lamina IX. The medial vestibulospinal tract primarily facilitates α and γ motor neurons innervating muscles of the neck that stabilize the head. Its principal functional role is to provide a stable platform for the eyes, and this system is intimately connected with the pathways that regulate eye movement.

The Cortical Motor Areas

The various regions of the cerebral cortex that are closely associated with motor function are interconnected by intracortical short association fibers. Since the various motor regions serve different facets of motor control, a brief review of these interconnections is necessary to have a basis for understanding the physiology of the region.

INTRACORTICAL CONNECTIONS

Somatotopically related areas of each of the three motor areas of the cerebral cortex, M-I, SMA, and PMA are precisely interconnected. The short association axons that make these connections arise from pyramidal cells of cortical layer II.[6] The SMA and PMA are reciprocally interconnected with one another, and both independently provide reciprocal connections to M-I (Fig. 7.8). M-I not only receives afferents from these "motor" areas, but also from the "sensory" areas: area 3a of S-I and areas 5 and 7 of the postcentral gyrus.

Pyramidal cells of cortical layer III in each of the motor areas also make somatotopically appropriate connections with the same motor areas in the contralateral hemisphere. These interhemispheric connections are most numerous between the areas corresponding to the trunk and proximal extremities. No interhemispheric connections are made between the areas representing the most distal part of the extremities, the hands and the feet. This relationship apparently facilitates side-to-side coordination between the anti-gravity and postural muscles while allowing for relatively greater independence of the prehensile hands and feet.

THALAMIC CONNECTIONS

In addition to the intracortical connections described above, the motor areas of the cerebral cortex receive fibers from three noncortical areas: the cerebellum, the basal ganglia, and peripheral sensory systems. All three of these systems are connected with the cerebral cortex through thalamic nuclei. Most of the details of these connections will be postponed (see Chapter 8). However, a few general observations are in order here.

Specific nuclei in the thalamus make extensive reciprocal connections with the motor areas of the cerebral cortex (Fig. 7.8). The **ventral lateral nucleus (VL)** and the **ventral anterior nucleus (VA)**, both in the ventral nuclear group of the thalamus, make the most extensive connections with the cortical motor areas. The ventral lateral nucleus has recently been divided into three subnuclei, the **pars oralis (VL$_O$)**, the **pars caudalis (VL$_C$)**, and **area X**. The ventral anterior nucleus has been divided into two subnuclei, the **magnocellular (VA$_{MC}$)** and the **parvocellular (VA$_{PC}$)** nuclei. While the nomenclature is confusing, the distinctions are based on cytological differences and the very specific connections these nuclei make with functionally separate areas of the motor cortex. A brief discussion of these connections will serve to make two important points. First, each of the three motor areas of the cerebral cortex (M-I, PMA, and SMA) has its own thalamic connections. Second, pathways from the cerebellum and basal ganglia each have separate feedback loops that converge on the motor cortex.

The basal ganglia connect with the motor cortex through VL$_O$ and VA$_{PC}$. These connections are primarily to SMA with a few connections to M-I from VL$_O$. The influence of the basal ganglia on eye movements is affected by VA$_{MC}$ projections to area 8, the frontal eye fields. In contrast, cerebellar pathways project to the motor cortex through three closely related nuclei. The primary motor cortex, M-I, receives most of these projections after they have passed through VL$_C$ and VPL$_O$. The PMA receives its thalamic projections from area X and a smaller contribution from VL$_C$. The physiological importance of these loops will be discussed in Chapter 8.

The nonspecific intralaminar and reticular nuclei of the thalamus also contribute afferent axons to the motor regions of the cortex. The **centromedian nucleus (CM)**, the largest of the intralaminar nuclei, makes reciprocal connections to wide areas of the frontal cortex. It also makes reciprocal contacts with the caudate and globus pallidus nuclei of the basal ganglia. Other connections related to the intralaminar and reticular nuclei of the thalamus and the motor areas of the cerebral cortex are not well understood, nor are their functions. Current speculation suggests that the nonspecific thalamic nuclei regulate the overall excitability of the cerebral cortex, a hypothesis that will be discussed in Chapter 8.

---

[6]Recall that the pyramidal cell in the cerebral cortex is named for its shape, not for any relationship it may have to the "pyramidal tract" (see Chapter 5).

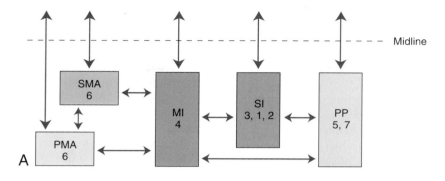

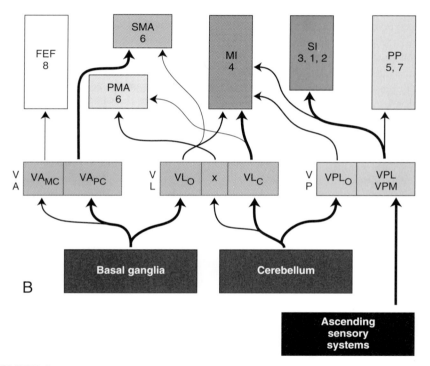

FIGURE 7.8.

A, Each of the three motor control regions of the cerebral cortex are somatotopically interconnected as depicted in this schematic illustration. Note the general convergence of signals onto M-I, not only from traditional "motor" areas, but also from parietal areas that are usually considered "sensory". *Arrows* at the top depict callosal fibers crossing to the contralateral hemisphere. **B**, The principal connections between the thalamic nuclei and the sensorimotor areas of the cerebral cortex are illustrated in this schematic drawing. Note that the basal ganglia and the cerebellum have separate gateways through the thalamus to the cortex. Note also that areas 4 and 6 receive afferents from several thalamic nuclei. Abbreviations: FEF, frontal eye fields; SMA, supplemental motor area; PMA, premotor area; M-I, primary motor cortex; S-I, primary sensory cortex; PP, posterior parietal cortex; VA, ventral anterior thalamic nucleus (composed of VA_{MC} and VA_{PC}); VL, ventral lateral thalamic nucleus (composed of VL_O, VL_C, and area X); VP, ventral posterior thalamic nucleus composed of VPL, VPL_O, and VPM). The numbers refer to the Brodmann areas.

Functional Responses of Cortical Motor Areas

Ablation experiments in animals have provided valuable information about cortical function. In these cases, animals are first trained to perform specific motor acts. Then a region of the cerebral cortex is removed. By analyzing deficits in the animal's ability to perform the previously learned motor act, one often gains insights into cortical function. Such experiments can be of great value when the lesions are made to mimic those that occur naturally in humans following cerebral injury.

Experiments involving awake, unanesthetized primates trained to perform specific motor tasks have also contributed greatly to our understanding of the functional relationships between the various motor areas of the cerebral cortex. In experiments such as these, one can record from single cells while the animal is performing a volitional act. This has provided unprecedented information about the functional organization of the cerebral cortex. The following paragraphs summarize some of this information.

PREMOTOR AREA

The PMA in the human is about six times larger than M-I. In spite of its size, the PMA contributes less to the CBS tract than M-I, and electrical stimulation does not usually produce muscle movement unless the stimuli are much more intense than effective stimuli for the SMA and M-I. What then is the role of the PMA in motor activity? A totally satisfactory answer to this question is not available, but available data suggests that the PMA is necessary to prepare M-I for the impending motor act.

Clever experiments using trained monkeys have provided certain insights into PMA function. In one set of experiments, a monkey was trained to rest his arm, between trials, in a specified location. Several targets were placed at various locations within reach. One of the targets would be lit up, indicating a "ready" signal, but the monkey was not allowed to touch it until a second "go" light came on. At the "go" signal, the monkey reached for the illuminated target to get his reward. The same paradigm was also established using auditory clues instead of visual clues.

While the monkey was performing his tasks, electrodes were placed in various parts of his motor cortex and the responses of individual cells recorded. In the PMA, many cells significantly increased their firing rate between the "ready" and the "go" signals, but only if the anticipated movement was in a specific direction (Fig. 7.9). Surprisingly, these PMA neurons usually decreased firing once the movement actually began. Experiments like these suggest that many

FIGURE 7.9.

A, The functional role of PMA cortical cells was studied in freely behaving monkeys. The animals were trained to rest their arm at a specific starting position (1) between trials. A light would come on at one of the other stations (2), the "ready" signal. Only when a second light at the same station appeared (3), the "go" signal, would the monkey reach out (4) and touch the illuminated panel to extinguish the lights and receive his reward. The various time periods between "ready" and "go" were randomized as shown below. **B**, Recordings of the activity of single PMA cells were made as the animals were performing this task. At the bottom of each of the four figures, each line represents one trial and each dot indicates the moment the neuron fired. At the top of each figure is a composite histogram of all trials, showing the number of action potentials versus time. All the records are aligned on the left at the "ready" signal and on the right on the beginning of movement. To keep all records the same physical length, data in the center needed to be discarded because of the randomization of the trials. Note that the PMA cells increased their firing rate after the "ready" signal only if the arm was to move to the right. The same cell was inhibited if the movement was to the left. The response was the same whether the signals were auditory or visual. These PMA cells did not participate in the actual motor act, since they were inhibited just before the arm began to move.

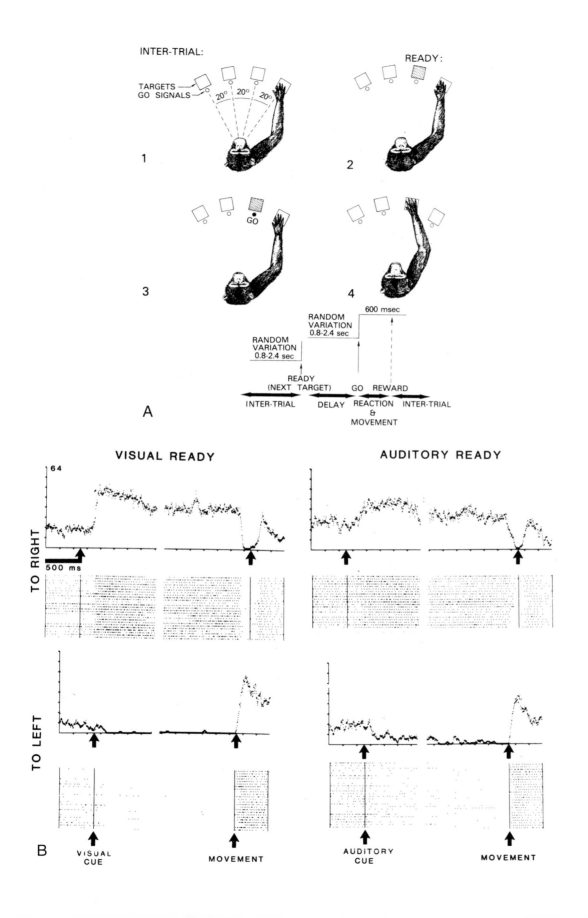

PMA neurons are anticipating a specific complex motor act. Their activity seems to be facilitating multiple motor columns in M-I, preparing them for action. Lesions in humans involving PMA do not cause paralysis, but slow complex limb movements. The slowing is apparently related to the inability to facilitate M-I in preparation for anticipated motor commands.

SUPPLEMENTAL MOTOR AREA

The SMA is necessary to coordinate complicated, bimanual motor acts. Stimulation of neurons in the SMA elicit complex movements, movements often involving the entire hand or arm or, in some cases, even postural movements of the whole body. Lesions in the SMA in humans result in motor apraxia [G. *a,* negation + G. *pratto,* to do] (loss of motor skill without paralysis) rather than paralysis.

Controlled SMA lesions in trained monkeys have cast some light on the role of the SMA in motor acts. A unilateral ablation of the SMA limits a monkey's ability to perform complex bimanual tasks. This ablation causes both hands to operate together, mirroring each other's movement, as if the intact SMA were controlling both hands. For example, in one classic experiment, a monkey was presented with a raisin lodged in a hole in a transparent plexiglass plate. To retrieve the raisin, the normal monkey simply pushed it through the hole from the top with its finger, catching it beneath the plate in its palm. After unilateral removal of the SMA, the monkey pushed the raisin from both sides simultaneously and was unable to retrieve the reward.

If the ipsilateral SMA and PMA are both removed, an entirely different apraxia appears. In another well-known experiment, a monkey with such a lesion was presented with an apple slice that was behind a small plexiglass barrier. The monkey could easily reach around the barrier to retrieve the apple with the hand ipsilateral to the lesion (remember the motor cortex controls the contralateral body), but would attempt to thrust his hand through the plexiglass with the arm contralateral to the lesion. If the lesion involved only M-I, the monkey could accomplish the task with either hand, although the hand contralateral to the lesion was necessarily weak and clumsy.

Experiments of this sort show that the SMA and PMA are not necessary for *motivating* the monkey to perform the motor act, nor are they necessary for the monkey to *understand* how to accomplish the act, for he would do just fine with the good arm. And, of course, the monkey was not paralyzed, for he could thrust his arm into the plexiglass barrier repeatedly. Somehow, the SMA and PMA are necessary to translate the knowledge of how to accomplish a motor act into a specific sequence of motor commands— in other words, *translating strategy into tactics.*

Experiments on humans have given some support to this interpretation of the monkey experiments. In humans, cerebral blood flow can now be measured noninvasively, and it has been shown that local cerebral blood flow increases as local neuronal activity increases. If a person is asked to tap a finger, cerebral blood flow increases over the hand area in S-I and M-I, indicating that both the motor and sensory areas have become more active (Fig. 7.10*B*). If instead of simple finger tapping,

---

FIGURE 7.10.

Changes in regional blood flow during motor activity in the human have shed some further light on the role of the SMA and PMA in motor acts. **A**, During isometric muscle contraction of the fingers, regional blood flow increased only over the contralateral M-I hand area of the cortex. **B**, During simple, repetitive finger tapping, blood flow increased in the hand areas of both M-I and S-I. **C**, A sequence of complex finger movements produced a bilateral increase in blood flow in the SMA as well as a contralateral increase in M-I and S-I. The subject had to concentrate on the act to do it correctly, since the sequence was difficult to coordinate, as opposed to the simple repetitive "mindless" act used in **B**. **D**, Blood flow increased bilaterally in the SMA but not in M-I and S-I if the subject simply thought through the motor act used in **C** but did not actually move his fingers. **E**, If the spatial reference for the motor act was outside the body, blood flow increased bilaterally in the SMA and contralaterally in the PMA, M-I, S-I, and the posterior parietal cortex. In this case, the subject was asked to draw a spiral pattern in air.

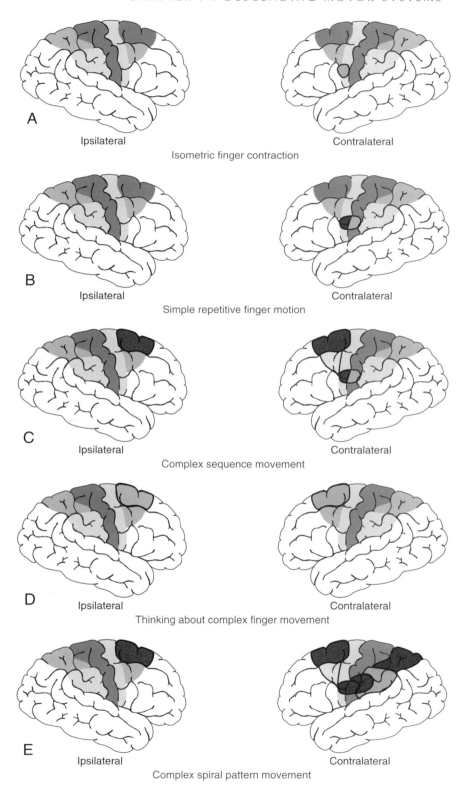

A   Ipsilateral   Contralateral
Isometric finger contraction

B   Ipsilateral   Contralateral
Simple repetitive finger motion

C   Ipsilateral   Contralateral
Complex sequence movement

D   Ipsilateral   Contralateral
Thinking about complex finger movement

E   Ipsilateral   Contralateral
Complex spiral pattern movement

the subject is asked to perform a complex sequence of movements with the fingers, blood flow in the SMA increases as well as in S-I and M-I (Fig. 7.10*C*). Finally, if the person is asked simply to think through the complex sequence without actually moving the fingers, blood flow increases only in the SMA (Fig. 7.10*D*). Experiments such as these not only suggest a role for the SMA in preparing for complex acts that require some planning, but also show that the SMA is not involved in very simple repetitive acts that require no "conscious" thought or preparation.

POSTERIOR PARIETAL AREA

The posterior parietal area is an enigmatic region of the cerebral cortex. In spite of its large size, only about 9% of the CBS fibers originate from it. Stimulation of the posterior parietal cortex does not result in consistent motor activity. Its motor functions seem to be related to motivation or interest, qualities of behavior that are difficult to quantify and thus difficult to study. Humans with lesions to the posterior parietal area display a behavior known as **neglect**. They seem *unaware* of the side of the body contralateral to the lesion. Although there may be no demonstrable lesion to the sensory system, such patients will ignore tactile stimuli presented to the neglected side. Similarly, they will ignore visual stimuli. This behavior is particularly evident if the lesion is to the nondominant hemisphere (see Chapter 12). Such a patient, when asked to draw a clock, will draw an egg-shaped figure and place all 12 numerals on one side of the figure, seeming to be unaware of the rest of the visual space (Fig. 7.11). Motor acts are similarly neglected, although there is no demonstrable paralysis.

Single unit recordings from trained monkeys performing specific tasks have provided some data that support the view that the posterior parietal cortex is an essential link between motivation and action. For example, some posterior parietal neurons in area 5 fire when the monkey reaches for something interesting, such as food, but do not fire during similar arm movements that are not associated with motivationally interesting objects. Similar neurons have been observed in area 7, but they are associated with eye movements that are directed toward interesting objects in the visual field.

PRIMARY MOTOR AREA

Anatomical and physiological studies consistently show that activity in cortical neurons related to motor acts begins in widespread areas of the cortex and converges on the primary motor area just before movement takes place. The activity of M-I cells influences small groups of motor neurons, giving this area more direct control of motor activity than any other cortical region. Although it is tempting to assume that there is a direct relationship between a single cortical pyramidal cell and a single motor neuron in the brainstem or spinal cord, in fact *populations* of motor neurons are controlled by *populations* of cortical neurons.

Populations of cortical neurons in M-I encode the *direction* of movement. In one particularly interesting study, an awake monkey was trained to push a lever with his arm in one of eight possible directions. For movement in each direction, action potentials were recorded from neurons in the precentral cortex. Each neuron fired most during movement in one of the eight directions, but all neurons increased their firing rates for about half of the directions and decreased their rates for the other half (Fig. 7.12). To gain insight into the behavior of all of the neurons at once, a single vector diagram was constructed for each of the eight directions. Each diagram represented the relative activity of all of the neurons during a given movement. Summing all of the vectors of each neuron produced a summed vector for the population of neuron that closely matched the actual direction of movement.

The *force* of muscle contraction is encoded by the firing rate of M-I neurons. This has been shown in many ways. For example, the activity of cortical neurons increases as the load placed on the muscle increases, although the length of contraction remained the same for each trial. In other experiments, if the load were increased in the middle of a muscle contraction, the firing rate increased to accommodate the increased load. How the nervous system computes the anticipated force required is an interesting unanswered question. We all know, however, that the brain can be easily fooled, for each of us has experienced the disorientation associated with lifting a package

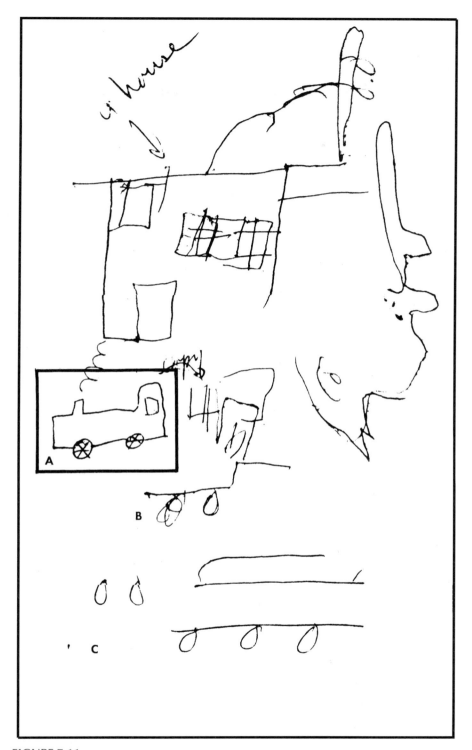

FIGURE 7.11.

Patients with certain lesions to the right posterior parietal cortex are not cognizant of the left half of their world. When they attempt to draw figures, as in this example, they exhibit what is called a constructional apraxia. The figure in the box labeled *A* was drawn by the examiner. The patient's attempts to copy the figure are labeled *B* and *C*. Attempts by the patient to draw a house or the examiner are at the top of the figure.

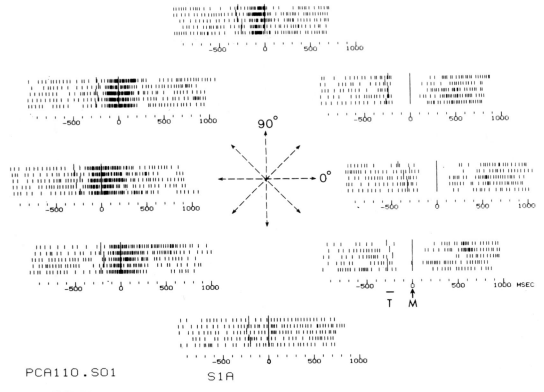

FIGURE 7.12.

Many cortical neurons are activated when a limb is moved voluntarily in a particular direction. To study this, monkeys were trained to push, on command, a lever in one of eight directions. This figure shows the firing patterns of one cell during movement in each of the eight directions. Five trials are shown for each direction; each line represents one trial and each dot within the line represents one action potential. The abscissa is in milliseconds relative to the beginning of movement at 0. This neuron increased its firing for about half of the directions and was inhibited for the other half. The firing rate for this individual neuron increased most when movement was directed between 135° and 180°. These data can be summarized as a series of vectors (*closed arrowheads*), one for each direction of movement, as shown at the center of the figure. A composite vector (*open arrowhead*) can be computed to represent the preferred direction for this particular neuron.

that weighs substantially less than what was expected.

Not all cortical neurons encode force. Some seem to be encoding the *velocity* of applied force. The speed with which one moves an appendage is under voluntary control, so it seems appropriate that velocity-encoded neurons would exist. They have in fact been observed in M-I. However, these neurons seem to exert their influence indirectly through the red nucleus. In one study only 10% of corticospinal neurons studied encoded velocity information, but 70% of the rubrospinal neurons did.

INTEGRATION OF MOTOR COMMANDS

As we have seen from the preceding discussion, the descending motor tracts exert their control over the final muscle contraction through multiple parallel pathways. It is useful to ferret out some of the differences in how these parallel tracts affect motor activity.

The corticospinal tract primarily regulates the most discrete muscle contractions of the most distal muscles. This is reflected in both the anatomy and the physiology, for the monosynaptic corticospinal connections with motor neurons exist predominantly on the distal motor neurons. Similarly, the lowest-threshold cortical units command the most distal muscles. Furthermore, corticospinal neurons are most active during the most delicate and precise finger movements. These same neurons may even decrease their firing rate if a strong, forceful gripping action is required. Rubrospinal neurons are very similar to corticospinal neurons. In general, the rubrospinal tract innervates proximal muscles in preference to distal muscles, and velocity in preference to force.

Stronger, sustained, more forceful muscle contractions under cortical control are more closely associated with reticulospinal neurons that synapse more densely on the motor neurons supplying proximal muscles of the extremities and muscles of the trunk. The vestibulospinal tracts, especially the lateral vestibulospinal tract, also predominantly affect proximal muscles. The vestibulospinal tract is unique because it is not subject to direct cortical regulation. Due to its close association with the peripheral vestibular apparatus, the vestibulospinal activity is most closely associated with one's orientation to gravity. Collectively, the reticulospinal and vestibulospinal systems may be considered *postural,* since they establish a background level of activity that resists the effects of gravity and establish the organism's orientation in space.

The Motor Systems as a Control System

The application of control system theory to physiological systems has provided important insights into the normal function of many physiological systems as well as providing a basis for understanding the ways in which these systems fail. Analyzing the motor systems in terms of control system theory is a useful exercise, for it establishes a framework against which clinical observations of patients with motor system lesions can be interpreted.

With controlled systems, there is a defined relationship between the output of the system and its input. If there is no communication between the output and the input, the system is said to be an **open-loop** system (Fig. 7.13). If, however, the input to the system is modified in some way by the output, the system is said to be a **closed-loop** system. Closed-loop control systems provide a *dynamic* mechanism through feedback regulation by which an output variable is regulated, within certain limits, around a desired value.

Closed-loop control systems fall into two classes: **feed-forward** and **feedback** systems. Feedback systems measure the **controlled variable** (output) in the environment and compare it with the desired value, the **set point**. If there is a discrepancy, an appropriate correction is applied to the input variable *after* the error has been detected. Feed-forward control systems anticipate the effect environmental disturbances will have on a system and apply corrective action *in advance* of a measured error in the controlled variable. Physiological motor systems have characteristics of both feed-forward and feedback control systems. Feed-forward systems will be considered in Chapter 8.

FEEDBACK CONTROL SYSTEMS

The feedback control system can be modeled with six elements: a **sensor**, the **set-point** signal, the **comparator**, the **effector**, the **controlled variable**, and the **error signal**. For example, in one's home, *temperature,* the controlled variable, is regulated by a *heating system,* the effector. A *thermometer* (the sensor) senses the temperature, which is compared by the *thermostat* with the *desired temperature* (set point) (Fig. 7.14). The thermometer continuously monitors the building temperature, while the thermostat continuously compares the present temperature with the set point. If the temperature differs from the set point, the comparator sends an error signal to the heating-cooling system, which starts the furnace if the building is too cool (or, if the building is too warm, shuts off the furnace).

Simple Controllers

With a simple thermostat, the error signal is either on or off. In other words, the thermostat can signal that the temperature is either below the set point or above it. It cannot signal the amount of deviation (see Box 7.1). A simple on-

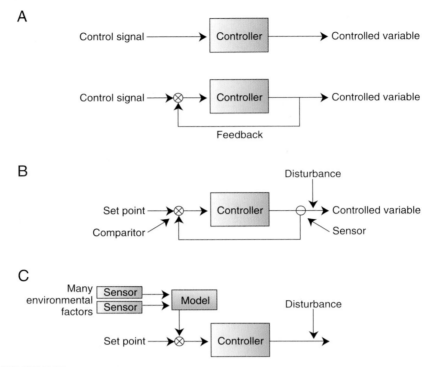

FIGURE 7.13.

A, Control systems are classified as open loop (*top*) if the controlling signal is not modified by the output of the system. They are closed loop (*bottom*) if some portion of the output is used to modify the control signal (feedback control). **B**, Feedback systems modify the control signal after an error in the controlled variable has been detected. **C**, Feedforward systems modify the control signal in response to events that are anticipated to affect the controlled variable. The anticipated effects are usually computed by comparing the existing state of the entire system against an idealized model.

off controller like a thermostat must not switch states the instant the temperature deviates from the set point, because the effector needs time to function. Therefore, a delay (called **hysteresis** [G. *hysteresis,* a coming later]) is built into the comparator, allowing the temperature to deviate from the set point by a substantial amount before the effector is activated. Because of the hysteresis, the controlled variable will oscillate within defined limits.

Proportional Controllers

With more sophisticated comparators, the error signal is **proportional** to the degree of deviation from the set point. In these systems the effector is also proportional. Using our furnace example, the *rate* at which heat is added to a cold building is a function of the magnitude of the error signal. Proportional controlled systems can regulate a variable much more closely than a simple on-off system, and they can be more responsive to a wider range of disturbing forces (see Box 7.2).

Gain.   The speed at which a proportional control system corrects a disturbance in the controlled variable is determined by the **gain** of the system. Gain is simply the amplification factor in the system that determines the magnitude of the applied correction relative to the magnitude of the error signal. In our heating system example, if the error signal were 10 (arbitrary units) and the furnace delivered 1000 kcal/hr, the gain of the system would be 100. If the furnace delivered 2000 kcal/hr, the gain would be 200. A system with a higher gain can respond more quickly to a disturbance.

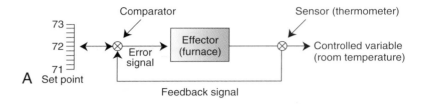

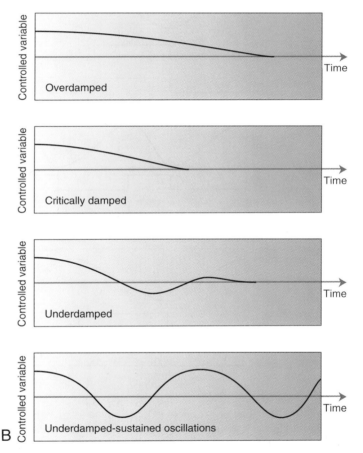

FIGURE 7.14.

A, Block diagram of a feedback control system with all the major elements labeled modeled as a typical heating system that uses feedback regulation to control building temperature. **B**, Proportional control systems oscillate if the gain is too great. The tendency to oscillate is expressed as a damping factor. The gain of overdamped systems is not optimal, for they respond slowly to disturbances. The gain of a critically damped system is set just below the point where oscillations first appear. This brings the system to the set point in the minimum time with no oscillations. In an underdamped system, the gain is set so high that the system overshoots the set point, causing oscillations. The oscillations gradually diminish in amplitude until the system becomes stable at the set point. Systems in which the gain is too high are unstable and oscillate indefinitely.

BOX 7.1.

Simple Control Systems

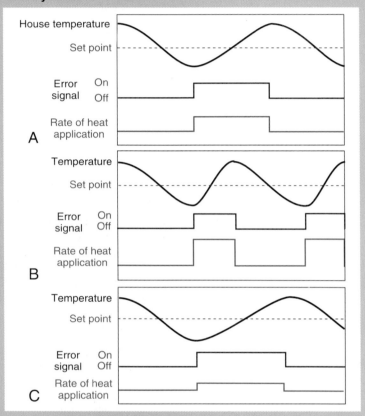

A simple nonproportional control system is characterized by an error signal that is either on or off. Using a home furnace as an example (*A*), the house loses heat at a steady rate according to how well the home is insulated (and other obvious factors we will ignore in this example). Contacts within the thermostat close when the house temperature falls too far below the desired temperature (set point). When closed, the contacts act as a switch that turns on the furnace. The furnace adds heat to the house at a steady rate. When the air temperature rises a certain amount above the set point, the thermostat contacts open and the furnace is shut off. The cycle repeats.

A simple system like this suffers from the fact that the response of the thermostat lags behind the actual room temperature. It cannot signal that the temperature has deviated from the set point until that deviation has exceeded a certain limit. This lag in response is called **hysteresis** [G. *hysteresis,* a coming later]. The system oscillates above and below the set point. Note that increasing (*B*) or decreasing (*C*) the size of the furnace (increasing or decreasing the system gain) does not solve the problem of oscillation.

BOX 7.2.

Proportional Control Systems

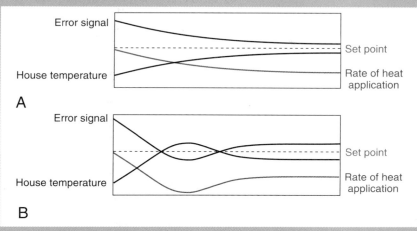

A

B

Proportional control systems can regulate an output variable very closely because the system responds to the magnitude of the disturbance. Consider a sophisticated heating system. Such a system would have an electronic thermometer that provides a voltage that is a continuous function of temperature and an electrical heating element that supplies heat in proportion to the applied electrical current. Given a large initial disturbance, the error signal is large. Consequently the controller supplies a large electrical current to the heater, which generates a large amount of heat. As the difference between the controlled temperature and the set point diminishes, the error signal diminishes proportionately and so does the amount of heat generated by the heater. Eventually the system approaches a steady state whereby a small residual error signal causes the system to generate heat at a rate that just matches the rate of heat loss (*black line*). If the rate of heat loss changes, the system will adapt itself to that change and generate heat at a new rate that matches the new heat loss. Well-designed proportional control systems do not exhibit the oscillations that are characteristic of simple on-off systems.

At any given gain the response of the system is related to the magnitude of the error signal. Consider our furnace control system. If the gain is fixed at 100, when the error signal is 10, the furnace delivers 1000 kcal/hr. As the building becomes warmer, the error signal diminishes. If the error signal drops to 2, then the furnace delivers only 200 kcal/hr. The key feature of a proportional control system is that the correction applied to the output variable becomes more delicate as the variable approaches the set point. This is quite similar to bringing an automobile to a safe stop. At 60 mph, one needs to press on

the brake firmly to rapidly slow the vehicle. When approaching the stop sign, however, one gradually reduces the pressure to avoid smashing the passengers into the windshield.

Damping Factor. Another important characteristic of any proportional control system is the **damping factor**. The formal definition of the damping factor is mathematical and is unnecessary for this discussion. But one can gain an intuitive understanding from the following description of the behavior of a proportional control system under various gains. If the gain of a proportional control system is a little larger than optimal, the controlled variable will follow a series of oscillations that get progressively smaller with time as the system is returned to its set point (Fig. 7.14). A system that responds with one or more oscillations is said to be **underdamped**. Underdamped systems quickly bring the controlled variable close to the set point, but overshoot the mark. The overshoot requires another correction in the opposite direction that, depending on the gain, may miss again, necessitating another correction. The system oscillates slightly, but eventually converges on the set point, where it stabilizes.

If the gain of the system is optimal, the system approaches the set point as quickly as possible without oscillating. The system is said to be **critically damped**. If the gain is reduced further, the system takes longer than the optimal time to achieve a steady state, and it is said to be **overdamped**. Most proportional control systems are operated slightly underdamped to maximize the speed while minimizing the oscillations to acceptable levels.

Proportional systems can become unstable if the gain is too large. As one increases the gain of an underdamped system, each correction applied by the controller causes greater and greater swings in the output variable. Eventually the system enters a state of **sustained oscillations**. Everyone has experienced this in an auditorium equipped with a public address system. Sound from the loudspeakers is received by the microphone (after a conduction delay across the auditorium). If the audio amplifier is properly adjusted, the system is stable and the audience hears the program. If the amplifier gain is at the critical point, one may hear a slight echo effect. If the gain is increased slightly, the audio system will begin to oscillate, which is that high-pitched squeal so irritating to audience and speaker alike.

INTERPRETING THE MSR AS A FEEDBACK CONTROL SYSTEM

The muscle stretch reflex operates as a feedback control system. All the elements of control systems can be identified. The nuclear bag receptors detect changes in muscle length. Feedback signals from these receptors are transmitted to the spinal cord. The spinal cord contains circuitry to compare the feedback signals with supraspinal command signals (set point) to control the muscle that is the effector. The controlled variable is muscle length. Lengthening the muscle causes the muscle spindle receptors to send a signal to the spinal cord, which returns a control signal to the muscle restoring the original muscle length. The MSR acts as a proportional control system, because the error signal from the receptors *varies in proportion to the change in length* of the muscle. Furthermore, the control signal—action potentials generated in α motor neurons—also varies in proportion to the size of the error.

Under normal circumstances, the gain of the MSR is set so that the system is *slightly underdamped*. If one elicits the MSR from the patellar tendon, the quadriceps muscle contracts moderately in response to the stretch; relaxes; and contracts slightly again before returning to its resting length. The leg makes a single visible jerk, but the extra small contraction can be seen under laboratory conditions (Fig. 7.15).

Certain lesions to the central nervous system change the gain of the MSR control system. Under these pathologic conditions, the system *behaves as an underdamped control system.* When the muscle is stretched, the leg swings more quickly and farther than normal. Instead of returning to its resting length after a single oscillation, the muscle will contract several times, and the leg may visibly swing two or three times. Muscle stretch reflexes that respond in this way are **hyperactive**. Patients with hyperactive reflexes have jerky, **spastic** [G. *spastikos,* drawing in] movements.

If the CNS damage is extensive, the gain of the MSR may be so great that the system becomes unstable. When this occurs, numerous oscillation in the MSR can be evoked, a response known as **clonus** [G. *klonos,* a tumult]. The rhythmic muscular contractions usually cease after several oscillations if the muscle is not otherwise stimulated. However, clonus can frequently be **sustained** if the muscle is placed under a slight stretch. For example, the physician can sharply flex the patient's foot at the ankle and then hold the flexed position; a steady, 5- to 7-Hz beating of the foot—sustained clonus—can be felt and often seen. Patients who have clonus can

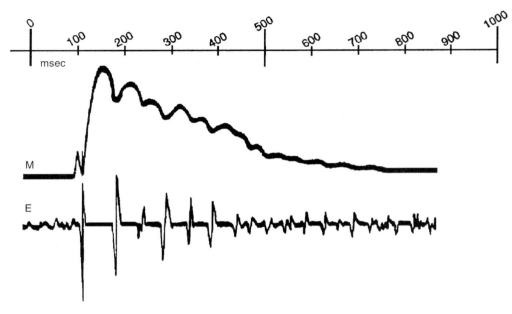

FIGURE 7.15.

The electrical and mechanical responses from the quadriceps muscle demonstrate clonus in this record. The muscle tension (*M*) and the EMG (*E*) are shown after a single tendon tap that occurred at the first small deflection in *M*. There followed a series of electrical discharges that are reflected in a series of waves of tension: clonus. In this case clonus was not sustained, but lasted approximately 600 msec.

be considerably inconvenienced. Walking becomes very difficult, since each step can provoke a rhythmic beating of the affected limbs. Even propelling a wheelchair across rough pavement can cause enough passive limb movement to engage the MSR, which in turn evokes clonus.

Lesions to the CNS affect the dynamic properties of the MSR more dramatically than the static properties. Hyperactive reflexes are very sensitive to the velocity of the muscle stretch. That is why the rapid stretching of the muscle caused by a sharp blow from the reflex hammer is so effective in displaying spasticity. The velocity sensitivity of spasticity can also be shown in other ways. For example, if one slowly pronates and supinates the arm of a patient, only a slight increase in resistance to passive movement will be detected. However, if the limb is rapidly moved, one will feel, in spastic patients, a sudden, brief "catch" in limb movement as the MSR is activated. This "catch" does not occur in normal patients. The rapid increase in tension or "catch" in the muscle is momentary because the

reflex increase in tension is great enough to invoke the **clasp knife reflex**, a reflex that causes massive inhibition of the α motor neurons innervating the muscle (see Chapter 6).

The characteristic changes in the MSR following lesions to the CNS are caused primarily by alterations in the excitability of motor neurons. Intracellular recordings from α motor neurons have shown that when the spinal cord is effectively transected,[7] the motor neurons become hypopolarized. This generalized facilitation of α motor neurons makes the motor neuron pools more excitable to all types of facilitation—that from remaining descending axons as well as to the I_A afferents of the MSR. Hypopolarization in effect increases the gain of the MSR feedback control system. The exact source of the motor neuron facilitation is not known. It is known that motor neurons are innervated by inhibitory in-

---

[7]This was done by cooling the spinal cord rather than actually cutting it in two. Since the cooling was reversible, the spinal cord could be rewarmed. Doing so restored the normal polarization to the motor neurons.

terneurons. Descending motor tracts probably tonically facilitate these inhibitory interneurons, creating a resting background inhibition on both α and γ motor neurons. Consequently, removal of descending motor tracts that facilitate the inhibitory interneurons *disinhibits* the motor neurons (Fig. 7.16).

Clinical Manifestations of Lesions to the Descending Motor Systems

Differentiating between lesions to the peripheral and central nervous system is a fundamental skill of medical practice. This is most easily accomplished by differentiating between upper and lower motor neuron afflictions. The concept of the "lower motor neuron" was presented in the previous chapter. A description of the "upper motor neuron" follows.

UPPER MOTOR NEURON

Lesions to the descending motor systems cause a unique set of symptoms. This constellation of symptoms is so characteristic that the concept of the **upper motor neuron (UMN)** has evolved, a concept that binds together several different but intimately interconnected anatomical entities, all involved with motor control. Before presenting the salient features of the UMN, it will be useful to recapitulate the clinical manifestations of LMN lesions (Table 7.1).

LMN lesions are characterized by a flaccid

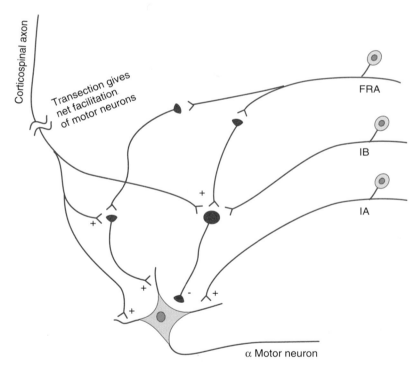

FIGURE 7.16.

One possible explanation for the development of hyperreflexia after spinal transection hypothesizes motor neuron disinhibition by interneurons. In this schematic, some of the influences on the α motor neuron are illustrated (the γ motor neuron is not shown for clarity). The α motor neurons receive many competing synapses, inhibitory and excitatory. The CBS tract, acting both directly and through interneurons, causes both inhibition and facilitation of α motor neurons. Removal of the CBS tract tips the balance in favor of greater facilitation.

Table 7.1. Comparison of the Principal Features of the Upper and Lower Motor Neuron

Lower Motor Neuron	Upper Motor Neuron
Flaccid weakness or paralysis	Spastic weakness
Decreased or absent MSR	Increased MSR with or without clonus
Signs of muscle denervation: fasciculations, fibrillations, profound atrophy	No signs of muscle denervation
Muscles affected singly or in small groups innervated by a common nerve or spinal root	Muscles affected in large groups, organized by quadrants or halves of the body
FRA reflexes normal	Some FRA reflexes reversed (signs of Babinski and Bing), others absent (abdominal and cremasteric reflexes)
Sensory patterns (a) stocking & glove or (b) follow dermatomes or fragments of dermatomes	Sensory patterns affect quadrants or halves of the body

weakness or paralysis, decreased muscle stretch reflexes, and signs of muscle denervation (fasciculations, fibrillations, and atrophy). Frequently, muscles are involved individually rather than in groups. The distribution of weakness may be patchy. Patterns of sensory loss that accompany LMN symptoms follow either the dermatomes or fragments of the dermatomes according to the individual spinal or peripheral nerve involved. The "lower motor neuron" represents a set of anatomical structures bound together by clinical observations. It is not a specific anatomical structure. Lesions to peripheral nerves, spinal roots, α motor neurons, and muscle can all give lower motor neuron signs.

In contrast to LMN lesions, damage to the "upper motor neuron" produces the following symptoms: **spastic weakness**, **hyperactive reflexes**, **reversal of certain FRA reflexes**, and the **absence of signs of muscle denervation**. Muscles are affected as groups, involving an entire extremity or half the body. They are never affected individually. The patterns of sensory loss are similarly distributed, involving quadrants or halves of the body. Losses can never be associated with specific dermatomes. Lesions that cause UMN symptoms invariably involve damage to the **corticobulbospinal tract**, the **rubrospinal tract**, and the **reticulospinal tracts**. As a practical matter, in human patients none of these tracts can be lesioned in isolation,[8] and one should consider these three tracts as an integrated UMN system. Damage to other parts of the motor system (see Chapter 8) does not produce these symptoms.

---

[8]Isolated lesions to the corticospinal tract have been performed in various animal species. The results of these experiments have reinforced the general conclusion that upper motor neuron signs are not attributable to a single descending motor pathway.

Weakness

The most prominent impairment that can be attributed to lesions involving the UMN is *weakness*. The usual term to describe voluntary muscle weakness is **paresis** [G. fr. *paritemi,* to let go], to be distinguished from a complete loss of voluntary motor control, **plegia** [G. *plege,* a stroke]. Accompanying the weakness, patients experience clumsiness and lack of precise motor control. The lack of precise motor control is most prominent in the distal extremities and is most closely correlated with damage to the corticospinal tract.

Hyperactive Reflexes

The increase in the muscle stretch reflex is the most characteristic sign of damage to the upper motor neuron. No other lesion to the nervous system gives this sign, so its importance cannot be overstated. As discussed above, the increase in the MSR is caused by a hypopolarization of the motor neurons following the loss of descending motor tracts. Since this brings the motor neurons closer to threshold, less facilitation from any source is necessary to cause them to fire. Therefore, in the quiescent state, there is little change in resting muscle tone, but once the muscle is stretched, even slightly, a rapid reflex contraction follows. It follows that, during a simple movement, the antagonist muscles will reflexively contract inappropriately as they are stretched. This reflex opposition to the desired movement makes the muscles seem "stiff" and spastic. The spasticity exacerbates the clumsiness and weakness.

Reversal of Certain FRA Reflexes

Several reflexes in addition to the MSR are commonly tested to verify suspected damage to the descending motor control systems. One such cutaneous FRA reflex is elicited by scratching the sole of the foot along its lateral margin, from the heel toward the toes. In normal individuals, the toes reflexively plantarflex. If there has been damage to the UMN, particularly the CBS tract, the toes extend and flare (Fig. 7.17). The reversal of this FRA reflex is called the **Babinski sign**, named after the Polish physician who first

described it. This valuable sign may be masked in people who are very ticklish, or who have unusually painful sensations from the bottoms of their feet. In such cases, one can use a pin to stab the dorsum of the foot over the extensor hallucis longus tendon. The normal reaction is to extend the foot at the ankle, away from the pin. The abnormal response indicating a UMN lesion, the **Bing sign**, is to flex the foot into the pin.

Two other FRA reflexes, the **abdominal reflex** and the **cremasteric reflex**, are less commonly used to confirm a UMN lesion. Both reflexes are normally present, and they are absent if there has been damage to the descending motor systems. The abdominal reflexes are evoked by lightly scratching each quadrant of the abdomen with a pin. Normally there is a reflex contraction of the underlying muscles. The cremasteric reflex is invoked by gently stroking the medial surface of the thigh with a cotton swab. In the male, one can see the ipsilateral testicle withdraw as the cremasteric muscle reflexively contracts. Of course this response is not visible if the scrotum is cold and the testicles are already withdrawn, nor can it be seen in the female.

Lack of Muscle Signs

UMN lesions are, by definition, lesions to the central nervous system, sparing the α motor neurons of the brainstem and spinal cord. Thus the muscle, although it may be paralyzed, remains innervated. Consequently, fasciculations, fibrillations, and profound atrophy—all prominent signs of LMN lesions—do not appear following UMN lesions.

UMN SYMPTOMS BY REGION OF THE CNS

Although the constellation of signs (spastic weakness, hyperactive MSR, reversed FRA reflexes, and absence of the signs of muscle denervation) are consistent signs of UMN lesions, there are other important characteristic signs associated with this lesion—signs that are dependent on the location. The differences in the clinical picture of the various UMN lesions depend on the *different crossing patterns* of the motor and sensory tracts. Further distinctions can be drawn based on the *location of nearby struc-*

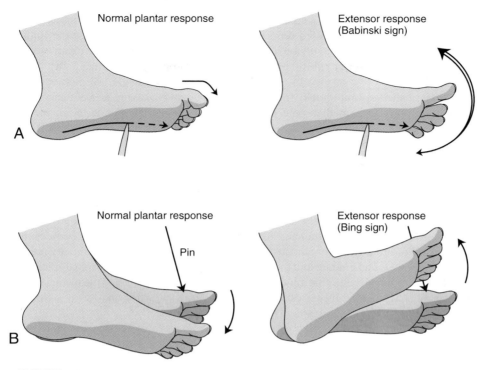

FIGURE 7.17.

   A, The Babinski sign indicates a lesion to the UMN, i.e., all of the descending, flexor-biased pathways. Scratching the lateral margin of the sole of the foot from the heel toward the small toe normally elicits a plantar (flexion) response. If the UMN has been damaged, the toes extend and flare. This reversal in the normal plantar response is the Babinski sign. **B**, A useful adjunct to the Babinski sign is the Bing sign. Stabbing the dorsum of the foot with a pin normally causes one to reflexively extend the foot at the ankle away from the pin. If the UMN has been lesioned, the foot is flexed into the pin.

tures. These distinctions are discussed in the following paragraphs.

Spinal Cord Lesions

   Complete transection of the spinal cord produces three dramatic signs below the level of the lesion: (*a*) loss of voluntary movement, (*b*) total anesthesia, and (*c*) a temporary period of areflexia, followed by permanent hyperreflexia.

   The loss of voluntary movement is due to the complete severance of motor commands descending from the brainstem and cerebral cortex. The level of transection determines the extent of the motor loss. Cervical transections will, of course, affect all four extremities, a condition known as **quadriplegia** [L. *quadri-*, four + G. *plege,* stroke]. Lesions below the cervical en-largement produce **paraplegia** [G. *para,* beside + G. *plege,* stroke]. If the lumbar cord is spared, and only the sacral region or conus is involved, only a partial paralysis of the lower extremities may be evident.

   Anesthesia over the body below the site of spinal cord transection is due to the loss of the dorsal columns and the ALS. With all sensory tracts cut, no sensory signals can reach cortical levels from the affected area. Many patients eventually develop abnormal sensations, **paresthesias** [G. *para,* beside, + G. *aisthesis,* sensation], from the affected areas. These are almost always unpleasant, usually a burning sensation. There is no satisfactory explanation for this phenomenon; however, it seems likely that a reorganization of synapses occurs in the brainstem and spinal cord that affects the thalamic targets of ALS transmission cells. This hypothesis is similar to the suggested reorganization following dorsal root avulsion (see Chapter 5) that produces similar painful paresthesias.

The areflexia, called **spinal shock**, following acute spinal injury is immediate and profound. There is a total flaccid paralysis below the site of the lesion that involves both the voluntary and involuntary functions. The MSR is absent from all muscles; the bladder and bowels become atonic; genital reflexes are absent; and autonomic functions (vasomotor tone, sweating, etc.) are lost. The areflexia usually subsides after a few (1 to 6) weeks in the human. Following the period of areflexia, spontaneous reflex emptying of bowels and bladder and the vasomotor reflexes are usually the first functions to reappear. Later muscle tone increases, especially in the flexor muscles and sphincters. Later still, the exaggerated hyperactive muscle stretch reflexes associated with UMN lesions appear. Complex alterations in reflexes below the level of the lesion may continue for several months.

The mechanisms of spinal shock are not understood. Some evidence suggests that loss of the reticulospinal and vestibulospinal tracts is the most important factor. This is supported by the observation that lesions more rostral than the pons do not produce a period of areflexia, but the basic explanation remains elusive.

Hemisection of the spinal cord, the **Brown-Séquard syndrome**, although not common, is an instructive lesion. Imagine a hypothetical patient who has suffered a perfect hemisection of the left spinal cord at T8.[9] Consider the findings of a neurological examination performed after the period of spinal shock (Fig. 7.18). The patient would be **monoplegic** [G. *monos,* single, + G. *plege,* stroke], having no voluntary motor functions of the left leg. Furthermore, one would observe **hyperactive reflexes** in the left leg and the FRA reflexes on the left would be reversed; the Bing and Babinski signs would be present and the left abdominal reflexes would be absent. If the patient were male, the cremasteric reflex would also be absent. These motor disturbances are directly related to the loss of all of the descending motor tracts on the left: the lateral corticospinal tract, the rubrospinal tract, the medullary reticulospinal tract, the lateral vestibulospinal tract, the medial vestibulospinal tract, and the medial corticospinal tract. Motor control and reflexes from the right leg are normal.[10]

The pattern of sensory losses allows one to locate the most rostral site of the lesion more precisely than the pattern of motor losses because the sensory losses will be absent below the last functional spinal segment. This level can be mapped to a dermatome. For the current example, pain and temperature sensations below approximately T8 on the right will be lost because the ALS axons on the left were sev-

---

[9]By common convention, T8 refers to the eighth thoracic segment of the spinal cord, not to the spinal column.

[10]In theory there ought to be some impairment of the right leg due to the loss of the medial corticospinal tract, since axons in that tract cross the midline at their level of termination. In practice, any impairments are not detectable by ordinary means.

FIGURE 7.18.

Left, This schematic diagram of the CBS tract and the ascending sensory tracts shows how a hemisection of the spinal cord produces the Harlequin pattern of sensorimotor losses. Only the contralateral CBS tract is shown for clarity. *Right,* This diagram illustrates the distribution of signs one would see in a patient who suffered a hemisection of the left spinal cord at T8. Sensory: Ipsilateral proprioception, vibratory senses, and fine tactile perception would be lost below the level of the lesion. Contralateral pain and temperature sensation would be lost below the lesion. This Harlequin-like distribution of sensory signs is called sensory dissociation. The sensory signs are the best indicators of the level of the lesion. MOTOR: Ipsilaterally and below the lesion, the MSR would be hyperactive and the Babinski (*arrow pointing up*) and Bing signs would be present. Contralateral to the lesion, MSRs and plantar signs (*arrow pointing down*) would be normal.

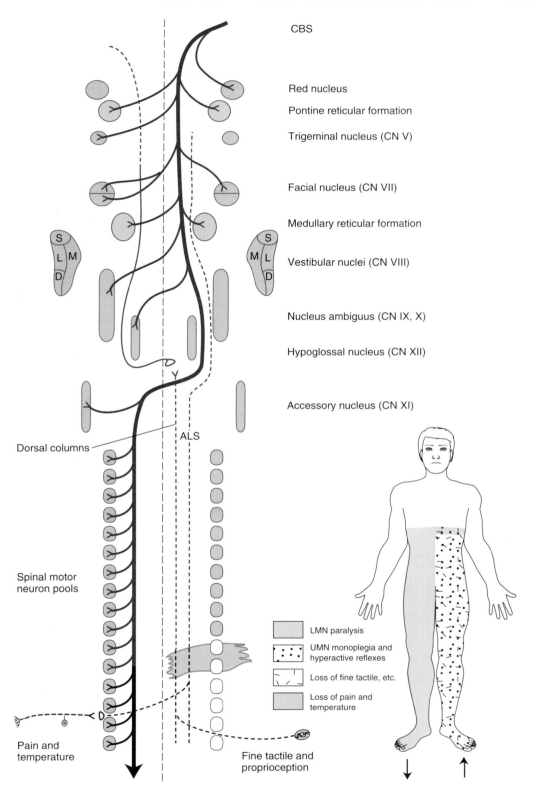

CBS

Red nucleus

Pontine reticular formation

Trigeminal nucleus (CN V)

Facial nucleus (CN VII)

Medullary reticular formation

Vestibular nuclei (CN VIII)

Nucleus ambiguus (CN IX, X)

Hypoglossal nucleus (CN XII)

Accessory nucleus (CN XI)

Dorsal columns

ALS

Spinal motor
neuron pools

Pain and
temperature

Fine tactile and
proprioception

LMN paralysis

UMN monoplegia and
hyperactive reflexes

Loss of fine tactile, etc.

Loss of pain and
temperature

ered. Since the ALS crosses at the level of ori-gin, deficits associated with its loss are con-tralateral to and below the side of the lesion. In contrast, proprioception and fine tactile sensa-tions would be lost at the same level and below, but ipsilateral to the side of the lesion because the dorsal columns remain ipsilateral through-out their course in the spinal cord. This dissoci-ation of pain and temperature versus fine tactile perceptions is an important sign found in spinal injuries.

As an exercise, write down the expected find-ings observed in a patient who suffered an in-farct[11] [L. *in-farcio,* pp. *-fartus,* to stuff into] in the territory of the anterior spinal artery.

Brainstem Lesions

Lesions to the descending motor pathways in the brainstem produce a different distribution of symptoms than do spinal cord lesions. One should note, however, that brainstem lesions are far more likely to be fatal because so many vital functions are packed into such a small area. These functions are described in Chapter 9. For now it is sufficient to know that critical respira-tory and cardiovascular centers reside in the brainstem and these centers are quite likely to be destroyed by most pathological processes af-fecting the brainstem.

---

[11]This word is so commonly misused that one should take special note of its meaning. **Infarct:** *An area of dead tissue resulting from the arrest of circulation in the artery supplying the part.* The defini-tion in no way suggests the process by which the circulation was ar-rested, and the word lacks medical precision unless qualified in some way to suggest either the artery involved or a description of the tis-sue killed. For example, *anterior myocardial infarct* is precise and meaningful; *cerebral infarct* is neither, unless one is attempting to describe the effects of decapitation.

Brainstem lesions produce sensory and UMN deficits that are codistributed over the same side of the body, contralateral to and below the le-sion. This is because both sensory systems and the corticospinal tract cross the midline below the level of the middle medulla. You will recall that the ALS crosses at the level of its origin in the spinal cord and that the medial lemniscus crosses at its origin at the most caudal regions of the medulla. Since the corticospinal tract crosses the midline at the medulla-spinal cord transition, most of the principal sensory and motor areas affected by brainstem lesions will produce con-tralateral signs.

The nuclei associated with the cranial nerves are located in the brainstem and they are fre-quently affected by brainstem lesions. These nuclei are discussed in Chapter 9. In this chap-ter it is worth noting that several cranial nerves have motor nuclei that supply voluntary mus-cles. A brainstem lesion that involves one or more of these nuclei will produce an ipsilateral LMN paralysis, since the α motor neurons in the cranial nerve motor nucleus will be lost. Be-cause of the compact nature of the brainstem, lesions involving motor nuclei of the cranial nerves almost always involve the descending motor tracts as well. Therefore, brainstem le-sions commonly present with an *ipsilateral LMN paralysis* of some motor function in the head combined with a *contralateral UMN hemiparesis.* This **crossed paralysis** is charac-teristic of brainstem lesions.

Consider an idealized lesion in the left brain-stem that affects the facial motor nucleus, the medial lemniscus, the ALS, and the CBS tract (Fig. 7.19). Such a patient will display a com-

---

FIGURE 7.19.

The most diagnostic feature of brainstem lesions is the crossed distribution of motor losses. This diagram illustrates how a brainstem lesion at the level of the facial nucleus produces these patterns. Since in this example the facial motor nucleus is lesioned, de-stroying its motor neurons, an LMN paralysis of the ipsilateral facial muscles follows. The CBS tract is also lesioned because it lies very close to the facial nucleus in the brain-stem. Its loss removes its influence on all structures below, causing a UMN paresis and the Babinski sign on the contralateral side. Since both the medial lemniscus and the ALS at the level of the facial nucleus convey contralateral sensory information, sensory losses will follow the UMN motor losses.

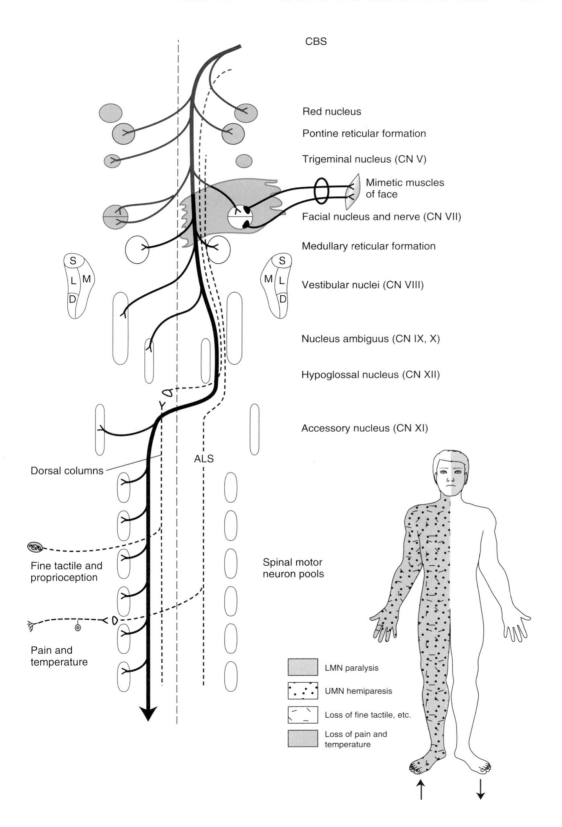

CBS

Red nucleus

Pontine reticular formation

Trigeminal nucleus (CN V)

Mimetic muscles
of face

Facial nucleus and nerve (CN VII)

Medullary reticular formation

Vestibular nuclei (CN VIII)

Nucleus ambiguus (CN IX, X)

Hypoglossal nucleus (CN XII)

Accessory nucleus (CN XI)

Dorsal columns

ALS

Fine tactile and
proprioception

Spinal motor
neuron pools

Pain and
temperature

LMN paralysis

UMN hemiparesis

Loss of fine tactile, etc.

Loss of pain and
temperature

plete paralysis of the mimetic muscles of the face on the left due to the loss of the facial motor nucleus. He will also have a spastic (UMN) hemiparesis on the entire right side of the body because of the loss of the corticospinal fibers. Such a patient will not be totally paralyzed on the right because the rubrospinal and reticulospinal tracts on the left, which have previously crossed the midline, can provide some motor control to the right side. Sensory losses will include a loss of pain and temperature (ALS), proprioception, and fine tactile sensations (medial lemniscus) from the right side of the body. There will be other specific abnormalities affecting cranial nerves below the level of the lesion. Details of those findings are discussed in Chapter 9.

Lesions Involving the Internal Capsule and Cerebral Cortex

Lesions to the internal capsule are commonly caused by strokes of the lateral striate arteries (see Chapter 1). These are very narrow and leave the parent artery at right angles, factors that make them a site for both occlusive and hemorrhagic strokes. Because the CBS tracts are compressed together in the posterior limb of the internal capsule, it is not uncommon for capsular lesions to produce pure contralateral motor signs. If so, then the lesion is almost certainly in the internal capsule, contralateral to the side of the hemiparesis (Fig. 7.20). However, nearly all tracts entering and leaving the cerebral cortex are found in the internal capsule. The optic radiations (see

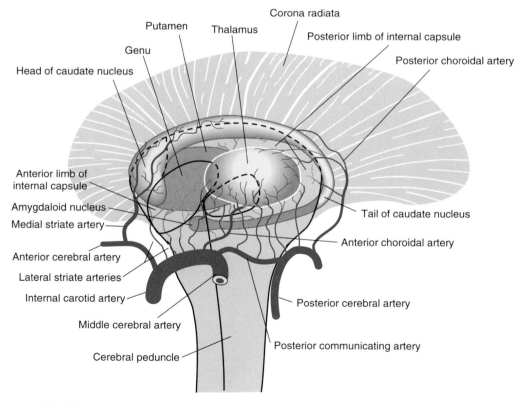

FIGURE 7.20.

Lesions in the internal capsule are often infarcts in the territory of the thalamostriate arteries. The various branches of the anterior and middle cerebral arteries and their relationship to the internal capsule are shown.

Chapter 11), the fiber tracts that connect the lateral geniculate nucleus with the visual cortex, are particularly close to the CBS tract in the internal capsules. If these fibers are included in the internal capsule infarct, partial blindness[12] will accompany the hemiparesis.

Infarcts involving the motor areas of the cerebral cortex can be in the territory of either the anterior or the middle cerebral artery. Middle cerebral artery infarcts produce motor and sensory losses that predominantly affect the contralateral face and arm while sparing somewhat the contralateral leg. Anterior cerebral artery infarcts do the opposite; they primarily affect the contralateral leg while sparing to some degree the contralateral arm and face. The explanation for these patterns lies in the distribution of the relative arteries (see page 4 and Chapter 1).

Infarcts involving the motor regions of the cerebral cortex rarely affect only those areas immediately adjacent to the central sulcus. Other cortical areas with nonmotor functions are also frequently involved, and documentation of these nonmotor functional losses is frequently helpful in allowing one to be able to localize the lesion to the cerebral cortex, even if motor losses may be minimal. These other functions are discussed in detail in Chapter 13.

STROKE

Strokes are the most common cause of UMN symptoms, and they represent the third most common cause of death in the United States. Strokes account for about half of all neurological disease. They are the consequence of cerebrovascular disease that interrupts blood flow to part of the brain with subsequent ischemia and infarction. Every physician will sooner or later be confronted with patients exhibiting the symptoms of cerebrovascular disease and therefore should be able to properly diagnose and evaluate this common malady.

The diagnosis of stroke depends on establishing the **time course** and the **distribution of symptoms**. The symptoms of stroke develop

rapidly over the course of a few seconds to a few hours. Few other disease processes develop so rapidly. Therefore, determining from the history the progression of the illness is essential in making the diagnosis of stroke. Furthermore, because stroke is primarily a vascular disease, the neurological deficits must be attributable to neuronal structures that lie within the territory of a single artery. This information is gathered from the neurological examination.

There are three principal causes of strokes: **thrombosis** [G. *thrombosis,* a curdling], **embolism** [G. *embolos,* a wedge or stopper], and **hemorrhage**. Establishing the type of stroke is important both for the initial treatment of the patient and for the follow-up care that is directed at reducing or eliminating the risk factors that the individual patient carries for stroke. The following paragraphs very briefly outline the major clinical features of the different types of stroke (Table 7.2).

Thrombosis

Cerebral thrombosis, the development of a blood clot in a cerebral vessel, is caused by the presence of an atherosclerotic plaque in an artery. Eventually the plaque causes degeneration of the vessel wall and damage to the endothelium that attracts platelets and fibrin to the damaged site. The thrombus slowly grows in size until eventually—usually during a period or reduced blood pressure—the lumen closes and the distal circulation ceases. This sequence of vascular events can be directly correlated with a sequence of neurological events that are characteristic of **atherothrombotic infarction**[13] or **thrombotic stroke**.

***Transient ischemic attacks (TIAs)* precede thrombotic stroke** in about 40% of patients. TIAs are temporary neurological deficits caused by disturbed neural function that can be related to the territory of a single artery. The deficit usually

---

[12]To be precise, a contralateral homonymous hemianopsia results. Visual field lesions are discussed in Chapter 11.

[13]Unfortunately it is common practice to call "atherothrombotic infarction" simply "infarction." This usage is unfortunate because it is neither accurate nor precise (see footnote 11). This is not the only example of imprecise language having penetrated the medical lexicon. I can only plead with students to understand how the term is commonly used, but never to use it incorrectly themselves.

Table 7.2. Comparison of the Principal Types of Stroke

Type	Location	Time of Occurrence	Consciousness
Thrombosis	Occlusion at atherosclerotic plaque	Periods of decreased activity (during sleep)	Preserved
Embolism	Displaced clot occludes downstream artery	Periods of activity	Preserved
Hemorrhage	Rupture of vessel wall	Anytime	Stupor or coma

clears within a few minutes, but always within 24 hours. The TIAs are presumed to be the result of the sloughing of small emboli from a thrombus that, once stuck in a vessel, quickly dissolve, allowing blood flow to be restored. TIAs may herald the eventual thrombotic stroke by hours or months, or the expected stroke may never appear.

Frequently (in about 20% of patients) there is a *stepwise development of thrombotic strokes* that progresses over a period of several hours to a few days. In these situations, some neurologic deficits appear, then there is a period without further deterioration, followed by expansion of the neurological signs. This sequence of events is known clinically as a *stroke in evolution,* and when the pathologic process has stabilized, *a completed stroke.*

Finally, most thrombotic strokes (about 60%) develop during sleep. These patients awake paralyzed. They may be unaware of their condition and get out of bed, only to discover their impairment by falling to the floor. The probable cause for the preponderance of nocturnal onset is that blood pressure is lowered during sleep. This may allow the vessel to narrow enough so that the lumen, compromised by the presence of the plaque, closes. Neuronal death occurs within five minutes of occlusion but is not made evident until the patient awakes. The patient usually does not lose consciousness, particularly not when he awakens.

Embolism

An embolism is a clot that has broken free from a thrombus and lodged in a remote part of the circulation, occluding it. **Cerebral embolisms** cause a very rapid development of neurological signs that usually do not progress. Only very rarely is there a prodromal TIA, an event that is so characteristic of the thrombotic stroke. Cerebral embolism usually occurs during times of activity, and consciousness is usually preserved. The source of the embolism is almost always the left side of the heart. Heart diseases that lead to the production of emboli include *atrial fibrillation, myocardial infarct,* and *defective* or *artificial heart valves,* particularly the mitral valve. Myocardial infarction frequently causes an akinetic region to develop in the heart wall. A thrombosis can develop within the nonmobile pocket of the ventricle. Emboli break off from the thrombosis, showering the vascular bed with clots. Artificial or damaged heart valves are also a frequent source of clot formation.

Hemorrhage

Intracerebral hemorrhage is associated with the sudden onset (minutes to hours) of neurological symptoms. Frequently this is associated with severe headache. Unlike the occlusive strokes discussed above, wherein consciousness is preserved, intracerebral hemorrhage causes stupor or coma that may progress with time (see Chapter 13). These patients are usually hypertensive, but not necessarily so, since **aneurysms** and **arteriovenous malformations** may bleed spontaneously under normal pressure.

Risk Factors

There are five important risk factors for stroke: **hypertension, hyperlipidemia, cigarette smoking, diabetes mellitus**, and **heart disease**. *Hypertension* damages the wall of small arterioles, making them susceptible to the development of atherosclerotic plaques, which in turn narrow the vessels further, exacerbating the hypertension. The plaques become sites of thromboses. Of course, hypertension itself can lead directly to arterial rupture and subsequent cerebral hemorrhage. The successful treatment of hypertension is the single most important factor in reducing the incidence of stroke (from 20 to 40%, depending on the study). *Hyperlipidemia, cigarette smoking,* and *diabetes mellitus* all accelerate the development of atherosclerosis. *Heart disease* is the most important risk factor for cerebral embolism. Since approximately 25% of the cardiac output reaches the brain, fragments from thrombi that develop in the left heart have a high probability of becoming cerebral emboli.

CASE STUDIES

Mr. R. B.

HISTORY: Mr. R. B. is an 80-year-old white male who fell this morning while in the bathroom. His wife found him on the floor, awake and alert,

but unable to move his right arm and leg. Mr. R. B. has been hypertensive (220/110), but this has been well controlled with medication (150/90). He states that recently he has had some minimal difficulty maintaining his balance. A few days before he had dropped his fork during dinner and had had some difficulty picking it up again. The clumsiness passed, and neither he nor his wife thought much about it.

NEUROLOGICAL EXAMINATION, JANUARY 11, 1986:

I.   MENTAL STATUS: Mr. R. B. is able to converse and he can give a reasonable accounting of this morning's events. He is aware of his surroundings and of his physical condition. He is a little confused at times.

II.   CRANIAL NERVES:

CN VII   There is no movement from the lower right side of the face. The right labial-nasal fold is quite flattened relative to the left side. He can raise his eyebrows on both sides approximately equally.

CN V   He reports that pin prick seems dull and cotton "feels funny" on the right side of the face but normal on the left side.

{There are other cranial nerve signs in this case; all point to involvement of the right side. Specifics of cranial nerve examinations will be deferred until Chapter 9}

III. MOTOR SYSTEMS: There is a nearly complete paralysis of the right arm, with only a few flickers of motion noticed in the biceps and triceps. The right leg can be elevated off the bed to a height of four fingers at the ankle. Strength on the left side seems normal. The left leg can be raised about 45°.

IV. REFLEXES: The MSRs are slightly brisk (+) from the right triceps, biceps, brachioradialis, quadriceps, and gastrocnemius. MSRs are normal on the left. The Babinski and Bing signs are present on the right and absent on the left.

V. CEREBELLUM: Not tested.

VI. SENSORY SYSTEMS: Sensory testing (pin and cotton) on the right side shows an ill-defined area where there is a "different feeling" compared with the left side, which the patients reports as feeling normal. The area of abnormal sensations clearly affects the right arm, shoulder area, and chest. The patient's reporting is unreliable over the right leg. Sometimes he detects differences between the two sides and other times he does not. He is unable to consistently delineate the boundaries on the right leg where sensory perceptions change.

VII. ANCILLARY DATA:

Computerized tomography (CT) of the head was ordered. It was normal.

SUBSEQUENT COURSE: Mr. R. B. was admitted to the rehabilitation unit of the hospital on January 14, 1986. At that time there had been some improvement in the strength of the right side of his body, but he could not stand without support, nor could he bring his right leg into a forward motion. A month later, on February 14, 1986, Mr. R. B. was walking well with a cane and was beginning to negotiate steps. There was still considerable weakness and clumsiness in his right arm and hand. At the time of his discharge 2 weeks later, his right arm was clearly spastic with very brisk reflexes (++) and assumed a "triple flexion" posture at rest. His walking had improved further, and the reflexes from the right leg were unchanged (+). The Bing and Babinski signs were still present on the right and absent on the left.

Comment: The neurological diagnosis depends on establishing the time course and localizing the site of the lesion based on the distribution of symptoms. The history and neurological examination provide the physician with the essential clinical facts. One interprets these facts in the light of the anatomy and physiology of the nervous system to establish a preliminary anatomical diagnosis. This diagnosis is then confirmed, refined, or disproved by ancillary studies that serve to augment the clinical facts. Finally, one attempts to determine the pathological process to determine the best course of treatment and establish a reasonable prognosis. These principles can be applied to the case of Mr. R. B.

Several important facts are immediately evident from the history. First, the time course of the symptoms is rapid; he collapsed suddenly in the bathroom after arising. Second, there is some evidence that there may have been prodromal symptoms; he had a minor transient bout of clumsiness: dropping his fork at dinner.

The neurological examination brings out other essential features of this case. Mr. R. B. is conscious. He is a little disoriented, but his fundamental mental capabilities are intact. Another important finding is the distribution and extent of the physical signs of sensory and motor losses. The sensory and the motor losses coexist over the same parts of the body, which include the face, arm, and leg, all on the right.

Having determined the clinical facts of the case, one must interpret these facts in accordance with known anatomy and physiology. The distribution of facial paralysis (see Chapter 9) and the increased MSRs on the right all point to a UMN lesion. Since the involvement of the facial nerve is ipsilateral to the involvement of the arm and leg, the lesion must be above the level of the facial nucleus, otherwise there would be a crossed palsy (see page). Furthermore, the face and arm are more paralyzed than the leg. This suggests that the lesion must be in the left cerebral cortex rather than in the internal capsule because of the distribution of the cerebral vasculature relative to the somatotopic organization of the cortex.

At this point one has established a neurological deficit that can be related to a single area of the brain that is perfused by a single artery. After a possible prodromal event, the onset of symptoms was rapid and did not progress. The patient is conscious and speaking sensibly. At this time one would entertain a diagnosis of thrombosis of the rolandic branch of the superior division of the middle cerebral artery (Fig. 7.21).

Having arrived at this conclusion, one would want to rule out cerebral hemorrhage, because treatment of cerebral hemorrhage is considerably different from treatment of cerebral thrombosis.[14] For this reason, a CT scan of the head was ordered. CT examinations within about 24 to 48 hours of an ischemic stroke (embolism or thrombosis) are normal, whereas the extravascular blood from cerebral hemorrhage is immediately visible. A CT scan will demonstrate an area of edema following ischemic strokes only after about 48 hours. In the case of Mr. R. B., the CT examination was normal, confirming the initial diagnosis of atherothrombotic infarction.

FURTHER APPLICATIONS

- Consider the neurological signs and symptoms one would expect to observe if Mr. R. B.'s

---

[14]Treatment is beyond the scope of this text. Please refer to any of the clinical texts listed in the references at the end of this chapter.

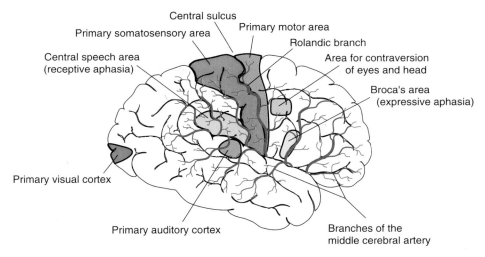

FIGURE 7.21.

The typical divisions of the middle cerebral artery and the principal functional regions served by its various branches.

thrombosis had occurred in the anterior cerebral artery.

- How would the history be different if Mr. R. B. had had a cerebral hemorrhage of the middle cerebral artery?

- Could Mr. R. B. have an embolism rather than a thrombosis? What is the difference between these two types of stroke? How does the clinical picture differ between them?

Ms. C. R.

Ms. C. R., a 39-year-old female and avid runner, was in good neurological health until January of 1992. At that time she fell while running and broke her left clavicle. Six weeks later, after the clavicle had healed, she resumed running. On her first day back, she fell again, scraping both her hands and her knees. By September, she had fallen frequently in her apartment building while ascending and descending the stairs. She states that at times her left leg feels like a piece of wood. Recently, she has noticed that her left hand has felt puffy and weak and has performed somewhat clumsily. She has no family history of neurological diseases.

NEUROLOGICAL EXAMINATION, SEPTEMBER 2, 1992:

I.   MENTAL STATUS: Normal

II. CRANIAL NERVES: All normal

III. MOTOR SYSTEMS: Ms. C. R. walks with a limp, slightly dragging her left side. She is not ataxic. She hops well on the right foot, but not on the left. She can do deep knee bends. On direct strength testing, her left arm and leg are weaker than the corresponding right extremities. There is definite drift of the left arm when she holds her arms outstretched.

IV. REFLEXES: The MSRs from all extremities are brisk, with those from the left greater than those from the right. Two to four beats of unsustained clonus can be elicited from the left ankle. The Babinski sign is present on the left, absent on the right.

V. CEREBELLUM: Rapid alternating movements of the left extremity are somewhat uncoordinated. Finger-to-toe and heel-to-shin maneuvers are performed reasonably well bilaterally.

VI. SENSORY SYSTEMS: Appreciation of pin prick, light touch of cotton, vibration, and joint position are all within normal limits. The Romberg sign is not present.

Comment: This case is very similar to that of Ms. P. J. (see Chapter 11). This patient's illness has been progressing slowly but relentlessly for several months. She has definite UMN signs, but there are no sensory deficits. At this point, Ms. C. R.'s difficulties can be explained by a single lesion. Given the patient's age and the general nature of her illness, the diagnosis of multiple sclerosis (see Chapter 11) was entertained, but without clinical evidence of exacerbations and remissions, or of multifocal lesions, that diagnosis remained tentative.

Ms. C. R. was examined again on August 19, 1993. At that time her weakness had spread to all four extremities. Her hands had become so weak that in July she had to quit her job.

NEUROLOGICAL EXAMINATION, AUGUST 19, 1993:

I. MENTAL STATUS: Normal.

II. CRANIAL NERVES: Normal.

III. MOTOR SYSTEMS: Ms. C. R. is now in a wheelchair. She is very unsteady when standing and requires support. When she attempts to walk, she lurches and would fall were help not available. Her upper extremities are very spastic, her lower extremities somewhat less. There are marked fasciculations in her tongue, occasional fasciculation in her right deltoid and over her lower ribs on the left. There is gross atrophy of the muscles in both hands and definite atrophy of the muscles in the shoulder girdle.

IV. REFLEXES: The MSRs from all four extremities are very hyperactive (++). There is sustained clonus from her left ankle. Unsustained clonus can be demonstrated from her jaw and right ankle. The Babinski and Bell signs are present bilaterally.

V. CEREBELLUM: Not testable.

VI. SENSORY SYSTEMS: Ms. C. R. is able to perceive pin prick, cotton wisp, vibration, and joint position normally.

Comment: Ms. C. R. has gotten considerably worse since her last examination. Her UMN signs have progressed to involve all four extremities. In addition, there are a number of LMN signs. In spite of the fact that her motor difficulties have progressed to the point that she is bound to a wheelchair, Ms. C. R. has a normal sensorium.

The presence of LMN signs is incompatible with the diagnosis of multiple sclerosis. The combination of UMN plus LMN signs with normal soma-tosensory findings is characteristic of amyotrophic lateral sclerosis (ALS), *a degenerative disease that affects both the α motor neurons and the descending motor tracts.*

ALS is a progressive disease of unknown etiology and is without effective treatment. Its diagnosis is based on the involvement of the voluntary motor systems with findings of simultaneous UMN and LMN signs. All other neurological systems remain intact. Patients remain fully sentient and mentally competent. The UMN signs are caused by the degeneration of the CBS tract, the rubrospinal tract, and the medullary reticulospinal tracts (Fig. 7.22). The LMN symptoms are brought about by the loss of the motor neurons in the ventral horn and the motor nuclei of the cranial nerves. Eventually the patient is left with a total, flaccid paralysis involving all voluntary motor systems except the extraocular eye movements and the urinary sphincters. The relentless progress of the disease leaves the individ-

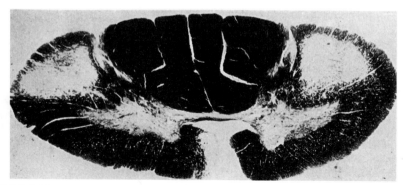

FIGURE 7.22.

   This section of a cervical spinal cord was obtained from a patient who died from ALS. Severe degeneration of the lateral white columns is evident from the almost total loss of myelin from the lateral white columns. Not evident in this preparation is the loss of motor neurons from the ventral horns.

ual, intact in every other respect, imprisoned in a paralyzed body. With many exceptions, most persons with ALS die within 5 years of the initial diagnosis. Patients can live a normal life span with this disease if they are willing to exist in a totally dependent state, connected to a respirator. The prominent mathematician and physicist Stephen Hawking made most of his professional contributions while strapped helplessly to a wheelchair.

FURTHER APPLICATIONS

- After the first examination, the patient's symptoms can be explained by a single lesion. Where would you place that lesion?

- Carefully review the UMN and LMN signs in this case. As the disease progresses, which symptoms predominate?

- Consider the ethical questions this disease imposes on the physician. With respiratory assistance and proper nursing care, patients can live for many years with this disease, albeit in a totally helpless state. In this helpless state, their intellectual functions remain entirely intact. Those patients who choose not to take that course face death by aspiration or asphyxiation. Many choose suicide. How would you determine the limits on a physician's duty to ALS patients?

SUMMARY

The **corticobulbospinal (CBS) tract** originates in a wide area of the cerebral cortex (Brodmann areas 4, 6, 3, 1, 2, 5, 7, 39, and 40) and descends to the brainstem and spinal cord. The axons that terminate in the brainstem may be separately named the **corticobulbar tract**, while those that descend to the spinal cord are called the **corticospinal tract**. CBS axons terminate in the red nucleus, the reticular formation, the voluntary motor nuclei of the cranial nerves, and the spinal cord. Most of these connections cross the midline before terminating. Some structures in the brainstem are bilaterally innervated. There is also a small component of the corticospinal tract that is uncrossed. CBS axons terminate primarily on interneurons, not motor neurons. They ultimately affect both α and γ motor neurons. In the spinal cord, most CBS axons facilitate flexor motor neuron pools. Some CBS axons terminate directly on α motor neurons, primarily those innervating the most distal muscles.

The **rubrospinal tract** originates in the red nucleus. Its axons immediately cross the midline and descend through the brainstem and spinal cord. Rubrospinal axons terminate on interneurons and facilitate primarily flexor motor neuron pools of the proximal musculature. The red nucleus receives afferents from the cerebellum and the motor areas of the cerebral cortex via corticobulbar axons.

The **reticulospinal tracts** originate in the pontine and medullary reticular formations. Both tracts terminate primarily on interneurons that affect the voluntary motor neuron pools. The medullary reticulospinal fibers primarily facilitate flexors, while the pontine reticulospinal fibers primarily affect extensors. Both systems receive afferents from the spinal cord, the cerebellum, and the motor regions of the cerebral cortex.

The **vestibulospinal tracts** originate in the vestibular nuclei, the lateral vestibular nucleus giving rise to the lateral vestibulospinal tract, while the medial vestibular nucleus contains the cells of origin for the medial vestibulospinal tract. The medial tract descends only to cervical levels, innervating primarily the muscles that stabilize the head. The lateral tract descends to all levels of the spinal cord, innervating extensor motor neurons. Both tracts terminate primarily on interneurons, but the lateral tract also terminates extensively on α motor neurons directly.

There are three motor areas in the cerebral cortex, the primary motor cortex (**M-I**), the supplementary motor area (**SMA**), and the premotor area (**PMA**). All three areas are somatotopically organized and are reciprocally interconnected through short association axons. The motor areas of the cortex receive ascending connections from the **ventral anterior (VA)** and **ventral lateral (VL)** thalamic nuclei. These nuclei serve as intermediate connections between the motor cortex and the cerebellum and basal ganglia.

Cells in the PMA seem to be necessary to facilitate specific sets of neurons in M-I required for a particular motor act. They fire before the evolving movement has been initiated. The SMA seems to be necessary to coordinate bimanual motor tasks. Unilateral lesions to the SMA produce mirror movements between the extremities. M-I provides the most direct cortical control of the motor neurons. SMA and PMA activity seems to converge on M-I neurons, and the latter are connected quite precisely to individual motor units in the voluntary motor neuron pools. M-I neurons appear to code for the force, velocity, and direction of the motor task.

The **upper motor neuron (UMN)** is a concept that contrasts with the concept of the **lower motor neuron (LMN)**. Certain CNS lesions produce a stereotyped constellation of symptoms that constitute a working definition of the UMN. They are (*a*) spastic paresis, (*b*) hyperactive muscle stretch reflexes (MSRs), and (*c*) the reversal of certain flexion reflex afferent (FRA) reflexes. The CBS tract, the rubrospinal tract, and the medullary reticulospinal tract share many physiological functions associated with the UMN. Lesion of these tracts is closely correlated with UMN signs.

The MSR has the properties of a proportional feedback control system. The controlled variable is muscle length. Alterations in muscle length are fed back to the spinal cord, where an error signal is generated that alters muscle tension to restore the original length. Damage to the UMN effectively increases the gain of the MSR control system, causing it to be underdamped. This is manifested clinically as hyperactive MSR responses, muscle spasticity, and clonus.

Strokes are the most common neurological problem and the third leading cause of death in the United States. Strokes are the result of cerebrovascular disease and are divided into three classes. **Atherothrombotic infarcts** represent the development of a thrombus within a cerebral vessel at the site of an atherosclerotic plaque. **Cerebral embolisms** occur when a clot breaks away from a remote thrombus and migrates into the cerebral vasculature. The most common site for the development of emboli is the left heart. **Cerebral hemorrhage** occurs when a cerebral vessel, aneurysm, or arteriovenous malformation ruptures or leaks fresh blood into the tissue of the brain or its ventricles.

Amyotrophic lateral sclerosis is a disease that affects both the UMN and the LMN. It is characterized by CNS lesions that are limited to the descending motor systems (CBS, rubrospinal, and reticulospinal tracts), sparing the sensory systems and higher cortical functions. Simultaneously, LMN symptoms also develop due to the loss of motor neurons in the brainstem and spinal cord. There is no effective treatment of this disease.

FOR FURTHER READING

Adams, R., and Victor, M. *Principles of Neurology.* New York: McGraw-Hill, 1993.

Binder, M. Spinal and Supraspinal Control of Movement and Posture. In *Textbook of Physiology,* Ch. 26, edited by Patton, H. D., Fuchs, A. F., Hille, B., Scher, A. M., and Steiner, R. Philadelphia: W. B. Saunders, 1989, pp. 563–581.

Brodal, A. *Neurological Anatomy In Relation to Clinical Medicine.* New York: Oxford University Press, 1981.

Brooks, V. *The Neural Basis of Motor Control.* New York: Oxford University Press, 1986.

Carpenter, M. B. *Core Text of Neuroanatomy.* Baltimore: Williams & Wilkins, 1991.

Fetz, E. Motor Functions of the Cerebral Cortex. In *Textbook of Physiology,* Ch. 28, edited by Patton, H. D., Fuchs, A. F., Hille, B., Scher, A. M., and Steiner, R. Philadelphia: W. B. Saunders, 1989, pp. 608–631.

Haines, D. *Neuroanatomy: An Atlas of Structures, Sections, and Systems.* Baltimore: Urban & Schwarzenberg, 1991.

Nieuwenhuys, R., Voogd, J., and van Huijzen, C. *The Human Central Nervous System; A Synopsis and Atlas.* Berlin: Springer-Verlag, 1981.

Rowland, L. *Merritt's Textbook of Neurology.* Philadelphia: Lea & Febiger, 1989.

Waddington, M. *Atlas of Cerebral Angiography with Anatomic Correlation.* Boston: Little, Brown, 1974.

8 Basal Ganglia and Cerebellum

The basal ganglia and the cerebellum are elements of the motor system that control and regulate motor acts as they are evolving. The descending motor systems, described in Chapter 7, affect the motor neuron pools directly. As we have seen, many cortical neurons code for the force and velocity of muscle contraction. Consequently, neuronal activity in these systems is closely correlated in time with the actual motor act. In contrast to this, much, but not all, of the activity of neurons in the basal ganglia and cerebellum occurs before any movement has begun. The basal ganglia are an important link between the *idea of movement* and the *motor expression of that idea*. The cerebellum shapes the motor act as it is conceived and as it is executed. It does so by regulating the *timing* of contractions among synergistic muscles across multiple joints.

Disturbances in the function of the descending motor systems and the spinal cord result in paresis or paralysis. The basal ganglia and the cerebellum only modulate the activity of the descending motor systems; therefore their loss cannot result in paralysis. Lesions to the basal ganglia cause disturbances in the initiation or cessation of a motor event. The loss of the cerebellum disrupts the coordinated activity of multiple muscle groups.

Nuclei of the Basal Ganglia

The basal ganglia are a set of nuclei that serve as a functional bridge between the telencephalon and the diencephalon. Some of these nuclei are telencephalic derivatives, while others develop out of the diencephalon. Information from the cerebral cortex enters telencephalic basal ganglion nuclei, is passed into diencephalic regions of the basal ganglia, and then is sent to thalamic nuclei. From the thalamus, the information is returned to the cerebral cortex.

Our understanding of the functional role that the basal ganglia play in motor control is changing rapidly. Originally thought to be a purely motor system, the basal ganglia are now recognized to play a role in certain cognitive functions also. The exact nature of this cognitive role is not known. In part for these reasons, there is legitimate disagreement among authors about the structures that properly constitute this area of the brain. For the purposes of this book the basal ganglia consist of the following nuclei: the **caudate nucleus**, the **putamen**, the **globus pallidus**, the **amygdaloid complex**, the **nucleus accumbens**, the **substantia nigra**, and the **subthalamic nucleus**.[1] All but

---

[1]Many authors include the claustrum in the basal ganglia. This enigmatic nucleus has major connections with the cerebral cortex. It also has reciprocal connections with nonspecific nuclei of the thalamus (e.g., cetromedian), but apparently does not have connections with specific thalamic nuclei. Therefore, it seems prudent at the present time not to include the claustrum with the basal ganglia.

Some authors prefer to exclude the amygdaloid complex from the basal ganglia because it appears to play no role in motor control. However, the amygdaloid complex is closely related to the caudate nucleus, developing from the same telencephalic anlage, and therefore, on developmental criteria, it seems desirable to include it in the basal ganglia. Recently, nonmotor functions have been attributed to the neostriatum, calling into question the idea that it is purely a motor control center. Therefore it seems probable that, with further research, the amygdaloid complex will be seen as an extension of the caudate nucleus in both ontology and physiology.

The nucleus accumbens has traditionally been considered part of the septal nuclei. Recent studies have shown that this nucleus shares many similarities with the caudate nucleus and putamen, including axonal projections into the globus pallidus and substantia nigra. Many authorities now consider it more appropriate to include this nucleus in the neostriatum.

the last two nuclei constitute the largest structure of the basal ganglia, **the striatum**. It is divided into three parts, the **neostriatum**, the **paleostriatum**, and the **archistriatum**. The striatum is divided by the internal capsule. The caudate nucleus lies medial to the internal capsule, while the putamen and globus pallidus lie nestled in the notch on the lateral side, between the anterior and posterior limbs. The substantia nigra and subthalamic nucleus are the nonstriatal structures of the basal ganglia.

NEOSTRIATUM

The neostriatum, a derivative of the telencephalon, consists of the **caudate nucleus**, the **putamen**, and the **nucleus accumbens** (Figs. 8.1 and 8.2). Both cytoarchitectural and physiological studies have provided evidence that these nuclei are in fact a single nucleus divided by the internal capsule. At the most anterior portion they are merged; only a few threads of the anterior limb of the internal capsule pass between the caudate nucleus and the putamen. The anterior portion of the caudate nucleus lies between the internal capsule and the lateral ventricle. It is enlarged relative to the remainder of the nucleus and therefore is called the **head**. Inferior to the head of the caudate nucleus is the **nucleus accumbens**. The rest of the caudate nucleus wraps around the lateral aspect of the internal capsule, forming the **body** and **tail**. The caudate nucleus forms part of the lateral wall of the lateral ventricle and follows it into the temporal lobe, where the caudate nucleus terminates at the **amygdaloid nucleus**, just deep to the uncus. The putamen lies lateral to the internal capsule, between its anterior and posterior limbs.

Cells in the neostriatum are immunoreactive to several neurotransmitters and neuroactive peptides. Most of the neurons in the striatum (more than 90%) have spiny dendrites. Different sets of these spiny neurons are immunoreactive to **γ-aminobutyric acid (GABA)** and **substance P (SP)** or **enkephalin (ENK)**. GABA is the principal neurotransmitter, while the neuroactive peptides are probably co-released with the transmitter. Another class of neurons in the neostriatum is the aspiny interneurons, so called because their dendrites lack spiny projections. Various sets of these neurons are immunoreactive to GABA or **acetylcholine (ACh)** and **neu-**

ropeptide Y (NPY) or **somatostatin**. The function of the neuroactive peptides is not known, but at this stage of our understanding it is clear that the neostriatum is not a biochemically homogeneous structure. This anatomical heterogeneity probably reflects considerable functional heterogeneity as well.

Neurons that stain for the various neuroactive peptides are unevenly distributed within the neostriatum, being collected together as islands or **patches** of immunohistochemically similar neurons that are isolated areas within a continuous **matrix** (Fig. 8.3). The patch-matrix pattern seems to represent a basic division within the total organization of the basal ganglion system. The matrix receives afferents primarily from M-I and S-I, and to a lesser degree from the frontal, parietal, and occipital lobes. The patches are innervated primarily by the prefrontal lobe and certain limbic structures (see Chapter 12). The patches themselves are heterogeneous, since patches that stain for one neuroactive substance do not necessarily overlap with patches that stain for another. All of this anatomical complexity has led to the hypothesis that the neostriatum performs several physiological functions and that these functions are isolated both anatomically and neurochemically within the neostriatum.

PALEOSTRIATUM

A derivative of the diencephalon, the paleostriatum is the **globus pallidus**, a nucleus that lies between the putamen and the internal capsule.[2] The globus pallidus consists of two separate nuclei, the **lateral (GPl)** and **medial (GPm)** divisions. They are separated by a thin layer of axons. Each nucleus has its own set of connections. The GPm is functionally associated with the reticular division of the substantia nigra. Most neurons within the pallidum are immunoreactive for GABA, which they probably use as a neurotransmitter.

ARCHISTRIATUM

The **amygdaloid nuclear complex**, a division of the telencephalon that lies deep to the un-

---

[2]The term "lentiform nucleus" has historically been applied to the combined structures of the putamen and globus pallidus. This union cannot be defended on anatomical or physiological grounds and therefore serves no useful function.

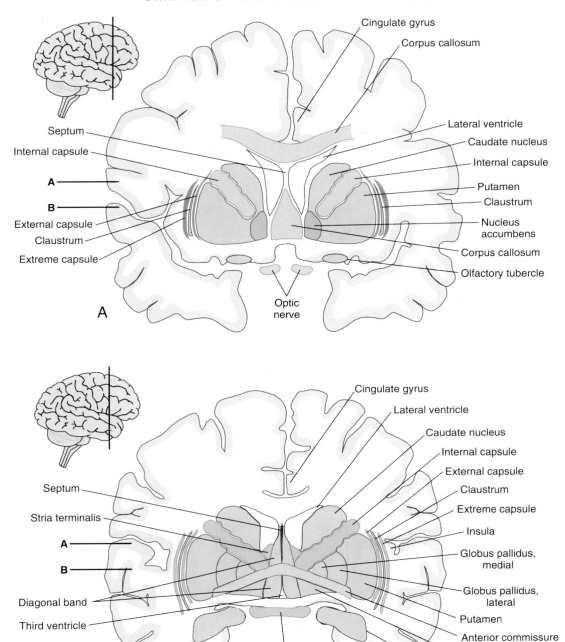

FIGURE 8.1.

A, A frontal section through the anterior portion of the basal ganglia. Note how threads of the caudate nucleus, putamen, and nucleus accumbens pass through the anterior limb of the internal capsule. Letters *A* and *B* on the left side correspond with the levels of the sections in Figure 8.2. **B**, This frontal section, at the level of the anterior commissure, is more posterior than **A**. Note the relationships between the septum, the diagonal band, and the nucleus basalis and the putamen and globus pallidus. Letters *A* and *B* on the left side correspond with the levels of the sections in Figure 8.2.

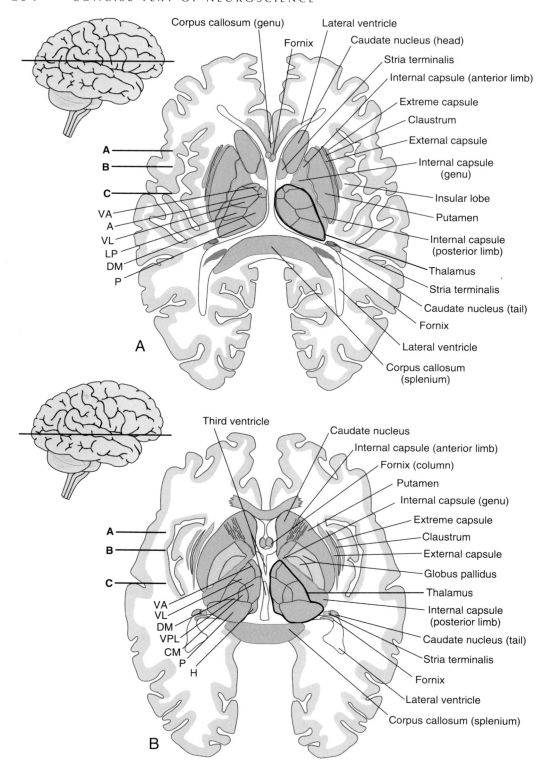

Corpus callosum (genu)
Lateral ventricle
Fornix
Caudate nucleus (head)
Stria terminalis
Internal capsule (anterior limb)
Extreme capsule
Claustrum
External capsule
Internal capsule (genu)
Insular lobe
Putamen
Internal capsule (posterior limb)
Thalamus
Stria terminalis
Caudate nucleus (tail)
Fornix
Lateral ventricle
Corpus callosum (splenium)

A
B
C
VA
A
VL
LP
DM
P

A

Third ventricle
Caudate nucleus
Internal capsule (anterior limb)
Fornix (column)
Putamen
Internal capsule (genu)
Extreme capsule
Claustrum
External capsule
Globus pallidus
Thalamus
Internal capsule (posterior limb)
Caudate nucleus (tail)
Stria terminalis
Fornix
Lateral ventricle
Corpus callosum (splenium)

A
B
C
VA
VL
DM
VPL
CM
P
H

B

cus in the temporal lobe, represents the archistriatum. This group of cells is divided into several subnuclei that make connections with diverse parts of the nervous system that are not directly related to motor functions. The nonmotor functions of this region will be considered in Chapter 12.

SUBSTANTIA NIGRA

The **substantia nigra**, although a nucleus of the mesencephalon, is considered by many to be a part of the basal ganglia because of its rich connections with the striatum (Fig. 8.4). This nucleus is divided into two regions, the **sub-**

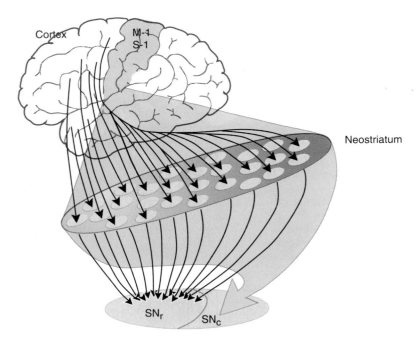

FIGURE 8.3.

The matrix, or background substance, of the neostriatum is punctuated with patches of neurons that have different neurochemical properties. Some of the patches are immunoreactive for enkephalin, while others are reactive for substance P. Neurons in the matrix are immunoreactive for a variety of other neurochemicals (see text). The patches and matrix have discrete, restricted connections with both the cerebral cortex, from which they receive afferents, and the globus pallidus and substantia nigra, to which they project.

FIGURE 8.2.

A, Horizontal section through the basal ganglia and the thalamus illustrating the relationships of these structures to the internal capsule and the lateral ventricle. Letters *A* and *B* on the left side correspond with the levels of the frontal sections in Figure 8.1; *C* corresponds with the level of the frontal section of Figure 8.4. **B,** This horizontal section is slightly inferior to **A.** Letters *A* and *B* on the left side correspond with the levels of the frontal sections of Figure 8.1; *C* corresponds with the level of the frontal section of Figure 8.4. Thalamic nuclei: *A,* anterior; *CM,* centromedian; *DM,* dorsal medial; *H,* habenula; *LP,* lateral posterior; *P,* pulvinar; *VA,* ventral anterior; *VL,* ventral lateral; *VPL,* ventral posterior lateral.

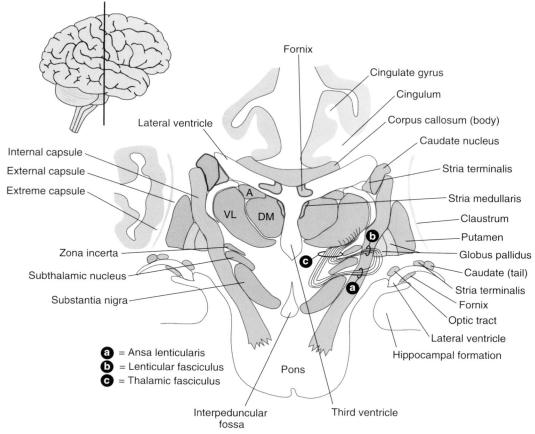

Fornix

Cingulate gyrus

Cingulum

Corpus callosum (body)

Caudate nucleus

Stria terminalis

Stria medullaris

Claustrum

Putamen

Globus pallidus

Caudate (tail)

Stria terminalis

Fornix

Optic tract

Lateral ventricle

Hippocampal formation

Lateral ventricle

Internal capsule

External capsule

Extreme capsule

Zona incerta

Subthalamic nucleus

Substantia nigra

A

VL   DM

ⓐ = Ansa lenticularis
ⓑ = Lenticular fasciculus
ⓒ = Thalamic fasciculus

Pons

Interpeduncular fossa

Third ventricle

FIGURE 8.4.

This frontal section through the diencephalon illustrates the relationship between the subthalamic nucleus, the zona incerta, the substantia nigra and the internal capsule. Note that the internal capsule becomes the cerebral peduncle as it enters the pons. The globus pallidus and putamen lie lateral to the internal capsule, and the caudate nucleus lies medial to it and dorsal to the thalamus. Also note the position of the tail of the caudate nucleus and its relationship to the stria terminalis and fornix.

stantia nigra pars reticulata **(SNr)** and the **substantia nigra pars compacta (SNc)**. This nucleus received its name because it is black in fresh cadaver brains. The black substance, **melanin**, is a byproduct of **dopamine** metabolism (Figs. 8.5 and 8.6). Neurons in the SNc produce large amounts of dopamine and by axoplasmic transport carry it to the neostriatum, where it is used as a neurotransmitter. It may have other metabolic effects as well. Most of the dopamine-producing cells of the SNc also produce **cholecystokinin (CCK)**, a neuroactive peptide. In addition to the striatum, these dual-transmitter neurons project to the amygdaloid complex, the nucleus accumbens,

and the prefrontal cortex. Neurons in the SNr, as well as those of the GPm, project to the thalamus and probably secrete **GABA** as a neurotransmitter, since most of the neurons are immunologically reactive to GAD (glutamate decarboxylase), an enzyme necessary for the synthesis of GABA.

SUBTHALAMIC NUCLEUS

The **subthalamic nucleus (STN)**, a derivative of the diencephalon, lies just medial to the internal capsule and ventrolateral to the thalamus proper (Fig. 8.4). Cells in this nucleus are immunoreactive to **glutamate** and may use it as a neurotransmitter.

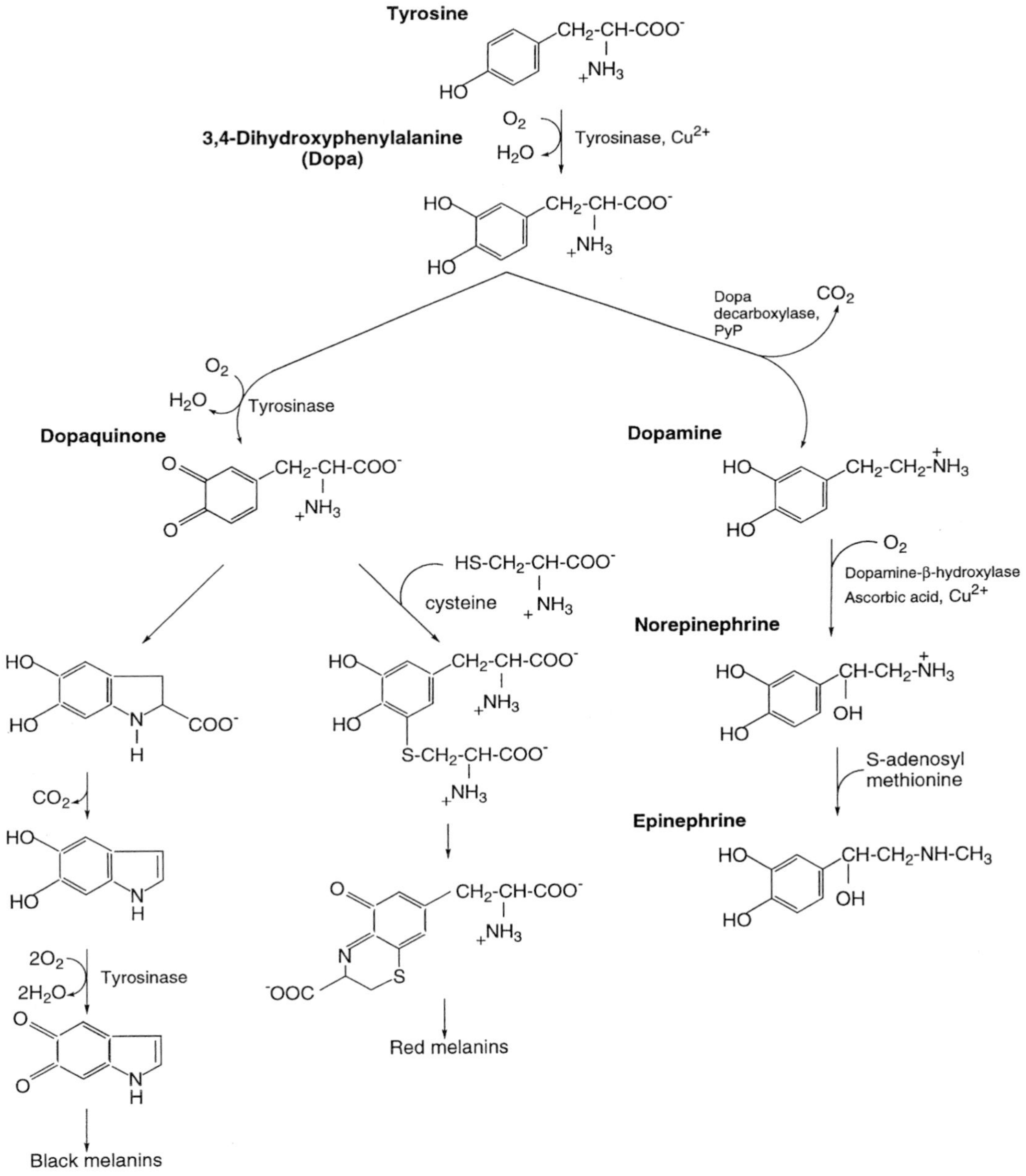

FIGURE 8.5.

The metabolism of tyrosine.

Connections of the Basal Ganglia

The general pattern by which the basal ganglia are interconnected with other parts of the nervous system is quite simple (Fig. 8.7). The most remarkable feature is that nearly the entire cerebral cortex projects to the neostriatum. For the most part these connections maintain the same spatial relationships in the neostriatum that exist in the cortex; the frontal lobe projects to the anterior caudate (the head) and putamen, while the temporal lobe projects to the tail. The

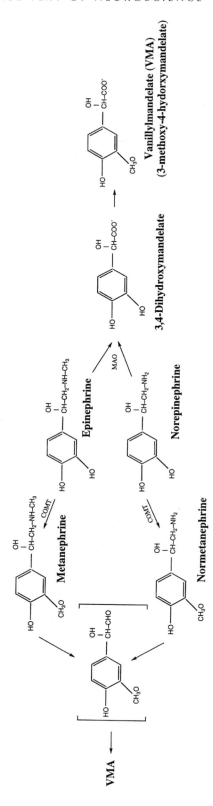

FIGURE 8.6.

The degradation of epinephrine and norepinephrine.

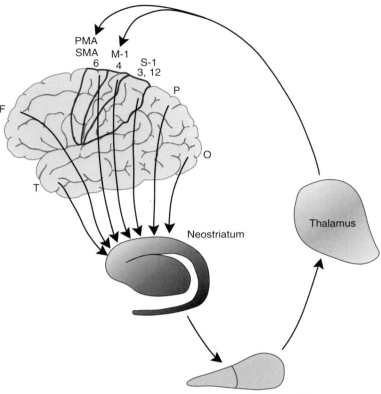

FIGURE 8.7.

This schematic diagram emphasizes the principal path through the basal ganglia. Note that the system forms a loop. Information originates from nearly the entire cerebral cortex. After passing through the basal ganglia and thalamus, the modified information is returned primarily to the specific motor areas of the cerebral cortex. *F,* frontal lobe; *O,* occipital lobe; *P,* parietal lobe; *PMA,* premotor area; *SMA,* supplemental motor area; *T,* temporal lobe.

neostriatum, in turn, projects to both nuclei of the globus pallidus. The GPm, in turn, sends axons to the ventral anterior (VA) and ventral lateral (VL) nuclei of the thalamus. These specific motor nuclei of the thalamus project, not to the entire cerebral cortex, but primarily to the premotor area (PMA) and supplementary motor area (SMA) of area 6 (see Chapter 7). Thus the basal ganglia are a central link in a part of the motor system that translates the desire to move into action. Information about the desire is collected from widespread areas of the cerebral cortex, modified by the basal ganglia, and returned specifically to the premotor and supplementary motor areas of the cerebral cortex, areas that are closely related to the final motor act.

The connections of the basal ganglia and the probable neurotransmitters are summarized in Figure 8.8.

AFFERENT CONNECTIONS

The neostriatum is the principal receptive area of the basal ganglia. Most afferent fibers arise in the cerebral cortex, but not all areas of the cortex project equally to the neostriatum. For example, the head of the caudate nucleus is much larger than the body or tail, reflecting the fact that the frontal lobe projects more densely to the caudate nucleus than do the other lobes. Most extensive are the connections from the primary motor area, M-I, which sends axons bilat-

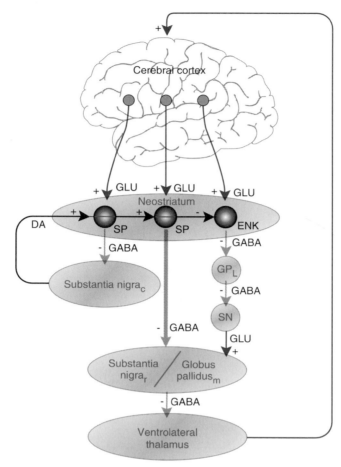

FIGURE 8.8.

A simplified schematic illustrating the principal connections of the basal ganglion. The probable neurotransmitters are indicated along with their apparent action (facilitation or inhibition). Note that the SNr and the GPm are functionally considered to be the same nucleus. This is similar to the situation with the caudate nucleus and putamen, which are indicated here simply as the neostriatum. GLU, glutamate; SP, substance P; ENK, enkephalin; GABA, γ-aminobutyric acid; DA, dopamine; SNc, substantia nigra pars compacta; GPl, lateral globus pallidus.

erally to the putamen, while the premotor area projects ipsilaterally to both the caudate nucleus and the putamen.

Cortical afferent fibers terminate on the spiny neurons of the neostriatum, and these same neurons are the source of efferent axons. The cortical projections are well organized and maintain a topographic relationship between the cerebral cortex and the neostriatum. Striatal neurons are excited by the cortical afferent fibers that terminate on the heads of the spines.

The neostriatum, particularly the putamen,

receives important projections from the substantia nigra pars compacta (SNc). These axons synapse on the necks of the spines and the dendritic shafts of the spiny neurons (Fig. 8.9). They also synapse on the aspiny interneurons. Their action in both cases appears to be inhibitory, mediated by the transmitter dopamine. The location of the nigral synapses on the necks of the spines suggests that they can selectively short-circuit excitatory postsynaptic potentials (EPSPs) originating on the heads, much the same way that inhibitory synapses on the base of

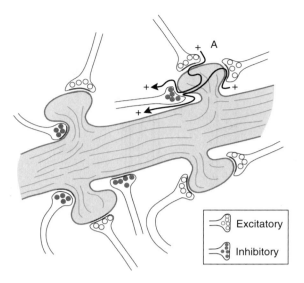

Excitatory

Inhibitory

FIGURE 8.9.

The spines on dendrites not only greatly increase the surface area of the dendrite (and thus the number of synapses it can receive); they can also act as miniature dendrites themselves. In this illustration, the excitatory synapses (*open vesicles*) make contact on the heads of the spines, while the inhibitory synapses (*closed vesicles*) contact the necks of the main dendritic shaft. Placed at the neck of a shaft, an inhibitory synapse can short-circuit the excitatory currents of synapses located on the head (see spine *A*).

dendrites can short-circuit EPSPs from synapses located more distally. The inhibitory action of dopamine is exerted through a G protein–coupled receptor. It is now clear that there are at least three and possibly as many as five different dopamine receptors in the neostriatum, and neurons expressing different dopamine receptors project to different areas of the basal ganglia. The specific functions performed by these different dopamine systems remain to be determined.

EFFERENT CONNECTIONS

The principal source of axons leaving the basal ganglia is the GPm and the SNr (Fig. 8.8). The GPm/SNr efferent axons terminate in VA_{PC} (the parvicellular subnucleus of the ventral anterior nucleus) and VL_O (the pars oralis of the ventral lateral nucleus), specific thalamic nuclei that project primarily to the SMA in area 6 of the cerebral cortex. VL_O also makes a small connection with M-I (Fig. 7.8). These neurons are inhibitory (GABA) to their thalamic targets. The thalamic neurotransmitters are not known, but they are probably excitatory to the cerebral cortex.

INTRINSIC CONNECTIONS

The nuclei of the basal ganglia are interconnected by three principal pathways (Fig. 8.8). The largest pathway arises from spiny neurons in the neostriatum containing substance P. Most of these axons, as previously mentioned, project to the GPm and SNr. A second set of spiny neurons, those containing enkephalin, project to the GPl. These neurons innervate the subthalamic nucleus, which returns fibers to the GPm. The third pathway consists of axons that originate from substance P–containing spiny neurons in the caudate and putamen and terminate in the SNc. These axons are not collaterals of the axons that terminate in the GPm and SNr, but are a separate group. The dopaminergic SNc neurons project back to the neostriatum.

Neurology of the Basal Ganglia

The basal ganglion is an area of great neurological importance because several common diseases have been correlated with specific lesions to this area. Damage to the basal ganglia produces movement disorders or **dyskinesias** [G. *dys,* bad + G. *kinesis,* movement]. Dyskinesias are motor disorders that imply some loss of voluntary control and regulation. They fall into two classes: those that result in spontaneous movements (**hyperkinesia**) and those that result in a poverty of movement (**hypokinesia**). Hyperkinesia is expressed as **involuntary spontaneous movements**. Hypokinesia is just the opposite, the **inability to initiate voluntary movement** in spite of the knowledge and will to do so. It is important to note that the motor system is otherwise intact. Dyskinesia differs from paralysis or paresis in two major respects. First, unlike paralysis, dyskinesia involves no dysfunction of the upper or lower motor neuron systems. Consequently there is no weakness. Neither is dyskinesia an apraxia, for apraxia is a deficiency in planning or executing a complex motor act. Apraxia follows lesions to the cerebral cortex. In contrast, dyskinesia is a *deficiency in the **initiation** or **cessation*** of complex motor acts. One's ability to conceptualize the task is unimpaired.

HYPERKINETIC DYSKINESIA

Involuntary spontaneous movements are described by several clinical terms: **chorea, athetosis, dystonia, ballismus,** and **tics.** Although the terms imply discrete pathophysiological phenomena, most authors now feel that they describe variations of a single symptom and that all hyperkinesias arise from a common set of basal ganglion lesions.

Chorea [G. *choros,* a dance] is a rapid, jerky type of involuntary movement that represents *fragments of purposeful movement.* One's arm might suddenly abduct, as a fragment of a more complex act of abduction and extension, as if reaching for an object. The motion is without conscious intent on the part of the patient; therefore it is a spontaneous movement. People with chorea find that having an extremity suddenly move on its own is disconcerting and embarrassing. To cover the embarrassing moment, they may consciously incorporate that spontaneous movement into a purposeful act, such as adjusting one's glasses. This may make their constant activity appear fairly normal to the ca-

sual observer, but such people appear very restless. Chorea has been associated with atrophy of the neostriatum.

Athetosis [G. *athetos,* without position or place] is a continual uncontrolled writhing. One spontaneous movement blends into the next in a ceaseless, sinuous, and purposeless motion. Usually the spontaneous movements involve the hands and face, but all parts of the body can be affected. The movements are frequently combinations of alternating antagonistic motions such as supination/pronation, flexion/extension, or inversion/eversion. Athetosis is an extension of chorea. Chorea consists of discrete, independent movements. As choreiform movements become more frequent, they may take on the slower, more continuous form of athetosis. This progression is common in Huntington's disease (see below). In such cases, the term **choreoathetosis** is applied.

Dystonia [G. *dys,* bad + G. *tonos,* tension] represents an extreme form of athetosis wherein a joint is forced into a locked position for a long period of time. This may result in a fixed abnormal posture with the muscles in painful spasm. Alternatively, the muscles may slowly change their state while antagonistic groups are in great tension, working against each other. These spasms may force the body into contorted positions. There are no grossly detectable lesions associated with dystonia. Therefore, most authorities assume that the lesion is a biochemical deficiency in the striatum. Two hereditary forms of dystonia have been identified, one dominant and the other recessive. Nonhereditary idiopathic forms also exist.

Ballismus [G. *ballismos,* a jumping about] is a violent involuntary movement involving the proximal muscles that results in a flinging of the extremities. This dramatic condition is self-limiting and evolves into a form of chorea or athetosis. Ballismus is caused by a discrete lesion to the subthalamic nucleus contralateral to the affected side. The lesion leaves the globus pallidus and its efferent connections intact. The subthalamic nucleus is excitatory to the GPm; therefore its loss would decrease the facilitation of the GPm. Since the GPm is inhibitory to the ventral lateral thalamus, loss of the subthalamic nucleus *disinhibits* the thalamus, thereby increasing its facilitation of the motor cortex, which results in the ballistic movements (Fig. 8.10).

Tics are a type of uncontrollable compulsive behavior. They are a quick fragment of a purposeful movement, making them similar to chorea, but they differ from chorea by being endlessly repeated in a stereotyped manner. Chorea consists of many different, randomly interspersed acts; tics are almost always the same movement. Tics may be sniffing, throat clearing, blinking, snorting, or even involuntary vocalizations. In Gilles de la Tourette syndrome, the spontaneous, involuntary vocalizations can take the form of barking like a dog or swearing.

HYPOKINETIC DYSKINESIA

There is only a single term, **bradykinesia** [G. *bradys,* slow + G. *kinesis,* movement], used to describe the hypokinetic movement disorders. Bradykinesia is characterized by a *poverty of movement.* Patients will maintain a fixed posture for unusually long periods of time. The normal, restless shifting of one's posture or the general fidgetiness is much reduced or absent. The facial expression is gaunt and fixed. The eyes seldom blink. When walking, the arms do not swing. With advanced disease, patients have difficulty initiating even the most routine acts, such as arising from a chair. Once walking one has dif-

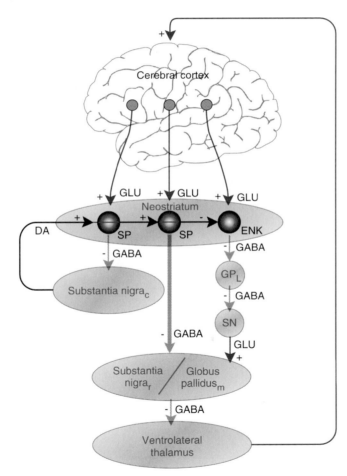

FIGURE 8.10.

Hyperkinetic dyskinesia is thought to involve the removal of the subthalamic nucleus from the basal ganglion circuitry. Here, ballismus is caused by a direct lesion to the subthalamic nucleus. Since it facilitates the SNr/GPm, a nucleus inhibitory to the ventral lateral thalamus, its loss disinhibits the thalamus, which creates greater cortical facilitation.

ficulty stopping, and often bumps into furniture or walls. Even spontaneous swallowing movements are diminished, causing one to drool, since the production of saliva remains normal.

Diseases of the Basal Ganglia

Several specific diseases are associated with each of the involuntary movements. The most common are discussed below.

HUNTINGTON'S DISEASE

In 1872 George Huntington reported on patients his father and grandfather had treated, all of whom shared the symptoms of **dementia** and **chorea**. These patients presented no symptoms at birth. In middle age they developed peculiar personality changes and choreiform movements. Their symptoms progressed until they were incapacitated and died. All patients had a parent who displayed the same symptoms, and Huntington went on to show the hereditary nature of the disease.

Dementia frequently precedes the development of involuntary movements, but just as frequently does not. The initial mental manifestations of the disease are usually alterations in personality. Patients may become irritable, impulsive, depressed, or violent. Later, obvious dementia develops, characterized by memory lapses and decreased attentiveness, progressing until the patient is incapacitated. Suicide is frequent, which may be due to the mental illness associated with the disease or because all Huntington's victims have witnessed their own future through their parent's suffering; it may be considered a rational act.

Abnormal choreiform movements usually first affect the hands and face. Frequently patients are able to mask the spontaneous movements by incorporating them into socially acceptable intentional acts. To the casual observer, the patient may appear unusually fidgety or restless. As the disease progresses, more of the body becomes involved. Eventually the patient is in constant motion. At the terminal stages the chorea evolves into athetosis or dystonia.

Pathology

The most consistent pathological finding in Huntington's disease is a gross wasting of the caudate nucleus and putamen. This can be visualized with computed tomography (CT) or magnetic resonance imaging (MRI) (Fig. 8.11) as an enlargement of the lateral ventricles due to the

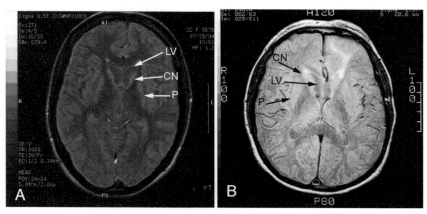

FIGURE 8.11.

A, T1-weighted MRI of a patient with Huntington's disease. This section is cut in the horizontal plane at the level of the basal ganglia. Note the large lateral ventricles (*LV*) and the atrophied caudate nucleus (*C*) and putamen (*P*). **B,** A T2-weighted MRI from a normal patient, taken in approximately the same plane as **A**. Compare the size of the ventricles, caudate nucleus, and putamen with the MRI in **A**.

atrophy of the caudate nucleus. The loss of neostriatal mass is closely correlated with the development of involuntary movements. Microscopically, the enkephalergic spiny neurons of the neostriatum seem to be the most affected, at least in the early stages of the disease (Fig. 8.12). Their loss serves to remove the neostriatal inhibition of the GPl, releasing its inhibitory influence on the subthalamic nucleus. This effectively removes the subthalamic nucleus from the basal ganglion circuitry. The loss of the subthal-

amic nucleus by increased GPl inhibition removes its facilitation of the GPm which, as we have seen in ballismus, releases the ventrolateral thalamus (see discussion of ballismus on page 262).

Also associated with Huntington's disease is a loss of cortical neurons, especially from layer 3. This results in some loss of cortical mass and a widening of the gyri, which can be visualized with MRI. The cortical atrophy is much less striking than the atrophy seen in the neostria-

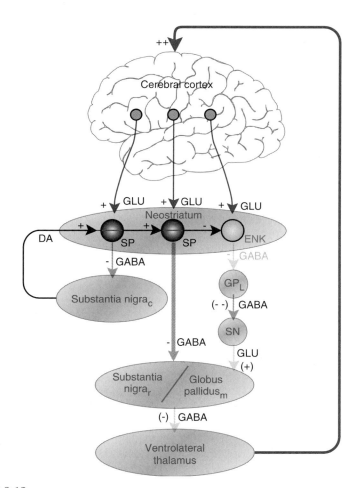

FIGURE 8.12.

Huntington's disease is a form of hyperkinetic dyskinesia. Early in the course of the disease, enkephalergic neurons in the striatum degenerate. These neurons project to and inhibit the GPl, a nucleus that is itself inhibitory to the subthalamic nucleus. Loss of the enkephalergic neurons of the neostriatum therefore enhances inhibition of the subthalamic nucleus, effectively removing it from the basal ganglion circuitry. In this regard Huntington's disease is similar to ballismus. The net result is a disinhibition of the thalamus, which results in increased cortical facilitation.

tum, being so slight that it does not seem to be able to fully account for the mental changes of the disease. Other types of dementia, such as Alzheimer's disease, are associated with more profound cortical atrophy and manifest certain mental disturbances not commonly associated with Huntington's dementia. For example, patients with Alzheimer's disease typically exhibit severe losses of memory, loss of language and calculation skills, apraxia, and agnosia. Huntington's patients are usually spared these difficulties. They are more likely to be afflicted by personality changes, simple forgetfulness, apathy, depression, and generally slowed thought processes. These differences have led some authors to suggest that Huntington's disease produces "subcortical dementia," while Alzheimer's disease represents a "cortical dementia." The concept of "subcortical dementia" implies a sparing of the cerebral cortex, an implication that is not absolutely correct, as MRI studies and autopsy material have consistently demonstrated cortical thinning in Huntington's disease. But the term also implies that the basal ganglia play an important role in cognitive function. This implication seems to be true for at least some parts of the basal ganglia. For that reason, distinctions between "cortical dementia" and "subcortical dementia" are probably useful.

Etiology

We now recognize that Huntington's disease is caused by an *autosomal dominant gene* located on the short arm of chromosome 4. Half of the children of an affected parent can expect to develop the disease. A study in 1932 showed that nearly all affected people in the eastern United States were the progeny of six individuals who emigrated to the New World in 1630 from one tiny village in Suffolk, England. Spontaneous mutations are rare. A 1915 study surveyed 962 patients. Only five were thought to be descended from unaffected parents and might represent spontaneous mutations.

In most individuals the disease is first recognized between the ages of about 35 to 40, however, various forms may be expressed as early as 4 and as late as 70. The disease progresses until death occurs 10 to 15 years after symptoms first appear. Due to the relatively late onset, most individuals do not know that they are affected until after their reproductive years, thus ensuring the transmission of the gene to the next generation. Although the specific gene responsible for Huntington's disease has yet to be identified, a marker close to it has been found. It is now possible to identify, with almost complete certainty, individuals who carry the Huntington's gene before they develop symptoms of the disease.

Treatment

To date there is no effective treatment for this disease. Genetic counseling is of some value to the children of afflicted persons. Not all who are at risk wish to avail themselves of this service, but many do.

PARKINSONISM

Approximately 1% of the U.S. population over 50 is afflicted with the illness that was described for the first time by James Parkinson in 1817. The disease that now carries his name usually begins between the ages of 40 and 70 and progresses inexorably to death in about 10 to 15 years. Parkinsonism is characterized by three signs: **bradykinesia**, **tremor**, and **rigidity**.

Perhaps the most prominent feature of parkinsonism is the 4- to 7-Hz tremor that almost always accompanies the disease. The tremor may begin in any extremity and once begun, tends to spread to the others. In the hand the tremor is quite characteristic, involving an alternating pattern of the thumb and fingers rubbing together. The more senior members of the neuroscience community say that it reminds them of ancient pharmacists making pills by rolling the ingredients between their thumb and fingers, so the term "pill-rolling" tremor is often used. The tremor is seen only while the extremity is at rest; it disappears during an intentional movement and during sleep.

Although the tremor is the most noticeable and emotionally disturbing aspect of the disease for the patient and his family, bradykinesia is the most debilitating. The bradykinesia of parkinsonism is a true slowing of movement. All movement for the parkinson patient is reduced to the same slow velocity. With normal persons, as the amplitude of a movement is increased, its velocity is also increased such that the total duration of the act is approximately a constant. The parkinson patient, however, cannot increase the velocity of movement. The duration becomes extended as the amplitude increases (Fig. 8.13). The parkinson patient is operating in slow motion.

Parkinson patients exhibit a special type of resistance to passive movement of the limbs known as rigidity. Rigidity is a very specific symptom and must not be confused with spasticity (see Chapter 7). Rigidity is caused by an inappropriate sensitivity of the muscle to stretching. Unlike the spastic patient, who exhibits very short-latency hyperactive muscle stretch reflexes, *the MSR is normal in the parkinson patient*. The sensitivity to stretching in the parkinsonian patient is due to a *long-latency polysynaptic* reflex pathway that probably includes the cerebral cortex (Fig. 8.14). To the physician, passively moving the patient's limb feels like bending a lead pipe, hence the rigidity of parkinsonism is frequently called **lead pipe** rigidity. Superimposed on the rigidity, one can often feel the tremor. This gives the sensation of two gears loosely meshing, as if one can feel each cog making contact. From this colorful description, the term **cogwheel rigidity** has evolved.

The diminution of cognitive function due to organic brain disease, **dementia**, is an inconsistent finding in parkinsonism. One study found 32% of afflicted patients demented, another only 8%. Unquestionably, some parkinson patients are demented, and in most of those one can demonstrate

white matter lesions with MRI. It is not settled whether or not the dementia is caused by the same factors that cause parkinsonism or is an independent development. In this regard it is noteworthy that the number of patients exhibiting the characteristics of both parkinsonism and Alzheimer's disease (a specific type of dementia that will be discussed in Chapter 13) is higher than would be predicted by chance alone. However, a causal relationship between parkinsonism and dementia, if one exists, has not been established.

Pathology

The most consistent pathological finding in patients with this disease is loss of the pigmented cells in the SNc (Fig. 8.15). This is accompanied by a depletion of other pigmented cells of the central nervous system: the locus ceruleus and the dorsal motor nucleus of the vagus. The SNc cells produce dopamine, a neurotransmitter that inhibits the target neurons in the neostriatum. These target neurons seem to be the substance P-containing spiny neurons of the neostriatum. Their inhibition releases the GPm and SNr, nuclei that inhibit the ventrolateral thalamus (Fig. 8.16). This removes thalamic drive from the motor cortex, producing hypokinesia. In addition to the inhibition of the substance P-containing spiny neurons, the SNc dopamine pathway seems to facilitate the enkephalergic spiny neurons. This greatly inhibits the GPl, which removes inhibition from the subthalamic nucleus (STN). Released in this way, the STN's facilitation of the

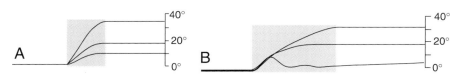

FIGURE 8.13.

Parkinson's disease is a hypokinetic dyskinesia. It results in an overall slowing of muscle action. The normal person, when moving an extremity, will increase the velocity of muscle contraction as the distance moved increases, so that the total action takes place in approximately the same amount of time. The parkinson patient is unable to do so. Contraction velocity remains constant. **A**, The elbow joint position of a normal person when asked to flex the elbow 10°, 20°, and 40°. Note that the slope changes for each trial, so the actual muscle action remains nearly constant. **B**, The same task performed by a person suffering from parkinsonism. Note that the slope remains fairly constant, so movement over the greatest distance takes much longer to complete than movement over a shorter distance.

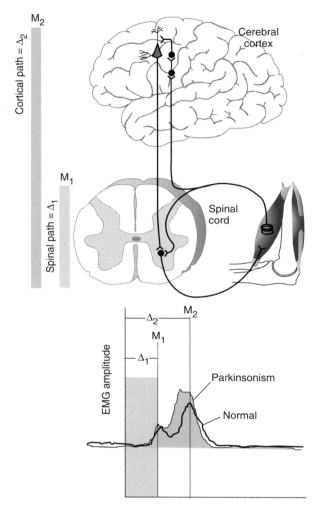

FIGURE 8.14.

Top, Simplified schematic of circuits activated by muscle stretch. The spinal path (Δ_1) is the MSR. The cortical path (Δ_2) is long latency and polysynaptic. *Bottom,* The electromyogram recorded from an intact individual displays two peaks in response to a single brief stretch, the first being labeled M_1 and the second M_2. M_1 results from the reflex contraction caused by the muscle stretch reflex (MSR), a reflex of spinal origin. M_2 is believed to involve cortical neurons because of its long latency. These recordings represent the M_1 and M_2 responses from a normal person and a parkinson patient. Note that the M_1 response is the same for both individuals, but the M_2 response is exaggerated in the parkinson patient. This hyperactive M_2 response to stretching is believed to be the basis for the rigidity associated with the disease.

GPm and SNr is increased, which causes further inhibition of the ventrolateral thalamus.

Etiology

The etiology of spontaneous SNc degeneration (idiopathic parkinsonism) has yet to be identified. Other non-idiopathic forms of the syndrome have been identified. For example, parkinson symptoms develop after various types of brain injury. **Head trauma**, particularly that associated with prize fighting, is one of the more preventable causes of parkinsonism. In one study, 50% of professional boxers were diagnosed near the end of their careers with "punch-drunk encephalopathy," a syndrome that frequently includes parkinson symptoms.

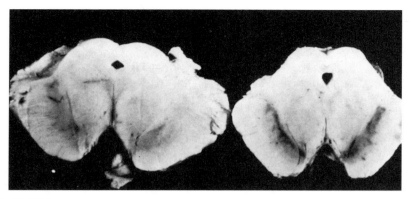

FIGURE 8.15.

Loss of the pigmented cells of the substantia nigra in parkinsonism is dramatically demonstrated at autopsy. Specimen on the *left* from a parkinson patient has very little color in the substantia nigra compared to the specimen from a normal patient (*right*).

Encephalitis lethargica, a viral infection that enveloped the world between 1914 and 1930, is of special importance. This deadly disease killed about 20% of its victims. Throughout the world, more people died of this disease than died from the hostilities of World War I. During the acute phase of the illness, many of the patients exhibited symptoms now attributable to basal ganglion pathology, including bradykinesia and chorea. A large number of the survivors developed a form of parkinsonism within a few months to as long as 25 years after recovering from the infection. Encephalitis lethargica has not been seen since 1930, but other forms of encephalitis and influenza are occasionally associated with parkinsonism. Although there is no evidence that ordinary influenza viruses cause parkinsonism, certainly some viruses are capable of selectively attacking SNc neurons, much like the polio virus can selectively attack the α motor neuron.

Certain **toxins** can injure the brain in a way that produces parkinson-like symptoms. Most notable are **carbon monoxide** and **manganese** poisoning. Very recently, a designer street drug, MPTP   (1-methyl-4-phenyl-1,2,3,6-tetrahydropyridine), was accidentally produced in an illicit kitchen laboratory. Those who were exposed to its effects developed a profound akinesia. In animal primates and in humans, it has been shown that this drug destroys nearly all the pigmented neurons in the SNc. The toxic agent is not MPTP itself, but a metabolite, MPP$^+$ (1-methyl-4-phenylpyridine), a free radical produced by

the oxidation of MPTP by monoamine oxidase. This is the first demonstration of a neurotoxic agent that is specific for SNc neurons. This discovery supports the hypothesis that certain toxic agents, be they viruses or chemicals, can produce parkinsonism in humans.

To date no infectious agent has been consistently associated with idiopathic parkinsonism, nor has a hereditary etiology been established. The cause of most parkinsonism remains obscure. The observations listed above have been used by some, however, to advance the hypothesis that idiopathic parkinsonism is an environmental disease caused by a number of possible toxins, particularly those that can be metabolized into free radicals. Stated in a very abbreviated form, this hypothesis emphasizes the following observations: (*a*) approximately 50% of the pigmented neurons in the SNc disappear in humans during the normal course of aging and, in spite of this loss, few people show signs of parkinsonism; (*b*) patients who develop parkinsonism have lost approximately 66% of the pigmented neurons; (*c*) consequently, there appears to be only a small safety factor in the number of neurons needed for normal function in the SNc throughout life; and (*d*) contact with a toxic agent that kills as few as 15% of the SNc neurons during one's youth might reduce the overall population enough so that in later years too few neurons remain for normal function, and parkinson symptoms develop. As yet, no single toxic agent or group of toxins has been identi-

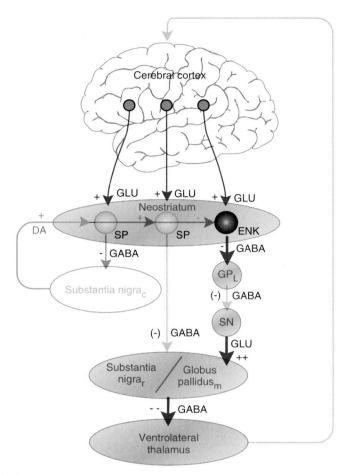

FIGURE 8.16.

Parkinsonism is a form of hypokinetic dyskinesia and is caused by loss of neurons in the SNc. These neurons appear to facilitate some and inhibit other neurons in the neostriatum. The most prominent effect seems to be the loss of inhibition on the enkephalergic neurons, which causes an increase in the inhibition they supply to the GPl. Inhibition of GPl results in disinhibition of the subthalamic nucleus, which facilitates the SNr/GPm nucleus. Loss of the dopaminergic neurons of the SNc also appears to result in a loss of facilitation to the substance P-containing neostriatal neurons. This reduces inhibition of the SNr/GPm. Therefore, activity in the SNr/GPm is increased by both mechanisms, resulting in profound inhibition of the ventral lateral thalamus. This reduces cortical drive.

fied that can explain the high incidence of this disease. Furthermore, there are difficulties with the hypothesis. For example, the hypothesis does not explain the higher incidence of the disease in the white races (2 to 4 times greater than for the nonwhite races). Nor does it explain the relatively constant occurrence of parkinsonism throughout the world. If environmental toxic agents were the cause of the disease, one would not expect a constant toxic exposure in all societies, since

some parts of the world are more polluted than others. Many important questions remain unanswered.

In 1817, when James Parkinson first described it, this disease appeared to be a single disease process with specific diagnostic criterion. The idiopathic form of the disease is known as **Parkinson's disease** or **primary parkinsonism**. Today, however, the picture is less clear. Many cases that resemble Parkinson's disease have a known cause. These non-idiopathic cases frequently respond differently to drug treatment and have a different clinical course. Other cases that resemble Parkinson's disease and are also idiopathic have additional features that segregate them from true Parkinson's disease. It appears that a number of different disease processes share the pathology of degeneration of the pigmented cells of the substantia nigra and the resulting clinical symptoms of bradykinesia and tremor. Since these variants have different prognoses and treatments, it is important to avoid labeling all of them simply "Parkinson's disease." The term **parkinsonism** or **parkinsonian syndrome** with appropriate qualifications is gaining favor.

Treatment

Once the connection between depletion of dopamine-producing cells and parkinson symptoms was made, replacement of the missing dopamine seemed logical. Indeed, ingestion of L-dopa[3] does ameliorate the symptoms in about 80% of the patients (Fig. 8.17). Unfortunately, L-dopa treatment does not curtail the relentless

loss of nigral neurons. The disease progresses, and consequently the treatment becomes less effective with time. Before L-dopa therapy was available, primary parkinsonism caused death in about 25% of patients within 5 years of the initial diagnosis. About 80% had died within 15 years. With L-dopa therapy, morbidity has been reduced by about 50%. Ultimately, with or without therapy, the patients become immobilized to the point where they are confined to bed, eventually to succumb to *complications of the bedfast state*.[4]

It is remarkable that L-dopa therapy is effective. One would hardly expect that flooding a system as complicated as the brain with a drug would approximate the natural release of a neurotransmitter at synaptic boutons in the neostriatum. The answer may lie in the observation that the firing of SNc cells is not correlated with any of the dynamic parameters of muscle movement. Alterations in firing rate tend to be much slower than any motor act. This has led to the hypothesis that dopamine may be acting like a neurohumoral agent or modulator of overall striatal excitability. Were that the case, systemic administration of L-dopa could be expected to perform, albeit crudely, the same function.

---

[3]Dopamine does not cross the blood-brain barrier. Its precursor L-dopa does and is effectively metabolized to dopamine (see Fig. 8.5).

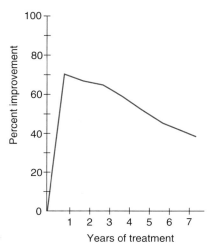

FIGURE 8.17.

The amount of improvement in parkinsonism symptoms with L-dopa treatment declines with duration of treatment.

Anatomy of the Cerebellum

In mammalian systems the cerebellum regulates posture and coordinates motor acts. The loss of cerebellar function produces disturbances of gait, balance, and stability and diminishes the accuracy of reaching motions. Its loss does not produce paralysis, paresis, apraxia, or deficits in the initiation or termination of voluntary motor acts. Cerebellar lesions also affect the ability to improve one's motor skills. In normal people, motor coordination improves with experience, a phenomenon known as motor plasticity. There is accumulating evidence that the cerebellum, not the cerebral cortex, is the site where this motor learning takes place.

---

[4]Many neurological diseases like parkinsonism are not fatal per se, but they immobilize the patient to the point that complications of immobility become the agents of death. These are frequently pulmonary infections secondary to the aspiration of mucus or saliva, urinary tract infections, or infections from decubitus sores.

GROSS ANATOMY

The cerebellum (Fig. 1.11) is a large, compact structure that is attached to the pons by three pairs of tracts, called the **cerebellar peduncles**: the **superior** (brachium conjunctivum[5]), the **middle** (brachium pontis), and the **inferior** (restiform body). The body of the cerebellum is divided into two hemispheres connected at the midline by the **vermis**. Like the cerebral cortex, the cerebellar cortex is deeply folded into **folia** to increase its surface area. Some of the folds are deeper than others, separating the cerebellum into various lobes. The **flocculus** and the **nodulus** are particularly noteworthy. Most of the cells of the cerebellum lie near the surface of the folia. Deep

in its core are a set of nuclei that contain most of the cells that project to other areas of the brain.

MICROANATOMY

The superficial cell-containing layer of the cerebellum consists of three parts. The deepest cortical layer is the **granule cell** layer, while the most superficial is the **molecular** layer. A single layer of cells, the **Purkinje cell** layer, lies between them (Fig. 8.18). There are five principal neuronal cell types in the cerebellar cortex. The cells and their probable transmitters are summarized in Table 8.1.

Purkinje cells have a large, goblet-shaped cell body and a gigantic, fan-shaped dendritic tree that extends toward the surface into the molecular layer, perpendicular to the long axis of the folium. Purkinje cell dendrites have nu-

[5]The names in parentheses are the traditional terms applied to the cerebellar peduncles.

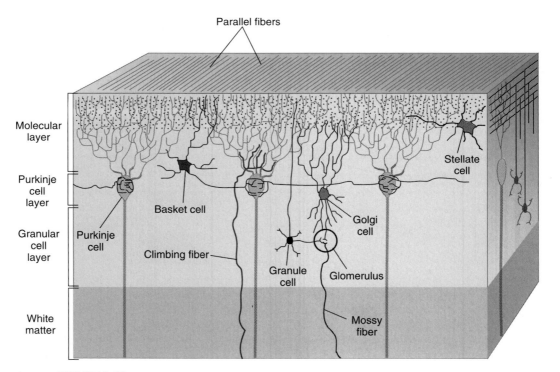

FIGURE 8.18.

The three-dimensional arrangement of the cerebellar cortex is shown in this schematic diagram. Note that the parallel fibers run perpendicular to the fan-shaped dendritic tree of the Purkinje cell. Basket cell axons run perpendicular to the parallel fibers and inhibit a row of Purkinje cells across the folium.

Table 8.1. Probable Neurotransmitters Associated with the Cerebellum

Synaptic Source	Target	Transmitter
Purkinje cell	Deep cerebellar nuclei	GABA (−)
Granule cell	Purkinje cell dendrites (via parallel fiber)	Glutamate (+)
Basket cell	Purkinje cell	GABA (−)
Golgi cell	Granular cell glomerulus	GABA (−)
Stellate cell	Purkinje cell dendrites	Taurine (−) ?
Mossy fiber	Granular cell glomerulus	ACh (+) ??
Climbing fiber	Purkinje cell	Aspartate

merous spines that increase the synaptic surface many fold. Axons from Purkinje cells are the only fibers to leave the cerebellar cortex. Most of them synapse in the deep nuclei; a few leave the cerebellum to synapse in brainstem nuclei. All make inhibitory synapses on their targets. The neurotransmitter is probably GABA.

Granule cells, the most numerous neuron type in the brain, are found in the granule cell layer. Axons from these cells leave the soma and project directly to the surface of the folia, where they bifurcate, sending the two branches along the folia and parallel to its surface for about 6 to 10 mm. These are known as **parallel fibers**. About 400,000 parallel fibers run through the fingers of the dendrites of a Purkinje cell. Each parallel fiber traverses the dendritic field of about 1000 Purkinje cells that are lined up in an orderly row along the length of the folia. The arrangement is much like wires running through the arms

of roadside telephone poles. Parallel fibers make excitatory synaptic contacts with Purkinje cells.

Basket cells lie in the deeper parts of the molecular layer. Their dendrites extend into the superficial region of the molecular layer, receiving excitatory synaptic input from the parallel fibers. The axon of the basket cell extends across the folia, at the junction of the molecular and Purkinje cell layers. As it traverses this interface, it extends collaterals at each Purkinje cell where the collateral envelops the Purkinje cell soma like a basket, making powerful inhibitory synapses. The inhibition is so powerful because the synapses are located on the Purkinje cell soma, very close to the trigger zone. The basket cell has the ability to temporarily stop all action potentials in the Purkinje cell.

Stellate cells are located in the molecular layer and receive excitatory contacts from parallel fibers. The stellate cell synapses with the dendrites of Purkinje cells, inhibiting them. Like basket cells, stellate cells send their axon across the folia, perpendicular to the parallel fibers. In contrast to basket cells, the inhibition produced by stellate cells is discrete, being directed at individual branches of dendrites. Stellate cells have the potential to electrically "prune" the dendritic tree, in effect temporarily removing limbs from the Purkinje cell, negating their influence.

Golgi cells lie slightly deep to the Purkinje cell layer. Their dendritic field extends into the molecular layer, where parallel fibers make excitatory synaptic contacts with them. The axon of the Golgi cell enters the granule cell layer and makes inhibitory synapses with dendritic extensions of the granule cells, extensions known as the **cerebellar glomerulus**. The glomerulus is a complex structure contained within a glial capsule. Mossy fibers (see below) also make synaptic contacts in the glomerulus.

One might expect to find a somatotopic map of the body represented in the cerebellar cortex, much as one finds maps in the primary sensory cortices and thalamic nuclei. Recently, convincing evidence has surfaced showing that the deep cerebellar nuclei are mapped to the body. However, no convincing maps of the cerebellar cortex have been discovered.[6]

AFFERENT CONNECTIONS

Although the cerebellum receives information from all parts of the nervous system, the entering axons make only three types of synaptic contacts in the cortex. The most numerous axons are classified by their synaptic endings; they are **mossy fibers** and **climbing fibers**. Both types of afferent fibers divide upon entering the cerebellum, sending one branch to the deep nuclei and the other branch to the cerebellar cortex.

The **inferior olive** is the source of all climbing fiber input into the cerebellum. Axons leave the inferior olive, cross the midline, and enter the cerebellum through the inferior peduncle,

from which they spread out to innervate the entire organ. On entering the cerebellum, climbing fibers divide; one branch enters a deep nucleus, while the other branch ascends to the cerebellar cortex, where it synapses with Purkinje cells. The projections from the inferior olive to the cerebellum are quite orderly and follow the longitudinal division outlined below for the three divisions of the cerebellum. Recently it has been discovered that the connections from the inferior olive to the deep nuclei are reciprocal.

It is not necessary to describe in detail all the afferent connections to the inferior olive. It will suffice to say that all parts of the sensory and motor systems appear to project into the inferior olive in an organized way. From the spinal cord, direct connections to the inferior olive are transmitted by the **spino-olivary tract**, a tract that courses in the ventral funiculus. Fibers from the dorsal column nuclei also reach the inferior olive after crossing the midline. Afferent fibers are also received from the **vestibular nuclei**, the **red nucleus**, the **superior colliculus**, and the **mesencephalic reticular formation**. Finally, axons from the sensory and motor areas of the **cerebral cortex** have wide projections into the inferior olive.

Mossy fibers are the most numerous axons entering the cerebellum. They arise from four principal parts of the CNS: the spinal cord, the vestibular nuclei, the brainstem reticular formation, and the deep pontine nuclei. The cortical branch of mossy fibers makes excitatory contacts with granule cells and Golgi cells in the glomerulus. These endings look like tufts of moss, giving these fibers their name. Mossy fibers excite granule cells, which in turn excite Purkinje cells via the parallel fibers. The extent of this excitation is remarkable. Approximately 600 granule cells are innervated by a single mossy fiber. Each granule cell, through its parallel fiber, synapses with about 1000 Purkinje cells. Therefore, assuming no overlap (not a reasonable assumption), approximately 600,000 Purkinje cells could be excited by a single mossy fiber. Even assuming considerable overlap, the amount of synaptic divergence in this system is impressive. The synaptic convergence on Purkinje cells is just as remarkable. Each granule cell receives synapses from about four mossy fibers, and each Purkinje cell receives about 80,000 synapses from parallel fibers. Again, as-

---

[6]Many textbooks offer illustrations with homuncular maps drawn on the surface of the cerebellum. These maps are based on recording evoked potentials from the cortical surface. There are many difficulties with this technique. The validity of these maps has recently been called into question by several authors. The issue is certainly not settled, but direct topographic mapping of the body onto the cerebellar surface now seems unlikely.

suming no overlap, this means that each Purkinje cell potentially receives information from approximately 320,000 mossy fibers.[7]

Parallel fiber synapses on Purkinje cells produce small-amplitude (0.02- to 4-mV) EPSPs at the Purkinje cell soma. These EPSPs are caused by Na^+-selective channels that open in response to the granule cell neurotransmitter, probably **glutamate**. The amplitude of the EPSPs summate as more parallel fiber synapses are recruited. If threshold is reached, an action potential is produced in the Purkinje cell. The synaptic activity from parallel fibers is very high, giving Purkinje cells a background firing rate of about 70 spikes per second.

Climbing fibers arise from the contralateral inferior olivary nucleus, dividing on entering the cerebellum into a deep branch and a cortical branch. The cortical branch further divides, sending collaterals into several lobes. Each collateral makes approximately 300 synapses on a single Purkinje cell. Each Purkinje cell receives only one climbing fiber. The climbing fiber receives its name because it contacts only the soma and basal dendritic branches, ascending the dendrite as if it were climbing a tree. The synapses are profoundly excitatory.

The action potentials produced by climbing fibers on Purkinje cells are dependent on the opening of Ca^{2+} channels, probably through the action of the neurotransmitter **aspartate**. The shape of the calcium-dependent action potential follows a *complex* time course and looks entirely different from the *simple* sodium-dependent spike associated with the parallel fiber EPSP. The complex, calcium-dependent spikes are not propagated through the cell or along the axon. Rather they act as powerful EPSPs. They are so powerful that a single climbing fiber calcium spike triggers an action potential. This is one of the very few obligatory synapses in the CNS of mammals. The calcium-dependent spikes initiated by climbing fiber input are much less frequent (only about 1 per second) than the sodium-dependent spikes initiated by the mossy fiber-parallel fiber input.

The importance of the calcium spike, however, probably has nothing to do with the action potential it initiates in the Purkinje cell. Complex calcium spikes are outnumbered by the simple sodium spikes by about 70:1. Therefore the effect on the stream of action potentials leaving the Purkinje cell is negligible. The real importance of the calcium spike is that it represents the inward flow of Ca^{2+} and that calcium flux can affect the metabolism of the Purkinje cell. Calcium ions are important intracellular messengers and can have a number of long-lasting effects (see Chapter 3). There is some evidence to suggest that the calcium-dependent spikes in the Purkinje cells are a critical component of the mechanism responsible for motor learning (motor plasticity) that occurs in the cerebellum (see page 287).

Very little is known about the third source of cerebellar afferents. These are the **monoaminergic axons** that arrive in the cerebellum from two sources. **Adrenergic** afferent fibers arise in the **locus ceruleus**, a small nucleus in the pons that is the principal source of adrenergic afferent fibers to all parts of the CNS. **Serotonergic** afferent fibers arise from the raphe nuclei of the brainstem. Both of these monoaminergic systems have major projections into the hypothalamus; their projections into the cerebellum are relatively slight. The physiological role of the monoaminergic systems is unknown.

EFFERENT CONNECTIONS

All axons leaving the cerebellar cortex originate in Purkinje cells. These axons terminate in nuclei located deep in the white matter of the cerebellum—the deep cerebellar nuclei. There are three pairs of deep cerebellar nuclei, the **fastigial**, the **interpositus**,[8] and the **dentate**. All contain cells whose axons leave the cerebellum through the cerebellar peduncles. Axons from these nuclei synapse in the thalamus, the brainstem motor centers, and the vestibular nuclei. The cells of the deep cerebellar nuclei are somatotopically arranged according to the muscles

---

[7]These numbers are based on observations from the cat. It is not possible to make such estimates from human material.

[8]In the human, the interpositus nucleus is usually divided into the **globose** and the **emboliform** nuclei. Two regions in the interpositus nucleus can be differentiated by cytological criteria. The anterior region is thought to be homologous with the emboliform nucleus, while the posterior region corresponds with the globose nucleus. In humans the globose nucleus is many times smaller than the emboliform. Most efferent axons from the interpositus nucleus arise in the anterior region, the analog of the larger emboliform nucleus.

they influence. The six nuclei are independently mapped, each having a complete representation in one body half (Fig. 8.19). This independent multiple representation of the body is particularly significant since each nucleus, and the corresponding portion of the cerebellar cortex that projects to it, appears to serve a different aspect of motor coordination. This arrangement divides the cerebellum into three independent motor control systems.

Functional Divisions of the Cerebellum

The cerebellum can be separated into three divisions based on the connections each division makes with the nervous system (Fig. 8.20). These functional divisions do not, with the exception of

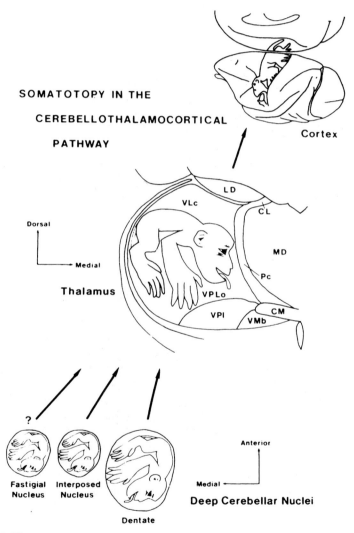

FIGURE 8.19.

The somatotopic organization of the monkey dentate, interpositus, and fastigial nuclei is schematically represented here and compared with the same organization in the thalamus and M-I cortex. Major thalamic nuclei that correspond with the human: *CM,* centromedian; *LD,* lateral dorsal; *MD,* medial dorsal; *VL$_C$,* ventral lateral pars compacta; *VPL$_O$,* ventral posterolateral pars oralis.

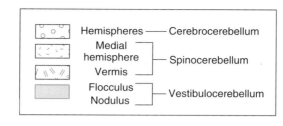

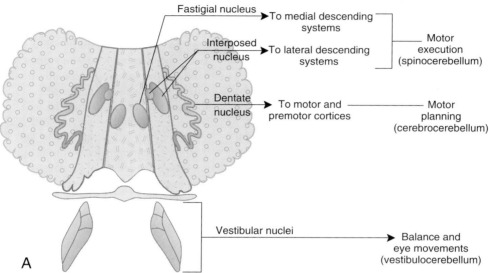

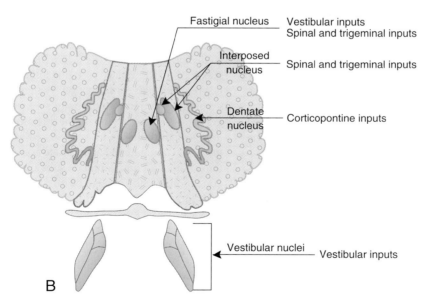

FIGURE 8.20.

The functional divisions of the cerebellum are shown here along with the principal deep nuclei to which their Purkinje cells project (**A**) and the areas from which they receive information (**B**).

the flocculonodular lobe, correspond with the prominent fissures that divide the cerebellum into lobes. The boundaries between the functional divisions are not precise and there is some overlap. Nevertheless, the functional divisions are much more useful than the gross anatomical divisions, both in understanding the physiology of the cerebellum and in interpreting its neurology.

VESTIBULOCEREBELLUM

The **vestibulocerebellum** is composed primarily of the **flocculus** and the **nodulus**. This **flocculonodular lobe** receives most of its afferent fibers from the vestibular nuclei, with a few coming from the vestibular nerve (CN VIII) directly (Fig. 8.21*A*). Axons from the vestibular nerve and the vestibular nuclei enter the cerebellum through the inferior cerebellar peduncle. They terminate as mossy fibers. Most of the axons from the Purkinje cells leave the vestibulocerebellum to synapse in the **fastigial nucleus**, but a few bypass this nucleus, leave the cerebellum through the inferior peduncle, and terminate directly in the vestibular nuclei.

Efferent fibers from the fastigial nucleus leave the cerebellum through the contralateral inferior peduncle, having crossed the midline within the cerebellum as an arch of fibers, the **uncinate fasciculus**. Most of the crossed fibers terminate in the lateral and inferior vestibular nuclei and the motor portions of the reticular formation. A few fastigial axons find their way to the thalamus (VL_C, VPL_O), the superior colliculus, and the cervical spinal cord (lamina IX). A few fastigial axons have ipsilateral connections, terminating in the lateral and inferior vestibular nuclei.

The vestibulocerebellum receives primary afferent fibers from the vestibular nerve (CN VIII). These axons originate in the labyrinthine receptors of the inner ear (see Chapter 10). Information about the orientation of the head with respect to gravity comes from the otolith receptors, while information about relative head motion originates in the semicircular canals. This information is used to maneuver the head with respect to body motion to provide a stable platform for the eyes. The vestibular input is also used to coordinate eye movements with head movements so that a stable image can be maintained on the retina. For example, while reading this page, if you move your head from side to side, your eyes will scan in a direction opposite to the head motion to keep the printed word fixed on the retina. The details of this system will be discussed in Chapter 9.

The vestibulocerebellum, through its extensive connections with the lateral vestibular nucleus, regulates gait and posture. The lateral vestibulospinal tract originates in the lateral vestibular nucleus, and this tract facilitates the motor neurons innervating the anti-gravity muscles. These connections are fast and direct, many of them making monosynaptic connections with α motor neurons. Alterations of the body's orientation to gravity are sensed by the otolithic organs in the inner ear and appropriate corrective signals are sent to the spinal cord to modify the activity of the motor neurons.

There are two principal symptoms associated with lesions to the vestibulocerebellum. The more obvious is a staggering, ataxic gait with a *tendency to fall toward the side of the lesion*. This ataxia is closely related to the inability to correlate proprioceptive information about the dynamics of body motion with the body's relationship to gravity. As one starts to fall, the direction and velocity of the fall has to be assessed relative to gravity to compute the appropriate corrective action. Without a functioning vestibulocerebellum, corrections are too slow to be effective.

Injury to the vestibulocerebellum also produces a characteristic spontaneous eye movement called **nystagmus**. This term describes an

---

FIGURE 8.21.

The principal connections of the three divisions of the cerebellum are summarized in these greatly simplified schematic diagrams (see text for details). *CBST,* corticobulbospinal tract; *DSCT,* dorsal spinocerebellar tract; *LVST,* lateral vestibulospinal tract; *MVST,* medial vestibulospinal tract; *VSCT,* ventral spinocerebellar tract.

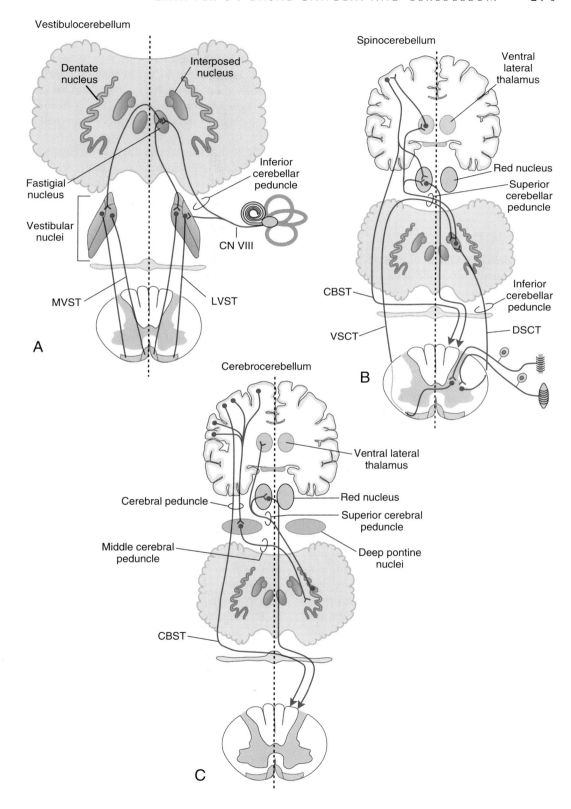

oscillating eye movement in which the eyes move slowly in one direction and rapidly in the other. Usually the eyes move together in the horizontal plane, but movement in almost any plane is possible. Nystagmus may be caused by several different lesions. Cerebellar lesions that produce nystagmus also frequently produce the sensation of the world spinning around, or **vertigo**. Most students experience nystagmus and vertigo upon lying in bed after an intemperate evening of entertainment. Nystagmus will be discussed in detail in Chapter 9.

SPINOCEREBELLUM

The **spinocerebellum** corresponds to the **vermis** and a small strip of the hemispheres located immediately adjacent to the vermis called the **intermediate hemisphere** (Fig. 8.21*B*). Most of the information reaching the vestibulocerebellum arrives from the spinal cord. Sensory information from the muscles and the lower half of the surface of the body reach the spinocerebellum over two principal fiber tracts, the **dorsal spinocerebellar tract (DSCT)** and the **ventral spinocerebellar tract (VSCT)** (Fig. 8.22).

The dorsal spinocerebellar tract carries very specific proprioceptive information to the spinocerebellum. Cutaneous information from touch and pressure receptors and proprioceptive information from the muscle stretch receptors and joint receptors enters the spinal cord over I_A, I_B, and group II primary afferent axons. As we have already seen (Chapter 6), the receptors associated with these axons have very restricted sensory fields, and thus they supply very discrete information to the CNS. In addition to the synaptic connections already discussed (Chapter 6), these axons also synapse in the ipsilateral **nucleus dorsalis (Clarke's column)**, a collection of large cells in lamina VII at the base of the dorsal horn (Fig.8.22). Axons leaving this nucleus collect, without crossing the midline, in the dorsolateral funiculus just inferior to the dorsal root entry zone and ascend the spinal cord as the DSCT, entering the cerebellum through the inferior peduncle, where they terminate as mossy fibers. Since the nucleus dorsalis exists only from about C8 to L3, the DSCT carries information only from the trunk and lower extremities.

An analogous system for the upper extremity consists of a few fibers from the fasciculus cuneatus that break away from the medial lemniscus and synapse in the **accessory cuneate nucleus**. Axons from this nucleus form the **cuneocerebellar tract** and join the DSCT to enter the cerebellum through the inferior cerebellar peduncle.

The ventral spinocerebellar tract originates in large "border cells" in lamina VII that border the dorsolateral boundary of the ventral horn. Axons from these cells cross the midline in the ventral white commissure of the spinal cord and collect in the ventrolateral funiculus (Fig. 8.22). In this location they ascend the spinal cord, pass through the brainstem, and enter the spinocerebellum through the superior peduncle. The superior cerebellar peduncle is a crossed tract, so axons of the VSCT ultimately terminate ipsilateral to the cells of origin. The border cells that give rise to the VSCT are innervated, polysynaptically, by flexion reflex afferents (FRAs) along with a few afferent fibers from muscle spindles and Golgi tendon organs. They also receive monosynaptic contacts from collaterals of all the descending motor tracts (corticospinal, reticulospinal, vestibulospinal, and rubrospinal).

It is clear that the VSCT conveys a qualitatively different type of information to the spinocerebellum than the DSCT. VSCT sensory information, for the most part, arises from sensory receptors with large receptive fields. It also returns to the cerebellum copies of the motor commands that are arriving at the motor neuron pools. In contrast, the DSCT carries direct information from the muscle and joint receptors, information that represents the actual effect the motor commands are having on limb placement.

Purkinje cells of the spinocerebellum send axons into the **ipsilateral fastigial** and **interpositus nuclei**. Cells located in the vermis project to the fastigial nucleus. Fastigial connections have been described above. The Purkinje cells in the intermediate hemisphere project to the interpositus nucleus. Axons leaving the interpositus nucleus leave the cerebellum through the superior peduncle. This tract is crossed, and consequently interpositus axons all terminate on the side contralateral to their origin. Most of the axons terminate in the red nucleus, and about

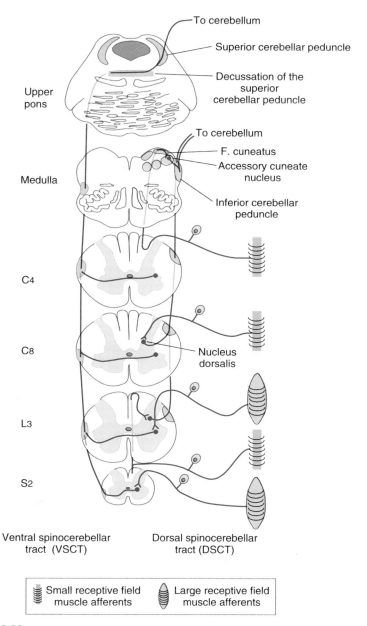

FIGURE 8.22.

This schematic illustrates the major features of the dorsal spinocerebellar tract (DSCT) and the ventral spinocerebellar tract (VSCT). Note that these tracts transmit qualitatively different information to the spinocerebellum. Note also the various crossing patterns that ultimately keep the information ipsilateral to the site of origin.

one-third pass through the nucleus and terminate in the thalamus (VL_C and VPL_O).

CEREBROCEREBELLUM

The **cerebrocerebellum** consists of most of the cerebellar hemispheres. It receives afferent connections indirectly from the cerebral cortex, through the intervening **deep pontine nuclei** (Fig. 8.21C). About 20 million axons from the deep pontine nuclei are sent into the cerebellum through each middle peduncle. Almost all of these axons cross the midline before entering. Within the cerebellum, they are distributed to the hemispheres, where they terminate as mossy fibers.

The deep pontine nuclei receive most of their afferent fibers from all four lobes of the cerebral cortex. The most massive projections come from the precentral and postcentral gyri. The most represented cortical regions are areas 6, 4, 3, 1, 2, and 5, which correspond with the principal motor and sensory areas. There are important projections from the primary visual cortex (area 17) as well. The corticopontine axons leave the cerebral cortex, descend through the internal capsule, and collect in the cerebral peduncle. Those from the frontal lobe lie in the ventromedial portion, while those from the remaining lobes lie in the dorsolateral part, separated from the former group by the CBS tract (see Chapter 7). These cortical projections onto the deep pontine nuclei are topographically precise and well organized.

Purkinje cells of the cerebrocerebellum project to the **dentate nucleus**. Axons leaving the dentate nucleus pass through the superior cerebellar peduncle and terminate in the contralateral VL_C and VPL_O nuclei of the thalamus. A smaller number of axons terminate in the contralateral red nucleus. In this respect, dentate projections mirror those from the interpositus nucleus.

OVERVIEW OF CEREBELLAR CONNECTIONS

The vestibular and spinal divisions of the cerebellum are arranged in a way that allows them to operate as a typical feedback control system (Fig. 8.23A). Both of these divisions receive sensory information concerning either body position or muscle action. The output affects the descending motor systems, either at the brainstem level (red nucleus, motor nuclei of the reticular formation, and the vestibular nuclei) or at the motor neuron pools of the spinal cord.

The cerebrocerebellum is arranged entirely differently (Fig. 8.23B). It receives most of its information from the motor, premotor, and sensory areas of the cerebral cortex. The sensory information from S-I probably conveys the present postural state of the body, while the motor information—from M-I, the SMA, and the PMA—relates to the motor command that *is about to be executed*. The output from the cerebrocerebellum feeds back to the red nucleus and M-I motor cortex, where it can affect the motor command *before the execution of the command*. In other words, the cerebrocerebellum is arranged to act like a feed-forward control system.

Feed-forward control systems modify the action of control systems before an action is taken. They do this by measuring environmental variables that may have an effect on how the control system as a whole operates. To do this they must have some internal representation or model of the system and how the environmental variables affect it.

Clinical Aspects of Cerebellar Function

Unlike the cerebral cortex and the basal ganglia, the cerebellum is an uncrossed organ. Lesions to the cerebellum have effects on the ipsilateral side of the body. This can be appreciated by considering the various crossing patterns of the afferent and efferent fiber tracts to the cerebellum. The DSCT carries ipsilateral information and enters the cerebellum over the inferior peduncle, an uncrossed tract. The VSCT, however, carries information from the contralateral side of the body, but it enters the cerebellum over the superior peduncle, a crossed tract. Similarly, the dentate efferent fibers leave through the superior peduncle, crossing the midline to innervate the red nucleus and the thalamus. Axons leaving the red nucleus immediately cross the midline before projecting to the spinal cord. Some of these crossing patterns are illustrated in Figure 8.21.

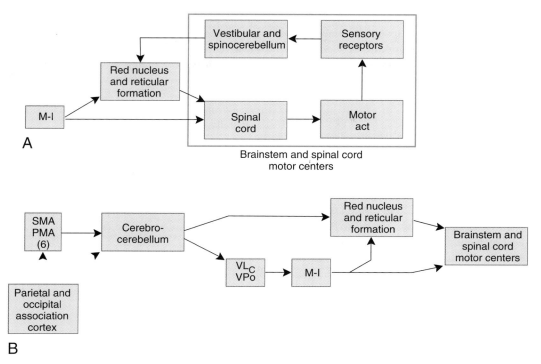

FIGURE 8.23.

A, Block diagram of a feedback control system labeled to illustrate how the spin-ocerebellum fits into such a scheme. Probably the cerebellum does not act strictly as a comparator (see Chapter 7), but uses sensory feedback information to adjust the timing of contractions among synergistic muscle groups (see text). **B,** This block diagram illustrates how the cerebrocerebellum lies between the sensorimotor-motor cortex and the primary motor cortex (M-I), and can act as a feed-forward controller. It is in a position to receive information about the impending motor act (from the PMA and SMA) and the current state of the body (S-I). It can adjust the developing motor command as it is actually being assembled (in M-I). The output from this system represents the input to the feedback system depicted in **A.** Missing from this diagram are the basal ganglia. They form a loop between the cerebral cortex and the VL. Note how both systems modify the motor act before it is implemented.

The function of the cerebellum is best appreciated by observing patients who have suffered cerebellar lesions. The motor disturbances are unique and are clearly distinguishable from disturbances associated with lesions to the basal ganglia or to the upper or lower motor neuron systems. A number of clinical terms are used to succinctly describe the effects of cerebellar dysfunction.

TIMING

Cerebellar lesions result in a **delay** in the initiation and termination of motor commands and affect the **sequencing** of individual muscle contractions. The velocity of motor acts is normal, as opposed to the situation with basal ganglia lesions, in which it is slowed (see page 267). Only the **timing** of the muscle contractions is delayed. This disruption of timing leads to a number of specific motor deficiencies. Perhaps the easiest to observe is the *inability to perform repetitive tasks* in which the timing between antagonistic muscle groups is critical. For example, a normal subject has no trouble rapidly pronating and supinating his hands. If the cerebellum is not functioning, this simple act becomes impossible, as the coordinated timing between the various muscle groups becomes disrupted (Fig. 8.24*A*). The inability to perform rapid alternating movements is called **dysdiadochokinesia** [G. *dys,* difficult; + G. *diadochos,* working in turn; + *kinesis,* movement].

ATAXIA

The effects of poor timing are also evident when multiple muscle groups across several joints are involved. This is often seen as **ataxia**, an uncoordinated, staggering, wide-based gait. If the patient is standing still, he will have a tendency to fall toward the side of the lesion (Fig. 8.24*B*). Timing derangements can also be demonstrated by having a patient touch the examiner's finger and then touching his own nose. The nor-

mal patient can make his finger travel a direct, straight course from one target to the other. If the cerebellum is dysfunctional, the finger travels a wandering course that deviates more and more from the straight line as the finger approaches the target (Fig. 8.24*C*). Performance *deteriorates as the motor act progresses.*[9] This is usually called **dysmetria**.

---

[9]The unsteadiness seen in target-directed movements is frequently referred to as "intention tremor." Superficially it looks like a tremor that is expressed only during muscle movement, in contrast to the "tremor of rest" that is associated with parkinsonism. The unsteadiness of motor acts due to cerebellar dysfunction is not rhythmic, nor is it of constant frequency or amplitude. True tremor has all of these characteristics and seems to be the result of continuous oscillatory behavior in central motor areas that is independent of external events. It is better to reserve the term tremor for the latter cases and simply describe "intention tremor" as what it is: dysmetria, a form of incoordination.

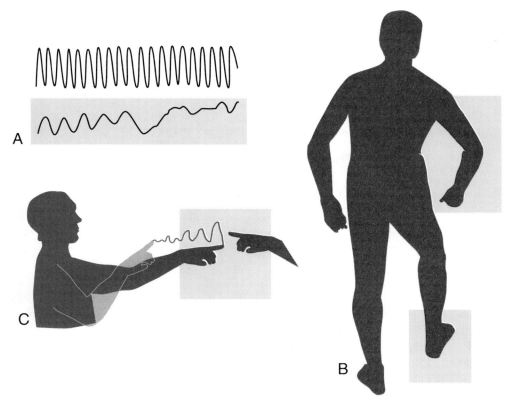

FIGURE 8.24.

A, These records indicate the position of a hand during rapid alternating pronation and supination. The upper record is from a normal individual, the bottom from one with a cerebellar lesion. **B**, The lack of coordination caused by cerebellar lesions may be revealed by a wide-based, staggering gait with a tendency to fall toward the side of the lesion. **C**, People who have cerebellar lesions have great difficulty projecting an extremity toward a target. Note that the deviations become greater as the motor act progresses.

DECOMPOSITION

The poor timing of motor acts that occurs with cerebellar lesions can be seen as a **decomposition of movement**. The patient, to complete the motor act, divides the synergistic components into separate acts in an effort to remove the necessity of precise timing between the components. In reaching for an object, he may first move the shoulder joint, then the elbow, followed by the wrist and fingers. This fails, of course, since each joint has antagonist muscles that must be coordinated, but decomposing the movement does improve performance somewhat.

REFLEXES

Other than observing the patient's motor performance, the most objective signs of cerebellar dysfunction are **pendular muscle stretch reflexes** and **hypotonia**. To visualize pendular reflexes, a limb must be allowed to swing freely, as the lower leg of a patient sitting at the edge of the examining table, when the tendon is struck by the reflex hammer. The MSR contraction occurs, but the distal extremity will swing back and forth several times like a pendulum, instead of damping quickly after one or two swings.

Pendular reflexes and hypotonia are related signs. Animal studies have shown that removal of the cerebellum or its deep nuclei reduces the facilitation of γ motor neurons. This loss of gamma drive in effect reduces the gain of the feedback control system, causing the MSR to be less effective in regulating muscle length. Muscle tone is reduced because the decreased facilitation of the γ motor neurons is reflected in a decreased facilitation of the α motor neurons (see Chapter 6). With less resting muscle tone, there is less resistance to passive movement after the MSR contraction has occurred. Therefore, a limb such as the lower leg will swing back and forth like a pendulum due to its own mass, not because of active muscle contractions.

Theory of Cerebellar Function

At this point one should note that, in spite of our fairly detailed knowledge of the anatomy and physiology of the cerebellar circuits, there is no comprehensive and coherent model of cerebellar function. However some ideas have been presented that seem promising and will be described here. One theory of cerebellar function is based on the observation that the principal function of the cerebellum seems to be the *coordination of muscle contractions across synergistic groups*. This is expressed by *regulating the timing* of individual muscle contractions.

ROLE OF PARALLEL FIBER EXCITATION

The parallel fibers in the molecular layer synapse with hundreds of Purkinje cells, but the synapses are not all active simultaneously. Because of the small diameter of parallel fibers, action potentials travel relatively slowly,[10] exciting one Purkinje cell after another in series. This means that a wave of Purkinje cell excitation spreads down a folium as parallel fibers are excited. The inhibitory basket cells and stellate cells are also excited by this same wave, but their axons run perpendicular to the parallel fibers. Therefore simultaneous with the wave of excitation, there is a wave of Purkinje cell inhibition that spreads across the folia, perpendicular to the excitatory wave. The Purkinje cells that are lined up along a beam of parallel fibers send axons into one of the deep cerebellar nuclei. There they synapse in an orderly manner across the nucleus that is somatotopically mapped to the body. *This means that parallel fiber input affects groups of muscles that are arranged sequentially along the body or an extremity* (Fig. 8.25A).

PURKINJE CELL SETS

A single row of Purkinje cells can be affected by different sets of parallel fibers. Therefore the same sequence of adjustments imposed on a motor act by Purkinje cell activity can be triggered by different sets of granule cell parallel fibers. If the parallel fiber-Purkinje cell synapses in the different sets are not identical, but some of the synapses have been modified by experience, then each set of parallel fibers could represent a slightly different timing pattern for the same

---

[10]Parallel fibers are unmyelinated and average about 0.2 μm in diameter. Such fibers conduct at about 0.5 m/sec.

group of synergistic muscles. The pattern selected would depend on the mossy fiber-granule cell activation (Fig. 8.25*B*).

ROLE OF BASKET CELL INHIBITION

Parallel sets of Purkinje cells in the cerebellar cortex represent different, but closely related, sequences of muscle coordination. Slight differences in sequencing can be established by slightly different connections the Purkinje cells make in the deep cerebellar nuclei. The radiating pattern of basket cell and stellate cell inhibition that spreads across the folia perpendicular to parallel fiber excitation probably represents the suppression of these competing parallel representations

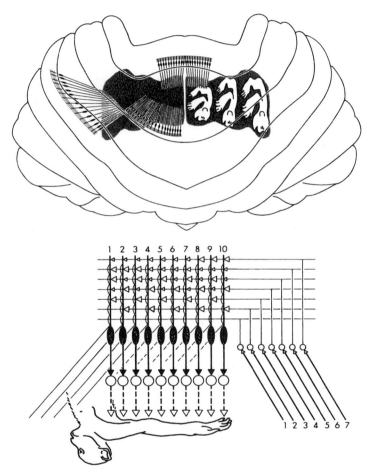

FIGURE 8.25.

A, This schematic depicts how a beam of parallel fibers selects a set of Purkinje cells the axons of which project across one of the deep cerebellar nuclei. This projection will automatically affect synergistic muscle groups because the nucleus is somatotopically organized. The timing is determined by the slow progression of excitation along the parallel fibers. **B**, The same beam of Purkinje cells can be selected by numerous parallel fiber pathways, each of which is excited by different sensory stimuli projected through the granule cell synapse. The various parallel fiber pathways do not all necessarily activate the Purkinje cells to the same degree. Here the effectiveness of the parallel-fiber Purkinje cell synapse is depicted by the size of the triangular symbol representing the synapse. Each of the seven parallel fibers shown here could represent a slightly different coordination of the same set of synergistic muscles, suggested here by the projection onto the monkey's arm..

of coordination sequences (Fig. 8.26). Therefore, as one Purkinje cell beam is excited, its immediate neighboring beams are inhibited.

CLINICAL CORROBORATION

These ideas of cerebellar function correlate well with two clinical observations. First, small lesions to the cerebellum produce few symptoms. Lesions must be quite large, involving most of a hemisphere, for example, before really obvious cerebellar symptoms appear. This is consistent with the hypothesis that there is a lot of functional redundancy in the cerebellar circuitry. As described above, if multiple repetitions of Purkinje cell sets represented only variations on essentially the same motor sequence, then moderate lesions could be absorbed with only very subtle changes in motor function observable. Second, clinical experience has always demonstrated that, unlike the cerebral cortex, there is no well-defined somatotopic organization to the cerebellar cortex. This observation is also consistent with the above hypothesis, because it is the cerebellar *output* that is somatotypically mapped through the deep nuclei, not the input as projected to the cortex. The input is redundantly represented on the cortical surface by the granule cell-parallel fiber path. The lack of a systematically arranged input map in the cerebellum has recently been demonstrated in animal experiments.

MOTOR PLASTICITY

Finally, there is a lot of evidence that the cerebellum is capable of modifying its function in relation to experience. This feature is called **motor plasticity** or **motor learning**. Motor plasticity is not a new observation. Everyone who has learned to ride a bicycle or play a musical instrument has personally experienced motor plasticity. Using bicycle riding as an example, when one is first learning to ride, one has to consciously think about every aspect of riding: balance, pedaling, braking, steering, etc. As long as one has to consciously think about the process, corrections are too slow to effectively correct errors. You fall off the bike a lot. But with practice, the motor skills necessary to ride a bicycle become more automatic. Eventually one does not think about the specific **tactical** motor acts

required to ride, but one's conscious efforts become directed toward the overall **strategy**, such as the best route to take. The execution of the motor tactics becomes unconscious. In fact, the transfer of motor tactics from the realm of the conscious to the realm of the unconscious is absolutely necessary if motor proficiency is ever to be achieved. Patients with cerebellar lesions cannot perform motor tasks unconsciously. Consequently they revert to consciously directing motor tactics, and that is why everything they do looks like a youngster learning to ride a bicycle.

Although the evidence is not complete, there is a consensus that the cerebellum is the part of the nervous system largely responsible for motor plasticity. For this to be true, there must be a mechanism in the cerebellum that can alter synaptic efficiency in response to experience. The climbing fiber synapse is believed to be this mechanism. Two lines of evidence support this view.

First, activating the climbing fiber synapse causes Ca^{2+} to enter the Purkinje cell. Calcium is a potent intracellular messenger. It can modulate the phosphorylation of proteins by activating protein kinase C or calmodulin. Once activated, these proteins can increase the production of neurotransmitter receptors, or signal the production of different receptors. Finally, some receptors can be directly phosphorylated, a process that can modulate their sensitivity to neurotransmitters. Therefore, although we do not know the exact effect the climbing fiber synapse has on the intracellular metabolism of the Purkinje cell, the potential exists for this synapse to have profound effects on the Purkinje cell's electrical behavior.

Second, climbing fiber activity is altered during a period of motor learning. In a classic experiment, monkeys were trained to move a lever into a target zone using only wrist flexion or extension. When the monkeys had learned the task, climbing fiber spikes in Purkinje cells occurred randomly and infrequently (about 1 to 2 Hz). If the load on the lever were changed so that a slightly different coordination between flexors and extensors was required to place the lever in the target zone, climbing fiber activity increased (to about 4 to 6 Hz) and was no longer random, but correlated with the lever displacement. When the new task was learned, climbing

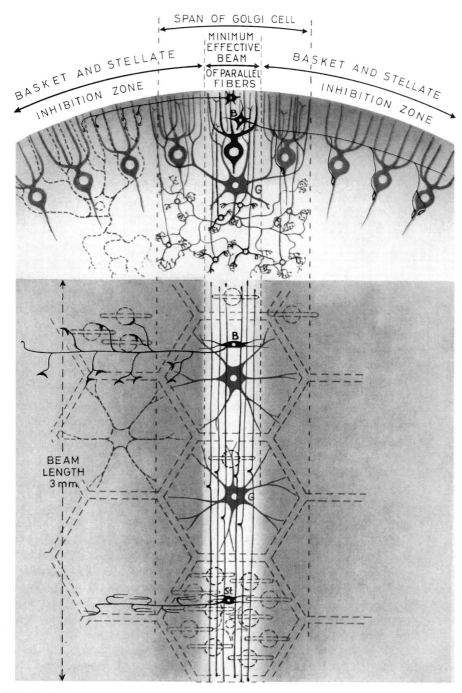

FIGURE 8.26.

The spreading wave of excitation by parallel fibers and inhibition by basket cells (and stellate cells) is schematically depicted here. The excitation arises in the center (*circle*), divides and spreads up and down in the figure. This wave serially excites Purkinje cells and basket cells. The axons of basket cells run perpendicular to the parallel fibers, creating a wave of inhibition (*left* and *right* in the figure). This inhibitory wave decreases the excitability of Purkinje cells that are not in the beam of parallel fiber excitation. The significance of the basket cell inhibition may be to deselect alternate, parallel coordinations for the same synergistic set of muscles (see Fig. 8.25).

fiber activity once again declined and was no longer correlated with lever position. Furthermore, simple spike activity (from parallel fibers) decreased as the learning period progressed and remained at the lower level after the new task was learned and the climbing fiber activity had returned to pre-learning levels.

CASE STUDIES

Ms. A. R.

HISTORY: Ms. A. R. is a 64-year-old female. In 1991 she first noticed a tremor in her hands. Over a period of a few months this spread to her left leg. She had weakness and awkwardness of the left side of her body. In February of 1992 she went to her family doctor, who sought a neurologic consultation because he thought that she had had a stroke. At that time, Ms. A. R. was feeling stiff all over and was having trouble initiating ambulation. She had tremors in all four extremities. Her countenance was expressionless, even when involved in animated conversation or telling a joke. She blinked rarely. She did not adjust her posture or make many subconscious movements when sitting in a chair.

She denied having had the flu during the great flu epidemic of the 1920s. She had no known occupational or social exposure to toxic levels of carbon monoxide, manganese, carbon disulfide, or MPTP. She had never been treated with the phenothiazines, reserpine, haloperidol, or similar drugs.

NEUROLOGIC EXAMINATION, 1992:

I. MENTAL STATUS: She is alert, witty, and conversant. She is orientated as to time and place. She can do serial subtractions by 7, she knows the current and past presidents back to George Washington. There are no memory deficits.

II. CRANIAL NERVES: All are normal. Although her voice is very soft, there is no indication of hoarseness.

III. MOTOR SYSTEMS: She has a shuffling gait. On turning, she stands in one place, shuffling her feet such that she turns about on a single spot as if standing on a pedestal (pedestal turning). She is stooped and holds her arms slightly adducted and flexed while standing. She does not swing her arms while walking.

There is an obvious 3- to 5-Hz resting tremor in all extremities. The left side is worse than the right side. On volition, the tremor decreases or even disappears. On examination, there is a well-developed cogwheel rigidity in both arms and legs. She is somewhat more rigid on the left side.

IV. REFLEXES: MSRs are all within normal limits and are symmetrical. There is no sign of spasticity, clonus, fasciculations, or atrophy. Plantar signs are flexor.

V. CEREBELLUM: Not tested.

VI. SENSORY SYSTEMS: Her perception of pin prick and cotton touches are symmetrical and within normal limits.

Comment: The history and physical appearance of this patient should provide one with enough information to arrive at the diagnosis of parkinsonism. The patient's initial complaint was a tremor in the hands that had shown progression over several years. The neurologic examination provides further corroborative evidence: cogwheel rigidity, shuffling gait, and poverty of spontaneous movements. The essential features of the parkinsonian state (bradykinesia, resting tremor, cogwheel rigidity, stooped posture with shuffling gait) are all well developed in this patient.

The original diagnosis of "stroke" is not defensible. The patient's initial complaint was tremor that had progressed inexorably over several months. Examination of the patient does not reveal a hemiparesis, loss of sensorium, or other deficits that can be attributed to an infarct within the territory of a single artery.

Once the initial diagnosis is made, one should question the patient concerning possible causative factors. Of particular importance are factors that can be modified or eliminated. Foremost among these are the antipsychotic drugs. The largest group, the phenothiazines, are potent drugs that have serious side effects. They work by blocking the postsynaptic dopamine receptors. As mentioned previously (see page 261), there are several different types of dopamine receptors; one type (D_2) is found primarily in the cerebral cortex, while others (D_1) are located primarily in the striatum. Many psychoactive drugs cross-react to more than one receptor and have the potential to produce motor disorders that mimic basal ganglion disease. These drug-induced symptoms include parkinsonism, spontaneous choreiform movements, and dystonia. Iatrogenic abnormal movements are called tardive dyskinesia. *Approximately 40% of patients on long-term antipsychotic medication develop tardive dyskinesia.*

Removal of the drug usually causes the symptoms to subside, but it may take months or years. In rare cases the effects may be permanent.

SUBSEQUENT COURSE: Ms. A. R. is currently taking Sinemet, a drug that combines **carbidopa** and **L-dopa** in a single pill. Carbidopa is a decarboxylase inhibitor that does not cross the blood-brain barrier. It prevents the conversion of L-dopa to dopamine in the systemic circulation while allowing the conversion to take place in the cerebral circulation. This combination has greatly reduced the side effects associated with taking L-dopa alone. Ms. A. R. is doing well. Her general motor performance is slower than normal, but she is able to walk unassisted and perform all the everyday functions of a normal life. She is living independently and has an active social life.

FURTHER APPLICATIONS

- Consider the criteria for the diagnosis of stroke. Are there any elements of this case that would legitimately lead one to arrive at the diagnosis of stroke?

- Ms. A. R. is not likely to have come into contact with MPTP, and inquiring about this in the history is mostly a matter of ritual on the part of the physician. However, contact with the other chemicals mentioned is quite possible. Which, in this case, would warrant special attention?

- L-Dopa treatment eventually fails to be effective in parkinsonism. Consider various hypotheses that would account for this failure. How does this differ from dopamine-responsive dystonia (see the case of Ms. T. L. below)?

- What are the side effects of L-dopa administration that are alleviated by the coadministration of carbidopa?

Mr. J. D.

HISTORY: Mr. J. D. is a 69-year-old male who first noticed a twitching in his right foot about 15 years ago (1975). This twitching became so severe that it interfered with his driving and he could not maintain a steady driving speed. His condition has deteriorated slowly over the years. He now experiences involuntary movements in all of his extremities and has an unsteady gait. He cannot dress himself, nor can he sit down without assistance. He is incontinent of urine, is frequently confused, and has memory lapses from time to time. He is quite irritable.

Mr. J. D. has six younger siblings, the youngest being 47 years old. Three of his siblings have similar difficulties. In 1932, his mother began to have memory lapses and she also became quite irritable and aggressive. Eventually her mental state deteriorated to the point where she was institutionalized in 1936 in the state hospital for the insane, since the family could not afford other care for her. She died in 1947, 15 years after her illness began.

NEUROLOGIC EXAMINATION, 1990:

I. MENTAL STATUS: Mr. J. D. is alert and cooperative. He is oriented to self, but does not know the day, date, or year. He does know the name of the current president but not his predecessor. He cannot do serial seven subtractions. He stated that he has four children, but in fact he has only two.

II. CRANIAL NERVES: Normal except for facial twitching and eye blinking.

III. MOTOR SYSTEMS: Mr. J. D. has spontaneous, involuntary movements in all extremities. Unable to sit still, he is continually crossing his legs, adjusting his clothing, or simply fidgeting. His face is alive with constant eye blinking, facial twitches, and nose wrinkling. His head is in constant motion.

Muscle tone seems normal in the upper extremities, but the lower extremities are slightly spastic. There is no rigidity and no tremor.

His gait is definitely ataxic, and he has a tendency to fall toward the right when he walks.

IV. REFLEXES: The MSRs are decreased (−2) in the upper extremities, but increased (+2) in the lower extremities. The distribution in both cases is somewhat worse on the right. Plantar signs are extensor in both feet.

V. CEREBELLUM: On finger-to-nose testing, Mr. J. D. demonstrates past pointing with the right hand; the left is normal. Rapid alternating slapping of his thighs with the palms followed by the backs of his hands is performed poorly on the right. He has difficulty running his right heel up and down the shin of the opposite leg. These tests are difficult to interpret because the involuntary movements mask the voluntary acts.

VI. SENSORY SYSTEMS: Mr. J. D. appears to have normal perception of pin prick and cotton, given his limited ability to cooperate and interpret the examination.

VII. ANCILLARY TESTS: A CT scan (1990) demonstrated enlarged anterior horns in the lateral ventricle due to a reduction in the size of the head of the caudate nucleus.

Comment: Mr. J. D. is in the later stages of Huntington's disease. He displays all of the characteristic spontaneous choreiform movements as well as the progressive dementia. The atrophied caudate nucleus, revealed by CT, confirms the diagnosis.

The cerebellum and the upper motor neuron system are not normally involved in Huntington's disease, but Mr. J. D. has definite signs that both systems are dysfunctional. The cerebellar signs are localized to the right. The increased MSRs in the lower extremities and the decreased MSRs in the upper extremities suggest a symmetrical lesion at the level of the cervical spinal cord. This could be caused by cervical stenosis, syringomyelia, amyotrophic lateral sclerosis (ALS), or some other problem. One would expect to find corresponding sensory signs, which may in fact exist, but cannot be demonstrated due to Mr. J. D.'s presumed inability to cooperate sufficiently to provide a valid examination of the sensory systems. In light of his primary neurological problem (Huntington's disease), these symptoms were not pursued. Unfortunately, having one neurological disease does not protect one from having another.

FURTHER APPLICATIONS

- How would cervical stenosis or syringomyelia explain the UMN and LMN signs seen in this case? Is multiple sclerosis a reasonable differential in this case?

- Could any of the diagnoses suggested above explain the cerebellar signs?

Ms. T. L.[11]

HISTORY, MARCH 1993: Ms. T. L. is a 42-year-old female who first began to have neurological problems when she was 4 years old. At that time she had trouble with her balance and fell down a lot. She was generally poorly coordinated. By the age of 5 her condition had worsened. Her left leg was quite rigid, being held almost constantly in a hyperextended position. Her other extremities were also somewhat rigid, but

the left leg was worse. A spinal puncture was performed and the CSF analysis was normal. To make it possible for her to walk in school, she was placed in braces.

By the age of 8 her rigidity was worse and she was having more difficulty walking. At this time surgery was performed on her ankle to "stabilize" it. An electromyogram was performed that was interpreted as normal. At the age of 13, she was experiencing such severe rigidity in both of her legs that she was walking only on her toes; she could not flex her feet at the ankle. Surgery was performed at this time to lengthen both Achilles tendons. This allowed her to walk somewhat better. Her condition progressed, however, so that by the time she entered high school, she required the use of a wheelchair.

She graduated from high school and entered college. Her condition deteriorated further, and she was forced to withdraw from school because of her physical condition. This led to her referral to a specialty clinic, where she underwent a complete neurological examination. No diagnosis was made at that time, and she was referred to a major university medical center, where another neurological evaluation was made. The doctors there determined that her problem was a psychiatric illness, and she was advised to seek psychotherapy, which she did (1971).

The psychotherapy made her feel responsible for her problems. She was told "you just aren't trying hard enough." Her mother, her principal care giver at the time, was told, "Quit helping your daughter so much. She will never get better until she has to do things for herself." Following the psychotherapy Ms. T. L. became very depressed. This was followed by feelings of anger, bitterness, and hostility toward the medical community. She removed herself from the "medical merry-go-round," and after this time sought medical attention only for routine physical conditions. At this time (1971) she had no neurological diagnosis for her condition.

By 1991 she was essentially bedfast. For the previous 4 years she had been attended by nurses from a local home care facility. She was very rigid. Her hands were continuously flexed and her arms pronated and extended. Her legs were locked in extension. She could not tend to her personal hygiene or dress or bathe herself. She could read, watch television, and converse on the telephone. Her symptoms were alleviated somewhat in the mornings and became progressively worse as the day wore on. By evening she was nonfunctional. During sleep her rigidity decreased significantly.

---

[11]I am grateful to Dr. S. R. White for bringing this case to my attention.

At this time she developed a hiatal hernia, for which her family physician prescribed metoclopramide (Reglan). Her condition immediately became dramatically worse. She could no longer even turn the pages in a book. She was having great difficulty chewing, swallowing, talking, and breathing. She was rigid to the point of nearly total immobility. The metoclopramide was discontinued, but her condition did not improve. Recognizing the relationship between metoclopramide and basal ganglion disease, her family physician arranged for her to be seen by a neurologist at major university medical center in another state. She felt at this time that she would "end up either in a nursing home or a cemetery."

Based on the history detailed above, the attending neurologist immediately prescribed carbidopa/L-dopa (Sinemet). Within a few hours her rigidity waned slightly but noticeably. Seven days later, she was mobile enough to begin physical therapy. Six weeks after entering the hospital and beginning a regimen of Sinemet, she went home. For the first time in her memory she was independent in her personal care. After 9 months, she was walking normally, had obtained a driver's license, and had gotten a job as an office manager at the same home care facility that had provided her with nursing care for the previous 4 years.

NEUROLOGICAL EXAMINATION, MARCH 1993:

I. MENTAL STATUS: Normal.

II. CRANIAL NERVES: All normal.

III. MOTOR SYSTEMS: She walks with an awkward but safe gait. She lifts her feet very high as she walks. She has a noticeable foot drop, which may be due to the bilateral Achilles tendon lengthening that was performed when she was a child. She gestures quite noticeably with her hands and arms as she talks. All four of her extremities seem "loose" and a little floppy as she walks or gestures, but these would not be particularly noticeable to the casual observer.

IV. REFLEXES: All MSRs are normal. Plantar signs are flexor.

V. CEREBELLUM: Not tested.

VI. SENSORY SYSTEMS: Perception to pin prick and cotton touch is everywhere intact.

VII. ANCILLARY STUDIES: An MRI revealed no abnormalities. Ms. T. L. agreed to participate in a research program and submitted to a positron-emission tomographic (PET) scan. This revealed active dopamine receptors in the striatum.

Comment: Ms. T. L. is suffering from a relentlessly progressing disease that affects her motor systems. The description that she provides in her history is that of dystonia, a basal ganglion disease of unknown etiology, pathology, and—until recently—treatment. It is common for people with this affliction, as in the case of Ms. T. L., to submit to numerous orthopedic procedures that are directed at restoring mobility. These measures are usually only of temporary benefit because of the progression of the disease process. What is remarkable in this case is that, in spite of two extensive neurological evaluations, Ms. T. L.'s dystonia remained undiagnosed.

Her family physician would probably not have prescribed metoclopramide if her condition had been properly diagnosed, because that drug is contraindicated with basal ganglion disease. Tardive dyskinesia and the development or exacerbation of basal ganglion signs are well-recognized side effects of this drug. In the case of Ms. T. L. the exacerbations were so severe that, in spite of her reluctance to become involved with the medical system again, she was persuaded to submit to re-evaluation. Fortunately, the attending neurologist recognized her symptoms and instituted appropriate therapy.

Dopa-responsive dystonia (DRD) is a recently recognized (1971) variation of the disease that is characterized chiefly by its immediate and dramatic responsiveness to L-dopa. Its prevalence is unknown but has been estimated to be as much as 10% of all cases of dystonia. It is hypothesized that the disease is caused by a defect in the release of dopamine from the terminals of the SNc neurons, for there is no pronounced atrophy of the substantia nigra, and PET studies have shown functional dopamine receptors in the striatum. Patients with DRD who have been treated with L-dopa have shown no decrease in the effectiveness of the drug over time (unlike in parkinsonism).

FURTHER APPLICATIONS

- Dopa-responsive dystonia has frequently been misdiagnosed as "juvenile parkinsonism," in part because DRD is sometimes associated with a tremor. Consider the similarities and differences between dystonia and parkinsonism and develop your own criteria for differentiating between these two diseases. Are there other causes of tremor?

- Consider the role of the family physician in this case. Had he not been alert to the connection between metoclopramide and the manifestations of basal ganglion disease, Ms. T. L.'s condition would probably still not be properly diagnosed. On the other hand, do you think that he should have recognized dystonia before he prescribed the metoclopramide?

- Consider the damage done in this case by the inappropriate psychotherapy. Imagine what your own feelings would be were you rigidly and painfully locked into an immobile body and told simply that "you are not trying hard enough." Consider how the care giver must have felt when she was told that she was contributing to the problem by helping too much.

SUMMARY

The basal ganglia are composed of a set of nuclei at the center of the cerebral hemispheres. They serve as a bridge between the cerebral cortex and the thalamus. Signals originating from all parts of the cerebral cortex pass through the striatum into the thalamus. From the thalamus they are returned, primarily, to the motor areas of the cortex. Several shorter loops exist. The subthalamic nucleus, for example, received fibers from the globus pallidus and sends fibers to it and to the substantia nigra. The substantia nigra receives axons from the caudate nucleus and putamen, to which it also projects.

The basal ganglia are essential structures of the nervous system for translating the conception of motor acts into action. Therefore, disturbances of basal ganglion function cause spontaneous movements or the opposite, a poverty of movement. These two extremes of basal ganglion disorders are exemplified by Huntington's disease and parkinsonism, respectively.

Parkinsonism is a biochemical disease of the basal ganglia, characterized by the loss of the dopaminergic neurons of the substantia nigra. Clinically, the disease produces bradykinesia, tremor, and cogwheel rigidity. Dementia is inconsistently associated with parkinsonism. Most parkinsonism is idiopathic, but it may be caused by head trauma, viral infection, and exposure to neurotoxins, most notably carbon monoxide and manganese. The symptoms of parkinsonism can usually be alleviated, at least partially, by the administration of L-dopa.

Huntington's disease is a genetic disorder that results in the early death of neurons of the neostriatum. Clinically it is characterized by dementia and chorea. Since it is caused by an autosomal dominant gene, 50% of the progeny of an afflicted individual may be expected to develop the disease. There is no effective treatment.

Dystonia is a disease of the basal ganglia of unknown etiology. It is characterized by nearly continuous contractions of various muscle groups. These contractions can lock joints into abnormal postures. These patients may experience rigidity of various extremities, the trunk, or the whole body. Unlike the parkinson's patient, the rigidity of dystonia is due to active contraction of antagonistic muscles. Some forms of dystonia respond to the administration of L-dopa.

The cerebellum is composed of a pair of hemispheres joined in the midline by the vermis. It has a superficial cortex that contains most of the cells and a set of four pairs of nuclei in the central core. The cerebellum is functionally divided into three divisions: the vestibular, the spinal, and the cerebral parts.

The vestibular and spinal divisions of the cerebellum are arranged as a feedback loop. They receive sensory information from the vestibular system and spinal cord and use that information to modify the activity of the descending motor systems after the commencement of the motor act. Output from the vestibular and spinal cerebellum synapses in the red nucleus, the motor areas of the reticular formation, and the vestibular nuclei.

The cerebrocerebellum is arranged as a feedforward loop. It receives sensory information from the cerebral cortex about current limb position, as well as information about the impending motor commands. Its efferent fibers project back to the thalamus and from there to the primary motor areas of the cerebral cortex, where it can affect the final development of the motor command.

Disturbances of cerebellar function result in a desynchronizing of activity among synergistic muscles. This is manifested clinically as decomposition of movement, dysmetria, dysdiadochokinesia, and ataxia.

FOR FURTHER READING

Adams, R. D., and Victor, M. *Principles of Neurology*. New York: McGraw-Hill, 1993.

Albin, R. L., Young, A. B., and Penney, J. B. The Functional Anatomy of Basal Ganglia Disorders. *Trends Neurosci* 12:366–375, 1989.

Alexander, G. E., DeLong, M. R., and Strick, P. L. Parallel Organization of Functionally Segregated Circuits Linking Basal Ganglia and Cortex. *Annu. Rev. Neurosci.* 9:357–381, 1986.

Alexander, G. E., and Crutcher, M. D. Functional Architecture of Basal Ganglia Circuits: Neural Substrates of Parallel Processing. *Trends Neurosci* 13:266–271, 1990.

Anderson, M. E. The Cerebellum. In *Textbook of Physiology,* Ch. 29, edited by Patton, H. D., Fuchs, A. F., Hille, B., Scher, A. M., and Steiner, R. Philadelphia: W. B. Saunders, 1989a, pp. 632–648.

Anderson, M. E. The Basal Ganglia. In *Textbook of Physiology,* Ch. 30, edited by Patton, H. D., Fuchs, A. F., Hille, B., Scher, A. M., and Steiner, R. Philadelphia: W. B. Saunders, 1989b, pp. 649–662.

Brodal, A. *Neurological Anatomy in Relation to Clinical Medicine.* New York: Oxford University Press, 1981.

Brooks, V. B. *The Neural Basis of Motor Control.* New York: Oxford University Press, 1986.

Carpenter, M. B. *Core Text of Neuroanatomy.* Baltimore: Williams & Wilkins, 1991.

Côté, L., and Crutcher, M. D. The Basal Ganglia. In *Principles of Neural Science,* Ch. 42, edited by Kandel, E. R., Schwartz, J. H., and Jessell, T. M. New York: Elsevier, 1991, pp. 647–659.

Eccles, J. C., Ito, M., and Szentagothai, J. *The Cerebellum as a Neuronal Machine.* New York: Springer-Verlag, 1967.

Gerfen, C. R. The Neostriatal Mosaic: Multiple Levels of Compartmental Organization in the Basal Ganglia. *Annu. Rev. Neurosci.* 15:385–320, 1992.

Ghez, C. The Cerebellum. In *Principles of Neural Science,* Ch. 41, edited by Kandel, E. R., Schwartz, J. H., and Jessell, T. M. New York: Elsevier, 1991, pp. 626–646.

Nieuwenhuys, R., Voogd, J., and van Huijzen, C. *The Human Central Nervous System; A Synopsis and Atlas.* 3rd ed. Berlin: Springer-Verlag, 1988.

Nygaard, T. G., Marsden, C. D., and Duvoisin, R. C. Dopa-Responsive Dystonia. *Adv. Neurol.* 50:377–383, 1988.

Peters, A., Palay, S. L., and Webster, H. d. *The Fine Structure of the Nervous System.* New York: Oxford University Press, 1991.

Rowland, L. P. *Merritt's Textbook of Neurology.* Philadelphia: Lea & Febiger, 1989.

Segawa, M., Nomura, Y., Tanaka, S., Hakamada, S., Nagata, E., Soda, M., and Kase, M. Hereditary Progressive Dystonia with Marked Diurnal Fluctuation: Consideration on Its Pathophysiology Based on the Characteristics of Clinical and Polysomnographical Findings. *Adv. Neurol.* 50:367–376, 1988.

Thach, W. T., Goodkin, H. P., and Keating, J. G. The Cerebellum and the Adaptive Coordination of Movement. *Annu. Rev. Neurosci.* 15:403–442, 1992.

Yahr, M. D. *The Basal Ganglia: Association for Research in Nervous and Mental Disease.* New York: Raven Press, 1976.

9 The Cranial Nerves

A thorough understanding of the anatomy and physiology of the cranial nerves is essential clinical knowledge. The cranial nerve nuclei reside at all levels of the brainstem. Because of this strategic location, deficits in cranial nerve function provide essential clues to the integrity of other brainstem functions, for example, the regulation of respiration, cardiovascular reflexes, and even consciousness. Furthermore, analysis of abnormal cranial nerve function often allows one to precisely localize intracranial lesions, an essential step in developing a neurological diagnosis. Consider just the analysis of eye movements and ocular reflexes during the neurological exam. The clinician's findings reflect not only the function of the three cranial nerves that innervate the muscles of the eyes, but also the portions of the cerebral cortex, the tectum, and the vestibular system that have connections to these cranial nuclei. Together they are an important window to the brain.

The concept of **neural modalities** and the associated nomenclature is presented in Appendix A (see Box A.1). This nomenclature arises from the developmental segregation of the neuroblasts into a series of columns (see Appendix A, Figs. A.9 and A.10). Each column represents a different neural function. Traditionally, the functional divisions of the cranial nerves are described in terms of these embryologically derived modalities. However, during development, many of the distinctions between the various modalities become blurred. Furthermore, none of the cranial nerves express all of the modalities. Therefore, the organizational description of the functional components of the cranial nerves can be usefully simplified to include only **afferent, efferent, somatic**, and **visceral**, an approach that will be used in this chapter. A summary table (Table 9.1) with the traditional classification schema is provided for reference purposes.

The afferent and efferent components of the cranial nerves are associated with nuclei that lie in the **alar** and **basal** plate regions respectively (see Appendix A). The division between these two embryonic regions is clearly marked in the brainstem, especially the medulla, by a shallow groove, the **sulcus limitans** (Fig. 9.1). In the medulla, an imaginary line extending from the sulcus limitans in a ventral lateral direction to the margin of the brainstem marks the basal-alar division. As one ascends the brainstem, this imaginary line becomes more horizontal. Ventral and medial to this sulcus lie the basal plate derivatives, the motor nuclei; dorsal and lateral are the alar plate derivatives, the sensory nuclei. The **visceral** cranial nerve components have nuclei that generally lie close to this imaginary line while the **somatic** components lie more medially (motor) or laterally (sensory).

Each of the cranial nerves will be discussed below, beginning with the simplest, the hypoglossal and ascending the brainstem in order. Because of their special importance, the auditory (Chapter 10) and optic (Chapter 11) nerves are treated in separate chapters. The olfactory nerve is treated in relation to the hypothalamus (Chapter 12).

Hypoglossal Nerve (CN XII)

The hypoglossal nerve is purely motor and served by a single nucleus in the medulla. The motor neuron somas of the **hypoglossal nerve**

Table 9.1. Summary of Cranial Nerves

Nerve	Component†	Function	Nuclei
I Olfactory	SVA[1]	Olfaction	Periamygdala and entorhinal cortex
II Optic	SSA[2]	Vision	Lateral geniculate nucleus
III Oculomotor	GSE	Oculomotion and elevation of eyelid	Oculomotor nucleus
	GVE	Pupillary constriction and lens accommodation	Edinger-Westphal nucleus
IV Trochlear	GSE	Superior oblique muscle	Trochlear nucleus
V Trigeminal	SVE	Muscles of mastication	Motor nucleus of V
	GSA	Sensory, light touch	Principal sensory nucleus of V
		Sensory, pain and temperature	Spinal nucleus of V
		Sensory, proprioception	Mesencephalic nucleus of V
VI Abducens	GSE	Lateral rectus muscle	Abducens nucleus
VII Facial	SVE	Mimetic muscles	Facial motor nucleus
	GVE	Lacrimation and salivation	Superior salivatory nucleus
	SVA	Taste, anterior two thirds of tongue	Nucleus of the solitary tract (rostral part)
	GSA	Sensory, pain, and temperature	Spinal nucleus of V

Table 9.1.—continued

Nerve	Component†	Function	Nuclei
VIII Acoustic/Vestibular	SSA	Balance	Vestibular nuclei
		Hearing	Cochlear nuclei
	SSE[3]	Regulate vestibular hair cell function	Periabducens nuclei
		Regulate cochlear hair cell function	Periolivary nuclei
IX and X Glossopharyngeal/ Vagus	SVE	Muscles of soft palate, pharynx and larynx	Nucleus ambiguous
	GVE	Salivation (IX)	Inferior salivatory nucleus
		Regulation of digestive organs	Dorsal motor nucleus of X
		Regulation of the heart	Nucleus ambiguous
	SVA	Taste, posterior one third of tongue (IX)	Nucleus of the solitary tract
	GVA	Sensory from viscera, baroreception, and chemoreception	Nucleus of the solitary tract
	GSA	Sensory, pain, and temperature	Spinal nucleus of V
XI Accessory	SVE[4]	Trapezius and sternocleidomastoid muscles	Accessory nucleus in spinal cord C2–C5
XII Hypoglossal	GSE	Intrinsic muscles of the tongue	Hypoglossal nucleus

†See Box A.1 for explanation of classification schema.

[1]The true olfactory nerve consists only of the axons of the bipolar receptor cells and is usually classified as SVA.

[2]The classification schema does not apply to the optic "nerve" because the "nerve" consists of second order axons that are myelinated by oligodendrocytes, facts that make it a CNS tract. In spite of this, most authors classify this "nerve" as SSA. The true cranial nerve of vision consists of axons of the retinal bipolar cells.

[3]Although these fibers are are not usually classified, the SSE designation is appropriate.

[4]The SVE classification is based on data that shows the accessory nucleus of the spinal cord to be a displaced fragment of nucleus ambiguous. This nerve has been classified as GSE by some authors and left unclassified by others.

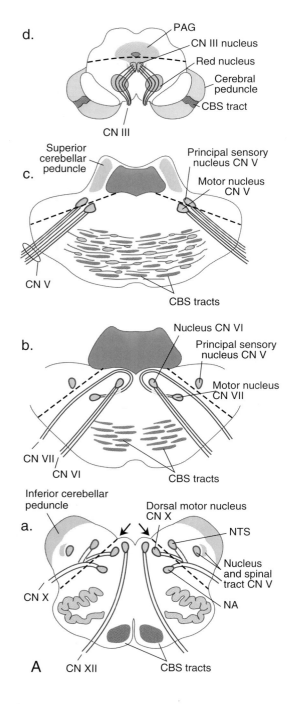

FIGURE 9.1.

A, The division between the sensory (alar plate derivatives) and motor (basal plate derivatives) is marked by the sulcus limitans which is most conspicuous at the medullary and lower pontine levels (*arrows*). The dashed line indicates the approximate line of demarcation between sensory and motor areas of the brainstem at various levels. **B,** This transparent view from the dorsal surface of the brainstem illustrates the location and extent of the cranial nerve nuclei and the internal course of their axons. The motor nuclei and nerves are on the left; the sensory nerves and nuclei are on the right. Not shown are the cranial nerve ganglia that are associated with all of the sensory nerves (except the mesencephalic division of V). The lines marked *a, b, c,* and *d* indicate the levels of the corresponding cross-sections in **A.**

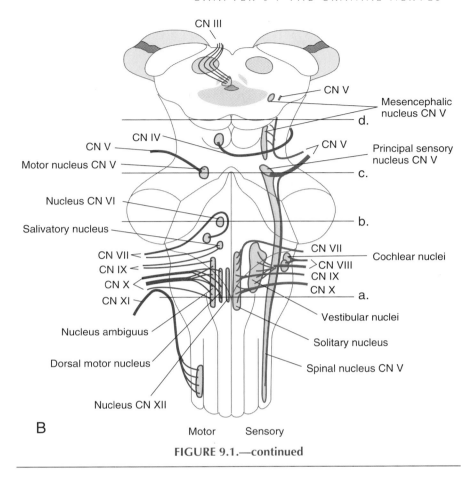

FIGURE 9.1.—continued

lie in the hypoglossal nucleus, a small, compact, tubular nucleus that lies very close to the midline in the caudal medulla, immediately ventral to the fourth ventricle (Fig. 9.2). The nucleus receives afferent connections from the *contralateral* corticobulbar tract. Hypoglossal axons leave the nucleus, take a ventral and slightly lateral course, and emerge from the brainstem between the medullary pyramid and the inferior olive. The hypoglossal nerve leaves the skull through the **hypoglossal canal**. Its axons innervate the *ipsilateral* tongue, supplying the **intrinsic muscles of the tongue** and the **styloglossus, hyoglossus, genioglossus and geniohyoid muscles**.

If the hypoglossal nucleus or nerve is damaged anywhere along its course, one may observe typical lower motor neuron findings in the ipsilateral half of the tongue. These include paresis or paralysis, depending on the severity

of the lesion. Fasciculations in the tongue are easily seen, followed in a few days by obvious atrophy. If one palpates the denervated tongue, the affected side feels soft and flaccid. When the patient is asked to protrude the tongue straight out, it will deviate toward the side of weakness. This is because the pair of genioglossus muscles each extend and adduct [L. *ad-duco,* to bring to] the tongue. When both muscles are active, the adductor forces from each balance and the tongue extends in the midline. If one genioglossus is paralyzed, its opposite member will adduct the tongue across the midline on extension.

If the corticobulbar tract is lesioned above the level of the CN XII nucleus, a paresis of the tongue contralateral to the lesion will follow. This will be of the upper motor neuron (UMN) type. Since it is impractical to test the muscle stretch reflex of the tongue, one can only ascer-

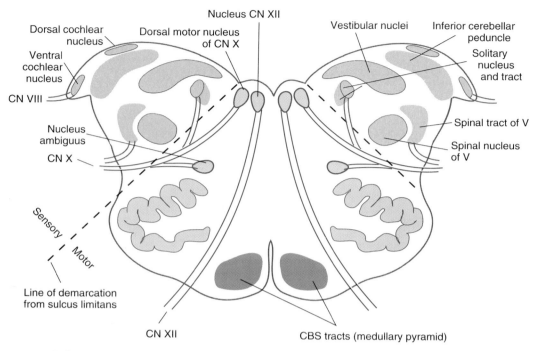

FIGURE 9.2.

A drawing of a cross-section of the medulla rostral to the obex showing the location of the nuclei associated with the vestibulocochlear, glossopharyngeal, vagus, and hypoglossal nerves.

tain the UMN nature of the lesion by the absence of lower motor neuron (LMN) signs in the tongue.

At this point one should note that deviation of the tongue *per se* has no localizing significance. If the nerve or nucleus is lesioned, then tongue deviation is toward the lesioned side. If the lesion is of the UMN type, the tongue deviates away from the side of the lesion. In either case, *deviation is toward the side of weakness.* Observing the presence or absence of fasciculations and atrophy as well as noting tongue muscle tone are essential in evaluating hypoglossal function.

Accessory Nerve (CN XI)

The accessory nerve is also purely motor and supplied by a single nucleus. The motor neurons are located in the **accessory nucleus**, a column of motor neurons that exist in the dorsolateral portion of the spinal cord.[1] Axons from the accessory motor neurons do not leave the spinal cord in the ventral root but exit the cord just dorsal to the dentate ligament, between the dorsal and ventral roots (Fig. 9.3). Accessory rootlets join one another as they ascend in the subdural space along the side of the spinal cord. They enter the cranium through the **foramen magnum** and briefly join the vagus nerve rootlets to leave the cranium through the **jugular foramen**. Immediately outside the skull, the accessory nerve separates from the vagus nerve and innervates the ipsilateral **sternocleidomastoid** and **trapezius** muscles.

Isolated lesions to the accessory nerve are not

---

[1] The accessory nucleus may be a fragment of the nucleus ambiguus that separated from the main nucleus during development and migrated into the spinal cord. If so, that would explain the unusual peripheral course of this nerve. This would also explain the "cranial" division of the accessory nerve that is often described in older texts. These "cranial" fibers are now considered by most authors to be part of the vagus nerve.

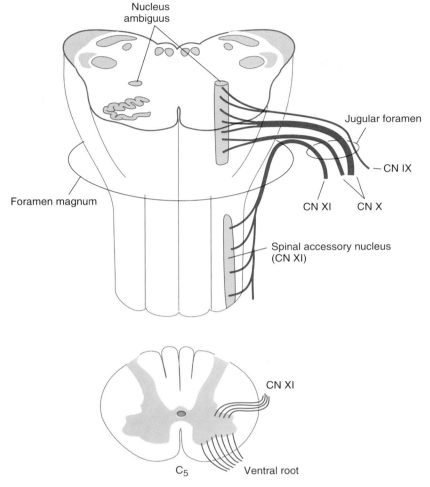

FIGURE 9.3.

The lower drawing shows the location of the accessory nucleus in the spinal cord and the separate egress from the ventral root its axons take when leaving the cord. Shown above is the peripheral course of the accessory nerve within the dura mater before it leaves the skull through the jugular foramen.

common. Its loss causes a LMN paralysis of the muscles it innervates. The sternocleidomastoid (SCM) muscle causes the head to rotate, thrusting the chin in a direction contralateral to the muscle. It also draws the chin toward the chest. Weakness of the SCM can be easily tested by asking the patient to rotate his head while the examiner resists the movement by holding the chin. The chin moves *away from the muscle being tested*. The trapezius is a major muscle of the back and shoulder. If paralyzed, shrugging the

shoulder, elevating the arm and outward rotation of the arm are all compromised on the ipsilateral side.

The accessory nucleus is innervated by axons in the corticospinal tract. Since it lies caudal to the pyramidal decussation, almost all UMN-type lesions involving the accessory nucleus occur contralateral to the side of paralysis. However, lesions to the corticospinal tract generally produce a hemiparesis, consequently the SCM and trapezius muscles are not affected in isolation and noting their involvement is generally not diagnostically significant.

Vagus (CN X) and Glossopharyngeal (CN IX) Nerve Complex

The vagus and glossopharyngeal nerves are separate in the periphery, but centrally they share the same anatomy and functions. All of the axons belonging to these nerves penetrate the brainstem dorsal to the inferior olive as a series of rootlets extending along the medulla the entire length of the inferior olive (see Fig. 9.1). Unlike the hypoglossal and accessory nerves, the vagus and glossopharyngeal nerves have several afferent and efferent components. Within the medulla, axons belonging to the various functional components are associated with five nuclei (Figs. 9.2 and 9.4).

EFFERENT COMPONENTS

Efferent axons of the vagus and glossopharyngeal nerves arise from three nuclei, the **nu-**

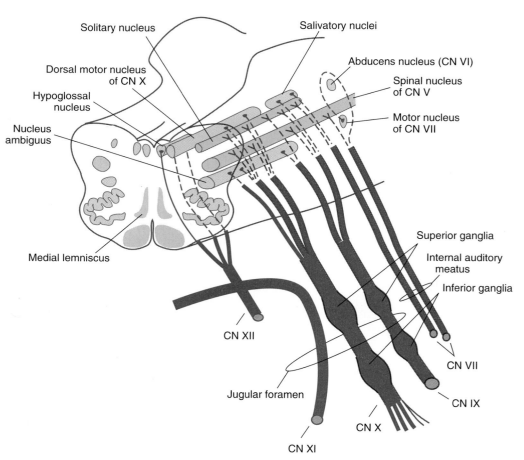

FIGURE 9.4.

Several of the lower cranial nerves are composed of axons of several modalities. Each modality is associated with a different nucleus in the brainstem. This schematic drawing of the medulla illustrates the relationships between the various nuclei, their axons, and the peripheral organization of those axons within the hypoglossal, accessory, vagus, glossopharyngeal, and facial nerves.

cleus ambiguus, the **dorsal motor nucleus of the vagus**, and the **inferior salivatory nucleus**.

Nucleus Ambiguus

The nucleus ambiguus is a large motor nucleus in the ventral lateral portion of the medullary reticular formation just dorsal to the inferior olive. It receives *bilateral innervation* from the corticobulbar tract. Axons leaving the nucleus ambiguus innervate the intrinsic voluntary muscles of the **larynx** (laryngeal nerve, a branch of the vagus), the constrictor muscles of the **pharynx**, the **palatoglossus** and **levator palati** (pharyngeal branch of the vagus), and the **stylopharyngeus muscle** (from the glossopharyngeal nerve), an elevator of the pharynx. Axons destined to join the glossopharyngeal nerve arise from the most rostral portions of the nucleus while those joining the vagus are from the caudal region.

Dorsal Motor Nucleus of the Vagus

The parasympathetic **preganglionic** axons of the vagus nerve arise from the dorsal motor nucleus of the vagus (DMN). The nucleus receives afferent fibers primarily from the hypothalamus and the nucleus of the solitary tract (discussed later in this section). Axons leaving the DMN are widely distributed throughout the viscera. They innervate all of the alimentary canal and the digestive organs, including the pharynx, esophagus, trachea, bronchi, stomach, liver, pancreas, small intestine, and most of the colon, as well as the heart.[2] Within the target tissue, short post-**ganglionic** axons synapse with glands or smooth muscle.

Inferior Salivatory Nucleus

Preganglionic parasympathetic fibers innervating the **parotid gland** arise from a separate nucleus, the inferior salivatory nucleus. This small nucleus contributes axons only to the glossopharyngeal nerve. They separate from the main nerve almost immediately after passing through the jugular foramen, forming the tympanic nerve.

It enters the temporal bone through the tympanic canaliculus and forms a plexus in the middle ear, innervating its mucous membranes. From the middle ear, the nerve continues as the lesser petrosal nerve, passing once again through the temporal bone and re-entering the cranium. It leaves the cranium for the last time through the foramen ovale and synapses in the **otic ganglion**. Postganglionic axons leave the otic ganglion to supply the **parotid gland**.

AFFERENT COMPONENTS

The cell bodies of the primary afferent axons of the vagus and glossopharyngeal nerves reside in peripheral ganglia. Each nerve has two ganglia, the superior and the inferior. Like the neurons in the dorsal root ganglia of the spinal nerves, the sensory neurons in the cranial nerve ganglia are pseudo-unipolar. In CN IX and CN X, the peripheral sensory axons arise from a variety of locations but synapse in the CNS in one of only two nuclei, the **nucleus of the solitary tract (NST)**, or the **nucleus of the spinal tract of V (STV)**.

Nucleus of the Solitary Tract

Most sensations from the viscera are not specifically carried to consciousness.[3] These sensations are, however, important for the regulation of many cardiovascular, respiratory, and gastrointestinal functions. Primary afferent axons of the glossopharyngeal and vagus nerves carrying visceral sensations arise from the same areas that are innervated by the efferent fibers of these nerves: the pharynx, larynx, trachea, bronchi, lungs, heart, and digestive organs. The cell bodies for these axons lie in the inferior ganglion of the two nerves.

As they enter the brainstem, the sensory axons destined for the NST form a prominent bundle called the **solitary tract**. Axons entering this tract bend in the caudal direction and descend in the medulla before terminating. Along its course, fibers leave the solitary tract and synapse in cells that line its lateral border, **the nucleus of the solitary tract** (see Figs. 9.2 and 9.4).

---

[2]There is good evidence that the nucleus ambiguus is the source of some parasympathetic axons, especially those destined for the heart.

[3]Visceral pain is brought to consciousness over axons of the sympathetic nervous system that enter the spinal cord (see Chapter 12).

Of particular note are axons that originate in specialized visceral receptors that are important in cardiovascular reflex regulation. Sensory axons in the glossopharyngeal nerve innervate the **chemoreceptors** of the **carotid body** and the **baroreceptors** of the **carotid sinus**. The sensory components of the vagus nerve innervate similar receptors in the **aortic arch**. The chemoreceptors detect oxygen tension in the blood while the baroreceptors measure blood pressure.

In the human, the glossopharyngeal is the principal nerve that conveys the sensation of taste. It innervates the taste buds in the posterior one third of the tongue, while taste buds from the anterior two thirds of the tongue are innervated by the facial nerve. A few taste buds are said to be located on the epiglottis and are innervated by the vagus nerve, but these degenerate in infancy. The various taste buds contain cells specialized for chemoreception. Sensory axons that innervate the taste buds enter the brainstem over the aforementioned nerves and synapse in the most rostral portion of the nucleus solitarius.

Neurons in the NST make extensive central connections. Locally, projections have been described to the **nucleus ambiguus**, the **dorsal motor nucleus of the vagus**, the **dorsal raphe nucleus**, and the **medullary reticular formation**. The reticular formation is important for several reflex functions including swallowing, vomiting, respiration, and cardiovascular regulation. Remote projections to the **hypothalamus** and the **amygdaloid complex** have also been described. These later connections are undoubtedly important for regulating many autonomic functions (see Chapter 12).

Rostral solitary neurons project directly to the ventral posteriomedial (VPM) nucleus of the thalamus as well as to the hypothalamus. The former nucleus projects to the primary sensory cortex where a gustatory area near the tongue representation in the primary sensory cortex has been identified in the human. This pathway is assumed to be responsible for the conscious sensations of taste. Projections to the hypothalamus probably serve as the afferent path for the feeding reflexes and perhaps some of the emotional affect associated with feeding.

Nucleus of the Spinal Tract of V

The glossopharyngeal and vagus nerves have a few axons that carry somatic sensations (i.e., from the body surface) to the central nervous system. These axons have their cell bodies in the superior ganglion of each nerve. The auricular nerve, which innervates a small portion of the concha, the external auditory meatus, and the external surface of the tympanic membrane, joins the vagus before entering the brainstem (see Fig. 9.14*B* and *C*). Sensation from the posterior one third of the tongue and the surface of the pharynx travel in the glossopharyngeal nerve. The central axons of the glossopharyngeal and vagus nerves become incorporated in the **spinal tract of V**. These axons terminate in the nucleus that lines the medial side of the tract, **the nucleus of the spinal tract of V**. The nucleus and spinal tract of V carry primary afferent axons from four cranial nerves, the trigeminal, the facial, the glossopharyngeal, and the vagus (see Figs. 9.4 and 9.14*C*). The central connections of this nucleus are described under the discussion of the trigeminal nerve.

CLINICAL CONSIDERATIONS

Peripheral lesions to the glossopharyngeal and vagus nerves are not common. Unilateral loss of the glossopharyngeal nerve results in paralysis of the *elevator muscle of the pharynx* and a loss of the *gag reflex* from the affected side. If the patient is asked to open his mouth and phonate, the paralyzed pharynx will droop in comparison to the intact side where the elevator muscles will draw the raphe palati into an arch.

Loss of the vagus nerve will paralyze the soft palate, which frequently causes the *uvula to tilt toward the intact side*.[4] Swallowing will be impaired due to the loss of the constrictors of the pharynx and the palatoglossus, a muscle at the base of the tongue that helps to initiate swallowing. During swallowing, food and fluids will tend to be regurgitated into the nasopharynx because of the general loss of tone in the pharynx and soft palate. More serious is the *paralysis of the vocal cord* on the affected side. This causes

---

[4]Remember, the tongue points toward the paralyzed side (nerve XII, hypoglossal nucleus) while the uvula points away from the paralyzed side (nerve X, nucleus ambiguus).

the vocal cord to collapse and partially occlude the air passage causing **hoarseness**. In the case of unilateral lesions, the hoarseness subsides in time. Bilateral lesions however, can be life threatening because the collapse of both vocal cords can severely impair air flow. The most common lesion occurs to the left recurrent laryngeal nerve. Its involvement can signal pathology in the thorax or mediastinum since this nerve hooks under the aortic arch. Aortic aneurysms, enlargement of the tracheobronchial lymph nodes, and mediastinal tumors can all cause hoarseness due to involvement of the left recurrent laryngeal nerve. Thyroidectomy and thymectomy are frequently complicated by injury to the laryngeal nerve.

Central lesions in the brainstem that involve the corticobulbar tracts do not produce very dramatic vagal or glossopharyngeal symptoms because the nucleus ambiguus is innervated bilaterally. Consequently, lesions above the brainstem that affect corticobulbar fibers will present minimal symptoms.

Brainstem lesions that directly affect the nuclei, however, produce the same lower motor neuron symptoms as direct lesions to the nerve in the periphery. For example, poliomyelitis is an infectious disease that kills motor neurons. It usually affects the lumbar spinal cord, but it may occasionally affect the brainstem, where it can produce weakness in swallowing, hoarseness, and tongue paralysis. Severe cases cause death. Direct lesions to the nuclei may also be produced by vascular lesions, the most common of which is *occlusion of the posterior inferior cerebellar artery* (Wallenberg's syndrome). Another lesion directly affecting the brainstem nuclei is *chronic bulbar palsy*. This is a degenerative disease causing degeneration of the voluntary motor nuclei of the lower cranial nerves (V, VII, IX, X, and XII).

Vestibulocochlear Nerve (CN VIII)

The vestibulocochlear nerve is mixed. Almost all of its axons are afferent, innervating the cochlea and the vestibular apparatus. However, there is a small efferent component that innervates both peripheral organs and serves to regulate the sen-

sitivity of the receptors and to modulate the afferent synapses. In the following section, only the vestibular apparatus and the vestibular division of CN VIII will be described. The cochlea and the central auditory pathways will be described in a separate chapter (see Chapter 10).

THE LABYRINTH

The vestibulocochlear sensory organ evolved from the lateral line organ of bony fish. This unusual receptor consists of specialized sensory cells that contain hairs extending from the apex of the cell into the slime on the surface of the fish. Whenever there is movement of the slime in which the hairs are embedded, the hairs bend and excite the **hair cells**. These cells can, therefore, detect any movement of water around the fish that disturbs the slime. The lateral line derivatives in mammals, the cochlea and the vestibular apparatus, detect sound pressure waves in air (hearing) and head movement (vestibular apparatus) by a mechanism that is essentially unchanged from that found in the fish—bending the cilia of the hair cells in response to fluid movement.

In mammals, the vestibulocochlear system consists of a **bony labyrinth** and a **membranous labyrinth**. The former consists of the cavities within the temporal bone that contain the latter. The membranous labyrinth consists of a tube within a tube. The outer tube is connective tissue that lines the bony labyrinth and forms a space that is filled with **perilymph**, a fluid that is similar in composition to CSF. The inner tube consists of diverse tissue including the sensory epithelia and is filled with **endolymph**, a fluid that is unique to the vestibulocochlear system. Endolymph is similar in composition to intracellular cytoplasm, but without the inclusions and cytoskeletal elements. It has a high K^+ concentration ($\sim$140 mM) and a low Na^+ concentration ($\sim$26 mM) (Table 9.2).

The labyrinth consists of four distinct regions (see Fig 10.1). The **cochlea** is the organ of hearing. The **utricle** is a large bulbous structure that lies at the base of three **semicircular canals**. The **saccule** lies between the utricle and the cochlea. The *endolymphatic space of all four structures is continuous,* but closed; it does not communicate with any other fluid space. The endolymphatic system is surrounded by a perilymphatic space that generally follows the contours of the endolymphatic system, except in the region of the utricle and saccule where it bulges to form a large cavity known as the **vestibule**. An oval hole, the **oval window** lies in the wall separating the vestibule from the middle ear cavity. It is normally filled by the footplate of the stapes (see Chapter 10).

THE VESTIBULAR SENSORY EPITHELIA

The utricle and the saccule each contain a sensory epithelium called a **macula**. The base of each semicircular canal is swollen, forming the **ampulla** that contains another sensory epithelium, the **crista** of the ampulla. The three cristae and the two maculae contain the neurosensory epithelium of the vestibular apparatus with its complement of hair cells.

Hair Cells

The hair cell is the sensory transducer in all parts of the vestibulocochlear system. This highly specialized cell gets its name from the hair-like processes that extend from its apical surface (Fig. 9.5). Vestibular hair cells have two

Table 9.2. Ionic Composition of Body Fluids (mEq/L)

	Plasma	CSF	Perilymph	Endolymph
Protein	6000–8000	10–38	75–100	10
K^+	20	12–17	15	140
Na^+	140	150	148	26
Cl^-	600	750	120	14–0
Mg^{2+}	1.0–3.0	2.0	2.0	0.9
Ca^{2+}	7.0	3.0	3.0	3.0

FIGURE 9.5.

A, There are two types of vestibular hair cells, labeled type I and type II. The principal difference between the two is the arrangement of the afferent axon terminal. For the type I hair cell, shown on the left, the axon terminal is in the form of a chalice, enveloping the entire cell, whereas the type II hair cell, on the right, has simpler, more typical afferent synapses on it. The afferent synapses are not highly vesiculated. Specialized densities in the hair cell, called synaptic ribbons, mark the active site of the afferent synapse. The efferent axons form highly vesiculated endings that have active sites on both the type II hair cell and on the afferent endings. The efferent synapses on the afferent chalice are not shown in this illustration. **B**, This is a cross-section near the base of several hair cell cilia. The kinocilium (k) is recognizable because of its characteristic 9 + 2 internal structure. The stereocilia have a much simpler internal structure of actin filaments. The central, darkly staining rootlet is visible in one. Note also that fine bridges connect adjacent stereocilia, ensuring that shearing forces will be distributed to all the stereocilia and that they will bend as a unit.

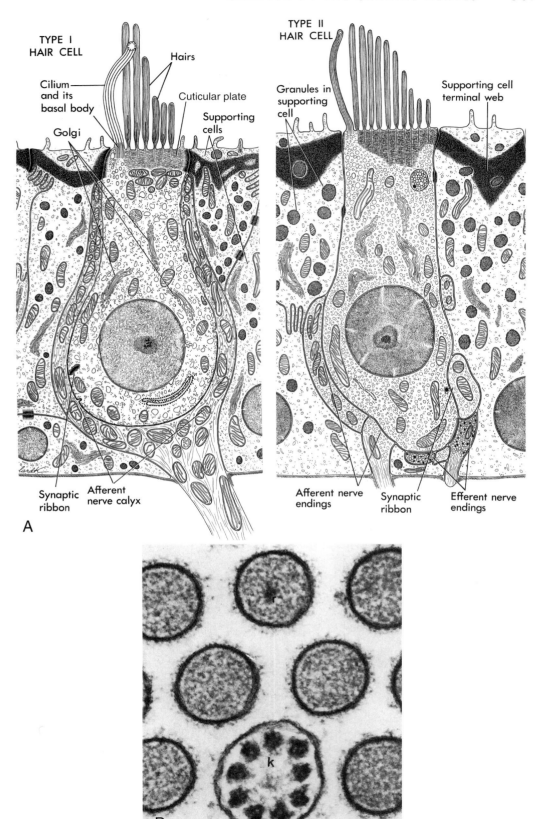

TYPE I HAIR CELL

Cilium and its basal body

Hairs

Cuticular plate

Golgi

Supporting cells

Synaptic ribbon

Afferent nerve calyx

TYPE II HAIR CELL

Granules in supporting cell

Supporting cell terminal web

Afferent nerve endings

Synaptic ribbon

Efferent nerve endings

A

B

types of hairs, a single **kinocilium** and between 50 and 110 **stereocilia**. The kinocilia are complex organelles, containing the 9 + 2 internal structure common to motile cilia. The stereocilia are much simpler. Their external limiting membrane is a continuation of the hair cell membrane. Internally, the stereocilia contain a dense network of actin filaments. At the base there is a spike-like structure, the rootlet, that extends into the **cuticular plate** at the apex of the hair cell. The stereocilia are arranged in rows. They are longest (about 100 μM) in the row adjacent to the kinocilia. Each subsequent row gets progressively shorter until the stereocilia are only about 1 μM long. This arrangement polarizes the hair cell anatomically.

The hair cells are also polarized physiologically. The stereocilia are stiff. A force directed perpendicular to the shaft causes the hairs to bend at the base where the stereocilium inserts into the cuticular plate. This bending causes changes in the ionic flux across the hair cell membrane. Bending the stereocilia toward the kinocilium hypopolarizes the entire hair cell and produces action potentials on the afferent axons. Bending the stereocilia in the opposite direction hyperpolarizes the cell. The ionic details of hair cell stimulation and synaptic activation will be discussed in Chapter 10.

The Maculae

The sensory epithelia of the saccule and utricle are called **maculae**. Each macula consists of numerous hair cells surrounded by supporting cells resting on a connective tissue base (Fig. 9.6). The surface of the macula extends into the cavity of the membranous labyrinth and is bathed by endolymph. The surface of the macula is covered with a gelatinous substance, the **otolithic membrane**. Calcium carbonate crystals **(otoliths)** are embedded on the surface, increasing the mass of the entire structure (Fig. 9.7).

Relative movement between the otolithic membrane and the surface of the hair cells is the essential macular stimulus, since this results in bending of the hairs. This relative movement is possible because of the inertia of the otolithic membrane. Whenever the head is moved **(linear acceleration)**, the macula moves with it because its base is firmly attached to the temporal bone. Inertia retards the movement of the otolithic membrane with its otoliths, resulting in the generation of a shearing force across the apex of the hair cells. This bends the hairs, causing an ionic current flow in the hair cells that initiates the release of neurotransmitter at the base of the cells. The terminal boutons of the afferent axons have neurotransmitter receptors that, when activated, produce action potentials that are carried to the CNS. *Linear acceleration is the adequate stimulus for the maculae.*

The hair cells of the maculae are not all arranged in the same direction of polarization. There are two levels of organization. First, the hair cells are lined up back to back along a dividing line, the **striola**, that runs approximately down the middle of the maculae, so that the direction of polarizations are 180° apart (see Fig. 9.6). Second, the striola bends about 90° across the surface of the maculae. This arrangement means that movement of the otolithic membrane in any direction in the plane of the macular surface will excite some hair cells and inhibit others.

FIGURE 9.6.

A, The macula of the saccule and the macula of the utricle are sensory epithelia containing both type I and type II hair cells. The cilia of the hair cells are embedded in a gelatinous otolithic membrane that has calcium carbonate crystals (otoliths) embedded into its surface. **B,** The size and shape of the otoconia varies across the surface of the saccule (*a*) and utricle (*b*). This size variation corresponds with the striola, the line that marks the interface between the hair cells of opposite orientations. The striola is also marked by large fenestrations in the gelatinous layer. An artist's rendition of the macula in cross section illustrates these points (*c*).

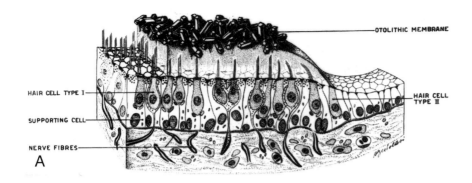

OTOLITHIC MEMBRANE

HAIR CELL TYPE I

HAIR CELL TYPE II

SUPPORTING CELL

NERVE FIBRES

A

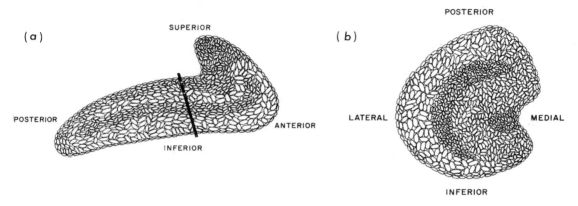

(a)

SUPERIOR

POSTERIOR

ANTERIOR

INFERIOR

(b)

POSTERIOR

LATERAL

MEDIAL

INFERIOR

(c)

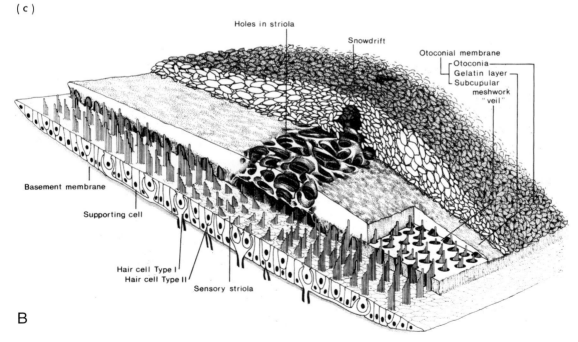

Holes in striola

Snowdrift

Otoconial membrane
Otoconia
Gelatin layer
Subcupular
meshwork
"veil"

Basement membrane

Supporting cell

Hair cell Type I
Hair cell Type II

Sensory striola

B

FIGURE 9.7.

Electron micrographs of the calcium carbonate crystals that lie on the surface of the otolithic membrane. **A,** A low-magnification SEM, illustrating the heterogeneous morphology of the crystals. **B,** A higher magnification.

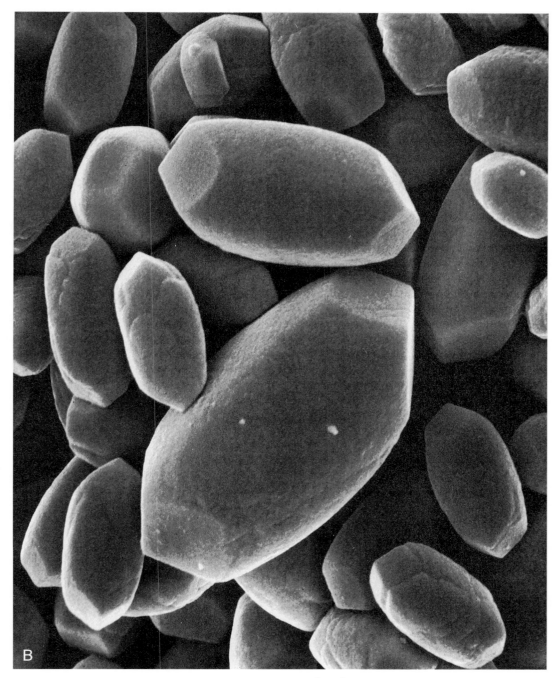

FIGURE 9.7.—continued

To further insure that hair cells will be stimulated by any linear movement of the head, the two macula are arranged perpendicular to each other. In the upright human, the saccule lies approximately in the sagittal plane while the utricle lies nearly horizontal. Because of its vertical position, the constant acceleration of gravity affects the otoliths of the saccule. If the head is tilted, the utricle can also be affected by gravity. Consequently, the maculae are very sensitive to static head position.

The Cristae

The sensory epithelium of the semicircular canals is the **crista ampullaris**. This structure consists of a small swelling of tissue containing hair cells and supporting cells resting on a connective tissue base (Fig. 9.8). The hairs are embedded in a gelatinous mass, the **cupula**, that extends into the endolymphatic space across the full diameter of the semicircular canal.

The semicircular canals are receptors of **angular acceleration**. The hair cells are sheared by the differential movement of the endolymph and the crista. To see how this works, consider one's head rotating in the plane of a semicircular canal. During the initial acceleration, the endolymph will lag behind the movement of the crista due to its inertia. This relative movement between the endolymph and the crista places pressure on the cupula, deflecting it and bending the hairs of the hair cells. The crista, therefore, *is stimulated by angular acceleration of the head in the plane of the semicircular canal.*

One should note that a constant angular *velocity* is not an appropriate stimulus for the semicircular canals. This is because there is friction between the endolymph and the wall of the canal. During rotation of the head at a constant angular velocity, the angular force will gradually be transferred to the endolymph. Eventually the fluid and the crista will be rotating at the same velocity and the shear across the cupula will cease. However, if the head is then brought to rest (negative angular acceleration), the en-

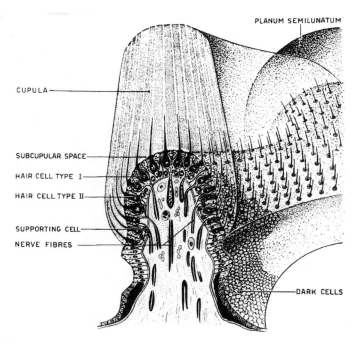

FIGURE 9.8.

The crista of the ampulla is the sensory epithelium found in the semicircular canals. Like the maculae, it contains both type I and type II hair cells and the cilia are embedded in a gelatinous membrane, here called the cupula. The cristae have no otoliths.

dolymph will tend to continue to flow through the semicircular canal. This movement of endolymph will again shear the hairs of the crista, but in the direction opposite of the original stimulus.

The three semicircular canals are oriented in approximately orthogonal planes (Fig. 9.9). The **horizontal** canal is inclined to the anatomical horizontal plane about $-25°$. The **anterior**[5] canal meets the sagittal plane anterolaterally at about 41° and the **posterior** canal intersects it at about 56°. This arrangement places the posterior canal on one side in nearly the same plane as the anterior canal on the other side while the two horizontal canals lie in the same plane.

The cupular hair cells, unlike those in the maculae, are all orientated in the same direction and, for each cupula, deflection toward the utricle will result in hypopolarization of all the hair cells. Deflection in the opposite deflection will hyperpolarize them. Therefore, stimulation of the hair cells in one canal will cause inhibition of the hair cells in the complementary canal on the contralateral side. Thus, just as in the maculae, for any given movement, some hair cells are stimulated while others are inhibited.

Vestibular Neurons

The primary afferent vestibular neurons are true bipolar neurons. Their cell bodies are found in two peripheral ganglia that lie in the internal auditory meatus. The peripheral axon of the vestibular neurons contacts the vestibular hair cells and is the post-synaptic element of the afferent synapse (see Fig. 9.5). Hair cells, when stimulated, probably release a neurotransmitter that hypopolarizes the afferent synaptic bouton of the vestibular neuron. The central axon of vestibular neurons enters the brainstem at the pontomedullary junction and nearly all bifurcate immediately after entering the CNS. The ascending branches synapse in the superior, medial, and lateral vestibular nuclei, while the descending branch synapses in the inferior and medial nuclei. A few enter the flocculus, nodulus, and uvula of the cerebellum where they terminate as mossy fibers.

Vestibular efferent axons arise bilaterally from a small collection of cells located just lat-

eral to the abducens nucleus. These axons travel with the vestibulocochlear nerve and terminate on the vestibular hair cells with highly vesiculated endings. The terminals secrete ACh and are excitatory to the vestibular hair cells. Their function is not known.

CENTRAL VESTIBULAR CONNECTIONS

The vestibular nuclei make connections with three functional areas of the nervous system, the **spinal cord**, the **cerebellum**, and the **nuclei of ocular motion**. The connections of the vestibular nuclei with the spinal cord have been discussed in Chapter 7. To recapitulate briefly, the lateral vestibular nucleus contains large neurons that project to all levels of the ipsilateral spinal cord as the lateral vestibulospinal tract. It is excitatory on extensor motor neuron pools. The **medial vestibular nucleus** gives rise to the **medial vestibulospinal tract**. These axons descend only to cervical levels, innervating motor neurons that serve primarily to stabilize the head. This action is necessary to provide a stable platform for the eyes. The connections of the vestibular nuclei with the cerebellum were discussed in Chapter 8. These contacts originate mostly from the **lateral vestibular nucleus** that makes extensive contacts with the **floccul-nodular lobe**. The connections with the nuclei of ocular motion will be discussed later in this chapter, after the external connections of these nuclei have been presented.

CLINICAL CONSIDERATIONS

One's sense of place in space is computed from three primary sets of information: vision, proprioception and the vestibular sense. The vestibular sense by itself, is incapable of providing adequate information to allow one to be stable with reference to the world. It must be supplemented by either vision or proprioception. This is the basis of the Romberg test (see Chapters 5 and 7). Inappropriate signals from the vestibular system can, however, disrupt one's sense of stability even when visual and proprioceptive signals are present and normal.

The principal symptoms associated with vestibular disfunction are **nystagmus** and **vertigo**.

---

[5]Formerly called the superior canal.

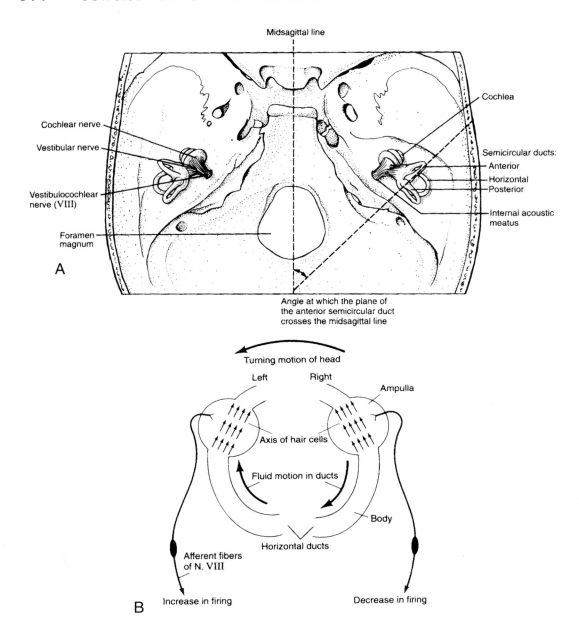

FIGURE 9.9.

A, The orientation of the membranous labyrinth in the temporal bone. Note that the arrangement of the semicircular canals pairs the anterior canal of one side with the posterior canal of the contralateral side because they lie in the same plane. The two horizontal canals lie the same plane and hence pair each other. Each member of the three pairs is polarized in the opposite direction. **B**, A schematic representation of the stimulation of a pair of semicircular canals. Since the polarization of the cristae are oriented opposite one another, excitation of one will ensure inhibition of the other. Note that the excited crista is on the side toward which the rotation is directed.

Nystagmus is a particular type of eye movement that is discussed below in relation to all types of eye movements. Vertigo is a technical term with a specific meaning. Patients will usually describe it as "dizziness," a common term that includes not only vertigo, but also syncope (light headedness or fainting associated with hypotension), weakness, and a number of other nonspecific conditions. Clinically, vertigo is the sense that the world is spinning around or that one's head or body is whirling. *It is an illusion of motion.* In response to the perceived motion, one will often lean against the perceived motion, as if being drawn by a magnet. Falling is common. In severe cases, the patient may involuntarily fling himself to the ground. Vertigo is usually accompanied by nausea, vomiting, tinnitus (a ringing in the ears), and deafness.

Benign positional vertigo is produced by the assumption of certain positions. It is thought to be caused by an otolith having become detached from one or more of the otolithic membranes of the saccule or utricle and thus able to float freely in the endolymphatic space. In certain critical positions, the otolithic crystals can interfere with the cupula in the posterior canal, giving rise to abnormal vestibular sensations. The patient is seized with instant vertigo that rarely lasts for more than a minute. The episodes will recur, however, until the crystals have become relocated or dispersed, which usually occurs in a few days or weeks. About 17% of the cases are brought about by head trauma which may be presumed to dislodge the otoliths. Benign positional vertigo is not associated with hearing loss.

Ménière disease is another type of vertigo. Unlike benign positional vertigo, Ménière disease is almost invariably accompanied by tinnitus and hearing loss as well as nausea and vomiting. Furthermore, the attacks are longer, lasting from a few minutes to several hours. This disease is probably caused by an imbalance between the production and reabsorption of endolymph, causing the endolymphatic space to expand (**endolymphatic hydrops**). The attacks of vertigo are believed to follow the rupturing of the membranous labyrinth that allows the potassium rich endolymph to contaminate the perilymph. This would depolarize the afferent axons and maintain the depolarization

until the rupture was repaired and the ionic milieu re-established. This probably kills hair cells also. Each attack, therefore, leaves the labyrinth, and the cochlea with which it communicates, more and more damaged. In cases of long duration, there is eventually complete deafness and total loss of labyrinthine function on the affected side. Most cases, however, recover spontaneously after a few years. The disease has been treated with some success by the introduction of an endolymphatic shunt that drains the endolymph into the CSF through a pressure sensitive valve. Other surgical approaches involve destruction of the labyrinth or the sectioning of the vestibular nerve.

Facial Nerve (CN VII)

The facial is a mixed nerve that is associated with four brainstem nuclei. Its axons leave the brainstem at the pontomedullary junction, just medial to the vestibulocochlear nerve (Fig. 9.10). The two nerves enter the temporal bone in the **internal auditory meatus**. Within the temporal bone, the facial nerve divides into three branches, the **chorda tympani**, the **superficial petrosal nerve**, and the **facial nerve** (motor component) (Fig. 9.11).

EFFERENT COMPONENTS

There are two efferent components to the facial nerve. The somatic axons originate in the facial nucleus and innervate the mimetic muscles of the face. The visceral, parasympathetic axons arise from the superior salivatory nucleus and mediate lacrimation and salivation.

Facial Nucleus

The main branch of the facial nerve innervates the voluntary **mimetic muscles** of the face: the **orbicularis oculi**, the **stapedius**, the **stylohyoid**, the **posterior belly of the digastric** and the **platysma**. The motor neurons are located in the **facial nucleus**, a collection of cells in the caudal pons (Fig. 9.11*B*). Axons leaving this nucleus take an unusual course, being directed dorsomedial toward the roof of the pontine tegmentum. They reach the abducens nu-

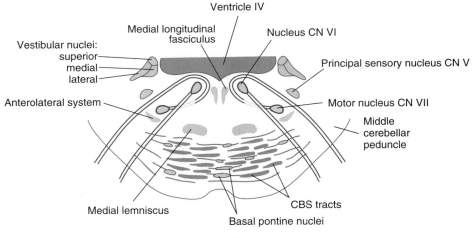

FIGURE 9.10.

A cross-section of the pons at the level of the abducens and facial nerves.

cleus and wrap around it to form the **internal genu** of the facial nerve. The axons course briefly in the rostral direction before proceeding in a ventral-lateral direction, passing between the facial nucleus and the principal sensory nucleus of V before leaving the brainstem at the cerebellopontine angle (see Fig. 9.11*B*). The facial nerve leaves the cranium through the internal auditory meatus of the temporal bone. Within the temporal bone, the facial nerve makes a sharp bend (**external genu**) before leaving the skull entirely by passing through the **stylomastoid foramen**.

The facial nucleus receives cortical motor commands via the corticobulbar tract. The termination of this tract within the nucleus is divided so that motor neurons innervating the *upper portion of the face (eyebrows and forehead) receive bilateral innervation* while the motor neurons innervating the *lower face receive only contralateral innervation* (Fig. 9.12). This arrangement allows one to differentiate lesions to the nucleus from those affecting only the corticobulbar tract: the former will produce an ipsilateral paresis of the entire face while the latter will affect only the lower face.

The facial nucleus also receives afferent fibers from several other sources. One source of particular importance is the reticular formation. Tactile and pain stimuli from the cornea are transmitted via the trigeminal nerve to the reticular formation

from the nucleus of the spinal tract of V. Both facial motor nuclei receive signals from the reticular formation that initiate *bilateral* contraction of the **orbicularis oculi**, the muscle that *closes the eyelids* (see Fig. 9.12). This is the **corneal reflex**. Although slow (latency > 40 msec), this reflex serves to protect and cleanse the eye of debris that could injure the cornea.

Since reflex closure of the eye is mediated by the reticular formation, other stimuli, particularly those that are novel and startling like flashes of light or loud noises, can also initiate bilateral eye closure. The unconscious continual blinking of the eye, important for maintaining proper corneal hydration, is probably also mediated by the reticular formation.

Superior Salivatory Nucleus

The facial nerve contains parasympathetic axons that arise from the **superior salivatory nucleus (SSN)**. The preganglionic parasympathetic axons innervate the **pterygopalatine** and the **submandibular ganglia**. These ganglia send postganglionic fibers to the **lacrimal glands**, the **mucous membranes** of the nose and oral cavity and the **sublingual** and **submandibular salivary glands** respectively (see Fig. 9.11).

The superior salivatory nucleus, like the inferior salivatory nucleus (ISN) is not a well-organized collection of neurons. Both nuclei consist

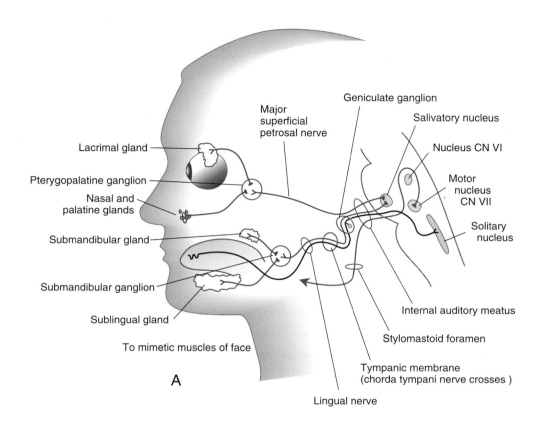

Geniculate ganglion

Major
superficial
petrosal nerve

Salivatory nucleus

Nucleus CN VI

Lacrimal gland

Motor
nucleus
CN VII

Pterygopalatine ganglion

Nasal and
palatine glands

Solitary
nucleus

Submandibular gland

Submandibular ganglion

Internal auditory meatus

Sublingual gland

Stylomastoid foramen

To mimetic muscles of face

Tympanic membrane
(chorda tympani nerve crosses)

A

Lingual nerve

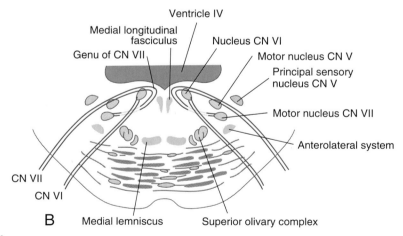

Ventricle IV

Medial longitudinal
fasciculus

Nucleus CN VI

Genu of CN VII

Motor nucleus CN V

Principal sensory
nucleus CN V

Motor nucleus CN VII

Anterolateral system

CN VII

CN VI

B   Medial lemniscus   Superior olivary complex

FIGURE 9.11.

   A, This figure depicts the course and distribution of the various components of the facial nerve. **B**, A drawing of the pons showing the internal course of the facial nerve and
its relationship to the abducens nucleus.

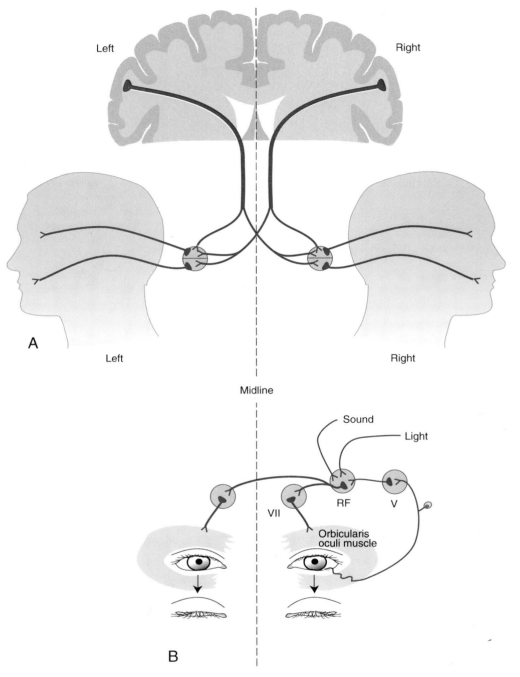

FIGURE 9.12.

A, This drawing depicts the dual innervation of the facial nucleus by the corticobulbar fibers. Note that the motor neurons innervating the mimetic muscles of the lower face receive only contralateral corticobulbar fibers while the motor neurons innervating the mimetic muscles of the upper face receive bilateral innervation. **B**, The corneal reflex involves two cranial nerves. Pain afferents from the cornea enter the CNS over the trigeminal nerve. Within the brainstem, secondary axons synapse in the reticular formation that, in turn, make bilateral contact with the facial nuclei. From the facial nuclei, motor axons innervate the orbicularis oculi muscles, closing the eyelid. The reticular formation also receives afferents from other modalities that can initiate lid closure.

of a few scattered cells in the dorsal lateral area of the reticular formation. Their axons progress directly in a ventral-lateral direction to join the main body of the facial (SSN) or glossopharyngeal (ISN) nerves. In the facial nerve, the parasympathetic fibers remain separated from the voluntary motor axons that form the main part of the facial nerve, and follow the motor root into the **internal auditory meatus** of the temporal bone. Since they lie between the acoustic nerve and the main facial nerve, they are frequently referred to as the **intermediate nerve**. The peripheral course of the intermediate nerve is complex and will not be described in detail. It will suffice to note here that at the external genu of the motor root of VII, the intermediate nerve divides, first giving off the **greater superficial petrosal nerve**. The remaining parasympathetic fibers follow the motor nerve for a short distance and then separate from it, joining the **chorda tympani**. The greater superficial petrosal nerve terminates in the pterygopalatine ganglion. The parasympathetic fibers of the chorda tympani terminate in the submandibular ganglion.

AFFERENT COMPONENTS

There are two afferent components of the facial nerve, neither of which are of great clinical significance.

Nucleus Solitarius

Axons that terminate in the **nucleus solitarius** convey the sensations of taste from the anterior two thirds of the tongue. These axons leave the tongue in the lingual nerve and later join the chorda tympani. As previously described, this nerve joins the main motor root of the facial nerve and follows it into the brainstem. The cell bodies of the sensory fibers lie in the **geniculate ganglion**, a collection of cell bodies found at the external genu of the facial nerve. The central connections of the nucleus of the solitary tract were described previously under the glossopharyngeal nerve.

Nucleus of the Spinal Tract of V

The second sensory component of the facial nerve carries somatic sensations from a small area within the pinna (Fig. 9.13). These sensory axons join the motor root of the facial nerve and follow its course through the temporal bone to the brainstem. Their cell bodies are also located in the **geniculate ganglion** and they synapse in the **nucleus of the spinal tract of V**.

CLINICAL CONSIDERATIONS

The importance of recognizing the differences between a paralysis of entire half of the face as opposed to only the lower quadrant has been previously discussed. To recapitulate, the motor neurons serving the upper quadrant of the face are supplied bilaterally by axons of the corticobulbar tract. This dual innervation makes them somewhat refractory to unilateral supranuclear (above the level of the facial nucleus) corticobulbar lesions. Motor neurons supplying the lower quadrant of the facial muscles have no dual innervation. They are supplied strictly by contralateral corticobulbar fibers. Supranuclear lesions will result in a marked weakness of the lower facial muscles contralateral to the lesion while sparing the upper face. For example, cerebral infarction in the territory of the middle cerebral artery, or infarction of the cerebral peduncle will cause marked weakness of the contralateral lower facial muscles that is readily seen as a flattening of the nasolabial fold and drooping of the corner of the mouth. A

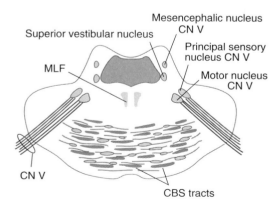

FIGURE 9.13.

A cross-section of the pons at the level of the motor and principal sensory nuclei of the trigeminal nerve.

voluntary grimace will be obviously unsymmetrical. Wrinkling of the forehead or raising of the eyebrows will be essentially normal. In contrast to this pattern, paresis or frank paralysis of the entire half of the face results from lesions that affect the facial motor nucleus directly, or the peripheral facial nerve.

Bell's palsy is an affliction of the peripheral portion of CN VII that causes complete paralysis of the mimetic muscles on the affected side. The patient cannot close the eye, wrinkle the eye brow, or smile on the side ipsilateral to the lesion. The corner of the mouth droops and the nasolabial fold is flattened. This disfigurement is very traumatic to the patient. Of great concern also is the loss of the blink reflex since the patient cannot close the eyelid. Drying, clouding, and ulceration of the cornea will occur due to the lack of corneal hydration normally supplied by lacrimation. Steps must be taken to protect and artificially hydrate the cornea.

Although the cause of Bell's palsy is unknown, most authorities believe that the nerve becomes compressed within the facial canal in the temporal bone, probably due to edema. Although the condition is dramatic, the rate of spontaneous recovery is very high. Approximately 80% of patients will recover within two months. If the nerve is severely damaged, complete regeneration of the nerve is required, which may take as long as two years. In such a case, the recovery is usually not complete. In severe cases, during regeneration, axons originating from motor neurons may cross over into the distal Schwann tubes of the parasympathetic fibers. After recovery, the patient upon smiling may shed a few tears due to this inappropriate innervation: the "crocodile tears" of Bell's palsy.

Trigeminal Nerve (CN V)

The trigeminal nerve is associated with four nuclei in the brainstem—one is motor, the remaining three are sensory. The motor axons innervate the muscles of mastication. The sensory fibers innervate the face and most of the internal surface of the nasal and oral cavities including the teeth.

EFFERENT COMPONENT

The **trigeminal motor nucleus** lies in the pons at the level where the trigeminal nerve leaves the brainstem. The nucleus lies in the lateral portion of the tegmentum, but medial to the principal sensory nucleus of V (Fig. 9.14). Motor axons from the nucleus leave the brainstem in the trigeminal nerve and become incorporated in the mandibular (V_3) division. Leaving the skull through the foramen ovale, V_3 innervates the **muscles of mastication**, the **tensor palatini**, the **tensor tympani**, the **mylohyoid** and the **anterior belly of the digastric** muscle. The motor nucleus receives bilateral connections from the corticobulbar tract. Consequently, supranuclear lesions do not produce dramatic signs of weakness in mastication.

FIGURE 9.14.

A, The principal sensory nucleus of the trigeminal (*N. V. sn. pr.*) and the nucleus of the spinal tract of V (NSTV) form a continuous structure from the pons to the spinal cord. For convenience, the NSTV may be divided into three parts: the oral (*sp. o.*), the intermediate (*sp. ip.*), and the caudal (*sp. c.*). **B,** A lateral view of the pons showing the descending course of the trigeminal fibers that synapse in the NSTV. **C,** In the head, the sensory receptors of the skin are supplied by the trigeminal nerve and the C_2-C_3 dorsal roots. The pattern of distribution is shown here. Note the boundaries of the three divisions of the trigeminal nerve and the division between the trigeminal and the cervical roots. Note particularly that the angle of the jaw is supplied by C_2 and that the anterior aspect of the pinna is supplied by VII. The dermatome pattern of pain and temperature fibers in the face are also arranged in an onion skin pattern centered about the mouth. **D,** Distribution of afferents in the spinal tract of V. Severing the spinal tract of V near the rostral pole (*A*) cuts nearly all the fibers, sparing the sensibility only around the lips (see

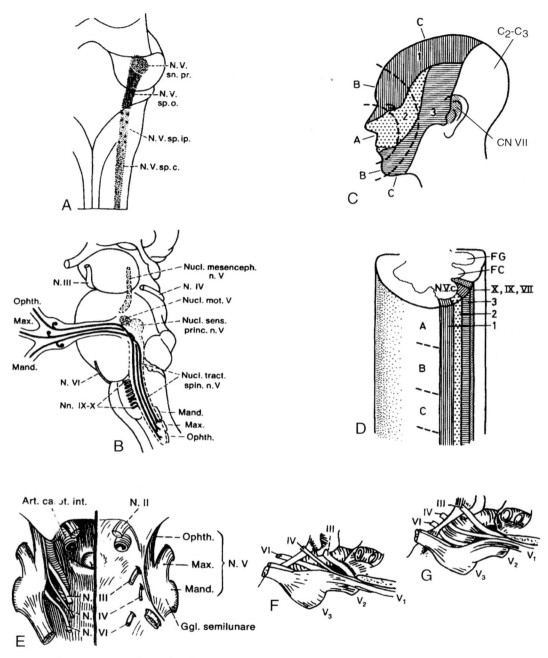

A

N.V. sn. pr.

N.V. sp.o.

N.V. sp. ip.

N.V. sp. c.

C

C₂-C₃

CN VII

B

N. III

Ophth.

Max.

Mand.

N. VI

Nn. IX-X

Nucl. mesenceph. n. V

N. IV

Nucl. mot. V

Nucl. sens. princ. n. V

Nucl. tract. spin. n. V

Mand.

Max.

Ophth.

D

FG

FC

N.V.c.

X, IX, VII

3

2

1

A

B

C

E

Art. carot. int.

N. II

Ophth.

Max.

Mand.

N. V

N. III

N. IV

N. VI

Ggl. semilunare

F

VI

IV

III

V₃

V₂

V₁

G

III

IV

VI

V₃

V₂

V₁

C). Sections at a lower level (B and C) spare an increasingly wider area. Note that fibers originating from the various divisions of the trigeminal and from the vagus, glossopharyngeal and facial nerves remain segregated and are organized laterally across the spinal tract of V. **E,** This diagram of the region of the cavernous sinus illustrates the relationships between the carotid artery, the three nerves of ocular motion, and the three divisions of the trigeminal nerve. The dura has been removed on the left and retained on the right. **F,** A lateral view of the cavernous sinus showing on the left the normal relationships of the structures. **G,** After an aneurysm of the carotid artery develops, the VI and IV nerves are displaced as shown. This also exerts pressure on the trigeminal nerve resulting in the ominous combination of facial pain and ophthalmoplegia.

AFFERENT COMPONENTS

Three nuclei are associated with the sensory functions of the trigeminal nerve, the **nucleus of the spinal tract of V**, the **principal sensory nucleus of V**, and the **mesencephalic nucleus of V** (Fig. 9.14).

Mesencephalic Nucleus of V

The mesencephalic nucleus consists of a thin line of pseudounipolar neurons that lie in the mesencephalon along the lateral margin of the fourth ventricle as it begins to narrow before becoming the aqueduct. These neurons are pseudounipolar and receive no synaptic connections. They appear to constitute a displaced peripheral ganglion. Their peripheral axons innervate the spindles and proprioceptors of the muscles innervated by the trigeminal motor root. There is some evidence that they may also innervate the muscle receptors of the extraocular eye muscles. The central axons form synapses in the motor nucleus and the principal sensory nucleus of V.

Nucleus of the Spinal Tract of V

The spinal trigeminal nucleus is a large structure lying at the lateral margin of the brainstem. It extends from the mid pons to the cervical spinal cord and is accompanied on its lateral aspect by the spinal tract of V, a collection of primary afferent axons that synapse in the nucleus (see Fig. 9.14). At its caudal pole it merges with the dorsal horn of the spinal cord. In the pons it merges with the principal sensory nucleus at the level where the trigeminal nerve enters the pons. Four cranial nerves have axons that synapse in this nucleus (X, IX, VII, and V), but the trigeminal nerve has by far the largest representation.

The sensory axons of the trigeminal nerve are distributed to the skin of the face, the mucous membranes of the oral and nasal cavities, and the base of the meninges over three divisions of the nerve: the **ophthalmic (V_1)**, the **maxillary (V_2)** and the **mandibular (V_3)**. There is essentially no overlap in the peripheral distribution of the three divisions and the patterns of innervation of the face are important features that need to be memorized (see Fig. 9.14C).

Principal Sensory Nucleus of V

The trigeminal sensory axons segregate on entering the pons, the largest axons terminating in the principal sensory nucleus while the smaller ones enter the spinal tract of V to synapse in the spinal nucleus of V. The former axons convey fine tactile sensations and proprioception. They have small receptive fields and are rapidly adapting. The latter carry pain and temperature sensations, have large receptive fields, and are slowly adapting. Many primary afferent trigeminal axons bifurcate on entering the pons, sending one branch to the principal nucleus and the other branch to the spinal nucleus. Such axons may account for the fact that electrical potentials related to tactile stimuli can be recorded from both nuclei.

The neurons of the principal sensory nucleus send axons to VPM in the thalamus. Most of the axons leaving the nucleus cross the midline and join the contralateral medial lemniscus. A significant minority (about one third) ascend in the ipsilateral medial lemniscus and terminate in the ipsilateral VPM. Axons from the nucleus of the spinal tract of V take an entirely different course. Very few axons appear to ascend to the thalamus. Most axons from the spinal nucleus appear to terminate in the reticular formation from which, one can presume, sensory information ascends to the thalamus as reticulothalamic fibers of the anterolateral system (ALS).

It is useful to note that the principal nucleus of V is functionally analogous to the dorsal column nuclei. Fine tactile and proprioceptive information is passed through it, and reaches the ventrobasal thalamus via a fast and direct pathway. The spinal nucleus of V, on the other hand, is analogous to the laminae in the dorsal horn of the spinal cord that receive pain and temperature information. Both the spinal and the trigeminal systems project to the ventrobasal thalamus by polysynaptic pathways. A minority of fibers proceed directly to the thalamus, but most synapse in the reticular formation from which the final thalamic projections arise. Thus the separation of fast versus slow–tactile versus pain pathways is preserved for sensations from the head.

Secondary fibers from the trigeminal sensory nuclei are also distributed locally to certain brainstem structures. These connections are important for local reflexes. The **corneal reflex**, as previously mentioned, is particularly important. The afferent limb traverses the ophthalmic division of the trigeminal nerve and synapses in the spinal nucleus of V. Secondary trigeminal sensory nuclei fibers innervate the facial nucleus indirectly through the reticular formation. From the facial nucleus, efferent axons innervating the orbicularis oculi muscle arise, closing the reflex path. The reflex relaxation of the masseter muscles and withdrawal of the tongue after biting one's tongue or cheek is presumably effected by similar connections involving the CN V and CN XII motor nuclei.

CLINICAL CONSIDERATIONS

The most conspicuous neurological problem associated with the trigeminal nerve is **trigeminal neuralgia** or **tic douloureux** [Fr., spasmodic movement + Fr., painful]. This condition is characterized by paroxysms [G. *paroxysmos,* to sharpen, irritate] of severe pain in the territory of the trigeminal nerve, usually the ophthalmic or mandibular divisions. The pain is intense, often described as burning, tearing, cutting, or stabbing. It lasts for several seconds. The pains occur frequently, day or night, and can be triggered by the slightest tactile (not painful) stimulus in specific areas of the face (trigger zones) within the territory of the trigeminal nerve.

Strictly speaking, trigeminal neuralgia is idiopathic, but symptomatic trigeminal neuralgia is sometimes associated with tumors, multiple sclerosis, and aneurysms that affect the trigeminal nerve. Although the disorder is not caused by seizures, it is frequently successfully treated with anticonvulsive drugs. Surgery is used to alleviate the pain in those patients for whom drug therapy fails. Several surgical procedures have been employed. Each is based on a different neuroanatomical principle. The first procedure was simply to sever the central axons of the nerve. This relieves the pain, but also anesthetizes the cornea. A severe keratitis [G. *keras,* horn + itis, inflammation] and ulceration of the cornea develop, usually with eventual loss of vision in the affected eye. To avoid that problem,

the trigeminal tractotomy was developed. This procedure takes advantage of the fact that the primary sensory axons of the trigeminal nerve are isolated in the spinal tract of V where they can be selectively cut (see Fig. 9.14*D*). Judicious placement of the lesion can spare sensation over much of the face while anesthetizing the trigger zone. The most recent surgical approach to trigeminal neuralgia is to stereotaxically direct a probe into the ganglion and carefully heat it with radiofrequency current. This procedure is said to destroy only the Aδ and C fibers, allowing some elements of touch and the corneal reflex to be preserved. Surgical therapy for trigeminal neuralgia, like surgery for any intractable pain syndrome, is often not very effective or long lasting and has many complications (see Chapter 5).

Because of its location in the region of the cavernous sinus, aneurysms of the internal carotid artery can exert pressure on the trigeminal nerve (see Fig. 9.14*E, F,* and *G*). Since the nerves of ocular motion (III, IV, and VI) also travel in the same area, one or more of them are frequently affected as well. Thus, as described in the next section, facial pain, diminished corneal reflex, and ocular palsy in any combination is ominous. Similar combinations of symptoms may arise from hypophyseal tumors that have expanded into the sinus.

Nuclei of Ocular Motion (CN III, IV, and VI)

Three cranial nerves, the **abducens (VI)**, the **trochlear (IV)** and the **oculomotor (III)**, innervate the **extrinsic muscles of the eye**, the muscles that move the eye in its orbit. Additionally, the oculomotor nerve also innervates the **levator palpebrae**, the muscle that lifts the eyelid. Also associated with the oculomotor nerve are parasympathetic fibers that regulate the size of the pupil and the shape of the lens.

THE EXTRINSIC MUSCLES OF THE EYE

Six muscles control the position of the eye in the orbit; the **medial** and **lateral rectus**, the **su-**

perior rectus and **inferior oblique**, and the **inferior rectus** and **superior oblique** muscles. The *medial* and *lateral rectus* muscles rotate the eye around an imaginary **vertical axis** (Fig. 9.15*A, B,* and *C*). Contraction of the medial rectus causes **adduction** [L. *ad-duco,* to bring to] of the eye while contraction of the lateral rectus causes **abduction** [L. *ab-duco,* to draw away]. The other eye movements are more complex because the remaining muscles do not pull perpendicular to any single axis during forward gaze. Since they lie medial to the sagittal axis of the eye, their force is exerted obliquely to all three axes. If the patient's gaze is straight ahead, contraction of the *superior rectus* muscle causes **elevation, medial rotation,** and **adduction.** Similarly, contraction of the *superior oblique* causes **depression, medial rotation,** and **abduction** of the eye. The *inferior rectus* causes **depression, lateral rotation,** and **adduction** while the *inferior oblique* causes **elevation, lateral rotation,** and **abduction** of the eye. The synergistic and antagonistic relationships between these muscles are summarized in Table 9.3.

When evaluating eye movements, it is convenient to test each muscle in isolation. The medial and lateral rectus muscles are isolated by adduction and abduction during direct forward gaze. One can isolate the remaining muscles by noting that the superior and inferior rectus muscles act as pure elevators or depressors if the eye is abducted about 23°, which corresponds to the angle of insertion of these muscles into the globe. The oblique muscles become pure elevators or depressors when the eye is adducted approximately 39°. Therefore, when testing the extrinsic eye muscles, elevation and depression of the eyes during left lateral gaze tests the two rectus muscles on the left and the two obliques on

the right and conversely upon right lateral gaze (see Fig. 9.15*D* and *E*).

THE NUCLEI OF OCULAR MOTION

The extrinsic eye muscles are controlled by motor neurons located in three pair of motor nuclei in the brainstem.

Abducens Nucleus

Motor neurons in the **abducens nucleus,** located near the midline of the pons at the internal genu of the facial nerve (see Fig. 9.10), send axons to the ipsilateral lateral rectus muscle over the **abducens (VI) nerve.** The nerve exits the basis of the pons near the midline at the pontomedullary junction. It courses in the posterior fossa until it enters the area of the cavernous sinus, leaving the cranium through the superior orbital fissure before innervating the lateral rectus muscle.

Trochlear Nucleus

Motor neurons in the **trochlear nucleus** innervate only the superior oblique muscle. The trochlear nucleus lies in the mesencephalic tegmentum at the level of the inferior colliculus (Fig. 9.16). The **trochlear (IV) nerve** leaves the roof of the brainstem, the only cranial nerve to do so, crosses the midline, and wraps around the lateral aspect of the cerebral peduncle before entering the region of the cavernous sinus. Like the abducens nerve, it leaves the cranium through the superior orbital fissure.

Oculomotor Nucleus

All of the other extrinsic muscles of the eye are innervated by motor neurons that reside in the

FIGURE 9.15.

The action of the extrinsic muscles of the eye on each of the three imaginary axes of rotation are represented (**A, B,** and **C**). Note that movement in any of the anatomical planes requires the coordination of several muscles. Elevation and depression of the eye can be accomplished by single pairs of antagonistic muscles if the gaze is shifted about 23 degrees. In **D,** the relationships of the superior and inferior rectus muscles (*left*) and the super and inferior oblique muscles (*right*) on lateral gaze are shown. **E.** The physician's view of the patient and the muscles being tested is shown.

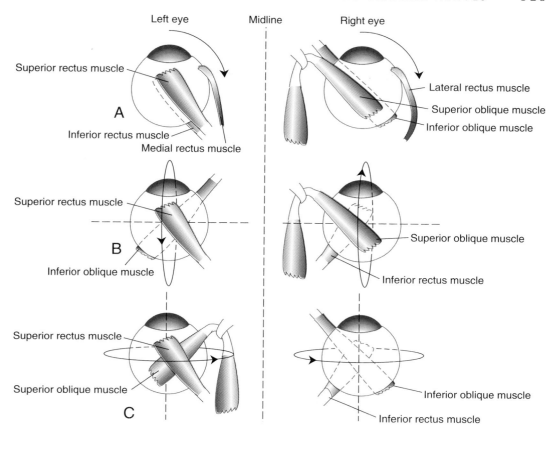

Left eye   Midline   Right eye

Superior rectus muscle

Lateral rectus muscle
Superior oblique muscle
Inferior oblique muscle

A

Inferior rectus muscle
Medial rectus muscle

Superior rectus muscle

Superior oblique muscle

B

Inferior oblique muscle

Inferior rectus muscle

Superior rectus muscle

Superior oblique muscle

C

Inferior oblique muscle

Inferior rectus muscle

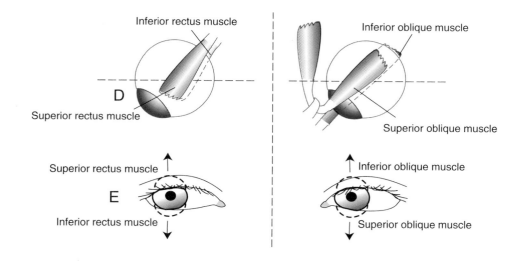

Inferior rectus muscle

Inferior oblique muscle

D

Superior rectus muscle

Superior oblique muscle

Superior rectus muscle

Inferior oblique muscle

E

Inferior rectus muscle

Superior oblique muscle

Table 9.3. Action of Extrinsic Eye Muscles

	Elevate	Depress	Medial Rotation	Lateral Rotation	Abduct	Adduct
Superior rectus	✓		✓			✓
Inferior oblique	✓			✓	✓	
Inferior rectus		✓		✓		✓
Superior oblique		✓	✓		✓	
Medial rectus						✓
Lateral rectus					✓	

oculomotor nucleus. This nucleus lies in the mesencephalon just rostral to the trochlear nucleus, at the level of the superior colliculus (see Fig. 9.16). The **oculomotor (III) nerve** leaves the nucleus and travels in a ventral direction, passing through the red nucleus and the medial portion of the cerebral peduncle before leaving the brainstem. Like CN IV and CN VI, it enters the area of the cavernous sinus and leaves the cranium through the superior orbital fissure. Once in the orbit, the oculomotor nerve divides into branches that innervate the superior, inferior, and medial rectus muscles and the inferior oblique.

Eye Movements

Clear vision requires stable congruence of the retinal images in the two eyes or diplopia (double vision) will result. Even transient diplopia will result in unclear vision. To prevent diplopia, the two eyes must be coordinated so that an image falls on complementary portions of the retina in each eye. If the image moves in space, the two eyes must track the object in a way that keeps the image appropriately fixed on the two retinas. This usually requires at least adduction of one eye and abduction of the other and usually requires compensation in the vertical plane as well. Since the eyes are located in the head, an unstable platform, head movement has to be taken into consideration in computing eye movements.

BRAINSTEM MOTOR CENTERS FOR EYE MOVEMENTS

Perhaps the most remarkable fact about the control of eye movements is that, unlike most voluntary motor systems, the motor neurons controlling the extraocular eye muscles are not directly regulated by motor areas in the cerebral cortex. Cortical control of eye movements is indirect, acting through brainstem visuomotor centers. The most potent motor commands directing eye movement come directly from brainstem nuclei. The **vestibular nuclear complex** plays a critical role in adjusting eye movements in response to head movement. The **superior colliculus** is a unique structure that receives a variety of sensory inputs and initiates motor commands to direct head and eye movements toward the stimulus. The **paramedian pontine reticular formation** and the **pretectal region** are integrating centers that direct horizontal and vertical gaze respectively. To understand how eye movements are initiated and regulated, these structures and their interconnections must be understood.

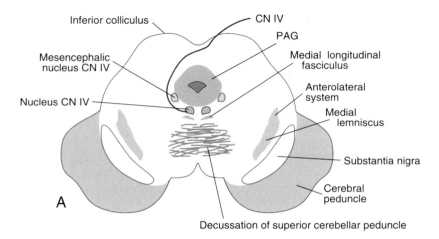

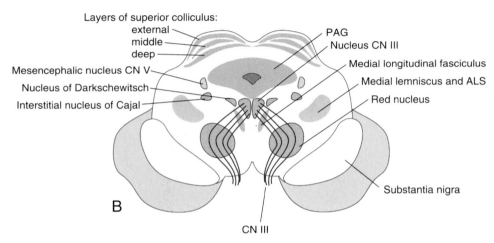

FIGURE 9.16.

A cross-section of the mesencephalon at the level of the trochlear nucleus (**A**) and at the level of the oculomotor nuclei (**B**).

The Vestibular Nuclei

The vestibular nuclear complex consists of four nuclei. These nuclei receive sensory information from the vestibular receptors about the movement of the head and its relation to gravity. Some of this information is used to regulate postural reflexes and stabilize the body relative to gravity. This is primarily the function of the lateral vestibulospinal tract. This information is also used independently to stabilize the head via the medial vestibulospinal tract (see Chapter 7).

The vestibular nuclei also control eye movements. Fibers originating in the medial and superior vestibular nuclei project to the nuclei of ocular motion, tectal and accessory oculomotor nuclei (see gaze centers below) over a fiber tract called the **medial longitudinal fasciculus (MLF)**. The MLF is a prominent collection of axons that courses the entire length of the brainstem. It is a paired structure that lies very close to the midline in the tegmentum just ventral to the fourth ventricle (see Figs. 9.2, 9.10, and 9.16).

Axons originating in the medial vestibular nucleus bilaterally innervate the abducens nucleus. Most of the axons also cross the midline

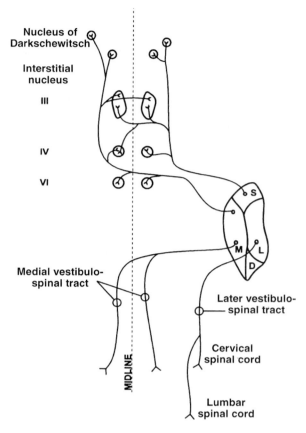

FIGURE 9.17.

A diagram illustrating the principal connections of the vestibular nuclei. Note that axons originating in the superior vestibular nucleus ascend on the ipsilateral MLF, making their contralateral connections by crossing at the level of the target nuclei. The axons originating in the medial vestibular nucleus cross the midline between the VI and IV nuclei and ascend on the contralateral MLF, making ipsilateral connections by re-crossing at the level of the target nuclei. D, descending vestibular nucleus; L, lateral vestibular nucleus; M, medial vestibular nucleus; S, superior vestibular nucleus.

and ascend in the MLF to innervate the contralateral trochlear nucleus, all parts of the oculomotor nucleus, and two accessory oculomotor nuclei, the interstitial nucleus of Cajal (INC) and the nucleus of Darkschewitsch (ND) (see Fig. 9.16). Axons originating in the superior vestibular nucleus ascend in the ipsilateral MLF and make similar connections (Fig. 9.17).

The Superior Colliculus

The superior colliculus comprises the tectum of the most rostral portion of the mesencephalon. It is a complex structure that is composed of three strata of cells and fibers, called the **superficial**, **intermediate**, and **deep** layers (see Fig. 9.16).

The superficial layer receives visual afferent fibers from both the *retina* and the *visual cortex* (areas 17, 18, and 19). The visual field is mapped onto the superior colliculus twice, once by retinal afferent fibers and once by cortical afferent fibers, creating overlapping maps. Therefore, certain cells in the superior colliculus receive information about the same visual space from both sources.

The innermost two layers of the superior colliculus receive information from nonvisual cortical areas. A special area in the frontal lobes, known as the **frontal eye fields** (part of area 8), is particularly important. *Auditory* information from the inferior colliculus also projects to cells in the intermediate and deep layers. *Somatosensory signals* reach the deep layers from several sources. A few fibers arise from lamina IV of the spinal cord, but the majority come from the dorsal column nuclei and the nucleus of the spinal tract of V (Fig. 9.18).

The efferent connections of the superficial layer project mainly to the pulvinar and these pathways will be considered further in the next chapter. The superficial layers also project to the pretectal nuclei (to be described below). There are probably no projections into the two deepest layers.

The intermediate and deep layers project to several structures. The small **tectospinal** tract leaves the deep layers, crosses the midline, and descends near the surface of the tegmentum. At the lower medulla, it becomes incorporated in the medial vestibulospinal tract and terminates in the cervical spinal cord. These fibers are important in directing head movements toward

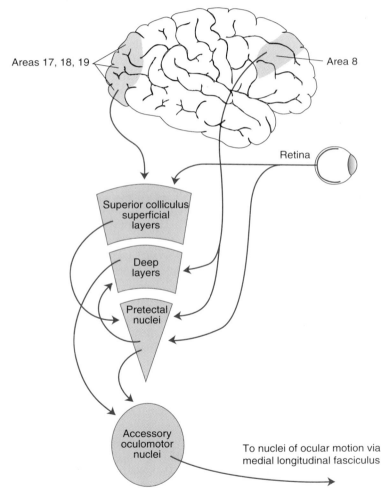

Areas 17, 18, 19

Area 8

Retina

Superior colliculus superficial layers

Deep layers

Pretectal nuclei

Accessory oculomotor nuclei

To nuclei of ocular motion via medial longitudinal fasciculus

FIGURE 9.18.

The superior colliculus plays a critical role in tracking eye movements. It serves to compare the actual visual target with the intended target (from cortical signals). It regulates the nuclei of ocular motion through its connections with the pretectal nuclei. These relationships are schematically illustrated here.

novel visual and auditory stimuli. **Tectobulbar fibers** terminate primarily in the **paramedian pontine reticular formation**. *There are no direct connections from the superior colliculus and the nuclei of ocular motion* (see Fig. 9.18).

The superior colliculus is a complex structure that is not well understood at this time. In many ways it is a dual structure being divided both anatomically and functionally. The superficial layer is primarily concerned with processing information from cortical and retinal sources. Its most prominent projections are to the pulvinar, a thalamic nucleus that projects to the cerebral cortex. This circular organization is reminiscent of the organization of the basal ganglia. In contrast, the deep layers receive information from noncortical sources and process mostly nonvisual information. Efferent connections from the deep layers project to brainstem and spinal centers.

The Pretectal Nuclei

A group of small nuclei located near the periaqueductal gray and adjacent to the oculomotor nucleus are known as the **pretectal nuclei** (see Appendix C, Fig. C.7). They receive afferent fibers from the retina, the visual cortex (areas 17, 18, and 19), the frontal eye fields (approximately area 8), and the superficial layers of the superior colliculus. Efferent fibers from the pretectal nuclei innervate the deep layers of the superior colliculus, the pulvinar, the interstitial nucleus of Cajal, the nucleus of Darkschewitsch, and the **Edinger-Westphal nucleus**. The latter nucleus mediates accommodation of the lens and constriction of the pupil through its connections with the ciliary ganglion (described below). Three pair of nuclei also found in this area, the interstitial nucleus of Cajal, the nucleus of Darkschewitsch and the **nuclei of the posterior commissure** are collectively known as the accessory oculomotor nuclei. They receive fibers from various parts of the pretectal nuclei and the vestibular complex and contribute fibers to the MLF. Axons from some of these nuclei directly synapse on oculomotor neurons.

The Gaze Centers

The deep layers of the superior colliculus and the vestibular nuclei send fibers into the **pontine paramedian reticular formation (PPRF)**, an area in the brainstem that has become known at the "horizontal gaze center." The PPRF is a collection of cells that lie between the abducens and the trochlear nuclei and lateral to the MLF. It receives afferent fibers from the deep layers of the superior colliculus and the vestibular nuclei. Horizontal and vertical eye movements are coordinated by different parts of this nuclear complex. Horizontal gaze is regulated by neurons in the PPRF and interneurons located adjacent to the abducens nucleus. Stimulation of the caudal region of the PPRF produces coordinated horizontally directed eye movements with very short latencies. Vertical eye movements can be elicited by stimulating the rostral portion of the PPRF and the **rostral interstitial nucleus of the MLF (RiMLF)**, a small collection of cells just ventral to the nucleus of Darkschewitsch. The RiMLF receives direct afferent fibers from the PPRF and the vestibular nuclei and projects directly to the oculomotor nucleus. The RiMLF is considered by many to be the functional "vertical gaze center."

TYPES OF EYE MOVEMENTS

The control of eye movements is quite different from the control of skeletal movements. Most skeletal movements are initiated and closely directed by conscious action. The regulation of skeletal movements by the muscle stretch reflex and cerebellar reflexes serves to enhance the precision of the consciously directed act. Eye movements are quite different.

Perhaps the most striking difference between skeletal and eye movements is the existence of two types of mutually exclusive eye movements, **smooth pursuit** and **saccades**. Smooth pursuit movements are also called **eye tracking** movements because they are only employed when the eye is tracking a moving object. For this reason they are considered *involuntary*. Smooth pursuit eye movements keep the image of a target of interest on the central 1° of the retina. This area, called the **fovea**, is specialized for high-resolution imaging. During smooth pursuit, the velocity of the eye movement varies with the velocity of the target, up to about 100°/sec.

In contrast to smooth pursuit, saccades are fast and of constant velocity, about 600°/sec.

They are also *voluntary* in the sense that one can consciously initiate the saccade and determine what object in visual space will be brought onto the fovea. One cannot, however, voluntarily set the saccade velocity.

You can very easily demonstrate these two types of eye movement to yourself. First, try to slowly shift your gaze between two objects in the room. You will note that it cannot be done. Your eye movements between the objects are broken into a series of short saccades. If, however, you focus on your finger and slowly move it about, you can follow it at very slow velocities with no interrupting saccades.

Saccades are used to quickly direct the gaze toward an object of interest. Smooth pursuit eye movements are then utilized to keep the eyes pointed at that object as it moves around. The superior colliculus plays a critical role in both types of movement.

Saccades may be initiated subconsciously by reflex. The stimulus for the reflex is the detection of novel objects in the peripheral visual field. The detection of novel objects is usually based on movement. Presumably this determination has great survival value for predators and prey alike. Once the novel object is detected, the superior colliculus computes the angular discrepancy between the fovea and the location of the object in the visual field and initiates a saccade to place the image on the fovea. Usually the head is oriented toward the object as well. The previously described structures and pathways between the superior colliculus, the spinal cord, the pretectal nuclei, and the nuclei interconnected by the MLF all participate in this task.

Saccades may be initiated by voluntary action, probably upon commands received by the superior colliculus from the frontal eye fields. Here the determination of what is novel or "interesting" is of cortical origin and transmitted to the superior colliculus by the corticifugal pathways previously described. The exact mechanisms are uncertain and some data are conflicting, but clinical data support this hypothesis. For example, electrical stimulation of the frontal eye fields causes contraversive (to the contralateral side) saccades and conversely, patients who have suffered damage to the frontal eye fields are unable to voluntarily initiate eye movements to-ward the side contralateral to the lesion. In bilateral lesions involving area 8, patients may be unable to voluntarily initiate saccades in any direction, although saccades initiated by reflex remain.

The pathways serving tracking eye movement are not understood. Some light has been shed on the subject by clinical evidence. In rare cases, patients may suffer a lesion to the occipital-mesencephalic fibers while the other visual pathways are spared. In these cases, patients may voluntarily move their eyes in all directions, but fail to track objects in space or maintain fixation on a desired stationary target. Probably, the superior colliculus, deprived of cortically derived visual map, is unable to assess the divergence of the identified "interesting" object in the visual field from the foveal image.

REFLEX EYE MOVEMENTS

Two important reflexes have evolved that utilize the two types of eye movements. The vestibulo-ocular reflex moves the eyes in response to head movement while the optokinetic reflex moves the eyes in response to visual stimuli. Both reflexes have important clinical applications.

The Vestibulo-ocular Reflex

When visually tracking or scanning an object, the specific muscle contractions required to keep the object centered on the fovea depend on both the static position of the head as well as any head movements that may occur. Dynamic head movements require modification of the muscle signals during the tracking maneuver. The vestibular apparatus detects both static head position and dynamic head movements and, through its central connections with the nuclei of ocular motion, adjusts the eye movements to compensate. This reflex movement of the eyes in response to vestibular stimulation is called the **vestibulo-ocular reflex** or **VOR**.

One can readily appreciate the sophistication of the VOR by tilting one's head while reading these lines. What had previously been a simple adduction/abduction movement of the eyes now requires movement around all axes. Information regarding head tilt, detected and measured by the saccule and utricle, is used to adjust the action of all six of the nuclei of ocular motion. The information is distributed to these nuclei by signals transmitted primarily over the MLF.

The VOR can compensate for dynamic movement as well as static head position. For example, if a man is sitting in a barber's chair and the barber turns him to the left, his eyes automatically rotate to the right.[6] Rotation of the head to the left stimulates the left crista of the horizontal canal and inhibits the right horizontal crista (see Fig. 9.9). The excitatory signals are passed from the left vestibular nuclei to the contralateral abducens nucleus that causes abduction of the right eye. The left oculomotor nucleus is also excited to adduct the ipsilateral eye. Similarly inhibitory signals are sent to the motor neurons of the right medial rectus muscle and the left lateral rectus. Most of the connections that affect these movements pass through the MLF. Similar circuits can be worked out for signals originating in the other crista.

Consider, again, the barber's chair. If someone is slowly rotated in the chair to the left, as we have noted above, the eyes slowly track an imaginary object to the right. If the rotation continues until the eyes have moved to the extreme right, they do not lock in place as one might first suspect, but saccade back to the gaze forward position. Then they slowly track to the right again. This slow tracking movement followed by a fast centering saccade in the opposite direction is called **nystagmus**. It is named for the *fast* component, so in our example, rotating someone to the left will produce a *left lateral nystagmus*.

When we take pity on our nauseated friend and stop rotating him in the barber's chair, the endolymph will continue to flow through the horizontal canal. Since the head is now stationary, the cristae will be stimulated in the opposite direction and one can observe a *right lateral post-rotatory nystagmus*. Post-rotatory nystagmus is a common phenomenon and very troublesome to pilots and astronauts.

The vestibulo-ocular reflex is an important tool for evaluating brainstem injury in uncon-

scious patients. When the patient is lying on his back, the horizontal canals lie nearly in the vertical plane. After having determined that there is no injury to the cervical spine, one simply rotates the head left and right to maximally stimulate the horizontal cristae. Normally, due to the VOR, the eyes will move together and in the opposite direction of the head. The action resembles a china doll equipped with weighted eyeballs that look forward no matter how the head is tilted. Hence this procedure is often called the **doll's eye maneuver**. If the eyes do not track together and react appropriately to head movement, a brainstem lesion is almost certain.[7] The value of this procedure lies in the fact that the structures necessary for the reflex to be intact, the vestibular system, the MLF, and the nerves and nuclei of ocular motion, are scattered from the medulla to the mesencephalon. Brainstem injury would almost certainly affect one or more of these structures and disrupt the VOR.

The Optokinetic Reflex

Brainstem and cortical visuomotor centers can also initiate oculomotor reflexes. For example, if an object is moving across the visual space, the visuomotor centers will initiate saccades to bring its image onto the fovea and then to maintain the image on the fovea as the object moves about in the visual space. This **fixation reflex** is more highly developed and elaborate in the human than in any other animal. The pathways for this reflex have not been definitively established, but it certainly requires the participation of the parietal and occipital cortex and the brainstem motor centers that have just been described. Fixation on a specific target also requires *consciousness* and *attention*.

The fixation reflex is responsible for a phenomenon sometimes called **railroad nystagmus**. A person sitting in a moving train will often fixate on a passing object, for example the utility poles that often line the tracks. As the object passes, his eyes having fixated on the object will track it until it has passes from view. Then the eyes will saccade back to the forward gaze position and fixate on the next pole. An obser-

---

[6]An apparatus similar to a barber's chair is the Bárány chair, named after the man who first used it to study vestibular nystagmus.

[7]Beware of patients with glass eyes!

vant traveling companion will notice nystagmus and clinically this term is usually called **optokinetic nystagmus (OKN)**. The phenomenon can be elicited by any repetitive visual stimulus. It is conveniently elicited during the neurological examination by a simple patterned ribbon drawn across the patient's visual fields. Failure to elicit the OKN in a patient with otherwise normal vision, usually suggests a *parietal lobe* lesion on the *side toward which the ribbon is being drawn.* This is probably due to the inability of the patient to maintain his attention on the moving pattern. Failure of the OKN also occurs following frontal lobe lesions. In these cases the fast component is lost, probably due to loss of the frontal eye fields.

Spontaneous Nystagmus

In contrast to rotational nystagmus and optokinetic nystagmus that are normal reflex responses to specific stimuli, nystagmus that is **spontaneous** *almost always indicates some underlying pathological condition.* Lesions to the labyrinth, the vestibular nerve, the brainstem, or the cerebellum all cause spontaneous nystagmus.

Unilateral peripheral lesions, those involving either the labyrinth or CN VIII directly, are manifested by nystagmus, vertigo, nausea, and unsteadiness with a tendency to fall toward the side of the lesion. If the nerve is affected directly, as happens with CN VII tumors (**acoustic neuromas**, described in Chapter 10), these symptoms will be accompanied by a diminished or total loss of hearing on the affected side. The vestibular symptoms are usually transitory, an observation that is usually explained by the ability of the CNS to compensate for the loss of the sensory signals from one labyrinth. Certain ototoxic drugs, particularly the **aminoglycoside antibiotics**, may destroy the vestibular hair cells bilaterally. In these cases, the patients may suffer a complete loss of vestibular function that results in a permanent loss of the vestibulo-ocular reflex. There is no spontaneous nystagmus because the absence of vestibular signals from the two sides is balanced. However, without the VOR, clear vision is impossible because the retinal image can no longer be stabilized in reaction to slight movements of the head.

Central lesions of the brainstem produce many of the same symptoms as peripheral lesions, but vertigo is usually absent. The nystagmus may be caused by any disease affecting the brainstem. Nystagmus derived from brainstem lesions is usually permanent since no compensation is possible if the brainstem nuclei are destroyed. *Cerebellar nystagmus* is frequently dependent on head position, the nystagmus being prominent in only one static head position. Otherwise, it may be indistinguishable from nystagmus of labyrinthine origin. In this regard, it is worth mentioning that the most common cause of nystagmus is alcohol intoxication, a condition that renders the cerebellum dysfunctional due to its extreme sensitivity to this drug. Experienced Saturday night bed pilots are quite familiar with the positional sensitivity of cerebellar nystagmus.

Ocular Reflexes

Clear vision also requires compensation of the lens and the pupil to adjust the eye for subject distance and brightness. These adjustments are undertaken automatically by certain of the ocular reflexes. All of these eye reflexes, tracking, accommodation, and pupillary constriction, are mediated through brainstem nuclei that are intimately interconnected. These relationships must be understood if one is to be able to relate impaired eye movements and ocular reflexes to localized brainstem pathology.

PUPILLARY CONSTRICTION

When light is shone into the eye, the constrictor muscle in the iris contracts, narrowing the pupil. Although it is the iris that contracts, by convention this is known as **pupillary constriction**. The constriction of the pupil is usually tested by shining the light from a small flashlight into each of the patient's eyes in turn and observing the pupils.[8] Normally, both pupils will constrict when the light is shone into one eye, the consensual light reflex.

---

[8]Of course if the patient is in a brightly lit room, shining a light into his eyes will not cause pupillary constriction for the pupils will already be constricted due to the ambient lighting.

The pathways for the light reflex are not known with certainty. It is generally agreed, however, that direct retinal afferent fibers synapse in the pretectal nuclear complex. Axons from the pretectal nuclei innervate both Edinger-Westphal nuclei, the contralateral connections having passed through the posterior commissure. These nuclei are the origin of preganglionic parasympathetic axons that leave the brainstem over the oculomotor nerves and innervate the ciliary ganglia, located on the globes. Post-ganglionic fibers then innervate the constrictor muscle of the iris (Fig. 9.19). Therefore the light reflex depends on two cranial nerves, the optic (that will be discussed further in the next chapter) and the oculomotor.

ACCOMMODATION AND CONVERGENCE

In addition to adjusting to changes in ambient light, the shape of the lens must be altered to focus on both near and distant objects. At rest, the lens is thin in order to focus distant objects on the retina. To focus on near objects it must become thicker, a process known as **accommodation**. When viewing near objects, if diplopia is to be avoided, the eyes must converge. Pupillary constriction also accompanies accommodation, an action that improves visual acuity by reducing the effect of spherical aberrations in the lens. Collectively these three actions, *accommodation, convergence,* and *constriction,* occur together when viewing close objects and are known as the **accommodation reflex.**

The mechanisms of the accommodation reflex are only partly understood. In this regard, the patient must have the ability to see and to cooperate with the examination. Therefore the occipital visual cortex must be functional; the frontal eye fields are not necessary. Probably signals from the occipital cortex are transmitted to the pretectal nuclei that, in turn, bilaterally innervate the **Edinger-Westphal** nuclei (see Fig. 9.19). As mentioned above, these nuclei supply the parasympathetic preganglionic neurons that innervate the ciliary ganglion by way of the oculomotor nerve. Postganglionic axons from the ciliary ganglion not only innervate the pupillary constrictor muscle, but also the ciliary muscles in the eye. The **ciliary muscles** are arranged such that their contraction shortens the suspensory ligament of the lens. This ligament nor-mally places tension on the lens, lessening its curvature, but when that tension is relieved by ciliary contraction, the lens passively thickens (see Fig. 9.19).

Accommodation ultimately relies on the *passive elastic* thickening of the lens after tension on it has been relieved. If the lens loses its elasticity, which it does with advancing age, it will no longer accommodate, even if all of the neurological apparatus for accommodation is functional. This well known middle age phenomenon begins about the fortieth year. The lens becomes an essentially rigid structure by age 60.

CASE STUDIES

Mr. A. M.[9]

HISTORY, MARCH 21,1990: Mr. A. M., a 44-year-old factory worker, was building shelving units in his garage late one Sunday evening when he noticed that his left arm and left leg felt a little weak. He decided to stop working for the evening, concluding that he was probably "overtired" and that it was time to turn in for the night. By the following morning, his left-sided weakness was much worse and, in addition, his right hand and leg felt numb and "tingly." Despite his difficulties, he resolved to go to work that morning because he didn't want to miss a friend's retirement party at lunchtime. He vowed to call his family doctor immediately upon his arrival to see if she could squeeze him in for an afternoon appointment.

Over the next several hours, Mr. M.'s weakness became progressively worse. At approximately ten o'clock, in the midst of his morning coffee break, he began to have difficulty speaking. He took a sip of his Diet Coke, but he had trouble swallowing and began to cough. Embarrassed, and visibly shaken, he explained to his colleagues that his mouth "wasn't working right." He nervously recounted the nature of the other difficulties that he had been having over the past two days. Disturbed by his noticeably slurred speech, a coworker volunteered to take him to the emergency room of a nearby hospital.

Mr. A. M. is a heavy long-time smoker (three packs of cigarettes a day for 25 years) and a former heavy drinker (two cases of beer per week). He states that he no longer drinks alcoholic bev-

---

[9]This case was kindly provided by Dr. Carl Marfurt of the Northwest Center for Medical Education, Indiana University School of Medicine.

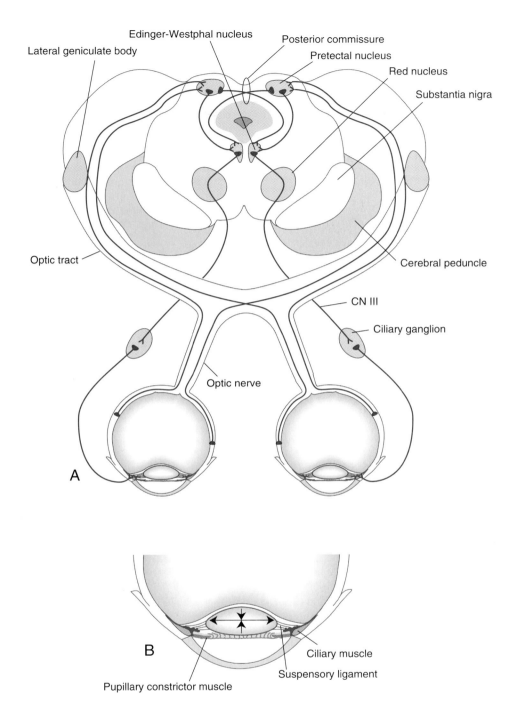

Lateral geniculate body

Edinger-Westphal nucleus

Posterior commissure

Pretectal nucleus

Red nucleus

Substantia nigra

Optic tract

Cerebral peduncle

CN III

Ciliary ganglion

Optic nerve

A

B

Ciliary muscle

Suspensory ligament

Pupillary constrictor muscle

FIGURE 9.19.

A, Probable pathways necessary for the pupillary light reflex. Note the afferent path is the optic nerve and the efferent path is the oculomotor nerve. Visual information from each visual hemi-field reaches both pretectal nuclei via fibers in the posterior commissure. Similar pathways also serve the accommodation reflex. **B**, The suspensory ligament of the lens places tension on the lens while the ciliary muscle is relaxed. This thins the lens, allowing it to bring distant objects into sharp focus on the retina. Contraction of the ciliary muscle relieves tension on the suspensory ligament and consequently the lens. If the lens is still resilient, it will thicken, bringing close objects into focus. If the lens is not resilient, it will not change its shape even if tension on the lens is relieved by ciliary contraction.

erages, but continues to smoke. He says that he weighs about 250 pounds and is 5′11″ tall. At the time of the present examination his blood pressure was 180/110 with a resting heart rate of 93. The remainder of the physical examination was otherwise unremarkable.

NEUROLOGICAL EXAMINATION

I. MENTAL STATUS: Normal

II. CRANIAL NERVES:

A. **Olfaction:** Not noted.

B. **Vision:** Full visual fields (see Chapter 11).

C. **Oculomotion:** Not noted (?).

Pupils: Pupils were of equal size. There was a consensual light reflex from each eye, accommodation was not noted (?).

D. **Trigeminal:** There was a decreased sensation to pinprick on the right side of the face. Responses to cotton and the corneal reflex were not noted.

E. **Facial:** The attending physician thought that the right nasolabial fold was somewhat flattened and that the mouth was slightly drooped on the right. The motility of the eyebrows was not noted (?).

F. **Audition:** Patient could hear the rubbing of fingers held close to the pinna. Hearing was grossly normal from each ear.

G. **Vagal-glossopharyngeal:** Voice was normal. The gag reflex, and the appearance of the pharyngeal arches and uvula were not noted (?).

H. **Accessory:** Not noted.

I. **Hypoglossal:** The tongue deviated to the right on protrusion.

III. MOTOR SYSTEMS: Muscle strength was moderately weak (3/5) in all four extremities. His gait was slightly ataxic and broad based. He was unable to perform a tandem walk. Some dysmetria was noted on finger-to-nose and heel-to-shin, bilaterally. There was some dysdiadochokinesia bilaterally, with the right worse than the left.

IV. SENSORY SYSTEMS: There was decreased sensation to pinprick on the entire left side of the body. Perceptions of vibration and joint position were intact bilaterally.

V. REFLEXES: All muscle stretch reflexes were 2+. There were bilateral Babinski, and Bing signs.

VI. ANCILLARY TESTS: An MRI of the head was ordered. It revealed that the basilar and right vertebral arteries were tortuous and dilated with a fusiform aneurysm of the mid- to upper basilar artery with an intraluminal thrombosis. No extravascular blood was demonstrated.

SUBSEQUENT COURSE

Early the next morning, Mr. A. M. experienced the sudden onset of vertigo, nausea, and vomiting. He also complained that he could no longer hear out of his right ear. Neurological examination revealed a spontaneous left-sided nystagmus. The next day, audiometry confirmed a total loss of hearing on the right.

Another MRI was ordered. In addition to the vascular abnormalities seen in the first MRI, it showed two areas of decreased signal (T1 weighted) in the pons at the level of the middle cerebellar peduncle. One area of low signal was in the midline, the other was to the right (Fig. 9.20A).

Two days later a cerebral angiogram was performed (Fig. 9.20B). It proved a marked enlargement of the basilar/right vertebral artery. There was also evidence of a fusiform basilar artery aneurysm with intraluminal thrombosis.

The aneurysm was considered inoperable. The patient was placed on anticoagulative drugs and given physical and occupational therapy. He was advised to stop smoking. Two months after discharge (three months after his initial symptoms), he was much improved. He was no longer smoking and had lost 30 pounds. His blood pressure was now 130/90 with a resting heart rate of 81. He still had a total hearing loss from the right ear, but his cerebellar deficits were greatly improved. He had a residual mild loss of pinprick sensation on the left side of his body. All MSR were 2+ and he had bilateral Babinski signs.

Comment: The process of placing limits on the physical extent of a lesion is deductive. One must decide what parts of the nervous system are not functioning properly by observing the capabilities of the patient. Then one must localize the lesion by picturing the anatomical relationships of the neurological systems at various levels of the nervous

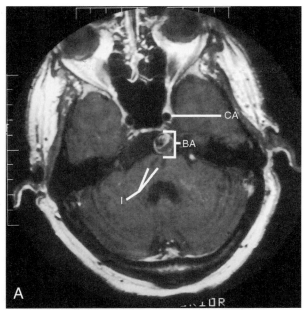

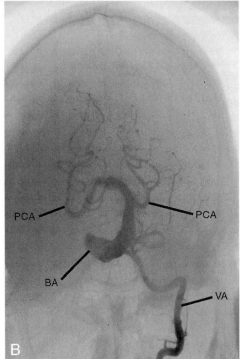

FIGURE 9.20.

A, This T1-weighted MRI from Mr. A. M. shows an enlarged basilar artery indicative of an aneurysm. In the middle and right pons, there is a large area of decreased signal indicating an infarct. This MRI was taken about 48 hours after his initial stroke. **B,** This arteriogram of Mr. A. M. reveals a tortuous and greatly enlarged basilar artery with intraluminal material suggesting a thrombus.

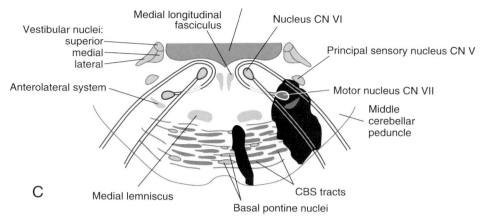

FIGURE 9.20.—continued

C, The extent of the lesion, as determined by deduction from the patient's signs and symptoms, is indicated. Compare with the actual lesion revealed by MRI.

system and figure out the location(s) where all the involved systems are in close proximity. A step-by-step analysis of this case will be discussed in detail to illustrate how one approaches a neurological diagnosis.

1.   The first step in evaluating a case such as this is to decide, approximately, the rostral-caudal level of the lesion. In general, cerebral lesions produce symptoms on the entire contralateral half of the body. Brainstem lesions produce crossed symptoms, one side of the head and the opposite side of the body. Spinal lesions spare the cranial nerves.

By the time that Mr. A. M. reached the emergency room, crossed symptoms were apparent. His right face drooped, the right nasolabial fold was flattened, and his tongue deviated to the right. There was decreased perception of pinprick to the right side of his face and the entire left half of his body. The loss of pain and temperature points to the ALS. Because the entire left half of the body and right half of the face is involved, the lesion must be above the neck and on the right, since the tract crosses at the level of origin in the spinal cord. The loss of pain and temperature on the right face points to the right spinal tract of V. The right facial weakness suggests involvement of the right facial nucleus.

2.   Once the approximate level of the lesion has been determined (in this case the brainstem), try to identify a small, specific structure that is dysfunctional. In this case, the spinal tract of V and the facial nucleus are implicated. The former is a long structure and runs the entire length of the medulla and pons. Its size limits its usefulness in localizing the lesion. The facial nucleus, however, is small

and compact; therefore, one should concentrate on proving or disproving its involvement.

If the right facial nucleus were lesioned, one would expect to observe a facial paralysis involving both the upper and lower face. Since the abducens nerve and nucleus are very close to the facial nucleus, one might also expect to observe a weakness of abduction in the right eye. If the paralysis only involved the lower face, then one would look above the level of this nucleus for a corticobulbar lesion. Unfortunately, in the case of Mr. A. M., the physician not only failed to observe the extent of facial weakness, but also did not report Mr. A. M.'s eye movements.

We can surmise with some certainty that the facial nucleus must be lesioned. If the left corticobulbar tract were lesioned above the facial nucleus, then the facial weakness and body numbness would be on the same side. This is because both tracts (anterolateral system and corticobulbospinal) would have crossed the midline before innervating the affected structures. Since this is contrary to the observed deficits, this hypothesis must be false. Therefore, either the lesion involves the facial nucleus or there are two lesions widely separated in space. The latter, although possible, is less likely than the former.

3.   Having decided that the lesion must be in the right brainstem involving the facial nucleus, one should decide what other structures may be involved based on the patient's disabilities. Do these other structures lie close enough to the facial nucleus to be involved in the lesion?

Mr. A. M. has increased muscle stretch reflexes (2+ MSRs) in all four extremities as well as bilateral

Babinski and Bing signs. These are classic upper motor neuron signs; therefore, the lesion affects the corticospinal tract (CBS) bilaterally. Note, however, that Mr. A. M. is not totally paralyzed. He can walk, but not well. He has voluntary control over all four extremities, but not normal control. Therefore, the CBS lesion is probably incomplete. At the level of the pons, the CBS is not collected together as a small compact bundle. Its fibers are widely separated in the basis of the pons. A small vascular lesion near the midline in the pons might easily damage some CBS fibers bilaterally, resulting in minimal UMN signs in all four extremities.

The patient clearly has signs of cerebellar involvement. There is bilateral dysdiadochokinesia and dysmetria, somewhat worse on the right. Such symptoms could be produced by a lesion of the cerebellum itself, but given the other conclusion already drawn, that seems unlikely. The basis of the pons, however, contains both the deep pontine nuclei and their axons that form the middle cerebellar peduncle. A midline lesion would be expected to injure these fibers, producing bilateral cerebellar signs. The fact that the signs are somewhat worse on the right suggests that the lesion may be asymmetrical, perhaps involving the right deep pontine nuclei and the right peduncle more than the left (remember that cerebellar signs are ipsilateral to the lesion).

The pattern of decreased sensation to pinprick and the dissociation between proprioception and vibration (intact), and pain and temperature (diminished), is of paramount diagnostic value. The loss of pain and temperature from the left body can be explained by the selective loss of the ALS in the brainstem on the right. In the pons, the ALS is located very close to the facial nucleus, and the loss of one would very likely be accompanied with the loss of the other. Since the ALS contains secondary fibers at this point, the information they are carrying originates from the contralateral side of the body. At this level the ALS has not yet joined the medial lemniscus, so preservation of proprioception, vibratory perceptions, and fine tactile sensations would not be unexpected. The spinal tract of V also lies very close to the facial nucleus. It consists of primary axons so its loss would produce decreased perception of pain and temperature on the ipsilateral side of the face, which is what was observed.

The problems Mr. A. M. has with swallowing are probably the consequence of the loss of corticobulbar fibers to the nucleus ambiguus. Since this nucleus receives bilateral supranuclear innervation, these losses would not be expected to be dra-

matic. Examination of the pharynx and evaluation of the gag reflex would have been helpful in analyzing this case.

Given the analysis so far, one would expect tongue deviation to the left as a result of the loss of the corticobulbar axons on the right. The observed right-sided deviation of the tongue could be due to left corticobulbar involvement with selective sparing of the corresponding fibers on the right. It seems more likely, however, that the deviation was recorded in error and that the tongue actually moved toward the left upon protrusion.

The physical limits placed on the location of the lesion by this analysis are illustrated in Fig. 9.20C.

4. After having localized the lesion as accurately as possible, one is then faced with determining its cause. A discussion of all the possible causes of Mr. A. M.'s difficulties is beyond the scope of this book. A few points are worth mentioning. Mr. A. M. is an overweight, heavy smoker with high blood pressure. These facts place him at increased risk for stroke. Given the relatively slow, step-wise development of his symptoms, one must entertain the diagnosis of thrombotic stroke in evolution (see Chapter 7).

If the diagnosis of stroke is correct, the lesion must be placed within the territory of a single artery. In the present case, one could conclude that occlusion of a paramedian *and* short right circumferential *branch of the* basilar artery *could account for all of Mr. A. M.'s initial symptoms. The second episode that resulted in vertigo, nausea, and deafness can be explained by a second thrombus that occluded the* right labyrinthine *artery. Its loss would destroy the cochlea, the vestibular apparatus, and the cell bodies of the vestibulocochlear nerve.*

5. Laboratory studies are an essential aid to neurological diagnosis. Such studies should be ordered to answer specific questions that cannot be answered by the physical and neurological examinations. In the present case, the initial diagnosis is infarction of the pons due to thrombosis of the penetrating arteries, but one must rule out the possibility that Mr. A. M. has a leaking aneurysm. The initial MRI was ordered for that purpose. It did not display any extravascular blood, an important observation since that rules out intracranial hemorrhage and allows one to proceed with anticoagulative therapy.

The MRI also revealed that Mr. A. M., in addition to the above mentioned risk factors, also has a tortuous basilar artery with a fusiform aneurysm. These represent further risk factors for stroke since atherosclerotic plaques and clots often form at sharp bends in arteries. Furthermore, aneurysms may leak or rupture, causing a hemorrhagic stroke (see Chapter 7). Note that the initial MRI did not reveal a pontine infarction. MRI, like CT, is unable to demonstrate pathological changes following stroke for about 24 hours.

When Mr. A. M. developed sudden new neurological deficits (vertigo and deafness), the possibility of an intracranial hemorrhage was again presented. The second MRI was ordered to make this determination. It showed no hemorrhage. By this time pathological changes could be imaged and the second MRI confirmed that which had previously been deduced from the clinical evidence, namely an infarct of the central basis of the pons in the territory of the paramedian branches of the basilar artery involving the facial nucleus and midline structures.

The arteriogram was ordered in order to evaluate Mr. A. M. for surgery. Many aneurysms can be clipped, sealed, or reinforced in order to prevent their later enlargement or rupture. In this case, the aneurysm was so large that it was deemed inoperable.

FURTHER APPLICATIONS

Restricted lesions of the pons are not common. More common, major branches of the basilar artery are occluded which produce large infarcts of the brainstem. For each of the arteries mentioned below, make a sketch illustrating the expected infarct and a list of sensory and motor symptoms one would expect to observe. You may want to review the brainstem circulation (see Chapter 1).

- Posterior inferior cerebellar artery (Wallenberg's syndrome).
- Anterior spinal artery at the level of the lower medulla
- Vertebral circulation at the level of the lower medulla
- Paramedian branches of the basilar or posterior communicating artery near the cerebral peduncle (Webber's syndrome).
- Make a schematic drawing of the ALS, the CBS, and the nuclei of the brainstem that shows how a single lesion could account for all the symptoms in this case.
- The emergency room physician skipped many parts of the neurological examination, especially the examination of the cranial nerves. How serious do you think these omissions were

in this case? Were these omission justifiable? Did the lack of information hinder the analysis of the case?

- The analysis presented above suggests that the direction of tongue deviation was recorded in error. Imagine yourself facing the patient during an examination. How easily do you think it is to make an error in recording left-right symptoms?

Mr. H. Y.

HISTORY, FEBRUARY 14, 1991: Mr. H. Y. is a 75-year-old white male. For the last 1–2 years he has been having trouble with his balance. He loses his balance very easily and tumbles backward. He senses no syncope or vertigo. He says that he cannot recover his balance when he starts to lean backward. He has had occasional trouble with seeing double.

His wife reports that his movements have, in general, slowed over the past two years. Recently he has had difficulty arising from chairs. This is apparently not due to weakness, but a general inability to get the process started. His speech has become soft and slurred.

NEUROLOGICAL EXAMINATION

I. MENTAL STATUS: Mr. H. Y. is somewhat apprehensive. He is oriented to place, time, and person. He has no memory deficits, recent or past. He could name the presidents from McKinley to the present and could recall three out of three items after 10 minutes.

II. CRANIAL NERVES:

A. **Olfaction:** Identified methyl salicylate independently from both nostrils (see Appendix B).

B. **Vision:** Full visual fields.

C. **Oculomotion:** Mr. H. Y. was unable to voluntarily direct his gaze downward. He had no problem looking from side to side. When his head was rotated by an assistant, his eyes would track in all directions (Fig. 9.21).

   Pupils: Pupils were of equal size. There was a consensual light reflex from each eye and both eyes constricted on accommodation.

D. **Trigeminal:** Sensations to pinprick and cotton were perceived the same from all

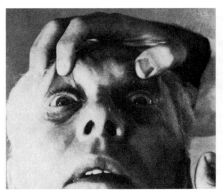

FIGURE 9.21.

This patient cannot voluntarily lower his gaze (*right*), but when his head is tilted upwards (*left*), the eyes move downward due to the vestibulo-ocular reflex, proving that the oculomotor nucleus, nerve, and connections through the MLF are intact.

three divisions bilaterally. Corneal reflex was present bilaterally. Jaw movements were symmetrical and strength seemed normal.

E. **Facial:** Face was symmetrical upon grimace. Eyes appeared normally hydrated. Salivary production seemed normal.

F. **Audition:** Weber and Rinne tests normal (see Chapter 10).

G. **Vagal-glossopharyngeal:** Voice was normal. Gag reflex present bilaterally. The pharyngeal arches were high and symmetrical; the uvula was centered.

H. **Accessory:** Sternocleidomastoid and trapezius muscle strength was normal and symmetrical.

I. **Hypoglossal:** Tongue was centered on extension. There was no atrophy nor any fasciculations.

III. MOTOR SYSTEMS: Mr. H. Y. presents himself with a marked retrocollis (the neck is hyperextended, causing the head to be thrust backward). He stands on a wide base. His gait is slow and tottering. When walking he leans backward and has great difficulty in correcting his balance. He falls quite frequently. During the examination he would have fallen were it not for the assistance he was offered. He has no obvious ataxia and his gait is not shuffling. He swings his arms somewhat when he walks.

There is a minimal increase in muscle tone in all of his extremities. The axial muscles are quite rigid, which accounts for his retrocollis. Upon examination, a definite cogwheel rigidity in both upper extremities was noted, greater in the left arm than in the right.

Mr. H. Y. exhibits very little spontaneous movement, his face is expressionless, and he seldom blinks.

IV. SENSORY SYSTEMS: Normal.

V. REFLEXES: All MSRs were normal. There was no Babinski sign.

VI. ANCILLARY TESTS: An MRI of the head was ordered. It demonstrated atrophy of the pontine tegmentum, but spared the basis of the pons. The aqueduct was enlarged. No abnormalities were noted in the cerebral cortex or the cerebellum.

Comment: Mr. H. Y. has bradykinesia, cogwheel rigidity, is expressionless, and has a slow tottering gait. These observations all suggest that he is suffering from parkinsonism. Several other observations, however, should alert one to the possibility that this case is a little different.

First, Mr. H. Y.'s chief complaint is falling. Parkinsonism is not usually characterized by falling. Second, the patient is dystonic. His posture is affected to the point that he assumes a retrocollic position. Parkinson patients have a kyphotic (stooped forward) posture. Finally, and perhaps most important to the correct diagnosis, Mr. H. Y. has little voluntary control of his eyes, particularly in vertical gaze; there is a definite supranuclear ophthalmople-

gia [G. *ophthalmikos*, **relating to the eye; G.** *plege,* stroke*].*

The term, supranuclear ophthalmoplegia, refers to a weakness of one or more of the extrinsic eye muscles (ophthalmoplegia) that cannot be explained by involvement of the motor neurons (i.e., supranuclear). Here, since eye movements could be proved in all directions during passive head movement (VOR), all of the nuclei of ocular motion and their nerves must be intact. Therefore, the disruption of voluntary vertical eye movement must originate in the structures that regulate the nuclei of ocular motion, rather than in the nuclei themselves.

Since the patient's ophthalmoplegia involves primarily downward gaze, one must consider a brainstem lesion involving the rostral interstitial nucleus of the MLF (RiMLF) and the pretectal nuclei, since these are the most important structures regulating vertical eye movements. These structures are not involved in parkinsonism.

In 1963, Richardson, Steel, and Olszewski described a syndrome now known as progressive supranuclear palsy. *It is characterized by supranuclear ophthalmoplegia affecting primarily vertical gaze, dystonia of the axial muscles especially affecting the neck and back, and unsteady gait with backward falling and diplopia. The pathology of this disease is characterized by a marked degeneration of the neurons within the globus pallidus, the substantia nigra, the pontine tegmentum, the periaqueductal gray, and neighboring pretectal nuclei. The cerebellum and the cerebral cortex are conspicuously unaffected.*

Mr. H. Y. displays all of the characteristics of supranuclear ophthalmoplegia and the diagnosis was confirmed by the findings of tegmental and periaqueductal gray (PAG) atrophy on the MRI.

FURTHER APPLICATIONS

- Tumors of the pineal gland are associated with abnormal eye movements. Consider the symptoms one would expect to see in a patient with such a tumor. How would they differ from the present case?

- Increased intracranial pressure is a major factor with pineal tumors. What are the special circumstances of the pineal tumor that creates this problem? The symptoms associated with increased intracranial pressure are discussed in Chapter 13.

- The MRI showed only degeneration of the pon-

tine tegmentum. What additional clinical signs would you expect to observe if the basis of the pons were also involved? Explain.

Mrs. P. F.[10]

HISTORY, November 21, 1993:   Mrs. P. F., a 44-year-old female, came to the emergency room at 9:45 PM complaining of a severe headache. She described the pain as sharp and shooting into the front of her head, then aching. She also says that her left eye "wants to stay shut." She has had this pain, on and off, for five days. Four days ago she saw her family physician who ordered a CT of the head. This shown no abnormalities.

Her physical exam was unremarkable. Blood pressure was 131/79 with a resting heart rate of 97.

NEUROLOGICAL EXAMINATION:

I.   MENTAL STATUS: She is alert and oriented to self, time and place.

II.   CRANIAL NERVES:

A. **Olfaction:** Identified methyl salicylate independently from both nostrils (see Appendix B).

B. **Vision:** Full visual fields; no papilledema (see Chapter 11).

C. **Oculomotion:** Mrs. P. F. has a weak upward and medial movements of the left eye. On forward gaze the eye is deviated down and lateral. There is a mild ptosis of the left eyelid.

   Pupils: The left pupil was larger than the right. The left eye did not react to light, either direct or consensual. The right pupil responses were normal. Accommodation was not tested.

D. **Trigeminal:** Sensations to pinprick and cotton were perceived the same from all three divisions bilaterally. Corneal reflex was present bilaterally. Jaw movements were symmetrical and strength seemed normal.

E. **Facial:** Face was symmetrical upon grimace. Eyes appeared normally hydrated. Salivary production seemed normal.

---

[10]This case was brought to my attention by Dr. Robert King, Memorial Hospital, South Bend, IN.

F. **Audition:** Patient is able to hear finger rubbing in both ears (see Chapter 10).

G. **Vagal-glossopharyngeal:** Voice was normal. Gag reflex present bilaterally. The pharyngeal arches were high and symmetrical; the uvula was centered.

H. **Accessory:** Sternocleidomastoid and trapezius muscle strength was normal and symmetrical.

I. **Hypoglossal:** Tongue was centered on extension. There as no atrophy nor any fasciculations.

III. MOTOR SYSTEMS: Strength is normal.

IV. SENSORY SYSTEMS: Normal.

V. REFLEXES: MSRs are all normal. Toes are down going.

Comment: In the present case, Mrs. P. F. had a third nerve palsy. She could not fully move her left eye upward or medially. The left pupil was larger than the right and it did not react to light. A consensual light reflex was present in the right eye, proving that she was not blind in the left eye. Third nerve palsy is frequently an ominous sign. In one study of 206 cases, 25% were caused by neoplasms and 18% by aneurysms. Since the CT four days earlier had shown no abnormalities (i.e., no neoplasm), a supraclinoid aneurysm of the carotid artery and/or the posterior communicating artery was suspected. Such vascular lesions cannot be visualized by CT. Therefore, Mrs. P. F. was scheduled for an arteriogram the next day. It revealed a small aneurysm of the posterior communicating artery. This was operated the same day without incident. Mrs. P. F. is doing well.

FURTHER APPLICATIONS

• Would it have been appropriate to order an MRI that evening rather than wait until the next day to do the angiography (angiography is not routinely available on an emergency basis; MRI is)? Would the MRI have shown anything of value? If it had, could the angiogram have been avoided?

• Review the course of the carotid artery. If the patient had presented with a sixth nerve palsy, where would you expect to find the aneurysm?

• The family physician in this case did not pursue the patient's symptoms after the CT proved to be negative. Was this reasonable or do you think that this represents an error on the part of the physician? If you believe this was an error, how do you think the medical community should handle such errors? How do you think you would feel if you were that physician? Discuss these issues among your friends and professors.

SUMMARY

Most of the cranial nerves are multimodal and are associated with more than one nucleus in the brainstem. The hypoglossal (XII) and accessory (XI) nerves are exceptions to this generalization, being purely motor nerves having their motor neurons located in correspondingly named nuclei. The hypoglossal nerve innervates the tongue while the accessory nerve innervates the trapezius and sternocleidomastoid muscles.

The vagus (X) and glossopharyngeal (IX) nerves should be considered together because their central connections are essentially identical; they differ primarily in their peripheral distribution. They have three efferent and two afferent components. The nucleus ambiguus is the largest motor nucleus and it supplies the voluntary muscles of the larynx and pharynx. The dorsal motor nucleus of the vagus is the principal source of preganglionic parasympathetic fibers to the viscera (heart, lungs, and gut). The inferior salivatory nucleus supplies the preganglionic parasympathetic fibers to the otic ganglion which supplies the parotid gland. The nucleus of the solitary tract receives sensory information from the chemoreceptors and baroreceptors that detect blood oxygen tension and blood pressure. It also receives signals from the taste buds of the posterior one third of the tongue. The glossopharyngeal is the principal nerve of taste. The nucleus of the spinal tract of V receives pain and temperature signals from a small portion of the external auditory meatus, the oral cavity and the posterior one third of the tongue.

The vestibular portion of CN VIII conveys sensations of angular and linear acceleration of the head to the CNS. Angular acceleration is detected by the semicircular canals while the maculae of the saccule and utricle are linear acceleration detectors. Vestibular afferents synapse in the vestibular nuclear complex and the cerebellum.

The facial (VII) nerve is a mixed nerve with both motor and sensory components. The facial nucleus supplies the mimetic muscles of the face and the *orbicularis oculi*. Motor neurons that innervate the portion of the face above the eyes are bilaterally innervated by corticobulbar axons. The motor neurons that innervate the lower portion of the face are innervated by contralateral corticobulbar fibers. The superior salivatory nucleus supplies preganglionic parasympathetic fibers to the pterygopalatine and submandibular ganglia that innervate the lacrimal glands and the sublingual and submandibular salivary glands respectively. The nucleus of the solitary tract receives afferents from the taste receptors of the anterior two thirds of the tongue. The facial nerve also makes a minor contribution to the nucleus of the spinal tract of V.

The trigeminal (V) nerve is also mixed. The trigeminal motor nucleus supplies the muscles of mastication. The nucleus of the spinal tract of V receives primarily pain and temperature signals from the face while the principal sensory nucleus receives fine tactile and proprioceptive information from the same area. The mesencephalic nucleus of V is a displaced ganglion, consisting of pseudounipolar neurons of the I_A trigeminal axons.

The oculomotor (III), trochlear (IV) and abducens (VI) are the nerves of ocular motion. The abducens and trochlear nuclei supply motor axons to the lateral rectus and superior oblique muscles respectively. The oculomotor nucleus supplies motor neurons to the remaining extraocular muscles and the *levator palpebrae*. Preganglionic parasympathetic axons originating in the Edinger-Westphal nucleus innervate the ciliary ganglion that, in turn, innervates the constrictor muscle of the iris and the ciliary muscle of the eye. They effect constriction of the pupil and accommodation of the lens respectively.

Disruptions of eye movements many be manifested as paralysis of one or more of the extraocular muscles or by malfunction of one or more of the eye tracking reflexes. Muscle paralysis usually indicates nerve or nuclear damage that can be diagnosed by carefully testing the various eye muscles independently. Disruption of saccades or tracking eye movements usually points to a lesion in the cerebral cortex or cerebellum. The optokinetic reflex consists of both the saccadic and tracking eye movements. Spontaneous nystagmus implies damage to the labyrinth, the brainstem, or the cerebellum.

FOR FURTHER READING

Adams, R., and Victor, M. *Principles of Neurology*, 5th ed. New York: McGraw-Hill, 1993.

Brodal, A. *Neurological Anatomy in Relation to Clinical Medicine*, 3rd ed. New York: Oxford University Press, 1981.

Carpenter, M. B. *Core Text of Neuroanatomy*, 4th ed. Baltimore: Williams & Wilkins, 1991.

Friedmann, I., and Ballantyne, J. *Ultrastructural Atlas of the Inner Ear*. London: Butterworth & Co, 1984.

Haines, D. *Neuroanatomy: An Atlas of Structures, Sections, and Systems*, 3rd ed. Baltimore: Urban & Schwarzenberg, 1991.

Nieuwenhuys, R., Voogd, J., and van Huijzen, C. *The Human Central Nervous System; A Synopsis and Atlas*, 3rd ed. Berlin: Springer-Verlag, 1988.

Rowland, L. *Merritt's Textbook of Neurology*, 8th ed. Philadelphia: Lea & Febiger, 1989.

Wilson-Pauwels, L., Akesson, E., and Stewart, P. *Cranial Nerves: Anatomy and Clinical Comments*. Toronto: B. C. Decker, 1988.

10 The Auditory System

The hearing apparatus is an extremely elaborate transducer that converts acoustic signals into coded action potentials suitable for analysis by the nervous system. The auditory system is divided conventionally into four areas: the **external ear**, the **middle ear**, the **inner ear**, and the **central auditory pathways**. Each part plays a specific role in the transduction, neural coding, and analysis of acoustic information. This chapter will describe the anatomy of the hearing apparatus, sound and sound measuring conventions, and the physiology of the transduction process. Finally, the medical problem of hearing impairment will be discussed.

Anatomy of the Hearing Apparatus

THE EXTERNAL EAR

The external ear consists of the **pinna** [L. *pinna,* a feather] and the **external auditory meatus** [L. *meo,* a passage]. The pinna collects and funnels sound into the meatus. In some animals, such as the bat, this structure is very elaborate and the animal has considerable control over directing its opening. Orienting the pinna helps localize the origin of the sound.

The external auditory meatus is a small tube leading through the temporal bone (Fig. 10.1*A*). It is bounded internally by the **tympanic membrane**, which is protected by the deep recess of the meatus. The external auditory meatus is a warm and moist environment that is conducive

to bacterial growth. External ear infections are common, but easily treated. If left untreated they can penetrate the tympanic membrane and involve the middle ear, a serious complication. Middle ear infections as well as a number of other conditions are discussed more fully later in the chapter.

THE MIDDLE EAR

The middle ear is an enclosed cavity in the temporal bone. The major chamber, the **tympanic cavity**, is connected with the **antrum**, air spaces in the **mastoid bone**, by a small opening called the **aditus** (Fig. 10.1*A*). The tympanic cavity is bounded on its lateral side by the tympanic membrane. On the medial side, there are two openings into the middle ear, the **oval** and the **round windows**. A fourth opening in the tympanic cavity is the **auditory tube** (eustachian tube) that extends into the posterior recess of the pharynx.

Three small bones, the **ossicles**, connect the tympanic membrane with the oval window. The **malleus** [L. *malleus,* a hammer] is attached to the tympanic membrane. The footplate of the **stapes** [L., a step] is inserted into the oval window. The **incus** [L., anvil] joins the malleus with the stapes. The three bones are articulated by tiny synovial joints. The footplate of the stapes fills the oval window and is held in place by ligaments. Vibrations of the tympanic membrane are faithfully registered as vibrations of the footplate of the stapes; therefore, anything that interferes with ossicular motion impairs hearing. For example, serous secretions from the lining

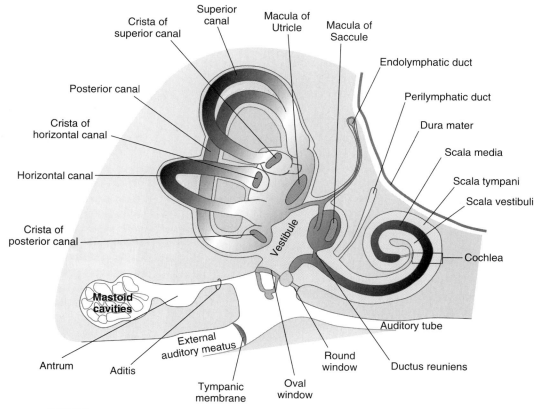

FIGURE 10.1.

A, This is a simplified illustration of the vestibular and cochlear structures within the temporal bone. The oval window opens directly into the vestibule from the middle ear cavity. The vestibule is filled with perilymph and is continuous with the labyrinth and the scala vestibuli of the cochlea. The scala vestibuli, in turn, is continuous with the scala tympani that ends blindly at the round window. The round window separates the scala tympani from the middle ear. Between the two scala lies the scala media, a space that communicates through the ductus reuniens with the other endolymphatic structures of the membranous labyrinth. **B**, An idealized drawing of a cross-section of the scala media, illustrating the principal anatomical structures.

of the tympanic cavity, puss from middle ear infections, or sclerosis of the ossicular joints (**otosclerosis**) all cause hearing impairment by restricting ossicular motion. Two striated muscles, the **stapedius** and the **tensor tympani**, are attached to the stapes and malleus respectively.

The auditory tube is a passageway through which the air pressure in the middle ear can be equalized with the external environment. This ensures that changes in air pressure between the middle ear and the environment do not become great enough to deform or rupture the tympanic membrane. Normally the auditory tube is closed, but opens easily with movements of the phar-

ynx. That is why yawning is helpful to airplane passengers during periods of rapid change in cabin air pressure. If the mucous membranes of the pharynx are irritated and swollen, it may be difficult or impossible to open the auditory tube, allowing a considerable pressure difference to develop across the tympanic membrane, a painful phenomenon well known to skin divers and air travelers.

The middle ear is another common site of infection since infectious agents can easily gain entry via the auditory tube. Once an infection is established, the auditory tube becomes inflamed and closes. Continuing bacterial growth pro-

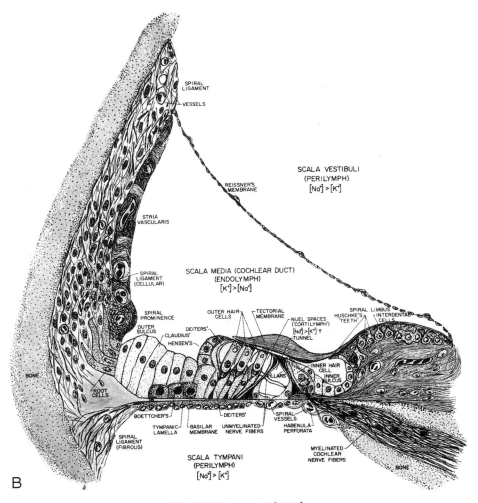

FIGURE 10.1.—continued

duces a large volume of puss that cannot escape the middle ear. If untreated, pressure increases; this not only causes great pain but may also result in the rupture of the tympanic membrane. Subsequent to infection, granulation tissue may envelop the ossicles, limiting their movement and impairing hearing. Occasionally, infectious agents gain entry into the inner ear through the round or oval window. Inner ear infections can produce permanent, ipsilateral hearing loss, ranging from a mild impairment to profound deafness. Finally, because of their direct connection with the tympanic cavity, the mastoid air spaces may become a repository for bacteria, resulting in chronic reinfection of the middle ear. In such cases, the **tegmentum tympani**, the thin bone that separates the mastoid air spaces from the cranium, can eventually become eroded, exposing the cranium to infection that may result in bacterial meningitis. Fortunately, modern antibiotic treatment of middle ear infections has nearly eliminated these disastrous sequelae to middle ear infections.

THE INNER EAR

The inner ear is the auditory portion of the membranous labyrinth (Fig. 10.1). The inner ear consists of the **cochlea** [L., snail shell], a spiral structure that is entirely enclosed by the petrous portion of the temporal bone. The cochlea is formed from three tubes or **scala** [L., stairway] that wind together in a circular pattern around a central core, the **modiolus** [L., the hub of a wheel]. The **scala vestibuli** begins in an opening from the **vestibule**, a large perilymphatic space (see Chapter 9) that contains the saccule. In the human, the scala vestibuli winds for 2¾ turns from the **base** to the **apex** of the cochlea. At the apex, it communicates with the **scala tympani** through a small hole, the **helicotrema** [G. *helix,* a spiral, + G. *trema,* a hole]. The scala tympani spirals from the apex to the base of the cochlea where it ends blindly at the membrane covering the round window. The scala tympani and scala vestibuli are filled with perilymph and are continuous with the perilymphatic space of the vestibular apparatus (see Chapter 9). Between them lies the **scala media**, an endolymphatic compartment that communicates with endolymphatic space of the vestibular apparatus through the **ductus reuniens**.

Scala Media

In cross section, the scala media is a triangular shaped structure, bounded at its base by the **basilar membrane (BM)**. Superior to the basilar membrane, **Reissner's membrane** separates the scala media from the scala vestibuli. The lateral wall consists of the **stria vascularis**. Lying on the basilar membrane is a complex structure, the **organ of Corti**, which consists of hair cells and a number of various types of supporting cells. In the human, the hair cells are arranged into three to five rows of **outer hair cells (OHC)** and one row of **inner hair cells (IHC)**. Two rows of pillar cells form the tunnel of Corti that separates the IHC from the OHC rows. Between the inner hair cells and the modiolus there is a recess, the inner sulcus. Extending from the upper lip of the inner sulcus, the **tectorial membrane (TM)** [L. *tego,* a covering] covers the surface of the organ of Corti. The hairs of the OHC are firmly anchored into the gelatinous material of the tectorial membrane. The hairs of the IHC are probably not firmly attached to it, but just brush its inner surface.

Cochlear Hair Cells

There are about 3500 inner hair cells and 20,000 outer hair cells in the human. They are similar to the type II vestibular hair cells (see Chapter 9) with a few specific differences. Most notably, the cochlear hair cells do not have a kinocilium; they retain only the basal body. Also, the stereocilia of the cochlear hair cells are considerably shorter than their vestibular counterparts. Like the vestibular hair cells, the cochlear hair cells are polarized such that a deflection of the stereocilia towards the basal body is hypopolarizing. All of the hair cells are polarized in the same direction so that movement of the hairs toward the stria vascularis is excitatory (Fig. 10.2).

Stria Vascularis

The lateral wall of the scala media, from Reissner's membrane to the spiral prominence, is lined with an epithelium, the **stria vascularis** [L., a stripe or furrow] (see Fig. 10.1). It is a highly vascularized, slightly pigmented tissue that is composed of three layers. The marginal layer (the layer in contact with the endolymph) is of ectodermal origin. The intermediate layer is a neural crest derivative and consists of melanocytes. The basal layer is derived from mesenchyme.

The principal function of the stria vascularis is to secrete the endolymph, a highly specialized fluid that fills the endolymphatic space of the entire vestibulocochlear system. As previously discussed, it is rich in K^+ (140 mEq) with a low Na^+ concentration (26 mEq) (see Chapter 9). The Nernst potential for potassium from the endolymph to the plasma is about +80 mV.[1] An electrical voltage that corresponds closely with this calculated figure, the **endolymphatic potential**, can be measured between the cochlear endolymph and the plasma. This correlation implies that the endolymphatic potential is dependent primarily on potassium ions leaking from the scala media to the plasma. This leakage apparently occurs only through the stria vascularis since the endolymphatic potential exists only in the scala media. It is nearly absent from the vestibular endolymphatic space,[2] a region that has no tissue

---

[1]The figure varies between 50 mV and 85 mV depending on the ionic concentrations used in the calculation. These figures vary somewhat between species and among the measurements of various research groups.
[2]A small 1- to 5-mV potential exists in the endolymphatic space of the vestibular apparatus.

FIGURE 10.2.

A scanning electron micrograph looking down on the reticular lamina of the organ of Corti. The tectorial membrane has been removed. Visible at the top of the figure are the hairs protruding from the inner hair cells (IHC) and below, three rows of outer hair cells (OHC). Between the IHC and OHC are the caps of the pillar cells that form the tunnel of Corti. Between the OHC lie the projections of the phalangeal cells (PC) that support and separate the OHC.

corresponding to the stria vascularis. Therefore, the stria vascularis not only elaborates the rich K⁺ environment of the scala media, but provides the conductive environment necessary for the production of the endolymphatic potential. The endolymphatic potential plays an essential role in hair cell transduction (see page 356).

The neural crest origin of the intermediate layer of the stria vascularis is of some practical significance. If the melanocytes of the intermediate layer fail to migrate into the stria vascularis, it fails to complete development during embryogenesis and soon degenerates. Consequently, neither the endolymph nor the endolymphatic potential is produced. For reasons that are not fully understood, the hair cells cannot sur-

vive in the absence of the endolymph and they soon degenerate following degeneration of the stria vascularis, producing total deafness in the affected ear.

Atrophy of the stria vascularis is frequently associated with disorders associated with aberrant melanocyte migration. For example, Dalmatian dogs, whose unique spotting pattern is due to melanocyte migration, are frequently deaf in one or both ears. Histological examination of the cochleas in these dogs has shown that they failed to produce a functional stria vascularis in the deaf ear. Melanocyte migration disorders also affect humans. People having **heterochromic irides** (eyes of different colors) or who have a white forelock are, like Dalmatian dogs, at risk for developing deafness early in childhood, a condition known as Waardenburg's syndrome (incidence 2/100,000) (Fig. 10.3). Deafness also frequently occurs in association with other pigmentation disorders such as **retinitis pigmentosa** (an inherited disease resulting in retinal degeneration), **hyperpigmentation** (leopard spot syndrome), **vitiligo** [L. *vitium,* blemish, vice] (white patches on the skin due to lack of pigment), **piebaldness** (patchy loss of pigmentation in the scalp and hair), and **albinism** [L. *albus,* white] (congenital lack of pigment in the skin and other normally pigmented parts of the body).

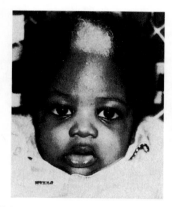

FIGURE 10.3.

Photograph of a patient displaying some of the physical characteristics associated with Waardenburg's syndrome. The approximate percentage of patients showing the various characteristics are indicated in parentheses: white forelock (20%); widely spaced eyes (90%); heterochromic irides (25%) (not obvious in this black and white photograph).

COCHLEAR INNERVATION

The auditory division of the eighth cranial nerve, like the vestibular division, has both afferent and efferent components. There are two morphologically different afferent neurons, identified simply as type I and type II. The cell bodies of both types of afferent neurons are located in the modiolus, the central core of the cochlear structure. The efferent somas are located in the superior olivary complex.

Afferent Neurons

There are about 30,000 afferent neurons in the human and about 95% of these are classified as type I. The type I neuron is a true bipolar cell, having a central and a peripheral axon. Both axons are myelinated over most of their length, as is the cell body. The distal axons leave the modiolus and enter the scala media through small perforations in the bone, the **habenula perforata** (Fig. 10.4), from which they precede directly to the inner hair cells. Each type I afferent fiber synapses with one IHC without dividing. Since there are many more type I axons than IHCs, each IHC receives approximately 10 afferent synapses. The afferent synapses on the IHCs are clustered at the base of the cell. The afferent neurotransmitter has not been identified, although there is evidence that it may be an amino acid.

The type II afferent neurons are unmyelinated. Their cell bodies reside in the modiolus and are apparently monopolar since a central axon has never been identified. The single distal axons accompany the type I axons into the scala media through the habenula. Once in the

FIGURE 10.4.

A, The innervation of the cochlea is schematically depicted as if one were looking down on the basilar membrane from above the hair cells. The *heavy lines* represent the type I afferent axons; the *thin lines,* the type II afferent axons. The efferent axons are shown as *dashed lines.* **B**, A more anatomically accurate oblique view of the cochlear innervation shows all fibers penetrating the bone of the modiolus at the habenula perforata (*HA*). Efferent fibers are shown crossing the tunnel of Corti between the rows of hair cells (*B*) and spiraling up the cochlea (*oS*). The type I (*I*) and type II (*II*) cell bodies are shown in the spiral ganglion. *iH,* inner hair cells; *oH,* outer hair cells. **C**, The efferent boutons (*e*) synapse directly on outer hair cells (*oH*) but make axo-axonic contact with afferent axons (*a*) that contact inner hair cells (*iH*).

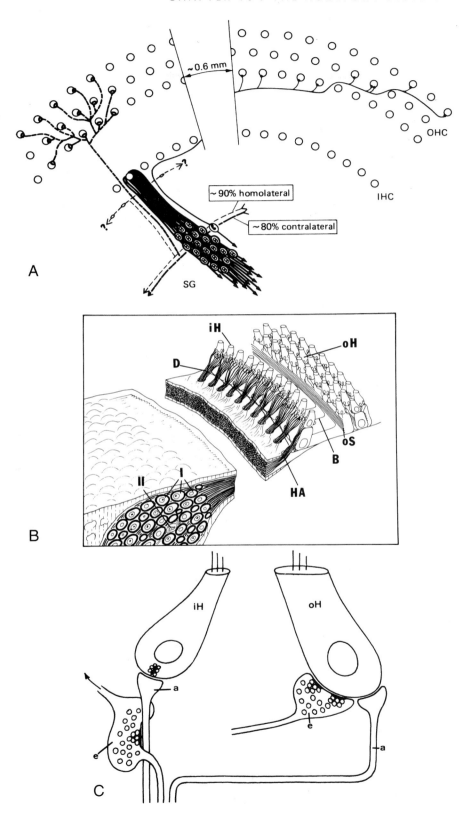

scala media the type II axons cross the tunnel of Corti to innervate only the OHCs. Among the OHCs, the type II axons divide, each axon sending collateral branches to innervate 10 outer hair cells (Fig. 10.4). Each OHC receives afferent synapses from about four type II axons. The afferent transmitter of the OHC is unknown.

Efferent Neurons

The efferent innervation of the cochlea originates in the superior olivary complex and arises from about 1500 neurons. The OHCs receive axosomatic synapses from efferent fibers, the majority of which originate in the contralateral superior olive. These synapses are large and highly vesiculated and are almost certainly cholinergic. In contrast to this, the IHCs do not receive efferent synapses. The type I primary afferent axons, however, do receive axoaxonic contacts from efferent fibers, just before the latter synapses with the IHCs. (Fig. 10.4)

The Nature of Sound

The auditory system is a highly specialized mechanical-to-electrical transducer. The process occurs in three steps. First, the middle ear acts as a *pressure to displacement* transducer, converting sound into a mechanical disturbance of the basilar membrane. Second, the cochlea performs a *mechanical frequency analysis* of the signal so that different hair cells will be maximally stimulated by different frequencies. Finally, the hair cells convert basilar membrane *displacement into action potentials*. In order to understand these processes, we must briefly examine the nature of sound and the conventions used to describe its physical characteristics.

SOUND IS A PRESSURE WAVE

Sound is the perception of pressure waves produced by mechanical disturbance in a conductive media which may be a gas, liquid, or solid. The energy source acts on the media by exerting a force such that some molecules in the media are compressed. When the force is re-

moved, the molecules in the media recoil. If the energy source is vibrating, a succession of compressions and rarifications occur among the molecules of the conductive media. The energy induced into the media by the source is propagated as a series of compressions and rarification wave fronts. A set of compressed molecules acts as an energy source to compress adjacent molecules within the media. In this way the wave of compression expands from the source like ripples spreading across a pond.

The physical properties of sound are characterized by **intensity** and **frequency**. Sound intensity is proportional to the *velocity* and the *amplitude* with which the stimulus compresses the transmission media, parameters that determine the amount of energy (power) that is transferred to the media. The intensity of sound (i.e., its power) is measured in watts. The frequency of sound refers to the number of compression/rarification events that occur per second. The unit of frequency, measured in cycles per second, is the hertz (Hz).

INTENSITY

It is inconvenient to express sound intensity in watts, partly because power levels of ordinary sound are very small (femtowatts per centimeter squared), and partly because the range of human sound perception spans 12 orders of magnitude. Therefore, a logarithmic notation is employed. The **bell (B)** is a unit of measure used to express the $\log_{10}$ of a ratio. For sound,

$$I_B = \log \frac{P}{P_0} \qquad \textbf{[Equ 10.1]}$$

where **I** is the sound intensity expressed in bells, **P** is the measured power in watts of the sound source, and **P$_0$** is the power in watts of an arbitrary reference standard. The bell, however, is an inconveniently large unit, so the **decibel (dB)** is used instead as:

$$I_{dB} = 10 \log \frac{P}{P_0} \qquad \textbf{[Equ 10.2]}$$

In practice, it is more convenient to measure sound in terms of **pressure** rather than power. The same relative scale can be used with the substitution of a pressure reference for the power reference. However, sound intensity is still *ex-*

pressed as a power ratio in spite of the fact that it is *measured* as pressure. This is relatively easy to do since power is proportional to the square of the pressure:

$$P = k \, p^2 \qquad \textbf{[Equ 10.3]}$$

where the intensity is measured as power (**P**) or pressure (**p**), and **k** is simply a proportionality constant. Hence:

$$I_{dB} = 10 \log \frac{P}{P_0} = 10 \log \frac{p^2}{p_0^2} \qquad \textbf{[Equ 10.4]}$$

and

$$I_{dB} = 20 \log \frac{p}{p_0} \qquad \textbf{[Equ 10.5]}$$

(see Table 10.1).

Since the decibel is a *relative unit of measure,* it is necessary to explicitly state the reference when expressing sound intensity. In acoustics, by agreement, the reference is 0.0002 dyne/cm², a pressure measurement that is usually abbrevi-

ated **SPL**.[3] This rather odd number is used because it approximates the threshold of human hearing at its most sensitive frequency (to be discussed below). Note that a change of 20 dB represents two orders of magnitude in power, but only one order of magnitude change in sound pressure.

FREQUENCY

The pressure/rarification waves of an acoustic signal repeat in time. The number of variations of the pattern per unit time is perceived as **pitch** and described as the **frequency** measured in hertz. If the oscillation pattern is perfectly regular over time it can be accurately described by the sine wave function (Fig. 10.5A). An acoustic signal that corresponds with a perfect sine wave is perceived as a pure, single tone. Pure tones are hardly ever produced by natural phenomenon, so most of

---

[3] 0.0002 dyne/cm² = 20 μPa and is equivalent to 10^{16} W/cm². The abbreviation SPL stands for "sound pressure level."

Table 10.1. Relationships between Decibels, Power, Power Ratios, and Pressure Ratios

dB	Power (W/cm²)	P/P_0	p/p_0
140	10^{-2}	10^{14}	10^{7}
120	10^{-4}	10^{12}	10^{6}
100	10^{-6}	10^{10}	10^{5}
80	10^{-8}	10^{8}	10^{4}
60	10^{-10}	10^{6}	10^{3}
40	10^{-12}	10^{4}	10^{2}
20	10^{-14}	10^{2}	10^{1}
0 (reference)	10^{-16}	1	1

Note: 1. A tenfold increase in the pressure ratio is registered as +20 dB because sound **intensity** is expressed as **power** and power is proportional to the square of pressure.
2. The range of sound intensity represented here approximates the range of human hearing intensity perception, although sounds at the highest levels indicated here would cause damage to the cochlea.

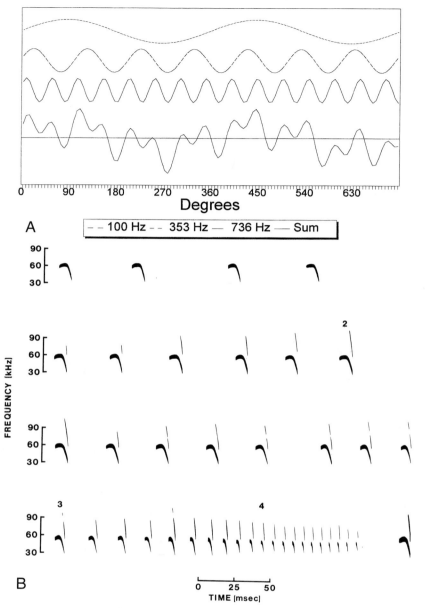

FIGURE 10.5.

A, Arbitrarily complex waveforms can be formed by the simple combination of pure sine waves of various frequencies. Conversely, the pure sine wave components can be extracted from an arbitrarily complex waveform by Fourier analysis. **B,** Vocalizations are complex auditory signals that cannot be described simply in terms of spectral and intensity components. The timing and amplitude of the various frequency components (the envelope) are critical to its overall perception. These sound spectrographs attempt to represent vocalizations in three dimensions: frequency versus time, with intensity crudely shown by the blackness of the trace. This figure shows a continuous sequence of vocalizations produced by a fish-catching echolocating bat during prey detection and capture. The search phase, shown on the top line, consists of a frequency-modulated chirp. Once the prey has been detected (second and third lines except the last three pulses), the character of the vocalization changes; a sharp high-frequency terminal component is added. The terminal attack phase is shown in the last three pulses in the third line and the last line except for the last pulse. Note the continuously changing envelope of the vocalization as the bat makes its final attack on the prey.

the sounds we hear are complex. *Arbitrarily complex sounds can, however, be synthesized from a combination of pure tones*. The inverse is also true: a complex waveform can be decomposed into a series of pure tone components. Mathematically, the process of decomposing a complex waveform into its fundamental pure tones is called Fourier analysis, named after the French mathematician who developed the method.

The sensitivity of the hearing apparatus to sounds of different frequencies varies among animal species. The human can hear sounds from about 20 Hz to about 20 kHz. This range is fairly typical for terrestrial mammals, although cats and dogs can hear sounds of somewhat higher pitch (up to about 50 kHz). A few animal species can hear very high frequencies, although at the expense of a reduced range at the lower end of the spectrum (Table 10.2).

DYNAMICS

A third parameter of sound analysis is not a physical measurement like intensity and fre-

Table 10.2. Range of Hearing for Common Animals

Species	Low (Hz)	High (kHz)
Human	20	20
Chimpanzee	100	20
Rhesus monkey	75	25
Cat	30	50
Dog	50	46
Chinchilla	75	20
Rat	1000	60
Mouse	1000	100
Guinea pig	150	50
Rabbit	300	45
Bat	3000	120
Dolphin	1000	130
Sparrow	250	12
Goldfish	100	2

Data from Fay, R. *Hearing in Vertebrates. A Psychophysics Databook*. Winnetka, Ill.: Hill-Fay Associates, 1988.

quency, but for human hearing (and probably all animals), it is critical for sound perception. Natural sounds do not instantaneously start and stop, nor do they maintain the same spectral (frequency) components and amplitude during the entire sound presentation. The timing, intensity, and spectral variability of animal vocalizations is called the **envelop** of the sound stimulus (Fig. 10.5*B*). Anyone who is familiar with bird calls, for example, must realize that the dynamic variations of the sound envelop are essential to one's overall perception and discrimination of the sound.

Sound Transduction

The inner ear is a lateral line derivative (see Chapter 9), and therefore is sensitive to displacement waves in a fluid that bend the stereocilia of the hair cells. Sound, however, is a *pressure* wave incapable of directly bending cilia. Therefore, the first step in sound transduction is the conversion of a pressure wave into a displacement wave, a process known as *far field to near field transduction*. This step is accomplished by the middle ear.

MIDDLE EAR TRANSDUCTION

The inner ear is a fluid filled cavity quite unlike the gaseous environment that conveys sound to us. The interface between a gas and a liquid represents a barrier to the transfer of acoustic energy from one environment to the other. A sound pressure wave in air is simply reflected off the liquid surface, because liquids are so incompressible compared with gases. Skin divers are quite aware of this phenomenon, for they can clearly hear sounds that are initiated within the water, such as the noise of an approaching propeller, but they are unable to hear the shouts of shipmates calling to them from the safety of the boat. Airborne sound simply bounces off water, transferring only about 2% of its energy in the process.

The middle ear is a transducer that efficiently transfers the airborne acoustic energy to the liquid environment of the inner ear. To understand how this is accomplished, consider a balloon you have filled with air while living in Denver, the mile high city. Since this is a thought experiment, we can consider this a perfect balloon, one that is simply a container and does not exert any forces on the gas inside it. In this situation the pressures inside and outside the balloon are equal. The skin of the balloon only acts as a separator. Take the balloon to sea level and its diameter will get smaller. The reason is simple enough. The increased atmospheric pressure at sea level will be exerted on the gas inside the balloon, compressing it. The size of the balloon reflects the amount of compression. Take the balloon to the top of a mountain and it will expand. In other words, changing the pressure outside the balloon will cause a *physical displacement* of its wall. The magnitude of the displacement is proportional to the magnitude of the pressure change.

The middle ear acts very much like this balloon. Most of its walls are rigid bone, hence inflexible, but the tympanic membrane is not. Changes in pressure outside the middle ear are reflected by a displacement of the tympanic membrane, just as pressure changes were reflected in changes in the size of our perfect balloon. In other words, *a change in pressure is transformed by the middle ear into a displacement*.

Tympanic membrane displacements are transferred to the oval window by the action of the ossicles. The efficiency of the transfer is affected by two parameters, the lever action of the ossicles and the size ratio of the tympanic membrane to the footplate of the stapes. The ossicles act as a series of levers with a mechanical advantage that magnifies the movement of the footplate of the stapes about 1.3 times greater than the movement of the arm of the malleus, a small but useful increase. The ratio of tympanic membrane area to the area of the footplate is about 17. Combined, these factors increase the force exerted on the footplate of the stapes by about 22 times over the force exerted on the tympanic membrane. Even so, the transfer of energy is not 100% efficient and when all factors are taken into account, approximately 67% of the acoustic energy falling on the tympanic membrane is transferred to the inner ear. This is, however, a great improvement over the 2% transfer one would expect in the absence of the middle ear.

THE MIDDLE EAR MUSCLES

The **tensor tympani** and the **stapedius** muscles contract in response to loud sounds and to many non-acoustic stimuli such as swallowing, closing the eyes, adjusting the pinna, and immediately before vocalization. Contraction of the middle ear muscles decreases sound transmission through the middle ear by about 10 dB, a modest effect. Most of the attenuation is caused by the action of the stapedius muscle; the tensor tympani makes almost no contribution.

The latency of the acoustic reflex contraction of the muscles varies between 150 msec and 25 msec, depending on stimulus intensity. It should be clear that because of this long latency, the acoustic reflex contraction of the stapedius cannot protect the inner ear from damage due to sudden loud noises. Nor can contraction of these muscles be of much protective benefit during prolonged loud sounds since the attenuation factor is only about 10 dB. Do these muscles serve any useful function? The answer may lie in the fact that the attenuation effect of middle ear muscle contraction is not uniform over the entire audio spectrum; the attenuation is most pronounced at frequencies below 1 kHz. Lower frequencies interfere with the perception of higher frequencies, a phenomenon called **masking**. The interference especially affects the frequencies in the 1 kHz to 5 kHz range, making them harder to hear. Sounds in the range of 1 kHz to 5 kHz are precisely the frequencies that carry the noises of rustling leaves, snapping twigs, and other sounds of high survival value. Contraction of the middle ear muscles, in effect, alters the spectral sensitivity of the auditory system, making it more sensitive to these threatening environmental noises. Coincidentally, these are the same frequencies that carry most of the auditory information in human speech. In humans, loss of the acoustic reflex, as for example in Bell's palsy, and the subsequent masking of critical frequencies by lower frequency sounds, causes speech to be less understandable.

COCHLEAR MECHANICS

The scala tympani and the scala vestibuli are a single continuous tube that is folded back on itself. Between them is the blind scala media.

All three tubes are normally coiled around each other like a snail shell, hence the term cochlea. To better visualize how this system works, it is useful to imagine this structure uncoiled with the scala media reduced to a thin membrane, the basilar membrane (BM), separating the other two scala (Fig. 10.6*A*).

In this uncoiled view, one can imagine the stapes vibrating in the oval window. As the footplate of the stapes moves into the scala vestibuli, the perilymph is *displaced* towards the helicotrema. A displacement occurs because the perilymph, being a liquid, is nearly incompressible. The pressure exerted on it must be relieved, in this case by an outward deformation of the round window. If the basilar membrane were a rigid structure, the perilymph displacement would have to pass the entire length of the scala vestibuli and return through the scala tympani in order to reach the round window. But the basilar membrane is not rigid; it is flexible. This means that the displacement of the perilymph caused by inward stapedial movement will result in a deformation of the basilar membrane towards the scala tympani. In effect, the entire length of the cochlea is short circuited and *the BM is deformed* in the process. When the stapes moves outward, the entire process is reversed.

Once the basilar membrane has been displaced at its base, the deformation moves towards the apex of the cochlea as a **traveling wave** (Fig. 10.6*B*). As the wave moves, its velocity slows, and as a result, the wave amplitude increases towards the apex, reaching a maximum at some point and then rapidly decreasing. The location of the peak *is frequency dependent:* low frequencies peaking near the apex of the cochlea while high frequencies peak near the base. The traveling wave not only displaces the basilar membrane, it also displaces the tectorial membrane. The differential movement of these two structures creates a shear across the stereocilia, bending them (Fig. 10.6*C*). This is the essential mechanical event that initiates hair cell hypopolarization.

COCHLEAR HAIR CELL TRANSDUCTION

The ionic mechanisms of hair cell transduction are considerably different from ordinary neurons. To understand cochlear hair cell transduction, one must first recall that the stereocilia

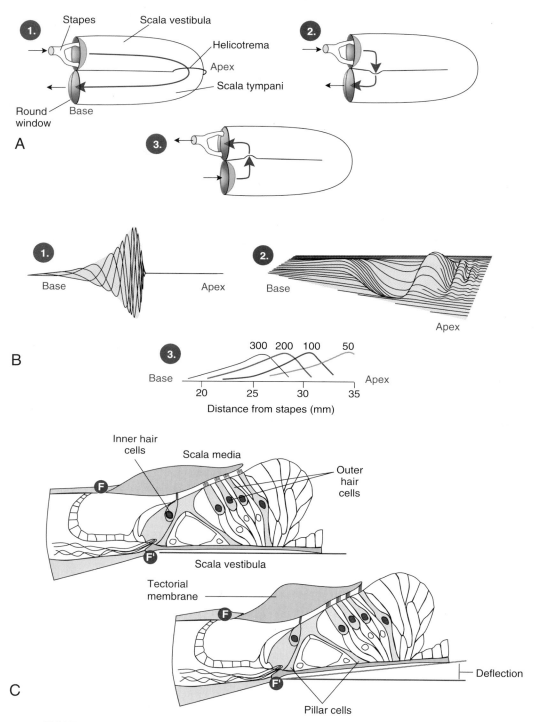

FIGURE 10.6.

    A, The cochlea is a coiled set of tubes that can be represented by a single tube folded back on itself. The scala media is depicted here as a simple boundary. *1,* If the basilar membrane were rigid, pressure caused by inward displacement of the stapes would be transferred the entire length of the cochlea to be relieved by outward displacement of

are in endolymph, an environment rich in K^+, while the hair cell body is surrounded by perilymph, a Na^+ rich fluid. If one assumes that the ionic concentration within the stereocilia is the same as in the soma, then the Nernst potential for K^+ from the endolymph to the intracellular space is -1.8 mV. This would create a strong inward gradient for K^+ entry into the hair cell (assuming a V_m of -60 mV) from the scala media through the stereocilia if potassium selective ionophores were present in the cilia.

Based on these considerations, the following hypothesis of hair cell excitation has been proposed (Fig. 10.7). First, mechanical deformation of the stereocilia opens potassium selective pores in the stereocilia, *allowing K+* to rapidly enter the cell from the scala media *through the stereocilia*. This initial entry of potassium hypopolarizes the entire hair cell, which causes the opening of voltage sensitive calcium gates in its *soma*. The rapid entry of Ca^{2+} ($E_{Ca^{2+}} = +275$ mV between perilymph and the cell) further hypopolarizes the cell and increases the intracellular $[Ca^{2+}]$. This entry of calcium opens calcium dependent potassium gates in the *soma*, a process that allows $K+$ to leave the cell soma, repolarizing it. The repolarization of the cell closes the calcium gates. The transient increase in intracellular calcium concentration is also the critical factor initiating the release of the neurotransmitter at the afferent synapse.

NEURAL CODING

Auditory information must be encoded as action potentials in a manner that preserves the *timing, frequency,* and the *intensity* of the stimulus. This can be accomplished in a number of ways. First, intensity may be conveniently coded as a function of impulse frequency. Second, intensity can also be encoded by the number of axons carrying the signal. Third, frequency can be coded by selectively exciting different axons each of which represents different frequencies, a strategy that seems to be an obvious outcome of the mechanical frequency tuning of the cochlea. And fourth, frequency coding can also be achieved by having the auditory neurons fire synchronously with each cycle of the stimulus. To varying degrees, all four strategies are used by the nervous system to represent auditory information.

If one simply records the intensity and frequency of sound stimuli that increase the firing of a primary afferent auditory neuron above its spontaneous level, a simple plot of frequency versus intensity is obtained (Fig. 10.8A). These plots illustrate two important characteristics of primary auditory nerve excitation. First, *any given axon*

---

the round window membrane. *2,* The basilar membrane is not rigid. Inward displacement of the stapes exerts a downward force on the basilar membrane that deforms it, transferring the pressure directly to the scala tympani and the round window membrane. This in effect short-circuits the system. Physiologically, most pressure changes within the cochlea are limited to the basal turn. *3,* This represents the reciprocal motion of the basilar membrane during outward displacement of the stapes. **B,** The displacement of the basilar membrane to a continuous pure tone stimulus is represented here. *1,* Each line represents the membrane position at periods of one-third of the stimulus period. The amplitude is greatly exaggerated with respect to natural movements of the basilar membrane. *2,* This is a three-dimensional representation of basilar membrane position at one instant during continuous pure tone stimulation. The amplitude of displacement is, as in the above illustration, greatly exaggerated. *3,* The envelope of basilar membrane displacement is depicted for stimuli of various continuous pure tone frequencies. **C,** Shearing of the hairs on the hair cells occurs during displacement of the basilar membrane and tectorial membrane. Each structure rotates around a different fulcrum (F and F'), a physical arrangement that creates a differential linear displacement between the base of the tectorial membrane and the reticular lamina of the organ of Corti. The hairs of the inner hair cells register this displacement because one end of the hair shaft is attached to the tectorial membrane while the other is embedded in the cuticular plate of the hair cell. The stereocilia of the outer hair cells are not embedded in the tectorial membrane, but apparently brush across its under surface during movement.

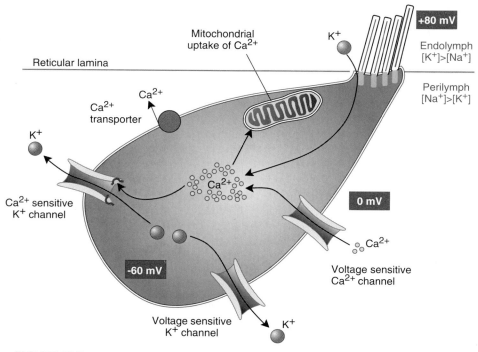

FIGURE 10.7.

This drawing illustrates the various electrical potentials in the cochlea that are important for transduction. Bending the stereocilia opens mechanically sensitive potassium pores. This initiates an **inward** flux of K⁺ due to the electrochemical gradient between the endolymph and the intracellular environment of the hair cell (see text). The increasing concentration of intracellular potassium hypopolarizes the cell, opening voltage sensitive calcium gates. This allows calcium to enter, further hypopolarizing the hair cell and triggering calcium sensitive potassium gates. The latter are located in the soma. When they open, K⁺ is able to leave the hair cell due to the electrochemical gradient between the hair cell and the perilymph. This restores the resting polarization to the hair cell.

can respond to a very wide range of frequencies, if the stimulus is loud enough. Second, *each axon is most sensitive to only one frequency*, its **characteristic frequency** (f_c). Plotting only the threshold of numerous primary afferent axons at their critical frequencies reveals the distribution of the spectral sensitivity of the animal's hearing (Fig. 10.8*B*). These charts reveal the animal's **hearing threshold curve** which agrees very well with the behaviorally determined auditory threshold of the animal. Such charts demonstrate that for any given frequency *at threshold, individual axons code for frequency*. This type of information coding is usually called **place coding**.

Another way of visualizing axonal responses to a stimulus is to plot a histogram of the number of action potentials versus time between spikes, an **interspike interval histogram**. Such a plot (Fig. 10.9*A*) reveals that the action potentials fire only during the rarification phase of the stimulus. Furthermore, each axon does not fire on every cycle, but, collectively, at least some axons will fire for each cycle of the stimulus. Integrated over a longer period of time, interspike interval histograms reveal that, *when stimulated well above threshold, primary auditory axons fire at some multiple of the stimulus frequency*. This type of information coding is usually termed **volley coding** (Fig. 10.9*B*).

A careful examination of the interspike interval histograms reveals that volley coding breaks down at frequencies much above 2 kHz. Therefore, for the higher frequencies, place coding predominates over volley coding whereas at the

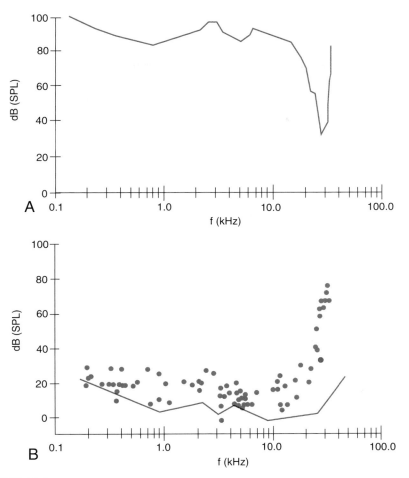

FIGURE 10.8.

A, This is a tuning curve for a primary afferent auditory neuron of the cat. The *line* represents the threshold of a single axon to auditory stimuli of various frequencies. Note that it is excited by a very wide range of frequencies (100 Hz through 35 kHz), although it is most sensitive to only one frequency, the characteristic frequency (here 28 kHz). Most auditory neurons have response curves with a shape similar to this. **B,** The threshold intensity at the characteristic frequency of a large number of neurons from the cat is plotted here as dots. The neuron depicted in **A** is indicated in red. The hearing threshold curve for the cat, as determined by behavioral responses, is indicated by the red line. Note that hearing threshold closely follows the sensitivity of the most sensitive neurons at their characteristic frequency.

lower frequencies, volley coding dominates. This dual representation correlates well with the mechanical selectivity of the cochlea. Low frequencies produce a traveling wave peak that is close to the apex of the cochlea, disturbing nearly the entire basilar membrane and firing almost all of the hair cells simultaneously. This makes place coding ineffective. Higher frequency stimuli cause the traveling wave to peak closer to the base, stimulating fewer hair cells in the process, making place coding more selective.

Neural Pathways

The central connections of the auditory system are among the most complex of all the sen-

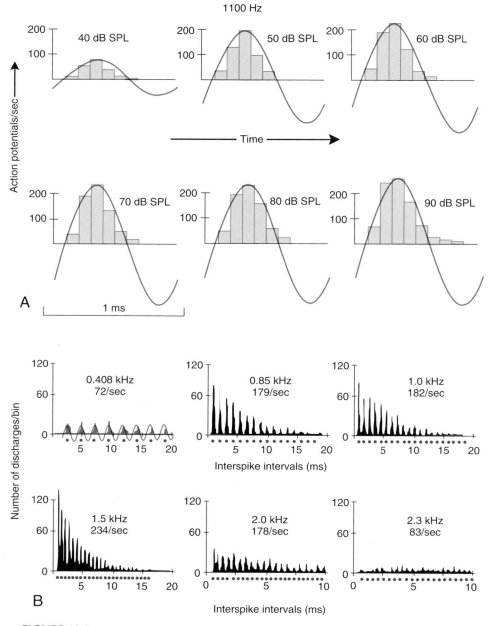

FIGURE 10.9.

   A, This is a response histogram of a single primary afferent axon driven at its characteristic frequency. The stimulus is superimposed over the histogram. Each bar in the graph represents the number of spikes collected during a time interval represented by the width of the bar. Note that spikes occur only during one phase of the sinusoid, even when driven at very high intensities. The excitatory phase corresponds to an outward movement of the stapes. Note also that the total number of spikes collected does not increase for intensities over about 60 dB. This means that intensity information at the higher intensities cannot be encoded by the number of action potentials carried by this axon. **B,** Interspike interval histograms taken from a single primary auditory afferent axon with a characteristic frequency of 1.6 kHz. It is driven at the various frequencies

sory systems. The ascending pathways are organized as two principal systems. The **core pathways** are carefully organized and mapped according to the characteristic frequency of the neurons, creating **tonotopic maps**. Tonotopic maps are maintained at all levels, and these core pathways are the fastest and most direct in the auditory system. The **belt pathways** are less carefully organized along tonotopic lines. Neurons in this system are not only frequency selective, but are also very sensitive to timing and intensity patterns in the envelop of the signal. Some parts of the belt pathways appear to be organized to preserve binaural interactions while other areas have cells that respond only to signals with complex envelops. Thus *the three critical parameters of auditory signal processing, frequency, intensity, and envelop, appear to have and to maintain separate information channels at all levels of the auditory system*. At many levels of the nervous system, and especially at the level of the cerebral cortex, the belt pathways physically surround the core pathway. The following account is greatly simplified and emphasizes the core pathways within the various nuclei of the auditory system (Fig. 10.10).

As discussed above, physiological studies have shown that each primary afferent fiber responds to a wide range of auditory frequencies, but has the lowest threshold for one specific frequency. Neurons in all of the auditory nuclei of the brainstem and cerebral cortex also have similar tuning curves, each responding to a wide range of frequencies, yet each with a single characteristic frequency. In general, and with many exceptions, the tuning curves of these neurons get narrower as one ascends the auditory system, making neurons at each successive level more frequency selective.

PRIMARY AFFERENTS

The central axon of the type I afferent neurons terminate in the cochlear nuclear complex that consists of three nuclei, the **dorsal cochlear nucleus (DCN)** the **anteroventral cochlear nucleus (AVCN)** and the **posteroventral cochlear nucleus (PVCN)**. After entering the brainstem, the primary afferent fiber bifurcates, the anterior branch terminating in the AVCN while the posterior branch enters the PVCN. Before terminating, the posterior branch divides, sending one collateral to synapse in the PVCN while the other enters the DCN where it terminates (Fig. 10.11).

THE COCHLEAR NUCLEI

The three cochlear nuclei give rise to three separate ascending pathways in the medulla, thereby converting the single information channel of the primary afferent fibers into a multichannel system. Undoubtedly each channel is specialized to convey specific attributes of the auditory signal and it is tempting to believe that each channel conveys one of the primary signal parameters. To some degree this is true, but unfortunately our knowledge of these attributes is too incomplete to fully justify that conclusion.

Axons from the DCN take the most direct route to higher auditory centers, crossing the midline as the **dorsal acoustic stria** without synapsing. The ascending portion of these axons form a tract known as the **lateral lemniscus**, the principal ascending auditory tract of the brainstem (see Fig. 10.10). Fibers in the lateral lemniscus ascend to the **inferior colliculus** where they terminate in an organized tonotopic pattern.

Axons originating primarily in the PVCN form the **intermediate acoustic stria**. Collaterals from these axons synapse in the ipsilateral and con-

indicated, all at 80 dB SPL. As in the previous illustration, the amplitude represents the number of spikes recorded at the time represented by the width of the bar. Note that for each frequency, the axon fires at an interval that is very close to the reciprocal of the stimulus frequency, i.e., the period of the stimulus (marked by the dots below the abscissae). Even though the axon will phase lock to the stimulus frequency, the number of spikes recorded is greatest at the characteristic frequency. Axons such as these show characteristics of both volley and place coding (see text).

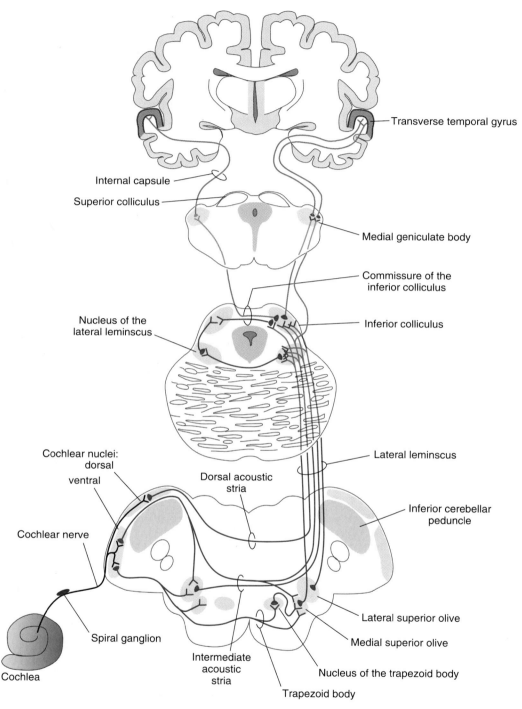

Transverse temporal gyrus

Internal capsule

Superior colliculus

Medial geniculate body

Commissure of the
inferior colliculus

Nucleus of the
lateral leminscus

Inferior colliculus

Cochlear nuclei:
dorsal

ventral

Dorsal acoustic
stria

Lateral leminscus

Inferior cerebellar
peduncle

Cochlear nerve

Spiral ganglion

Cochlea

Intermediate
acoustic
stria

Lateral superior olive

Medial superior olive

Nucleus of the trapezoid body

Trapezoid body

FIGURE 10.10.

The principal ascending auditory pathways.

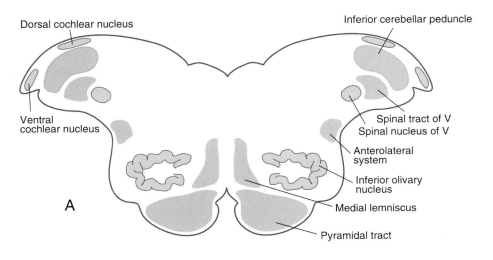

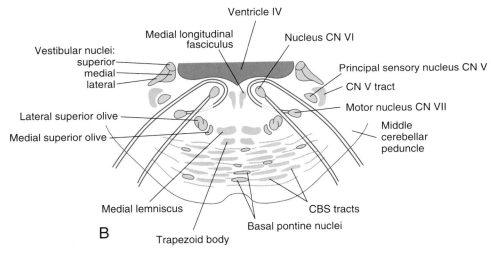

FIGURE 10.11.

A, Cross-section of the brainstem at the medulla showing the cochlear nuclei. **B**, Cross-section of the brainstem at the level of the superior olivary complex. Note the relationship between the fibers of the trapezoid body and the medial lemniscus.

tralateral **superior olivary complex** while the main branch joins the lateral lemniscus to ascend and terminate in the inferior colliculus.

Axons originating primarily in the AVCN form the most ventral of the acoustic striae, a tract known as the **trapezoid body**. It is located immediately dorsal to the basis of the pons. These axons develop several collateral branches that synapse bilaterally in the superior olivary complex and the contralateral **nucleus of the trapezoid body**. Axons of the trapezoid body do not join the lateral lemniscus. Third-order fibers originating in the superior olive ascend in

the lateral lemniscus to terminate in the inferior colliculus.

THE SUPERIOR OLIVE

The superior olivary complex is a collection of nuclei in the pons located at the level of the facial nucleus (Fig. 10.11*B*). The largest structures in the complex are the **lateral superior olive (LSO)**, the **medial superior olive (MSO)**. Surrounding this complex are a number of small nuclei simply called the **peri-olivary nuclei**. Medial to the superior olive is the nucleus of the trapezoid body (NTB). The superior olive and trapezoid nuclear complex are very prominent in carnivores, especially the cat, but the size of the superior olive is much reduced in the human.

The cells in the LSO and MSO have two large dendrites extending from opposite sides of the soma. One dendrite receives projections from the contralateral cochlear nuclei, the other from the ipsilateral cochlear nuclei. This arrangement allows the olivary neurons to compare auditory information from the two ears. This system is used to determine the **spatial localization** of sound based on the inter-aural differences in sound *intensity* (LSO) and *time* (MSO) between the two ears.[4] The sensitivity of MSO neurons to intra-aural arrival time differences of as little as 400μsec is remarkable. This represents a difference in path length between the two ears of approximately 5 in. The barn owl can locate and capture mice in total darkness, having as its only sensory clue the rustling sound generated by the mouse during foraging.

The superior olivary complex is also the origin of the efferent axons of the auditory nerve. This tract, the **olivo-cochlear bundle (OCB)**, contains two classes of efferent axons, those that terminate as large vesiculated boutons on the outer hair cells, the **medial efferent system**, and those that make axoaxonic synaptic contact with the type I afferent axons, the **lateral efferent system** (see Fig. 10.4).

Most of the 800 axons of the medial efferent system arise from the contralateral peri-olivary nuclei. These axons collateralize extensively in the cochlea and ultimately innervate the entire population of about 20,000 outer hair cells. In contrast to the medial efferent system, most of the neurons of the lateral efferent system originate on the ipsilateral peri-olivary nuclei. They too extensively collateralize, and form synapses on most of the approximately 35,000 afferent axons. They do not terminate on any of the hair cells.

The effect of stimulation of the lateral effer-

---

[4]The sound from a source that is not equidistant from the two ears does not stimulate the ears equally. The ear closest to the source will receive a more intense sound stimulus with a shorter latency than the opposite ear. The brain uses both of these clues, intensity and intra-aural time difference, to localize the sound source in space. Many animals have a mobile pinna that can be oriented toward the source, maximizing the intensity differences and therefore increasing the precision of localization.

ent system is remarkable; it causes the outer hair cells to change their length! Apparently the change in OHC length lifts the tectorial membrane slightly off from the IHC stereocilia, thereby reducing the amount they are bent by basilar membrane movement. Stimulation of the crossed OCB can reduce auditory sensitivity by as much as 26 dB. The effect of lateral efferent activity is not known, although it too is presumed to be inhibitory.

The physiological role that the efferents play in hearing is not known. Originally, it was hypothesized that the cochlear efferents sharpened the tuning curves of the primary afferent fibers, making them more frequency selective. This appears not to be the case. In trained monkeys, the only effect that has been demonstrated on auditory perception is a decreased *ability to make auditory discriminations in the presence of noise* after the OCB has been lesioned. Extracting an interesting signal from environmental noise is an important attribute of our auditory system. It is exemplified by our ability to listen to a single individual in a room full of talking people, the "cocktail party effect." In noisy environments humans with certain types of hearing losses have difficulty discriminating speech in noisy environments. There is evidence that this deficit is related to loss of cochlear efferent axons.

LATERAL LEMNISCUS

The lateral lemniscus is the principal fiber tract entering the inferior colliculus (Fig. 10.12). It carries fibers from the dorsal and intermediate acoustic striae and from the superior olivary complex. Many of these fibers send collaterals into a small nucleus, the **nucleus of the lateral lemniscus**. The commissure of the lateral lemniscus carries fibers from one nucleus to its contralateral partner. Other projections are to the inferior colliculus. Almost nothing is known of its function.

INFERIOR COLLICULUS

The **inferior colliculus (IC)** is a major auditory center (Fig. 10.12). Afferent fibers enter it from three principal areas, (*a*) the lateral lemniscus, (*b*) the contralateral inferior colliculus, and (*c*) descending connections from the ipsilat-

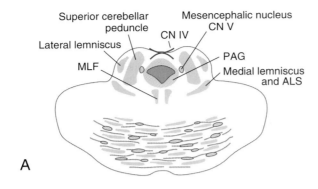

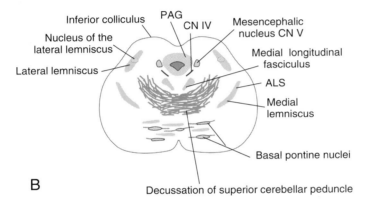

FIGURE 10.12.

A, Cross-section of the pons at the level of the trochlear decussation. Note the relationships between the lateral lemniscus, the medial lemniscus, and the superior cerebellar peduncle. **B**, Cross-section of the pons at the level of the decussation of the superior cerebellar peduncle. Again note the prominent lateral lemniscus and its nucleus.

eral auditory areas of the cerebral cortex. Efferents leave the IC to innervate (*a*) the ipsilateral medial geniculate nucleus, (*b*) the contralateral inferior colliculus, (*c*) the deep layers of the superior colliculus, (*d*) the pretectal area, and (*e*) the mesencephalic reticular formation.

The functions of the inferior colliculus remain obscure. Portions of the IC associated with the belt pathways are not rigorously organized tonotopically. Unfortunately, the underlying organizational plan is not apparent. Many cells in those regions respond to complex auditory signals. Unfortunately, the wide variety of response patterns precludes any generalizations at this time.

The tonotopic organization that is characteristic of the auditory system is represented in the IC. In spite of this well-organized tonotopic arrangement, it would be a mistake to consider frequency analysis the principal function of the IC. Although most cells in the IC respond to binaural clues, the binaurally driven cells do not appear to be arranged in a manner that would map the three-dimensional acoustic space. That function appears to reside in the superior colliculus where there are cells driven by both visual and auditory stimuli (see Chapter 9). Those cells are arranged so that maps of the acoustic and visual spaces are congruent. At noted in the previous chapter, the superior colliculus and its efferent projections are important in initiating movements to orient the head (and pinna if it is mobile) toward important stimuli in the visual/auditory space.

MEDIAL GENICULATE NUCLEUS

The thalamic nucleus of the auditory system is the **medial geniculate nucleus (MGN)** (Fig. 10.13). It is divided into four subnuclei, which receive mostly ipsilateral projections from the IC. The largest subnucleus, the lateral nucleus of the medial geniculate, is tonotopically organized. It projects to the ipsilateral primary auditory cortex (Brodmann areas 41 and 42) and constitutes the core pathway. The other MG nuclei receive belt projections from the inferior

colliculus and project to all auditory areas of the cerebral cortex.

The MGN also receives a particularly prominent descending collection of axons that arises from the auditory cortex. The ascending axons from the IC and the descending axons from the auditory cortex converge on the dendrites of geniculate neurons forming complex synaptic structures that have been dubbed "synaptic nests." This anatomical arrangement allows close regulation of MGN activity by the cerebral cortex.

AUDITORY CORTEX

In the human, most of the primary auditory cortex lies on the surface of the temporal lobe (Brodmann areas 41 and 42), buried within the lateral fissure (Fig. 10.14). This corresponds with the **transverse temporal gyrus (Heschl's gyrus)**. In most individuals, the left transverse temporal gyrus is considerably larger than the right, an asymmetry that can be seen on favor-

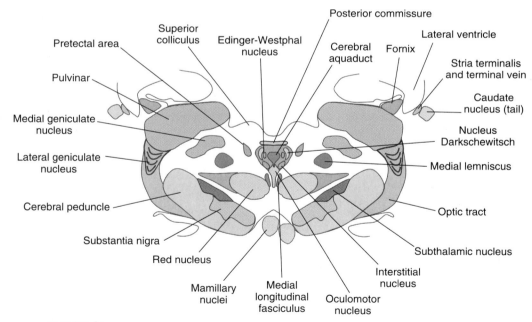

FIGURE 10.13.

Cross-section of the mesencephalon at the level of the superior colliculus. The accessory nuclei of ocular motion are depicted near the oculomotor nucleus. Note also the relationship of the medial geniculate nucleus to the lateral geniculate nucleus, the pulvinar, and the cerebral peduncle.

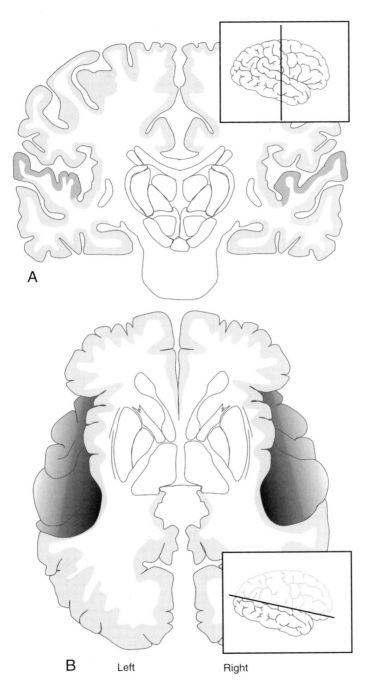

A

B   Left   Right

FIGURE 10.14.

Location of the primary auditory cortex in the human. **A**, A frontal section through the cerebral cortex locating the auditory cortex in gray on the surface of the temporal lobe. **B**, A view of the surface of the temporal lobe with the remainder of the cerebral cortex removed. The two auditory areas are usually asymmetric as shown here, with the left being somewhat more prominent than the right.

able magnetic resonance images. The auditory cortex receives ascending afferents from the medial geniculate nucleus. The auditory areas of the two hemispheres are interconnected by commissural fibers that connect each area with its contralateral homologue.

The core area of the auditory cortex (usually designated A-I) is prominently organized tonotopically.[5] Reflecting the pattern seen in other sensory and motor areas of the cortex, A-I is organized into vertical columns where the cells in all the cortical layers of a particular patch respond to the same characteristic frequency. Running perpendicular to the isofrequency bands one finds columns that represent binaural information. One set of columns responds if the stimulus is bilateral while another set prefers monaural stimuli. There is some evidence that there may be a third level of cortical organization representing a map of the three-dimensional auditory space.

The belt area of the auditory cortex surrounds A-I and has been subdivided into several areas. These have not been identified in the human although some anatomical equivalent probably exists. The specific functions of the belt areas remains obscure.

Studies of the cat have shown that the *cerebral cortex is not necessary for frequency discriminations* which are unimpaired after bilateral cortical ablation. However, *discriminations that are based on the timing and pattern of auditory events are severely impaired* after bilateral cortical ablation. For example, in one classic experiment cats were trained to recognize a Morse code pattern of low-high-low pitch tone pips as signaling a "safe" condition. When the pattern changed to high-low-high pips, the cat had only a few seconds to move before being shocked. After cortical ablation, the cats were unable to recognize the difference in the tone patterns to avoid the shocks. Remarkably, if any auditory cortex escapes the experimenter's knife, timing and pattern discriminations remains intact. Noteworthy is the fact that in the human, speech perception is a task that is critically dependent on timing and pattern discriminations and quite refractory to cortical infarction.

---

[5]This is based primarily on studies of the cat and monkey. Anatomical data on the human auditory cortex is sparse. In the absence of better data, conclusions based on animal studies will have to be projected to the human.

One particularly useful case has been reported that addresses this question directly. The patient suffered an infarction of the left auditory cortex. He had some transient difficulty with speech discrimination. Sixteen months later he suffered another infarction, this time of the right auditory cortex, a lesion that left him unable to comprehend speech. His hearing threshold as measured by pure tones was nearly normal. He was also able to localize sounds in space. The patient's own description of his situation was "I can hear you talking, but I can't translate it."

Clinical Evaluation of Hearing Disorders

There are many different types of hearing tests, many of which are very subtle and sophisticated. Those that are most relevant to primary care medicine will be discussed here.

SIMPLE HEARING TESTS

Two simple hearing tests are an essential part of the neurological examination. Both depend on the differences between **air conduction** and **bone conduction** of sound. Air conduction simply refers to the normal presentation of sound by air pressure waves impinging on the tympanic membrane. Sounds presented in that way are, as we have seen, critically dependent on the transducing function of the middle ear. Bone conduction refers to stimuli presented as oscillations imposed on the temporal bone. Vibrations presented in this way can set the basilar membrane into motion directly, bypassing the middle ear. This is less efficient than air conduction and bone conduction thresholds are about -35 dB with respect to air conduction thresholds.

Rinne's Test

For the office examination, a tuning fork (512 Hz is typical and corresponds with middle C on the piano) is set into vibration. Its stem is placed on the mastoid process in order to stimulate the cochlea by bone conduction. The intensity of the stimulus decreases gradually as the tuning fork's vibrations diminish. The patient is asked to signal when tone can no longer be heard. At that time the tines are placed near the pinna in order to stimulate the ear by air conduction. The patient with normal hearing should hear the tone for an additional 15 seconds since air conduction is 35 dB more sensitive than bone conduction. If the patient can hear the stimulus by bone conduction, the cochlea is grossly functional. If the patient cannot hear by air conduction below the bone conduction threshold, then the middle ear is dysfunctional.

Weber's Test

Rinne's test only determines the relative sensitivity of the ear to air and bone conduction. It is not very useful for determining partial hearing loss. Weber's test provides a useful adjunct. In this test, the base of the vibrating tuning fork is placed at the midline on the skull. The patient is asked to localize the sound. There are three possible interpretations. First, if hearing is equal from both ears, the sound will be *localized to the inside of the head*. Second, if the cochlea or auditory nerve are defective, sound will be *lateralized to the good ear* since one would expect a severe hearing impairment in the bad ear. Finally, if the middle ear is defective, the sound will be *lateralized to the bad ear*. This counter intuitive result is thought to be due to the fact that air and bone conduction interfere with each other in the good ear, partially canceling each other. Since only bone conduction is available in the bad ear, interference is not possible. Therefore, the cochlea in the bad ear receives greater stimulation and perception is lateralized to it. Obviously, an abnormal Weber's test cannot be interpreted in isolation, it must be augmented by the Rinne test. The Weber test can detect relative differences in hearing loss between the two ears.

PURE TONE AUDIOMETRY

Simple office hearing tests with tuning forks, while useful, cannot accurately evaluate hearing disorders. For that, more sophisticated auditory examinations are required. Describing all the methods of testing hearing is beyond the scope of this text, however, the family physician is frequently faced with interpreting the results of common hearing tests, so a few comments are appropriate. The most common hearing test is **pure tone audiometry**, which simply measures **hear-ing threshold**. To do this, the patient is fitted with a pair of calibrated earphones. Pure tones are presented at various intensities and frequencies; the patient simply indicates whether or not he hears the tone. The ears are tested separately.

Since individual hearing thresholds vary, data from thousands of young subjects with normal hearing have been averaged and an internationally agreed upon standard hearing threshold curve has been established (Fig. 10.15*A*). Several features of the normal hearing threshold curve are apparent. First, the threshold is not identical for all frequencies. Second, human hearing is most sensitive between about 500 Hz and 5 kHz. Third, although it is common to state that the normal frequency range of human hearing is from 20 Hz to 20 kHz, few individuals hear very well below 50 Hz or above 16 kHz.

Hearing loss curves are a convenient method of presenting a patient's audiogram. These charts represent the *difference between the **standard** hearing threshold curve and the **patient's** threshold hearing curve*. The data are presented as hearing *loss* in decilbels versus frequency (Fig. 10.15*B–E*). These charts are easy to read and interpret, because normal hearing is represented between 0 and −20 dB for all frequencies. One must understand, however, that the hearing loss curve represents a threshold measured against a relative standard. Hence the *absolute intensity* of the 0-dB reference in the hearing loss diagram *varies with frequency according to the standard threshold curve*.

Other audiometric tests measure the ability to understand speech in the presence of noise and recruitment (see page 374). These tests will not be described here, but the nonspecialist should be aware that they ought to be included in any evaluation of hearing impairment.

TYMPANOMETRY

The compliance (stiffness) of the tympanic membrane can be affected by any process that restricts or enhances it movement. Two processes are common, the accumulation of fluid in the middle ear and pathological changes in the ossicles. Middle ear compliance, often called impedance, is measured by bouncing sound waves off the tympanic membrane and analyzing the echo. If the middle ear is filled with fluid, the tympanic membrane will be very stiff and its

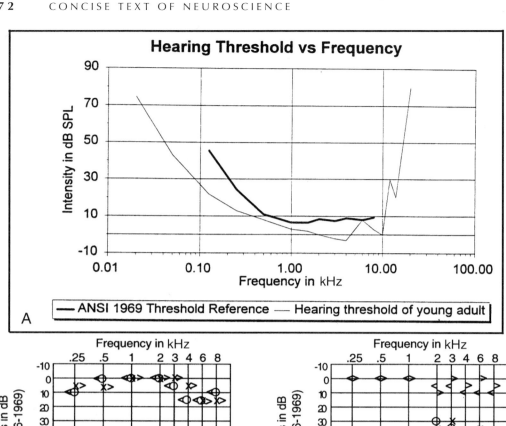

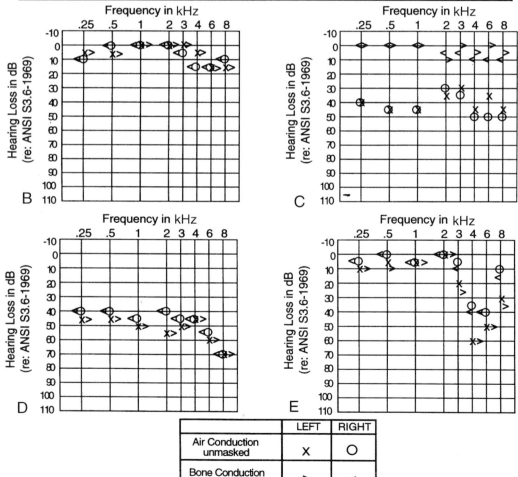

	LEFT	RIGHT
Air Conduction unmasked	X	O
Bone Conduction unmasked	>	<

compliance will be reduced. This is of practical importance in diagnosing serous otitis media (see page 375), and is now routinely employed in examining children. Tympanometry is also useful in diagnosing other, less common middle ear problems that will not be discussed here.

BRAINSTEM AUDITORY EVOKED POTENTIALS

The electrical events evoked in the brainstem by auditory stimuli can be recorded from surface electrodes under certain circumstances. In order to be measurable from remote sites, these signals must represent the nearly simultaneous action of a large number of closely spaced neurons. Even so, at the surface of the body these signals are so small that they usually cannot be extracted from the background noise.[6] Under special stimulus conditions, neural signals can be extracted from background noise by computer averaging techniques. Signal averaging is based on the principle that noise is random. Random noise, averaged over time, will result in a signal of zero amplitude. If a synchronized, repetitive signal is superimposed on the noise, it will not average to zero. Therefore a neural signal that is smaller in amplitude than the noise itself, can be extracted by the averaging process, provided the timing of the signal is known.

This averaging technique became clinically

---

[6] There are two exceptions to this generality. The EEG (electroencephalogram) and the ECG (electrocardiogram), can be recorded without averaging because the signals are unusually large and generated close to the recording electrode.

practical with the development of small, high-speed computers. Such a system can digitize and average multiple recordings in real time, extracting the neural signal in the process. This method can be applied to any situation where a repetitive, stereotyped neural response can be elicited thousands of times. It has been applied to the auditory system where calibrated clicks (square wave auditory signal) are presented to one ear, and electrical events recorded from an electrode place on the scalp. Typically, some 1000 to 2000 responses are averaged, resulting in a recording of a complex waveform (Fig. 10.16) known as a **brainstem auditory evoked response (BAER)**. The BAER is a powerful test because it does not require the cooperation of the patient. For example, threshold audiograms can be performed on infants where cooperation is impossible, yet early diagnosis of hearing impairment essential.

Within the first 10 msec of the presentation of the click, five negative peaks can be identified in the BAER recording. The electrical events that contribute to these peaks originate in the brainstem auditory structures. By noting the delay or absence of the peaks, one can quite accurately locate brainstem lesions.

Common Hearing Disorders

Hearing impairment is a common affliction. Approximately 22 million Americans are afflicted with a **hearing loss**, the most common

---

FIGURE 10.15.

A, The ANSI 1969 standard curve, used as an approximation of the normal sensitivity of human hearing, is shown as a heavy line. An audiogram from a young adult is superimposed on it. **B**, Hearing loss diagram of a person with normal hearing. A chart such as this represents the hearing threshold at various frequencies subtracted from the ANSI 1969 standard curve at the same frequency. The difference is represented as a hearing loss in dB. Normal individuals have no loss, and consequently the ideal hearing loss curve is a straight line at 0 dB. Thresholds from 0 to 20 dB are within the normal range, which is the case on this audiogram. The bone conduction threshold has been corrected 35 dB so that the two curves, if normal as in this case, are superimposed. **C**, Hearing loss curve illustrating conductive hearing loss from a child with serous otitis media. Note that hearing by bone conduction is greater than by air conduction. **D**, Hearing loss curve illustrating presbyacusis. In addition to the expected high frequency hearing loss, this patient also demonstrates an overall loss of hearing sensitivity at all frequencies. **E**, Hearing loss curve illustrating noise induced trauma, which is characterized by a sharp loss at 4 kHz with some recovery of sensitivity at 8 kHz.

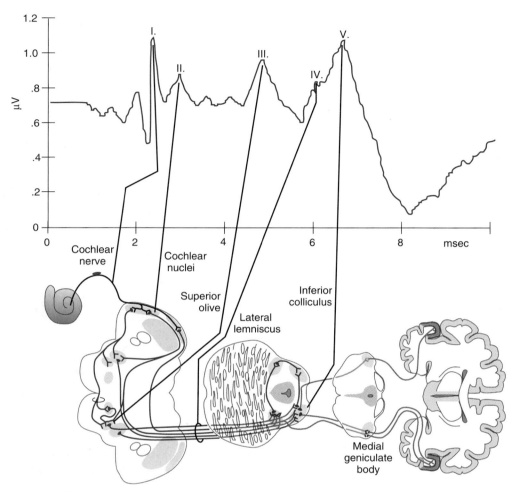

FIGURE 10.16.

Normal brainstem auditory evoked response (BAER). This illustrates only the first 10 msec, which corresponds to the responses obtained from the brainstem nuclei. Peaks I through V are indicated. Below the BAER, the presumed origin of the various peaks is indicated.

form of hearing disorder. Of these, about 6 million are affected severely enough to affect communication. Among school aged children, approximately 5% are affected by hearing loss and about 10% of these (0.57%) have hearing handicaps that require some form of special education. The U.S. Office of Education estimates that only about 18% of children with hearing loss are receiving appropriate support.

Frequently associated with hearing loss is **tinnitus** [L. *tinnio,* a jingling], the objective perception of sounds that originate in the ear. Tinnitus is frequently said to be a ringing sound, but it may be a whistling, pure high pitched tone, buzzing, roaring, or chirping. In fact, almost any type of noise has been reported. Tinnitus is very common, affecting about 37 million Americans. The cause of tinnitus is unknown and there is no effective treatment. For most people is it a nuisance that is tolerated.

More serious than moderate hearing loss and tinnitus is a phenomenon called **recruitment**. Normally, the perception of loudness increases with increasing intensity of sound. With certain types of hearing disorder, the increased perception of loudness increases at a greater rate than the actual increase in intensity. In effect, the dynamic range of hearing, normally about 120 dB, is reduced to as little as 60 to 80 dB. Louder sounds, particularly speech, collapse into a cacophony of garbled noise. Recruitment seriously affects communication. It is thought to be caused by a loss of cochlear efferent fibers. It is treated by hearing aids with nonlinear amplifying characteristics. They amplify low-intensity sounds but do not amplify high-intensity sounds. Some hearing aids even de-amplify high-intensity sound in a effort to collapse the dynamic range of the natural world to fit the patient's narrow dynamic range of hearing.

The location of lesions affecting auditory perceptions has a significant effect on treatment; therefore hearing disorders are frequently categorized according to the site of the lesion. What follows is a very brief discussion of the three principal classes of hearing impairment: **conductive**, **sensorineural**, and **central**.

CONDUCTIVE HEARING LOSS

If the ossicles become restricted or disarticulated, acoustic energy is less efficiently passed through the middle ear. Consequently, any lesion involving the middle ear produces a **conductive hearing loss**. Typically, all frequencies are affected, producing a relatively flat hearing loss curve, but at reduced sensitivity (see Fig. 10.15*D*). Determining conductive hearing losses is especially important because most causes are treatable.

Middle Ear Infections

Children commonly develop middle ear infections. These infections invariably cause significant pain that promptly brings them to medical attention. With modern antibiotic therapy, these infections are effectively treated and, as discussed earlier in this chapter , do not have the long-range implications on hearing that they once did.

Serous Otitis Media

Children are also subject to fluid accumulation in the middle ear, **serous otitis media**, a condition that is non-infectious and not painful. Consequently these children frequently escape medical attention. The accumulation of fluid, however, does cause hearing impairment, a condition with long-range consequences.

The cause of this condition is not known. Most authorities believe that, in children, the drainage of the middle ear is less efficient than in the adult because the auditory tube is both narrower and lies more horizontal in the immature head. Therefore, normal middle ear secretions may not be properly drained. They become concentrated and mucoid, restricting ossicular movement and impairing hearing. Because the condition develops slowly, the loss of hearing is insidious and is rarely detected by the parents.[7] Since there is no inflammation or swelling, the condition is not painful. Furthermore, routine otological examinations fail to reveal the condition for the same reasons. Quizzing the parents about hearing loss may elicit a revealing comment such as "she certainly does like to have the TV really loud." The alert physician may detect delayed or impaired language acquisition symptoms that suggest hearing difficulties. More frequently there are no clues. *The best routine examination is tympanometry, which will reveal the condition immediately.*

Treatment is straightforward. After the middle ear is evacuated, silicone rubber tubes are placed through the tympanic membrane, a procedure that provides the middle ear with an alternate drainage path. The tubes are left in place for several years, until the head has grown sufficiently to properly drain the middle ear cavity. Restoration of hearing is immediate and dramatic; children frequently awake from the anesthesia complaining of how loud everyone is talking.

About 5% of school-aged children have elevated hearing thresholds. Hearing impaired children have retarded speech development, delayed acquisition of language, and impaired social development and academic achievement. Most of the hearing impaired children do not receive any particular assistance in overcoming these difficulties since only about 10% of them have hearing difficulties severe enough to re-

---

[7]Young children cannot be expected to complain of hearing loss since they are too young to appreciate normal hearing; their present experience is normal to them. This is compounded by the fact that many children with this problem have yet to acquire spoken language.

quire special education. Although this author is not aware of any conclusive studies, it seems probable that most of the less severely hearing impaired children may have *treatable* hearing disorders such as serous otitis media. They can and ought to be relieved of the burden of hearing loss by timely (preschool) diagnosis and treatment.

Otosclerosis

One of the most common causes of adult onset deafness is a condition known as **otosclerosis**. This disease is characterized by pathological growth of bone at the margin of the oval window. The growth initially impedes the movement of the stapes, but eventually will fix it rigidly in place. Otosclerosis is inherited by an autosomal dominant gene with variable penetrance. Surgical removal of the stapes followed by its replacement with a Teflon prosthesis restores hearing in most patients.

SENSORINEURAL HEARING LOSS

Hearing loss associated with pathology in the cochlea or the auditory nerve is referred to as a **sensorineural hearing loss**. People with this class of hearing loss often benefit from the use of hearing aids.

Presbyacusis

The most common form of sensorineural hearing loss is **presbyacusis** (also spelled presbycusis) [G. *presby,* old + G. *akouo,* to hear], which simply means the loss of hearing with advancing age. The hearing loss primarily affects the high frequencies and correlates with a loss of hair cells in the first turn of the cochlea (see Fig. 10.15*D*). The process is progressive and apparently begins early in life, for few individuals over the age of about 20 have undiminished high-frequency hearing.

Ototoxicity

The class of antibiotic drugs known as the **aminoglycosides**, are well-known ototoxic agents. The exact mode of action is not known.

It is known that these antibiotics are concentrated in the endolymph, apparently by the action of the stria vascularis. Therefore, over a long period of time (days) extremely high concentrations develop. The very high concentrations of aminoglycosides in the endolymph are toxic to all hair cells, both cochlear and vestibular, although some drugs may affect one class more than the other (Fig. 10.17).

The aminoglycosides, like most antibiotics, are cleared from the blood through the kidney. Therefore, any disease process that reduces renal clearance potentiates the ototoxic effect of these drugs. Ototoxic antibiotics can be used with relative safety if they are delivered as a bolus or rapid infusion at high concentrations at long intervals (6 hours or more). Long term infusion of these drugs exacerbates the conditions that lead to high endolymph concentrations and hearing loss.

While the aminoglycoside antibiotics are among the most potent ototoxic agents, the physician must be alert to other iatrogenic causes of hearing impairment. High on the list are diuretics. Quinine is moderately ototoxic. Its use as an illegal and fairly ineffective abortifacient was often the cause of "congenital" deafness. Aspirin is a very mild ototoxic agent. The elderly who use aspirin to reduce the pain of arthritis may inadvertently take toxic doses.

Noise Exposure

Intense sound kills hair cells. The more intense the sound and the longer the duration of exposure, the greater the damage. Having said that, the cochlea is also remarkably resistant to damage by intense sound (Fig. 10.15*E*). Members of rock bands usually require several years of exposure during performances before they suffer the consequence of their art. Members of rock concert audiences usually return home with a feeling of fullness in their ears and a decrease in overall sensitivity. This phenomenon, the **temporary threshold shift**, resolves in about 24 hours. Occupational noise exposure is now a well-recognized risk and can be eliminated by the faithful wearing of ear protection devices.

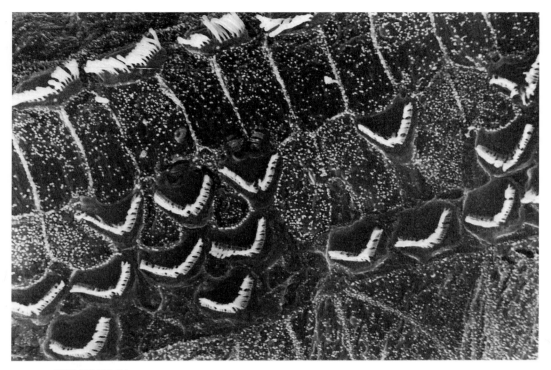

FIGURE 10.17.

Ototoxic drugs kill hair cells as in this example taken from a chinchilla treated with polymixin B. This scanning electron micrograph is taken from above the organ of Corti after the tectorial membrane has been removed. The regular pattern of outer hair cells seen in Figure 10.2 is clearly disrupted here. The reticular lamina remains intact; the absent hair cells being replaced by phalangeal cells.

Acoustic Neuroma

Schwannomas of the eighth nerve occur at the rate of about 1 per 100,000. They are called **acoustic neuromas** because their most common presenting symptom is unilateral deafness, even though research has shown that the tumors almost always arise from the vestibular branch of the nerve. As the tumor grows, it envelops and places pressure on the contents of the internal auditory meatus. If left untreated, it will expand into the posterior fossa and eventually involve the trigeminal nerve. In addition to unilateral deafness, patients frequently experience vertigo, unsteadiness of gait, facial pain, facial weakness and tinnitus. Gadolinium-enhanced magnetic resonance imaging is almost always able to confirm this diagnosis if the tumor is present. Surgical removal is quite successful if diagnosed promptly. Although hearing can seldom be preserved on the affected side, the facial nerve can be saved in the majority of cases.

Central Deafness

Central lesions are frequently called **retrocochlear lesions**. These hearing impairments are usually subtle. This is due to the number of midline crossings of the ascending auditory pathways. All auditory nuclei above the cochlear nuclei have prominent binaural representation. Clinically this means that the auditory system is very refractory to injury above the cochlear nuclei. In particular, unilateral cortical lesions have very little effect on auditory thresholds, the ability to localize sound, pitch perception, and as illustrated by the case described earlier in this chapter, even on speech perception. Sophisticated audiometry may tease out the exact nature of a central auditory deficit, but such lesions are usually refractory to treatment.

CASE HISTORY

Mr. R. W.

HISTORY, NOVEMBER 17, 1988: Mr. R. W. is a 45-year-old white male who went to see his family physician, Dr. J. P., for a life insurance examination. He had no specific complaints and stated that he felt in excellent health. During the course of the examination, Mr. R. W., who is an enthusiastic bird hunter, spoke of how good the hunting was this season. He is planning a trip to Texas next year to hunt wild turkeys. The physical examination was unremarkable for a man of his age.

NEUROLOGICAL EXAMINATION

I. MENTAL STATUS: Normal

II. CRANIAL NERVES:

   A. **Olfaction:** Patient identified methyl salicylate from both nostrils.

   B. **Vision:** Full visual fields. Fundi were normal.

   C. **Oculomotion:** Eye movements were unrestricted in all directions. There was no nystagmus. Pupils were equal and reactive to light, both direct and consensual.

   D. **Trigeminal:** Sensation was intact to cotton from all three divisions. Blink reflex was present bilaterally. Masseter muscles seemed of equal strength.

   E. **Facial:** Grimace was symmetrical.

   F. **Audition:** Rinne's test was normal for both ears but Mr. R. W. lateralized Weber's test to the right ear. He seemed to have some trouble hearing light finger rubbing from the left ear, but not from the right. Examination of the external auditory meatus revealed it to be clean and free of obstructions. The appearance of the tympanic membrane was normal.

   G. **Vagal-glossopharyngeal:** Voice was normal. The gag reflex was present bilaterally and the pharyngeal arches were fully elevated and symmetrical.

   H. **Accessory:** The drape of the shoulders was symmetrical. Strength of shoulder shrug was good and symmetrical as was head turning against resistance.

   I. **Hypoglossal:** Tongue extended in the midline and was grossly normal in appearance.

III. MOTOR SYSTEMS: Strength was normal and symmetrical from all extremities. Gait was normal.

IV. SENSORY SYSTEMS: Sensorium was present from all extremities to cotton touch and pin prick.

V. REFLEXES: MSRs all normal. Toes were down pointing.

Comment: Dr. J. P. was concerned that Mr. R. W. lateralized Weber's test to the right ear and that he also appeared to have difficulty hearing very soft sounds from the left ear. There was no sign of external ear obstruction or damage to the tympanic membrane. He has no specific complaints about hearing difficulties. Unexplained unilateral hearing loss in an adult suggests a diagnosis of acoustic neuroma. Therefore, Dr. J. P. sent her patient to the local audiology clinic for an evaluation of his hearing. Since the audiogram revealed a marked hearing loss on the left, this was followed by a BAER. The results of the two examinations are shown in Figure 10.18.

The 4-kHz notch seen in the hearing loss curve from the right ear is characteristic of noise-induced hearing loss. In this case, Mr. R. W., an enthusiastic hunter, evidently hunts without hearing protection. Since he is left handed, his right ear is exposed to the report. The left ear is protected somewhat by the acoustic shadow provided by the head and so shows no 4-kHz notch at this time. What the left audiogram does show, however, is a broadband decrease in auditory threshold both for bone and air conduction.

At this point Dr. J. P. ordered an MRI study that revealed the tumor (Fig. 10.19). The tumor was subsequently removed, sparing the facial nerve, but hearing was lost on the left. Mr. R. W. continues to hunt birds, but now wears a protective device in his right ear to avoid further noise-induced trauma.

FURTHER APPLICATIONS

Dr. J. P. is a very thorough and astute physician. This case emphasized the need to be complete in one's examination of the nervous system. Had the Rinne and Weber tests not been performed, it is quite likely that the tumor would not have been discovered until it produced symptoms associated with other cranial nerves.

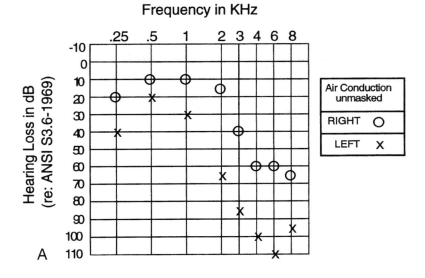

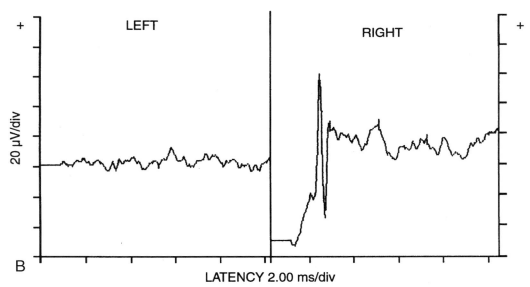

FIGURE 10.18.

A, Hearing loss curves for Mr. R. W. Although both ears demonstrate presbyacusis, the loss is much more prominent in the left ear. **B**, BAER for Mr. R. W. A 95-dB high-frequency click was used as the stimulus in each ear. Since this test emphasizes high-frequency hearing sensitivity, the left-right differences in Mr. R. W. are easily seen. This is not a normal BAER; compare it with Figure 10.16.

- What symptoms accompany well-developed acoustic neuromas? Which cranial nerves are likely to become involved and in what order?

- Had Mr. R. W. been right handed, would one have been justified in ordering the MRI after the hearing tests? (Assume that there was no evidence of noise-induced hearing loss in the right ear.) Defend your answer.

- Look up in a textbook of neurology the causes of unilateral hearing loss. Are any of these differential diagnoses relevant in this case? If so which? Was Dr. J. P. justified in pursuing a tumor so quickly or should she have performed

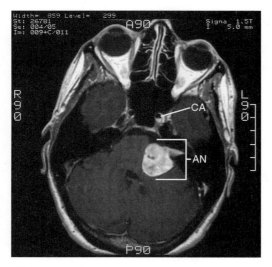

FIGURE 10.19.

MRI study (T1-weighted) illustrating a very large acoustic neuroma (*AN*) on the left. In this patient (not Mr. R. W.) the tumor has invaded the posterior fossa and displaced the brainstem. *CA*, carotid artery.

other tests to rule out other possible causes? Defend your answer.

SUMMARY

The auditory systems consists of the external, middle, and inner ear as well as the central auditory pathways. The external ear consists of the pinna and the external auditory meatus. It is separated from the middle ear by the tympanic membrane. The middle ear consists of two major spaces, the tympanic cavity and the antrum, connected together by the aditus. Three ossicles within the middle ear conduct sound from the tympanic membrane to the inner ear at the oval window. The middle ear is drained by the auditory tube.

The inner ear consists of the cochlea. Its perilymphatic portion consists of the scala tympani and scala vestibuli, structures that communicate with the vestibule. The endolymphatic compartment, the scala media communicates with the saccule through the ductus reuniens. Within the cochlea, Reissner's membrane separates the scala media from the scala vestibuli while the basilar membrane separates it from the scala tympani. The stria vascularis forms the lateral wall of the scala media. Hair cells within the scala media are the principal transducers of the auditory system.

The inner hair cells are innervated by approximately 10 type I afferent neurons that do not branch. The outer hair cells are innervated by about 4 type II afferent neurons that branch and innervate about 10 outer hair cells. Efferent axons originating in the contralateral superior olivary complex form axoaxonic synapses on type I afferent axons near the base of the inner hair cells. The also make axosomatic synapses on the outer hair cells. The central axon of the type I afferent neurons branch and synapse in the ipsilateral cochlear nuclear complex. From this complex, three secondary pathways arise and cross the midline. Axons of the dorsal acoustic stria form the lateral lemniscus and ascend in the brainstem to synapse primarily in the inferior colliculus. Axons of the intermediate acoustic stria send collaterals to synapse in the ipsilateral superior olivary complex before joining the lateral lemniscus. Axons of the ventral acoustic stria, better known as the trapezoid body, synapse bilaterally in the superior olivary complex. From there, tertiary axons join the lateral lemniscus to ascend to the inferior colliculus. Neurons in the inferior colliculus transmit auditory information to the medial geniculate nucleus from which information is transmitted to the primary auditory cortex (areas 41 and 42).

Sound is a pressure wave conducted in air. It is transformed into a displacement of fluid within the inner ear by the action of the middle ear. Within the inner ear, the translocation of fluids causes a displacement of the basilar membrane which, in turn, ultimately results in a shearing of the hairs at the apex of the cochlear hair cells. Shearing the hairs allows K^+ to enter the cell, hypopolarizing it. This hypopolarization opens voltage sensitive Ca^{2+} ionophores in the hair cell soma, allowing Ca^{2+} to enter the cell. The entry of Ca^{2+} initiates the release of a neurotransmitter at the hair cell-afferent axon synapse and also opens calcium dependent K^+ ionophores in the hair cell soma that repolarizes the hair cell.

Sound is separately coded into frequency, intensity, and timing elements by the action of the cochlea. Frequency information is place coded such that auditory neurons have a characteristic frequency at threshold. Place coding is primarily determined by the cochlea, which performs a mechanical Fourier analysis that distributes acoustic information to different hair cells based on frequency. Volley coding is another method of frequency coding that is most effective for the lower frequencies.

Hearing disorders are usually classified into one of three types, conductive, sensorineural, and central. Conductive hearing losses can usually be differentiated from sensorineural losses by the simple Rinne and Weber tests. All types of hearing losses can be quantified by pure tone audiometry.

Conductive hearing losses are due to mechanical disruption of middle ear function. The most important of these are serous otitis media, middle ear infections, and otosclerosis. Sensorineural hearing losses are caused by the loss of hair cells in the cochlea. Most common of the sensorineural disorders is presbyacusis, characterized by the progressive loss of high-frequency sensitivity with advancing age. Ototoxic agents are chemicals toxic to hair cells. Foremost among these are the aminoglycoside antibiotics, and they must be carefully administered to avoid iatrogenic hearing loss. Noise exposure is another hazard to normal hearing, being a particularly important cause of hearing loss in those frequently exposed by occupation or attendance at rock concerts. Central hearing losses are less common and are often subtle, especially if the lesion involves the auditory cortex. Tinnitus and recruitment are hearing disorders with uncertain etiologies. Recruitment can seriously disrupt communication and may be related to the loss of efferent innervation to the cochlea.

FOR FURTHER READING

Adams, R., and Victor, M. *Principles of Neurology,* 5th ed. New York: McGraw-Hill, 1993.

Brodal, A. *Neurological Anatomy in Relation to Clinical Medicine,* 3rd ed. New York: Oxford University Press, 1981.

Buser, P., and Imbert, M. *Audition.* Cambridge, MA: MIT Press, 1992.

Friedmann, I., and Ballantyne, J. *Ultrastructural Atlas of the Inner Ear.* London: Butterworth & Co., 1984.

Hallahan, D., and Kauffman, J. *Exceptional Children: Introduction to Special Education,* 6th ed. Boston: allyn & Bacon, 1994.

Nieuwenhuys, R., Voogd, J., and van Huijzen, C. *The Human Central Nervous System: A Synopsis and Atlas,* 3rd ed. Berlin: Springer-Verlag, 1988.

Northern, J. L., and Downs, M. P. *Hearing in Children,* 4th ed. Baltimore: Williams & Wilkins, 1991.

Wilson-Pauwels, L., Akesson, E., and Stewart, P. *Cranial Nerves: Anatomy and Clinical Comments.* Toronto: B. C. Decker, 1988.

The Visual System

The visual system consists of the eye and the CNS structures associated with visual imaging. The eye is a complex sensory organ that transduces electromagnetic energy in the form of photons into neural signals. CNS structures interpret these signals to create the visual perceptions we have of the environment in which we find ourselves. The visual system is of great neurological significance not only because vision plays such an important role in human behavior, but because disturbances to vision provide clues to a large number of neurological disorders. Experimental investigations of the visual system have also provided fundamental insights into the nature of cortical information processing.

Anatomy of the Eye

The eye is the transducing organ of the visual system. It consists of both passive and active elements. The passive elements collect photons in a manner that allows a visual image to be projected onto the retina. The retina contains the cells that convert photon energy into electrochemical signals that can be interpreted by the CNS. To understand how the nervous system organizes and interprets visual information, it is necessary to understand how the passive and active elements of the eye function.

STRUCTURE OF THE EYE

The eye is a spherical structure, approximately 24 mm in diameter (Fig. 11.1). The anterior surface is the **cornea** [L. *orneus,* horny], a transparent structure composed of collagen. The white, opaque **sclera** [G. *skleros,* hard] covers the remainder of the eye. In referring to the structures of the eye, the description "inner" refers to structures closer to the center of the globe, while "outer" refers to those closer to the sclera. The posterior inner surface of the eye is lined with the **retina**, the photosensitive epithelium. The **lens** [L. *lentil,* a bean] lies suspended between the cornea and the retina. Like the cornea, the lens is transparent and composed of collagen fibers. Behind the lens is the **vitreous cavity** [L. *vitreus,* glassy], which is filled with a clear gelatinous material, the **vitreous body**.

The **ciliary body** [L. *cilium,* eyelid] (Fig. 11.2) lies near the junction between the cornea and the sclera, on the inner surface of the eye. This complex structure evolves from the both the inner and outer layers of the optic cup. The inner layer forms the pigmented, highly vascular layer of the ciliary body. The **ciliary muscle** and suspensory ligaments of the lens develop from adjacent mesenchyme. Constriction of the ciliary muscles relieves tension on the lens, allowing it to thicken and increasing its refractive power (see Chapter 9).

The **iris** [G., rainbow] is a thin extension of both layers of the optic cup where they fuse at the anterior rim. The inner, pigmented layer gives the iris its characteristic color and makes it opaque. The iris divides the space between the lens and the cornea into the **anterior chamber** and the **posterior chamber** (Figs. 11.1 and 11.2). The unattached, medial edge outlines a hole known as the **pupil**. The size of the pupil

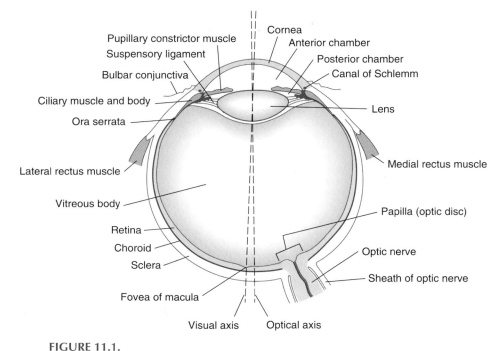

FIGURE 11.1.

A cross section of the human eye illustrating the major anatomical features.

can be adjusted by the action of the **constrictor** and **dilator muscles** of the iris (see Chapter 9).

The anterior and posterior chambers are filled with a clear fluid, the **aqueous humor**, that nourishes the avascular lens and helps to hydrate both the lens and the cornea (Fig. 11.2). Proper hydration of these structures is essential to the maintenance of their transparent state. Secreted by the ciliary body into the posterior chamber, the aqueous humor passes into the anterior chamber through the pupil. At the lateral margin of the anterior chamber, the collagen fibers of the cornea and sclera separate into a trabecular meshwork through which the aqueous humor can pass before it is collected in a particularly large cavity of this meshwork, the **canal of Schlemm**. From there it passes into the venous drainage of the sclera. It is replaced about every 2 to 3 hours. Regulation of the ratio of production to reabsorption of aqueous humor determines the **intraocular pressure**, normally between 15 and 20 mm Hg. If the balance is disrupted and production exceeds removal, intraocular pressure will increase, a condition known as **glaucoma** [G. *glaukos,* bluish green]. If untreated, glaucoma causes blindness by restricting blood flow to the retina.

OPTICS OF THE EYE

The eye is an optical instrument. Light entering the eye through the cornea is focused by several refractive structures to form an inverted image on the surface of the retina (Fig. 11.3). The most important of these refractive surfaces is the cornea, contributing about 42 of the 60 diopters[1] of total refraction of the human eye. The lens is also an important refractive element, providing about 18 diopters of refraction in the relaxed eye. It is held in place to the posterior by the viscous **vitreous humor** and radially by **zonal fibers** that attach it to the ciliary body. The lens is elastic, and its refraction can be increased by about 12 diopters through the action of the ciliary muscles. The elasticity of the lens decreases with age, beginning slowly in childhood, and progressing very rapidly after about age 45, a condition known as **presbyopia** [G. *presbys,* old + *ops,* eye]. By age 60 virtually no lens accommodation remains (Fig. 11.4).

The eye is normally a sphere about 24 mm in diameter (Fig. 11.5). **Emmetropia** [G. *emmet-*

---

[1]The unit of refractory power is the diopter (D), which is the reciprocal of the focal length of the lens measured in meters.

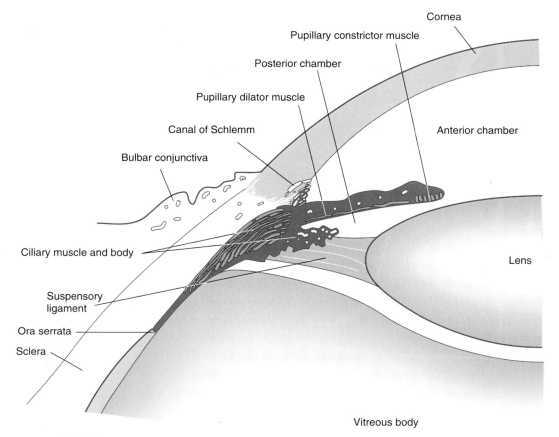

FIGURE 11.2.

The aqueous humor fills the posterior and anterior chambers. It is secreted continuously into the posterior chamber by the ciliary body. It is returned to the vascular space through the canal of Schlemm, which lies in the anterior chamber at the junction of the cornea and the sclera, adjacent to the ciliary muscle.

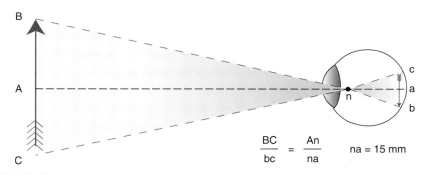

FIGURE 11.3.

There are four refractive surfaces in the eye: the two surfaces of the cornea and the two surfaces of the lens. This complex system can be simplified to the equivalent single lens system, the idealized eye. A ray diagram for such an idealized eye is shown here to demonstrate the inversion and reduction of the image on the retina. The nodal point, *n*, represents the effective focal point of all refractive elements combined. The size of the retinal image can be computed by solving the formula shown.

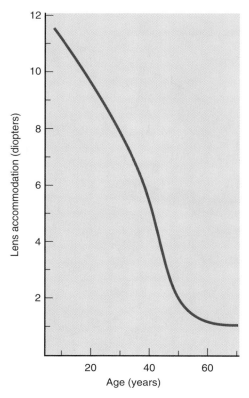

FIGURE 11.4.

The loss of lens elasticity with age, presbyopia, occurs throughout life, as shown here in this graph. The rate of change increases sharply at about age 40. By about the age of 60, the lens is essentially inelastic. The loss of elasticity is reflected in amount of diopter change the lens can accommodate.

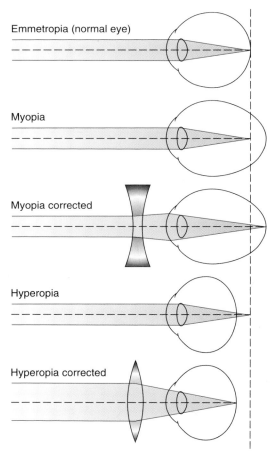

FIGURE 11.5.

This diagram illustrates how abnormalities in the shape of the globe affect vision and how simple lenses, placed in front of the cornea, bring the image to focus at the retina in the myopic and hyperopic eye.

ros, according to measure + *ops,* eye] is the term used to describe a normal, spherical eye in which distant objects are brought to focus on the retina while the ciliary muscles are relaxed. If the eyeball is elongated, the focal plane will be in front of the retina, a condition known as **myopia** [G. *myo,* to half close, + *ops,* eye]. Only objects close to the eye can be focused in the myopic eye, so these people are said to be "nearsighted." Myopia is caused by nonuniform growth of the eye and usually becomes noticeable at puberty. **Hyperopia** [G. *hyper,* over, beyond, + *ops,* eye] is the opposite condition, "farsightedness." The image falls behind the retina because the eyeball is flattened. This is an unusual congenital disorder that is apparent early in life.

Both conditions are easily corrected with appropriate lenses.

Since the cornea is the most refractive element of the eye, defects in its structure, especially the outer surface, have significant effects on visual acuity. **Spherical aberrations** in the image are produced if the lens curvature is not precisely spherical. Light passing through different parts of an imperfect lens will be brought to focus at different distances. In the normal eye, spherical aberrations of the cornea are not significant. However, **keratoconus** [G. *keras,* horn + *konos,* cone] is a condition in which the cornea, with time, gradually thins at the mar-

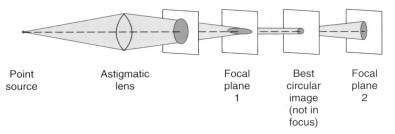

FIGURE 11.6.

An astigmatic lens brings a point of light to focus, not on a plane, but over a distance between two planes, as diagrammed here. At either of the two planes, a distant point is focused as an ellipse. The ellipses at the two planes are perpendicular to each other. Between the planes, the distant point is focused as a large circle. Good focus cannot be achieved at any distance from the lens.

Point source · Astigmatic lens · Focal plane 1 · Best circular image (not in focus) · Focal plane 2

gins, causing it to protrude from the center. This transforms the spherical cornea into a cone, creating spherical aberrations that seriously impair vision. The condition is progressive and is ultimately treated by corneal transplant.

Another aberration caused by corneal imperfections is **astigmatism**. Astigmatic lenses have different radii of curvature at different meridians around the lens (Fig. 11.6). This defect causes a point to be brought into focus over a distance between two different planes rather than as a point on a single plane. Astigmatism can be demonstrated by viewing a series of radially oriented lines (Fig. 11.7). If the curvature of the cornea is uniform, all lines will be in focus. If it is astigmatic, one line will be darker and in sharpest focus, while the line oriented perpendicular to it will be lighter and out of focus. Astigmatism is easily corrected with contact lenses, since the tear film between the contact lens and the cornea eliminates the defective anterior corneal surface as a refractive element, replacing it with the outer surface of the contact lens. Astigmatism can also be corrected with external lenses.

CELLS OF THE RETINA

Histologists identify ten layers in the retina composed of neuron cell bodies, neuron processes, and supporting cells (Fig. 11.8). Five types of neurons are distributed throughout these layers: **receptor cells**, **bipolar cells**, **ganglion cells**, **horizontal cells**, and **amacrine cells** (Fig. 11.9). The light-sensitive receptor cells are the outer-

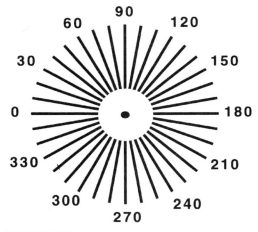

FIGURE 11.7.

The set of radiating lines in this Lancaster-Regan chart will appear of even density to an observer with normal vision. If astigmatism is present, some of the lines will appear darker and focused, while those perpendicular to them will appear lighter and out of focus.

most of the neural cells in the retina. All the other neural cells and the inner vascular layer lie juxtaposed between the light entering the eye and the visual transducers. This counter-intuitive structure results from the development of the neural elements from the inner layer of the optic cup.

There are two types of receptors, **rods** and **cones**, being named according to their general shape (Fig. 11.10). Both have an **outer segment** that consists of a series of folded membranes. In rods the folds separate to form independent

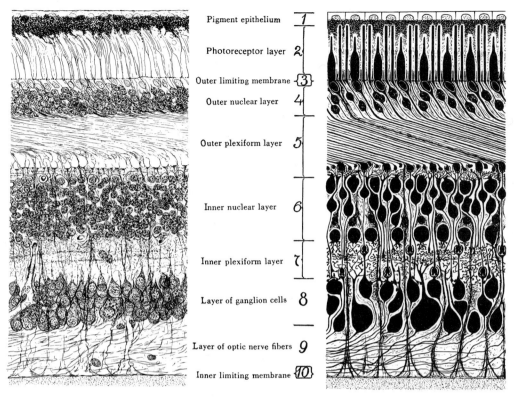

Pigment epithelium *1*

Photoreceptor layer *2*

Outer limiting membrane *{3}*

Outer nuclear layer *4*

Outer plexiform layer *5*

Inner nuclear layer *6*

Inner plexiform layer *7*

Layer of ganglion cells *8*

Layer of optic nerve fibers *9*

Inner limiting membrane *{10}*

FIGURE 11.8.

The ten layers of the retina, as defined by histology, are illustrated on the *left* as they appear in routinely stained preparations. On the *right* is a schematic representation of the cellular elements of the retina, produced at the same magnification as the section on the left. Note that the axons of the receptor cells in the outer plexiform layer are greatly extended at their base. The synapses between the bipolar cells and the receptor cells occurs at the junction of the outer plexiform layer and the inner nuclear layer. The synapses between the bipolar cells and the ganglion cells occurs in the inner plexiform layer. Note that in this illustration, light reaches the retina from the bottom of the figure and must pass through nine of the ten layers to reach the outer segments of the receptor cells before it can be detected.

disks, each disk being a closed, membrane-bound sack. In cones, the membrane folding of the outer segment is considerably reduced compared with rods, and the elaborate assembly of independent disks is missing. The photopigments necessary for phototransduction are located on the interior surface of the outer segment's membrane. The **inner segment** of the receptor is separated from the outer segment by a narrow stalk that contains a short cilium. The inner segment is rich in mitochondria but does not contain the nucleus. That organelle lies in a separate bulbous segment.

In the mammalian retina, the cones, numbering about 6×10^6, are most conspicuous in the **fovea** [L., a pit], where they constitute the only receptor (Figs. 11.1 and 11.11). The fovea is a small depression in the retina, no more that 700 μm in diameter, that occupies approximately the most central 2° of the retina. The fovea not only contains a single type of receptor cell, but these cones are the smallest in diameter of those found in the retina and the most densely packed. Cellular elements of the retina are minimized at the fovea, bringing the receptor cells closer to the inner retinal surface where they can be more di-

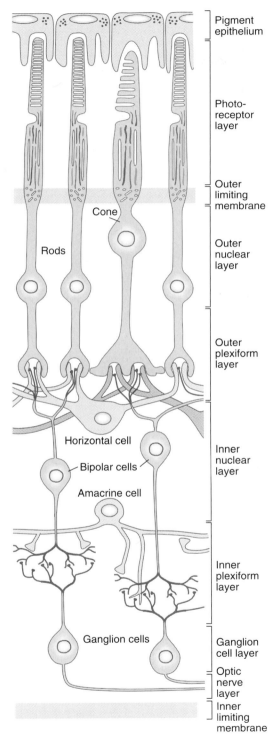

rectly affected by light. These structural modifications greatly enhance visual acuity from the fovea, hence it lies at the center of the visual fields, on the **visual axis**, an imaginary line running from the center of the cornea to the center of the fovea. The visual axis is congruent with the straight-ahead line of sight. Rods greatly outnumber cones, being about 100×10^6 in number. They are found in all parts of the retina except the fovea and are most numerous about 3 mm from its edge, about 10° to 15° lateral to the visual axis (Fig. 11.12)

The **bipolar cells** have, as their name implies, two processes. The outer process makes synaptic contact with numerous receptor cells in the outer plexiform layer. The inner process synapses with **ganglion cells** at the inner plexiform layer. The ganglion cells number about 1 $\times 10^6$ and are the cells of origin for all the axons that leave the retina as the **optic nerve**.[2] There are two types of interneurons in the retina. **Horizontal cells** make synaptic contact with receptors and bipolar cells in the outer plexiform layer, while **amacrine cells** synapse with the ganglion cells and the bipolar cells at the inner plexiform layer. Both types of interneurons serve to spread visual information laterally across the retinal surface.

The axons of ganglion cells traverse the inner surface of the retina and collect at one point, the **papilla** [L., a nipple] (or optic disk), where they penetrate the sclera to form the optic nerve. There are no receptor cells at the papilla, so no vision is possible from that portion of the visual field, known as the **blind spot**. No blindness is perceived because the blinded area is ignored by higher CNS structures and is never brought to perception. The blind spot can be demonstrated by observing, monocularly, an appropriately de-

[2]The optic nerve is a CNS tract by virtue of its developmental history and substantiated by the fact that its axons are myelinated by oligodendrocytes. However, by convention the pre-chiasmatic portion of the tract is called the optic nerve and the post-chiasmatic part the optic tract.

FIGURE 11.9.

This schematic illustration emphasizes the connections among the neural elements of the retina.

Note that the receptor cells synapse with bipolar and horizontal cells with chemical synapses. Some rods and cones form electrical synapses with each other.

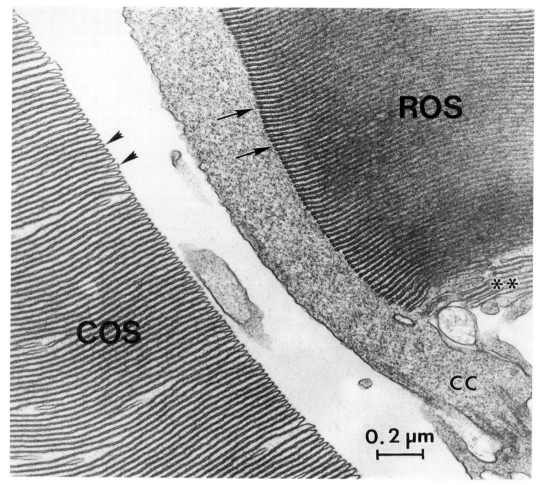

FIGURE 11.10.

Transmission electron micrograph of a rod outer segment (*ROS*) and a cone outer segment (*COS*) from an amphibian retina. Note the continuous lamellar organization and patency of the COS disk membranes (*arrowheads*), compared with the individual, closely spaced membranes of the ROS (*arrows*). *Asterisks* denote nascent disc membranes at the base of the ROS, emanating from the connecting cilium (*cc*). Scale bar, 0.2 μm.

signed test figure (Fig. 11.13). Blinded areas of the visual fields are known as **scotomas** [G. *skotos,* darkness] and may be natural, such as the blind spot, or due to disease. The ability of the brain to ignore and perceptually fill in blinded areas in the peripheral retina is truly remarkable. Individuals with very large scotomas may be totally unaware of their peripheral blindness until it is discovered during an examination of the visual fields. Scotomas in the foveal area cannot be ignored, as they severely interfere with visual perception.

The papilla is the terminus of the subarach-noid space that surrounds the optic nerve. Intracranial pressure is transferred directly through this space to the papilla. If intracranial pressure is elevated, the pressure constricts the axons of the optic nerve like a tourniquet, restricting axoplasmic flow. The axons, whose cell bodies lie in the retina, swell at the point of constriction, blurring the margin of the papilla. The increased intracranial pressure also produces papillary venous congestion. Both conditions can be seen during examination of the eye with an ophthalmoscope as a blurring of the disk margin,

FIGURE 11.11.

The fovea is an area of the retina specialized for high-resolution vision. It contains only small-diameter cones, and the inner layers of the retina are displaced laterally, as seen here in this histologic section of a monkey fovea, so that light is presented to the outer segment of the receptors with as little obstruction as possible.

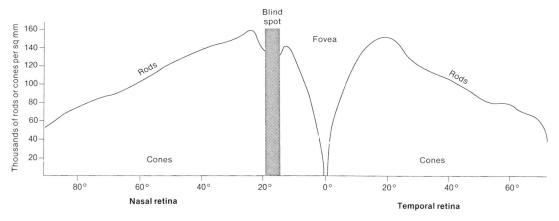

FIGURE 11.12.

The distribution of rods and cones is not uniform across the retina. Only cones are found in the fovea, while rods are almost exclusively found beyond about 10° from the visual axis. The density of receptors declines rapidly beyond about 20°.

papilledema. Papilledema is the cardinal sign of increased intracranial pressure.

Central Visual Pathways

Axons leaving the *temporal half* of the retina traverse the **optic nerve** to the **optic chiasm** [G., *chiasma,* two crossing lines], where they join the **optic tract** and project to ipsilateral structures. Axons leaving the *nasal half* of the retina cross the midline at the chiasm and terminate in the contralateral structures (Fig. 11.14). This arrangement means that all the axons in one optic tract carry information about the contralateral visual field, a pattern of hemispheric reversal with respect to the external world that is consistent with the motor and sensory systems. Axons of the optic tract terminate in three areas of the central nervous system, the **lateral geniculate nucleus**, the **superior colliculus**, and the **pretectal area** (Fig. 11.15). The trajectory through the lateral geniculate nucleus is the largest, most direct, and clinically most important pathway by which visual information reaches the cerebral cortex. A second pathway passes through the superior colliculus and the pulvinar before reaching the cortex. The third pathway from the retina does not reach the cerebral cortex. It terminates in the brainstem pretectal nuclei that regulate eye movements, lens accommodation, and pupillary size (see Chapter 9).

About 80% of the optic tract axons synapse in the **lateral geniculate nucleus (LGN)** (Fig. 11.16). The LGN is a laminated structure, having six layers. The ipsilateral fibers of the optic nerve terminate in laminae 2, 3, and 5, while the contralateral fibers terminate in the remaining laminae, 1, 4, and 6. The ventromedial laminae (1 and 2) contain large cells and are therefore called the **magnocellular** division, while the remaining (3 to 6) are known as the **parvocellular** region. Ganglion cell axons are distributed in the LGN according to their origin in the retina, thus establishing a retinotopic organization in this structure. Projections from the retina are not distributed to the LGN in proportion to their spatial distribution in the retina but by their receptor density. For example, the central 20° of the retina occupies about 10% of the retinal area, yet contains about half of the total number of receptors. The ganglion cells that receive receptor input from the central 20° send axons that synapse in about 65% of the LGN (Fig. 11.14).

There are about 1 million neurons in each LGN, all of which project to the ipsilateral occipital cortex (area 17) as the **optic radiations**. These axons fan out as they leave the LGN (Fig. 11.14). Consequently, some axons take a direct route to the occipital pole, while others take an indirect route through the temporal lobe before reaching the **calcarine sulcus**, where the direct fibers terminate in the superior lip and the indirect fibers terminate in the inferior lip of the calcarine gyri. The portion of the cerebral cortex that receives LGN axons is usually labeled V_1, or called the striate cortex (Brodmann area 17), to designate it as the primary visual cortical area. The expanded neural representation of the fovea found in the retina and LGN is maintained in the visual cortex (Fig. 11.14).

Most of the remaining axons of the optic tract terminate in the superior colliculus. In Chapter 9, we discussed the organization of the superior colliculus. To recapitulate briefly, the superficial layer receives both retinal and cortical visual information, the latter descending from V_1. Neurons in the superficial superior colliculus project both to the pretectal area and to the **pulvinar** [L., a couch made from cushions], a large nucleus of the thalamus. Neurons in the pulvinar project

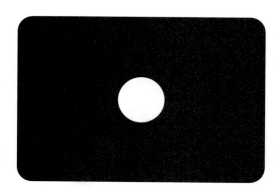

FIGURE 11.13.

To demonstrate the presence of the blind spot, hold this illustration at arm's length while closing the right eye. Focus on the white circle and slowly bring the page toward you. The crossed circle on the left will disappear at about 10 inches. To experience how the brain fills in scotomas, making them imperceptible, repeat the experiment, but now close your left eye and focus on the crossed circle. At about 10 inches, the white hole in the figure will disappear and be filled in with vertical lines.

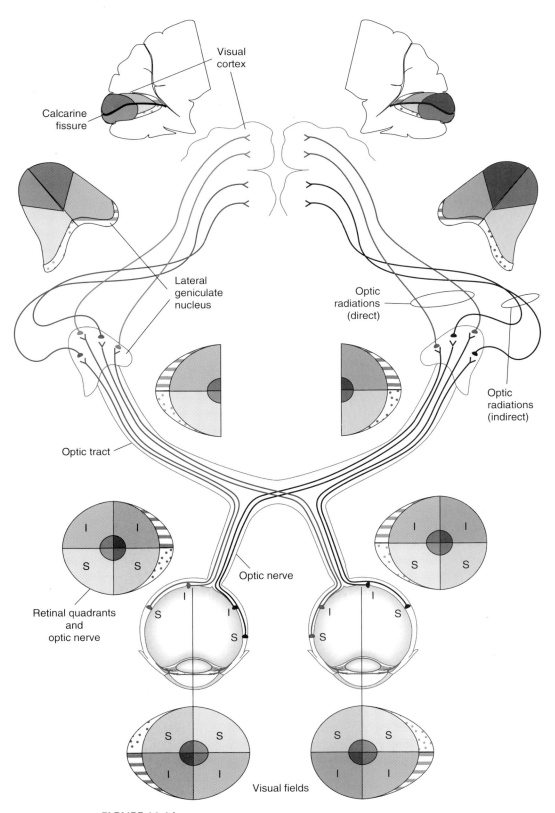

FIGURE 11.14.

The axons of the neurons that make up the visual pathways remain highly organized throughout their course, as depicted here. Note the reversal of visual fields on the retina due to the action of the lens. *I,* inferior visual field; *S,* superior visual field.

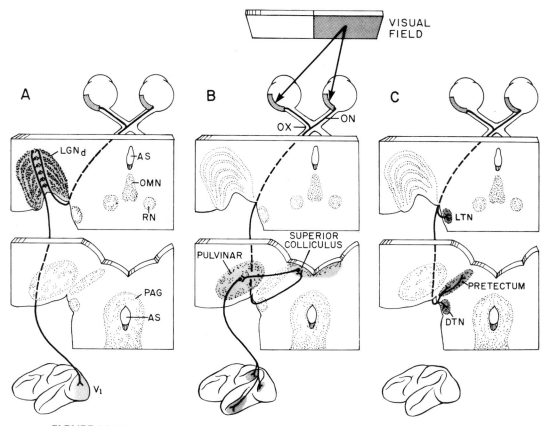

FIGURE 11.15.

The three sites of termination of ganglion cell axons are illustrated here. **A,** The principal pathway, comprising approximately 80% of the optic tract axons, terminates in the lateral geniculate nucleus (*LGN*). LGN axons project to the primary visual cortex (V_1). **B,** Most of the remaining axons of the optic tract terminate in the superior colliculus. One efferent path from the superior colliculus projects to the pulvinar. Pulvinar efferents terminate in visual centers of the cerebral cortex other than V_1. **C,** A few optic tract axons terminate directly in brainstem nuclei, primarily in the pretectal area. Abbreviations: *AS,* sylvian aqueduct; *DTN,* dorsal terminal nucleus; *LGN,* lateral geniculate nucleus; *LTN,* lateral terminal nucleus; *OMN,* oculomotor nuclei; *ON,* optic nerve; *OT,* optic tract; *OX,* optic chiasm; *PAG,* periaqueductal gray matter; *RN,* red nucleus.

to the cerebral cortex. Thalamocortical connections from the pulvinar, unlike those from the LGN, do not synapse in V_1.[3] Instead the pulvinar innervates a number of extra-striate cortical areas, including areas 18 and 19 and the temporal lobe, particularly the superior temporal gyrus. These regions are important for processing highly abstracted visual perceptions (see discussion below).

[3]Area 17 (V_1) is innervated by the pulvinar in the cat and probably other carnivora. Most authorities do not report connections from the pulvinar to V_1 in primates.

Phototransduction

The conversion of photon signals into neuronal signals, phototransduction, takes place in the receptor cells. It is a three-stage process involving first the isomerization of a pigment subsequent to the absorption of a photon, second a biochemical cascade, and finally alterations in a sodium ionophore that modulate ionic current within the receptor. Alterations in ion current

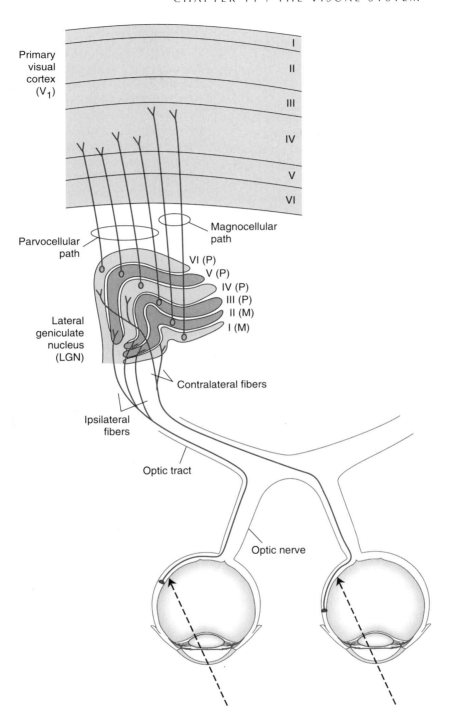

FIGURE 11.16.

Ganglion cell axons from each hemi-retina project to different layers of the LGN as illustrated here. Layers 1 and 2 of the LGN contain larger neurons than those in the remaining layers. This divides the LGN into magnocellular (*M*) and parvocellular (*P*) regions. Axons leaving the LGN remain segregated according to their magnocellular and parvocellular origins and synapse in different layers of the cortex.

flow, as we have seen in other neurons, lead to transmitter release and the transfer of information to other neurons.

PHOTOCHEMICAL EVENTS

The stacked disks in the outer segment of rods contain a membrane protein, **rhodopsin**, that is composed of two subunits. The larger fragment (384 amino acid residues) is a membrane-spanning protein called **opsin**. Attached to residue 296 is a smaller molecule called **retinal**, a six-carbon ring with a side chain (Fig. 11.17). In the dark, the side chain is bent at the 11th carbon atom. In this form it is called **11-*cis*-retinal**. If this molecule absorbs a photon, it undergoes photoisomerization, forming the straight chain version, **all-*trans*-retinal**. This subtle alteration in the retinal fragment unleashes a series of conformational changes in the opsin fragment (Fig. 11.18). These steps produce metarhodopsin II, the activated form of rhodopsin. Most of these changes occur in less than a millisecond, but the last transformation, from metarhodopsin II to metarhodopsin III, requires several minutes to accomplish. Ultimately, the last intermediate form of rhodopsin (metarhodopsin III) dissociates into opsin and all-*trans*-retinal. All-*trans*-retinal is subsequently reduced to **vitamin A** (all-*trans*-retinol), which is synthesized back into 11-*cis* retinal. It re-associates with opsin, completing the cycle. It is noteworthy that the metabolic substrate for this cycle is vitamin A, which cannot be synthesized by the human; it must be supplied in the diet. Consequently, dietary vitamin A deficiency leads to visual impairment because phototransduction cannot occur without the availability of the complete photopigment. If unresolved, the chronic deficiency of retinol in the outer segment leads to degeneration of the receptors, producing permanent blindness.

BIOCHEMICAL CASCADE

Metarhodopsin II, the active form of rhodopsin, initiates the second stage of the phototransduction process. The activated rhodopsin collides with **transducin**, a G protein, activating it by initiating the exchange of GDP for GTP within its α subunit (Fig. 11.18). The α subunit is immediately transferred to a **phosphodiesterase** that, when activated in this way, hydrolyzes cytoplasmic cGMP to 5'-GMP. Reduction in the concentration of cytoplasmic cGMP in the rod's outer segment releases bound cGMP from the cGMP-gated Na^+ ionophores. Dissociation of cGMP from the ionophore initiates the final stage in the phototransduction process, the closing the Na^+ channel in the outer segment.

This complex, multi-staged process seems cumbersome. How much simpler it would be if the photon directly affected the sodium ionophore! The principal advantage of this indirect photochemical process is the degree of amplification if affords. One molecule of rhodopsin can absorb only a single photon. But the resulting activated rhodopsin activates approximately 500 transducin molecules, each of which activates approximately 500 phosphodiesterase molecules every second. Therefore, *a single photon can hydrolyze some 250,000 cGMP molecules per second*. This biochemical amplification is directly responsible for the remarkable sensitivity of the retina to light. Elaborate experiments have shown that the human is capable of consciously detecting the absorption of a single photon by a rod.

IONIC CURRENTS

In the dark, receptor cells have a resting potential of about −40 mV. This relatively small value is due to a steady current flow through the cell, the **dark current**, that is carried by two

FIGURE 11.17.

Rhodopsin is a membrane protein located in the membrane of the outer segment disks (**A–C**). It consists of a large protein, opsin, and a covalently bound hydrocarbon, retinal (**D**). Retinal consists of a six-member ring with a nine-carbon side chain. In the dark, the side chain is bent (11-*cis*-retinal). After absorbing a photon, 11-*cis*-retinal undergoes isomerization that straightens the side chain, forming all-*trans*-retinal.

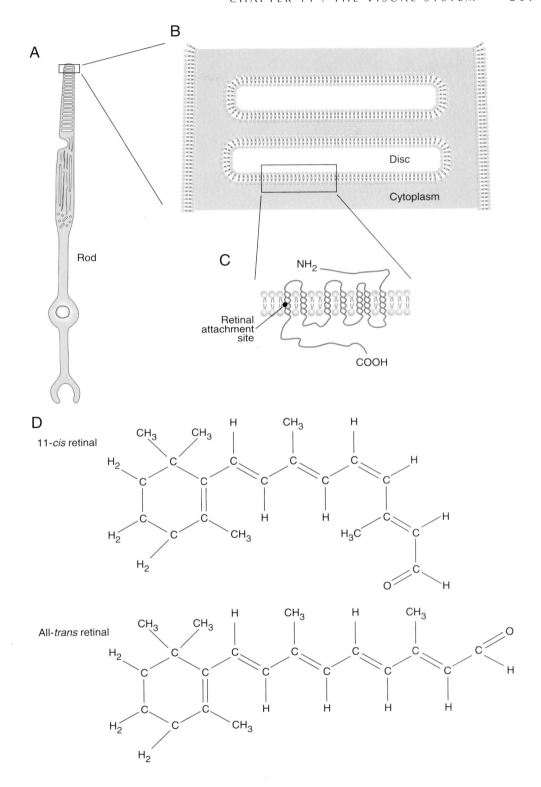

A   Rod

B   Disc   Cytoplasm

C   NH₂   Retinal attachment site   COOH

D   11-*cis* retinal

All-*trans* retinal

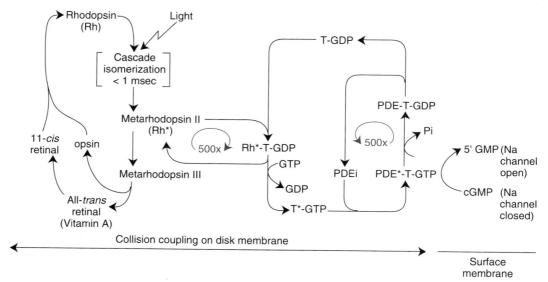

FIGURE 11.18.

This schematic summarizes the essential steps of the biochemical events that occur in the membranes of the outer segment of the photoreceptors. The absorption of a photon activates rhodopsin, allowing it to couple with transducin. Transducin is subsequently activated by the replacement of GDP with GTP, allowing its α-subunit to activate a phosphodiesterase. The activated phosphodiesterase catalyses the conversion of cGMP to 5'GMP. The loss of cGMP from the sodium ionophore closes it, reducing the inward flux of sodium.

ions (Fig. 11.19). First, Na$^+$ enters the outer segment through **cGMP gated ionophores** that are open as long as they bind cGMP. In the dark, the continuous inward flow of sodium hypopolarizes the receptor cell, producing the resting potential of −40 mV. This steady hypopolarization, in turn, continuously drives K$^+$ out of the cell at the inner segment. To maintain homeostasis between Na$^+$ and K$^+$, an ATP-dependent exchange pump restores the ionic milieu.

To recapitulate, if the cell is illuminated cytoplasmic cGMP concentration is reduced in the outer segment of the receptor. This favors the diffusion of cGMP from the ionophore into the cytoplasm. Without the cGMP bound to the ionophore, the sodium gate closes, reducing the inward flow of sodium ions. Thus, when light strikes the receptor, the sodium gates in the outer segment close, reducing the dark current, allowing the cell to become hyperpolarized due to the unopposed potassium efflux. This means that alterations in the number of photons absorbed in the outer segment directly modulate the dark current

in proportion to the intensity of the light falling on the receptor. Consequently, the ionic current flowing through the receptor is an analog of the photic signal; in effect, *the modulated dark current is the receptor potential in the photoreceptor cells.* As we have seen in other neurons, the flow of current through the cell affects the membrane potential, which in turn determines the degree of binding of vesicles at the presynaptic membrane (see Chapter 3). Consequently, in photoreceptors modulation of the dark current will regulate the release of neurotransmitter (probably glutamate) at the base of the rods and cones.

The dark current continuously flows in rods and cones, leaving these cells in a continual state of hypopolarization. This means that, unlike "ordinary" neurons, which release transmitter from the bouton as a discrete event in response to an action potential, *in photoreceptors there is a continuous release of neurotransmitter from the synapses,* even in the dark. Therefore, modulation of the dark current serves to modulate the release of neurotransmitter from the receptor

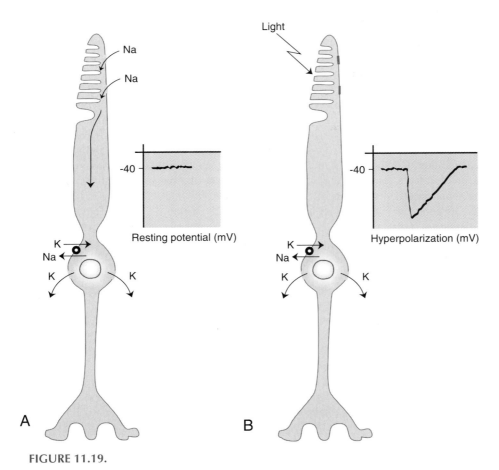

FIGURE 11.19.

A, In the dark, ionic currents are continuously flowing through the receptors. Sodium ions enter the outer segment through cGMP-gated ionophores. The current leaves the body of the cell as a potassium current. An Na+-K+ exchange pump maintains the ionic balance within the cell. B, Light striking the receptor closes the cGMP-gated sodium ionophore, reducing the sodium current. K+ ions continue to leave the cell, due to the equilibrium potential. This hypopolarizes the receptor cell.

cells. The modulation of neurotransmitter release would be of no benefit if the bipolar cells responded to the transmitter with action potentials. All of the subtle changes in transmitter concentration above and below threshold would be lost in the all-or-nothing response. Therefore, it is not surprising that, except for the ganglion cell, none of the retinal cells display action potentials. Instead, *all of the signaling within the retina is performed by graded potentials.* Only when the information must be conveyed a considerable distance from the retina is the visual information converted to a digital form (action potentials).

DIFFERENCES BETWEEN RODS AND CONES

The photochemical and biochemical events in rods and cones are believed to be essentially the same. Nevertheless, there are important physiological differences between the two receptors. Perhaps the most conspicuous difference it that rods are about 30 times more sensitive to light than cones. At very low ambient light levels, the cones are too insensitive to be stimulated, and vision is completely dependent on the rods. Vision mediated by rods under low-light conditions is called **scotopic vision** [G. *skotos,* darkness]. At high levels of ambient light, the rhodopsin in the rods is so completely converted to the all-*trans* state that the rods cannot contribute to visual perception. Vision mediated by cones under high ambient levels of light is called **photopic vision**. Together, the rods and the cones are able to detect light over about five orders or magnitude of light intensity The action of the iris adjusts the amount of light falling on the retina. It can affect the light intensity on the retina by about one order of magnitude. Combined, the rods, cones, and iris provide for a total range of sensitivity covering about six orders of magnitude.

As discussed earlier, acute vision is possible only if the image falls on the fovea. Since the fovea contains only cones, high-resolution imaging must take place under photopic conditions. In contrast to this, the highly sensitive vision occurs in the extrafoveal retina. Lateral to the fovea, the density of cones diminishes rapidly until, at the periphery of the retina, only rods are found. The rods are not only more sensitive to light than cones, but they are larger and less densely packed. Consequently, rod-mediated vision, although highly sensitive, is of poor resolution. One can demonstrate the sensitivity of rod vision on a dark night by viewing the Pleiades, a northern constellation of the winter sky. If one looks directly at this group of stars, one is hard pressed to count the seven sisters. This is because the fovea has no rods and the cones are not sensitive enough to detect the light from dim stars. By looking a few degrees away from the constellation, to direct the starlight onto the rod-rich perifoveal retina, all seven of the daughters of Atlas come into view.

The rods and cones also differ from each other in their spectral sensitivity. The differences arise from the incorporation of different opsin proteins with 11-*cis*-retinal, altering the absorption characteristics of the molecule. In human cones there are three different types of opsin molecules, segregated according to their absorption characteristics into blue, green, and red. Each cone produces only one type of opsin, so the signals transmitted to higher centers by the cones carry spectral as well as intensity and spatial information (Fig. 11.20). Although there is considerable overlap in their spectral sensitivity, particularly between the green and red cones, the combination of signals from these three types of cones is sufficient to account for our perception of colors.

Neural Coding in the Retina

The most obvious representation of visual information is as a collection of points or pixels of light, each pixel derived from an individual receptor cell. Each pixel would contain spatial, spectral, and luminance information about that one spot of the visual image. This would be much like the inverse of the color TV picture tube, in which the image is composed of thousands of tiny red, green, and blue dots of varying intensities. Attractive as this model appears, it is in fact too simple. The retina transforms the 100 million or so pixels of light into about 1 million packets that convey more than spatial, spectral, and luminance data. In the following paragraphs we will briefly look at some aspects of retinal function.

Information from receptors is transferred both radially across and trans-retinally through the retina (Fig. 11.9). The trans-retinal organization passes through the bipolar cells. As noted previously, in the retina there are well over 100 million receptor cells but only about 1 million ganglion cells. This means that considerable convergence takes place along the trans-retinal pathway. For example, in the cat fovea, approximately 200 receptor cells affect a single ganglion cell. In addition to convergence along the trans-retinal pathway, the horizontal and amacrine cells provide the mechanism for the lateral spread of information radially across the retina. This ensures that a single receptor cell can affect several adjacent ganglion cells.

One can record the action potential from a single ganglion cell in response to light stimuli applied to the retina. They respond to light presented to a restricted locus in the retina, which is the **receptive field** of the ganglion cell. One type of ganglion cell will respond by increasing its firing rate if one shines a small spot of light on the retina within its receptive field. If the size of the spot is increased, the ganglion cell response also increases (Fig. 11.21). If the size of the illuminating spot is increased beyond a certain point, the response of the ganglion cell begins to decrease. Beyond a certain size, further increases in the spot size have no additional effect on ganglion cell output. One can conclude from these observations that the responses of this ganglion cell represent

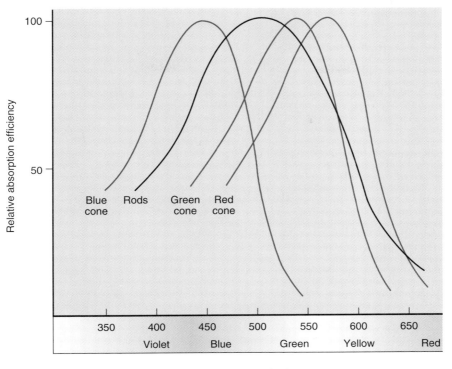

FIGURE 11.20.

The spectral absorption curves of the four photopigments are illustrated. The ordinate is a normalized relative scale; the four pigments do not all have the same absorption efficiency.

the influence of a large number of adjacent receptors. Those in the center of the illuminated field converge on and excite the single ganglion cell. Receptors outside this central core of synergistic receptors inhibit the ganglion cell since, beyond a certain spot size, increasing the size of the illumination reduces ganglion cell output. Another type of ganglion cell has been discovered that has inverse receptive properties; the center must be dark while the surrounded area is illuminated (Fig. 11.21).

From experiments such as these, it has become evident that the *pattern* of illumination that maximally excites this ganglion cell looks like a doughnut. The doughnut hole is the center, while the doughnut itself represents the surround. This **center-surround receptive field** is typical for almost all of the ganglion cells in the retina. Both types of ganglion cell, on-center and off-center cells, are about equally represented in the mam-

malian retina. The most effective size of the central spot varies with its location on the retina. In the cat fovea, it is about 30 seconds of a degree, while at the periphery it may be as large as 8°. By comparison, the size of an individual cone in the fovea is about 4 seconds.

One might well ask at this point what the advantage is of representing an image as a field of spot annuli as opposed to a field of simple pixels? In Chapter 4 we discussed the principle of **lateral inhibition** as a computational mechanism to enhance the detection of boundary conditions (see Fig. 4.12). The center-surround response characteristics of retinal ganglion cells is simply the visual manifestation of that same principle. The lateral inhibition manifested in retinal center-surround receptive fields enhances the boundary between the light and dark areas of the image, which are subsequently emphasized in the neural signaling of the ganglion cells. There-

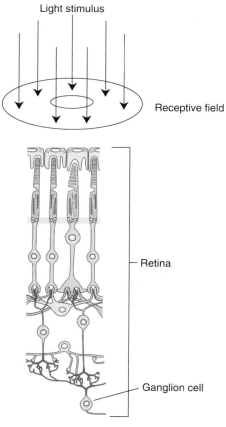

Light stimulus

Receptive field

Retina

Ganglion cell

Ganglion cell responses

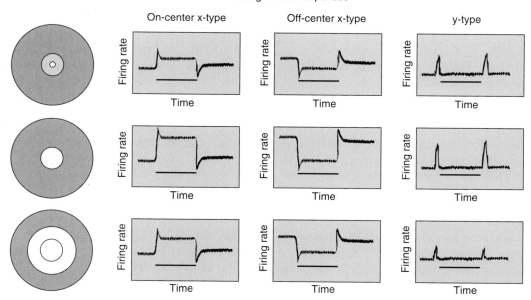

On-center x-type

Off-center x-type

y-type

fore, the *information conveyed to central visual structures carries information not inherent in the simple pixel representation of the receptors.*

Even though virtually all ganglion cells respond to the center-surround pattern, they can be subdivided into three classes based on their temporal responses to stimuli (Fig. 11.21). Classified in this way they are named **X**, **Y**, and **W** cells. In the most general terms, X cells synapse on the parvocellular cells in the LGN. They have small receptive fields (about 1°) and for the most part carry information from the central area of the retina. They respond during the presentation of light, roughly in proportion to its intensity. Y cells synapse in the magnocellular laminae of the LGN. They have larger receptive fields (about 3°) and originate, for the most part, in the peripheral portions of the retina. Y cells generally respond at the onset and termination of the stimulus and show a preference for stimuli moving quickly across the visual field. W cells are not as well characterized as the X and Y cells because they have unusual response requirements that do not correspond with the center-surround receptive fields of the X and Y cells. They have very slow conduction velocities. Many respond to general levels of illumination and are apparently important in regulating the iris. The most numerous central connections of W cells are with the superior colliculus and the pretectal area.

Central Processing of Visual Information

The visual system has been studied extensively for several decades. Although relatively little is known about information transformations in the LGN, certain features of information processing in the visual cortex have been uncovered that shed light on the overall neocortical function. Therefore it is worthwhile to briefly examine a few of the features of central visual processing, because it can serve as a framework for studying cortical function in general.

SERIAL PROCESSING IN THE VISUAL CORTEX

One can study the properties of cortical cells in the visual system in the same manner as one studies retinal ganglion cells, by presenting visual stimuli to the retina and recording the activity of the cortical cells. Just as with ganglion cells, one can determine the most effective stimulus shape and map their receptive fields. The most effective stimulus shape for individual neurons in the primary visual cortex, V_1, is different from the center-surround that is so effective in driving ganglion cells in the retina and LGN. In the visual cortex, neurons respond best to rectangular shapes. Neurons in V_1 also respond in accordance to other criteria; this characteristic has allowed them to be categorized into two broad classes: simple cells and complex cells.

A **simple cell** in V_1 responds best to a bar of light, oriented at a particular angle in the visual field. The bars must be presented within receptive fields that are also rectangular. The cells respond best to stimuli having excitatory centers and inhibitory surrounds, or vice versa (Fig. 11.22). The effective stimuli and receptive fields for simple cortical cells are entirely analogous to the spot annulus fields of the ganglion cells except for their shapes. One attractive explanation for the bar-shaped receptive fields of simple cells supposes

---

FIGURE 11.21.

Top, The response characteristics of a typical on-centered retinal ganglion cell. As the radius of the spot is increased, the firing rate increases, up to a certain spot size. Further increases in spot size reduce the firing rate of the ganglion cell, up to a point beyond which no further changes in ganglion cell response occur. *Bottom,* The response characteristics of on-centered and off-centered ganglion cells to various stimuli (light areas). The duration of the light presentation is indicated by the *bars* below the firing rate histograms. *Top group,* Small spot in the central field; *Middle group,* Spot filling the central field; *Bottom group,* Diffuse illumination covering both fields. Note that X cells respond more or less to the amplitude of the stimulus, whereas Y cells respond primarily to the presentation boundaries.

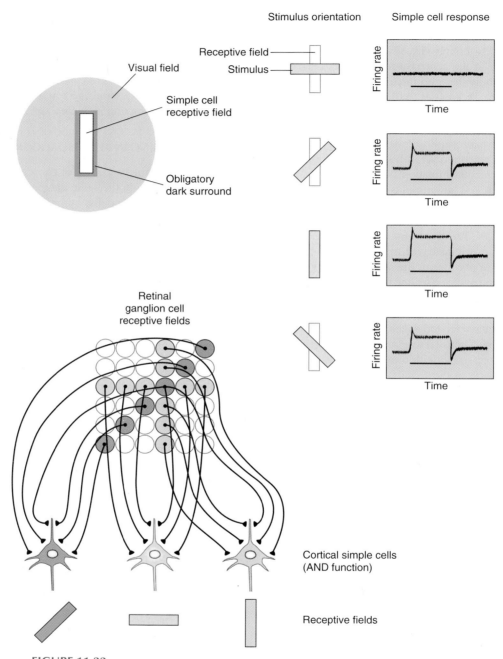

FIGURE 11.22.

Top, Simple cells have rectangular receptive fields. The response of the simple cells to light varies according to how well the stimulus matches the rectangular receptive field (*right*). *Bottom,* The receptive fields of LGN cells (*circles*) are represented with connections going to several simple cortical cells. It is proposed that the simple cell receptive properties are an AND function relating several LGN cells as shown.

that several geniculate neurons, with "typical" spot-annular receptive fields that overlap in the shape of a bar, converge on a single cortical cell. The simultaneous activity of all these converging geniculate cells would be required to cause the cortical cell to fire. This is the neural equivalent of a logical AND function (Box 11.1).

Complex cells have response properties similar to those of simple cells, but their response fields are much larger, approaching in some cases the size of the entire visual space. The stimulus for a complex cell is a properly oriented bar of light, but the bar may appear anywhere in the large receptive field of the cell (Fig. 11.23). In other words, *complex cells respond to an abstracted version of the stimulus that excites simple cells.* The quality that is abstracted, or generalized, is location within the visual space. The behavior of complex cells can be explained by assuming that all the simple cells of the same orientation, but with receptive fields in different parts of the visual space, synapse with a single complex cell. Each simple cell is capable of causing the complex cell to fire. This is the neural equivalent of a logical OR function (Box 11.1).

COLUMNAR ORGANIZATION OF THE VISUAL CORTEX

The visual cortex, like the other cortical regions we have discussed, is organized into functional columns (also see Chapter 5). One type of visual column is based on the origin of the LGN thalamic afferent axons. In layer 4 of the primary visual cortex, the cells in one functional column receive afferents only from LGN laminae 2, 3, and 5, while an immediately adjacent column receives afferents from LGN laminae 1, 4, and 6 (Fig. 11.24). Since these sets of LGN laminae receive visual information from the opposite eyes, the cortical columns to which they project also receive information from opposite eyes. In other words, the origin of the visual information alternates within layer 4 from the left to the right eye, across functional columns. These are called the **ocular dominance** columns. The pair of left-right ocular dominance columns is called a **hypercolumn**. Hypercolumns are organized across V_1 according to the retinotopic map that represents the visual fields (see Fig. 11.14).

Each hypercolumn is subdivided into a large number of **orientation columns**. Within each orientation column all the simple cells respond to bars of light of the same orientation in the visual field. Across a hypercolumn, essentially all possible angles are represented in separate orientation columns! Sprinkled among the hypercolumns are **blobs**, a term used to describe a cortical column that processes color information and does not have orientation requirements. Blobs are named for the blob shape of the axonal arborization of the thalamic afferents that arise from intralaminar neurons in the LGN and terminate in layers 2 and 3 of the cortex.

There is also a horizontal structure to the cerebral cortex. Pyramidal cells within layers 3 and 5 send axons horizontally across the cortex for several millimeters. These axons extend vertical arborizations at intervals that correspond to the width of hypercolumns. Physiological studies have shown that color blobs and orientation columns of the same angle are interconnected across hypercolumns. This horizontal structure provides an anatomical basis to account for observed interactions between the left and right ocular dominance columns (see below).

PARALLEL PROCESSING IN THE VISUAL SYSTEM

The visual system also processes information along separate parallel pathways. We have seen that four parallel channels of information, one scotopic (monochromatic) for the rods and three photopic (red, green and blue) for the cones, are extracted from the visual signal by the receptors. The parvocellular (X cell) and magnocellular (Y cell) pathways divide the visual space into a high-acuity path (X) with static properties and a low-acuity path (Y) that is very sensitive to movement. Finally, this visual information is segregated spatially according to position within the visual field. All of these channels are segregated at the level of the retina.

Further divisions occur at the level of the thalamus. Visual information passing through the LGN is directed to V_1 of the cortex. Paralleling this, a separate path through the pulvinar innervates all other visual areas in the cortex *except* V_1. Within the LGN and cerebral cortex, the parvocellular and magnocellular channels remain segregated. The parvocellular path synapses in

BOX 11.1.

Boolean Logic Systems

Boolean logic is a mathematical system that functions within the constraints of a base 2 number system. It is useful for working with digital electronic systems that have only two stable states, on and off. As we have noted earlier, the action potential is a digital signal, having only two stable states, and therefore boolean systems have been used to analyze some neural systems. The analysis of neural systems by boolean logic cannot be pressed too far, since the synapse is inherently not digital; it is analog. Nevertheless, the fact that digitally represented signals enter and leave a neuron allows one to use boolean concepts in analyzing some neural systems.

Digital electronic circuits are the basis of all digital computers. Digital signals interact in these systems at **gates** that perform specific boolean functions. At the lowest level, all digital circuits are formed by various combinations of three functions: **AND, OR,** and **NOT.** These functions can best be appreciated by building a "truth table" that illustrates all of the input-output relationships of a gate.

NOT: The NOT function is the boolean function of negation. The input is simply transformed into its opposite. The electronic gate known as the "inverter" implements this function. Its neural equivalent is the inhibitory interneuron.

NOT	
Input	**Output**
0	1
1	0

In ———▷o——— Out

AND: The AND function is the boolean function of simultaneity. The function is asserted if and only if all inputs are also asserted. The electronic "AND gate" performs this function. Simple AND gates have only two inputs, but there is no theoretical limit to the number. In some respects, neurons act like AND gates, since many simultaneous excitatory inputs are usually required to cause a neuron to fire.

AND		
Input 1	**Input 2**	**Output**
0	0	0
0	1	0
1	0	0
1	1	1

In 1 ———
In 2 ——— ———o——— Out

BOX 11.1.—continued

OR: The OR function is the boolean function of alternation. The function is asserted if any input is asserted. The electronic "OR gate" provides this function in computers. Neurons can also act something like OR gates. For example, a set of excitatory stimuli on a single dendrite can cause a neuron to fire. A different set of excitatory stimuli on a different dendrite can also cause it to fire; the two dendritic paths can act independently, like an OR function.

OR		
Input 1	Input 2	Output
0	0	0
0	1	1
1	0	1
1	1	1

In 1 ———\
In 2 ———⊃o——— Out

POSTSTRIATE PROCESSING

Feature abstraction continues beyond V_1 in the extrastriate visual areas. As many as 20 different retinotopically mapped areas have been discovered. In very broad terms, the anatomical organization of these areas seems to extend both the serial and parallel organizational schemes just described. At least two parallel information channels leave V_1 (Fig. 11.25). The *magnocellular (Y) pathway* appears to extract information regarding motion, interocular disparity, and spatial relationships. It proceeds serially through several individually mapped regions into the posterior parietal cortex. The *parvocellular (X) pathway* analyzes form, color, and interocular disparity. It projects in serial stepwise fashion to the deepest part of layer 4, while the magnocellular path synapses more superficially in layer 4. The parallel distribution of information is elaborated further by the output pattern from the cortical columns. The upper layers (4 and 2) of the cortex project to other cortical areas by intracortical association fibers. The deeper layers project to the superior colliculus, the pulvinar (from layer 5), and LGN (layer 6).

The temporal lobe. The *blob system*, a subset of the parvocellular system, seems to be exclusively associated with color analysis and may represent yet a third parallel path. These parallel pathways are not strictly separated. They interact at several levels.

The various extrastriate visual areas seem to individually abstract certain global attributes from the visual image such as shape, color, motion, and interocular disparity (which is necessary for depth perception). It now seems reasonable to hypothesize that *abstractions of the visual scene are analyzed and individually brought to conscious perception by anatomically separate parts of the cerebral cortex.* This hypothesis predicts that appropriately placed lesions would produce perceptual losses of specific attributes while other attributes would be preserved. This hypothesis is supported by some clinical data.

Lesions that produce a defect in recognition or meaning without losses in objective sensations are termed **agnosias** [G., ignorance; from *a + gnosis,* knowledge]. Agnosia is not a sensory loss in the sense that blindness or lack of vibratory sensations are sensory losses. Agnosia is the *inability to recognize or attach meaning to stimuli.*

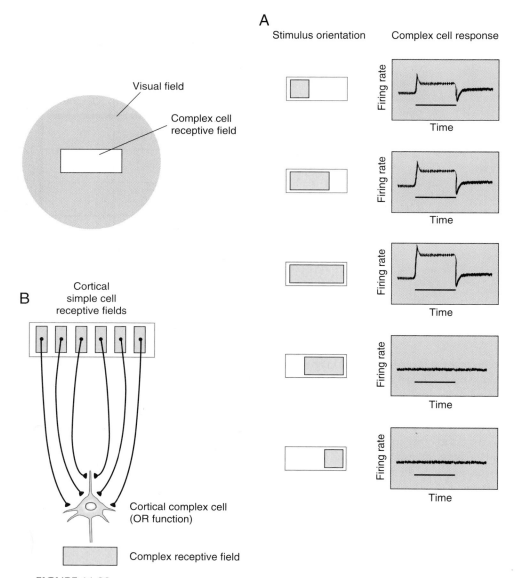

FIGURE 11.23.

Complex cells have large receptive fields as shown here (*top left*). **A**, Complex cortical cells have multiple stimulus requirements, some of which are generalizations of simple cell response requirements. In **A**, a properly oriented edge facilitates the cell, no matter where it is located in the response field, so long as the illumination is greater on the left than the right, otherwise the cell is inhibited. The edge must be properly oriented for the complex cell to respond, no matter how the field is otherwise illuminated. **B**, The synaptic drive for complex cells is thought to be derived from the output of many simple cells as depicted here, where simple cell receptive fields are depicted within the complex cell receptive field. It is hypothesized that any simple cell with appropriate response characteristics in the receptive field of the complex cell would be able to drive the cell (OR function).

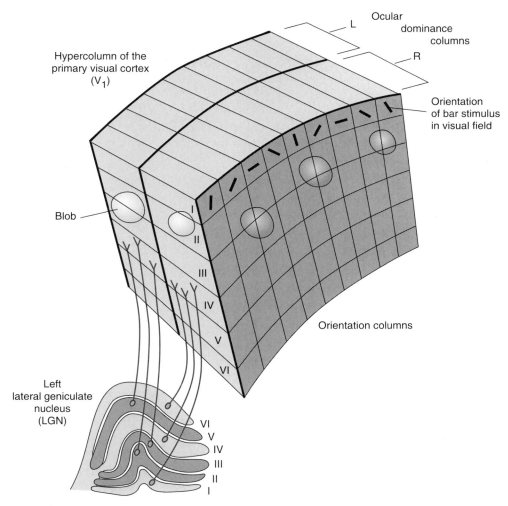

FIGURE 11.24.

Hypercolumns in the visual cortex consist of pairs of ocular dominance columns and numerous orientation columns. Interspersed in the hypercolumns are blobs—columns that are not orientation sensitive and appear to process spectral information. Each hypercolumn processes information from a single discrete spot of the visual field. Hypercolumns are arranged across the surface of the primary visual cortex in a retinotopic pattern.

Several visual agnosias have been correlated with specific cortical lesions. For example, **object agnosia** is the inability to recognize familiar objects by visual means alone. The objects are readily recognized by tactile manipulation, auditory cues, or even odor (if appropriate), but visual sensations do not make the object recognizable or meaningful to the patient. Visual recognition is not possible because the visual sensations, even though consciously perceived, hold no meaning to the patient. Object

agnosia can be very specific. For example, patients have been discovered whose object agnosia is limited to a single class, such as vegetables. One extreme, but well recognized, form of object agnosia is **prosopagnosia** [G. *prozapine,* countenance], the inability to recognize faces. Such patients cannot recognize people from their face, but have no difficulty recognizing them by body movement or the sound of their voice. Object agnosias are associated with bilateral lesions to the ventromedial portion of

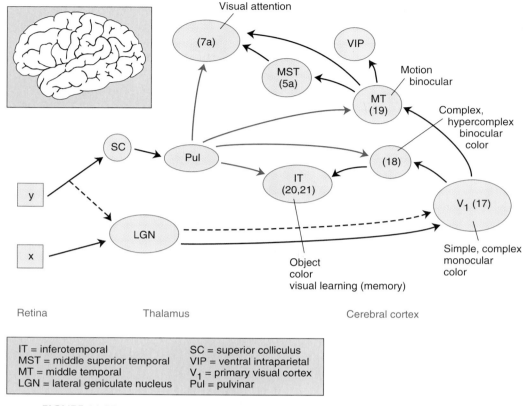

FIGURE 11.25.

  This schematic illustrates a few of the interconnections between some of the visual areas of the cerebral cortex and suggests some of the discrete functions served by each area (see text for details).

the occipitotemporal cortex (Brodmann's areas 20 and 21).

  In the monkey, the inferotemporal cortex is remarkable for its collection of neurons that respond only to the most esoteric of stimuli. Within it, single cells have been discovered that respond only to visual stimuli in the shape of a simian hand (the subject was a monkey) or the outline of a simian face! Such cells are encountered rarely and have been dubbed "grandmother cells," a term used apocryphally to suggest that certain cells may be so specific as to respond only to one's grandmother. Discovery of these cells has provided a physiological explanation for such bizarre syndromes as prosopagnosia.

  Achromatopsia [G. *a + chroman,* without color, *+ oasis,* vision] is the inability to make color hue discriminations. The patients remark that everything looks gray. Not to be confused with color blindness, which is a retinal defect due the absence of one or more sets of color-sensitive cones, achromatopsia is a type of agnosia. Color information is sent to the brain; it just has no meaning to the patient. Without meaning, discriminations cannot be made. Achromatopsia is associated with bilateral lesions to area 37 (perhaps also 18), which is a transition zone between the occipital and temporal lobes.

  Motion plays a critical role in visual perception. For example, many subprimates have considerable difficulty attaching meaning to nonmoving objects. A cat seems unable to pounce on the still mouse, but springs instantly the moment the creature makes the slightest twitch. Primates and humans are not quite so dependent on movement for visual perception. Nevertheless, a separate specialized region of the cerebral cortex is necessary in order to consciously perceive motion. Recently, experiments have been performed on monkeys that have shown that artificially stimulating a very small number of cells in this movement area can alter the monkey's *perception* of the direction a moving target is taking. In the human, a bilateral lesion of the posterior part of the middle temporal gyrus and the adjacent lateral occipital gyrus results in the inability to perceive motion. There is no other abnormality in visual perception. A description of this phenomenon, as documented by the experience of one patient, follows:

> The visual disorder complained of by the patient was a loss of movement vision in all three dimensions. She had difficulty, for example, in pouring tea or coffee into a cup because the fluid appeared to be frozen, like a glacier. In addition, she could not stop pouring at the right time since she was unable to perceive the movement in the cup (or a pot) when the fluid rose. Furthermore the patient complained of difficulties in following a dialogue because she could not see the movements of the face and, especially, the mouth of the speaker. In a room where more than two other people were walking she felt very insecure and unwell, and usually left the room immediately, because "people were suddenly here or there but I have not seen them moving." The patient experienced the same problem but at an even more marked extent in crowded streets or places, which she therefore avoided as much as possible. She could not cross the street because of her inability to judge the speed of a car, but she could identify the car itself without difficulty. "When I'm looking at the car first, it seems far away. But then, when I want to cross the road, suddenly the car is very near." She gradually learned to "estimate" the distance of moving vehicles by means of the sound becoming louder.[4]

Development of Cerebral Function

How are the synaptic connections of the cerebral cortex determined? Obviously, the principal

[4]Quoted from Zihl, J., von Cramon, D., and Mai, N. Selective Disturbance of Movement Vision after Bilateral Brain Damage. *Brain* 106:313–340, 1983, with permission.

organization is determined by the genetic code. The lateral geniculate nucleus is always connected by the optic radiations with the cerebral cortex of the calcarine fissure; the visual cortex is never seen straddling the lateral fissure. The ocular dominance columns are always arranged retinotopically. But what about the synaptic connections within and among the cortical columns? Are they determined genetically, or can experience influence the nature of the final synaptic arrangement of the cerebral cortex in a way that ultimately determines function? The columnar structure of the cerebral cortex has allowed a number of important experiments to be designed that have probed these questions.

It has been convenient for investigators to physiologically categorize cells in the visual cortex into one of seven classes, based on how strongly they are driven by either eye. The cells in category 1 are driven exclusively by the contralateral eye; those in category 7, by the ipsilateral eye. Category 4 represents cells driven about equally by both eyes. The other categories represent intermediate levels of lateralization. The degree that a cell is driven by both eyes is determined by the strength of the horizontal circuits that interlink the ocular dominance columns. If the horizontal connections do not develop, all the cells in a column will be driven monocularly.

If one deprives a newborn kitten of visual input to one eye by suturing the eyelids together, one can observe the effect monocular deprivation has on cortical development, for in the kitten, most of the cortical synapses are formed during the second through fifth weeks of life. After 3 months of monocular deprivation, kittens behave as if blind in the deprived eye, and sight is never restored during the life of the animal. The retina and LGN appear to function normally. In V_1, however, all the cells can only be driven from the normally stimulated eye. The synapses from the unstimulated eye have failed to develop. Three months of monocular deprivation in an adult cat does not seriously affect the binocular pattern.

The orientation columns are also affected by perinatal experience. Kittens raised in a visual environment consisting of vertical or horizontal stripes develop orientation columns in V_1 that correspond to the orientation of the visual stimulus. The wide range of orientation columns representing all angles, that is so characteristic of the normal V$_1$, is absent. One can only speculate how this adaptive transformation of cortical function affects what these animals actually perceive in their visual world.

The difference in effect of monocular deprivation between adult and neonatal cats suggests that there is a perinatal **critical period** during which cortical synaptic relationships are indeterminate. After the critical period, the synaptic structure of the cortex is sufficiently established to be permanent. *Experience is the force that ultimately determines adult cortical synaptic structure.* In cats, experiments have shown that this critical period extends for approximately the first 100 days (13 weeks) of life. In humans, the critical period cannot be established with certainty. Some insight is provided by children with **strabismus** [G. *strabismos,* a squinting], a condition in which the eyes do not track together. This is frequently caused by a weakness of one or more of the muscles of ocular motion. It causes **diplopia**, or double vision, since the two retinal images cannot be fused by movement of the eyes. If it persists, the diplopia is resolved by a neurologic suppression of one of the two images. If the condition begins during infancy and persists into the fifth or sixth year, vision is permanently lost in the suppressed eye, just as with the monocularly deprived kittens. Treatment during this critical period consists of forcing the child to use both eyes, usually by alternative eye patching. Good vision can usually be maintained in both eyes. If the underlying cause of the strabismus is not relieved, one eye will be suppressed after patching is discontinued. Later in life, if vision is lost in the good eye (perhaps through trauma), vision from the previously suppressed eye can often be restored.

Lesions to the Visual Pathways

Lesions to the visual pathways produce very characteristic deficits in the visual fields. These deficits are easily documented during the neurological examination and are extremely useful in reaching a proper diagnosis. For these reasons, the visual field losses associated with specific visual pathway lesions (Fig. 11.26) need to be memorized.

Visual fields are tested at the bedside or an office examination by a **confrontation test**. The examiner faces the patient, who covers one eye with his hand and fixates on the examiner's nose while the examiner places objects in the visual field of the open eye. Moving targets such as twitching fingers are sufficient in most cases. One should, at a minimum, determine that vision is present in all four quadrants of each eye separately and that the visual fields extend nearly 90° from forward gaze. If necessary, the visual fields can be accurately mapped by computer-assisted perimetry. It is customary to describe visual deficits in terms of the *visual fields,* not the retinal fields. The visual fields are, of course, the opposite of the retinal fields due to the inverting action of the lens.

The term to describe total blindness in one eye is **monocular blindness**. This can be caused by total failure of the *retina* or loss of the entire *optic nerve*. Both conditions are less common than partial monocular blindness, in which the retina has punctate lesions that leave scotomas in the visual fields, much like the blind spot produced by the papilla. **Glaucoma** is manifested by a shrinking of the visual fields from the perimeter in a more or less symmetrical pattern. If the optic nerve is the primary site of the lesion, the patients may experience an acute blurring or haziness of vision in one eye. In some cases there may be a scotoma. If the optic nerve is involved near the papilla, the disk margin may be swollen or blurred, a condition known as **papillitis**. It is differentiated from papilledema because the later does not usually produce visual disturbances.

A **homonymous hemianopia** [G. *homos,* same + *nomos,* law + *hemi,* half + *a,* without + *ops,* eye] is the loss of vision in half of the visual field divided by the vertical meridian. As the name implies, the same hemifield is affected in both eyes. A homonymous hemianopia may be **complete** or **incomplete**, depending on whether or not the entire half of the visual field is affected. It can be produced by almost any lesion more central than the optic chiasm. This very characteristic deficit is explained by the crossing pattern of the optic nerve axons at the optic chiasm. Since the axons arising from the nasal half

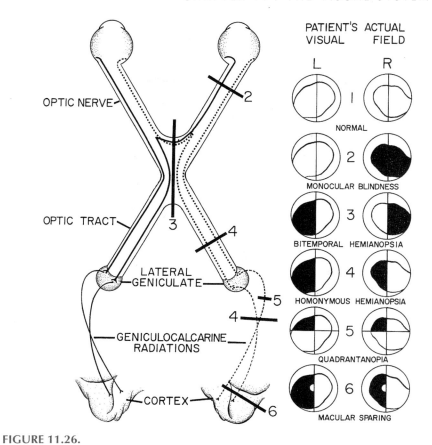

PATIENT'S ACTUAL
VISUAL FIELD

L   R

I
NORMAL

2
MONOCULAR BLINDNESS

3
BITEMPORAL HEMIANOPSIA

4
HOMONYMOUS HEMIANOPSIA

5
QUADRANTANOPIA

6
MACULAR SPARING

OPTIC NERVE

OPTIC TRACT

LATERAL
GENICULATE

GENICULOCALCARINE
RADIATIONS

CORTEX

FIGURE 11.26.

Lesions to the visual pathway produce various patterns of blindness as illustrated.

of one retina cross the midline to join the axons arising from the temporal half of the opposite retina, all of the axons in the optic tract receive light from the same visual hemifield. Lesions producing a homonymous hemianopia are contralateral to the visual field defect.

A complete homonymous hemianopia can be produced if the entire *optic tract, lateral geniculate nucleus, optic radiations,* or *primary visual cortex (V$_1$)* is destroyed. One cannot differentiate, with visual field data alone, which of these structures is affected if the homonymous hemianopia is complete. It is useful to know, however, that lesions to the optic tract and lateral geniculate are rare, but infarctions of the optic radiations are quite common. Furthermore, this is a large tract and most lesions are partial, producing an **incomplete homonymous hemianopia**. As discussed previously, about half of

the axons in the optic radiations take an indirect course from the LGN to the calcarine fissure, passing around the lateral ventricle in the temporal lobe.[5] Axons taking the indirect path terminate in the inferior lip of the calcarine fissure, and their loss produces a **superior quadrant homonymous hemianopia**. Loss of axons taking the direct path leaves the patient with an **inferior quadrant homonymous hemianopia**.

One can anticipate that a complete homonymous hemianopia would be produced if the occipital cortex were lesioned. As a practical matter, occlusion of the *calcarine artery,* a branch of the posterior cerebral artery, usually spares the macular portion of the visual field. This

[5]This portion of the optic radiations is often called Meyer's loop, but has also been referred to as Flechsig's loop, Archambault's loop, and Cushing's loop. In the face of such a multiplicity of candidates, one is hard pressed to justify any eponym for this structure.

macular sparing is probably due to perfusion of the occipital pole by anastomotic collateral circulation reaching it from the terminal branches of the middle cerebral artery.

Lesions to the optic chiasm may produce almost any type of visual field disturbance, since all of the optic fibers are present in that structure. The most characteristic deficit is a **bitemporal hemianopia** caused by cutting only the crossing fibers, sparing the uncrossed fibers. Such visual field deficits are not uncommon. They may be caused by pituitary tumors, aneurysm of the circle of Willis, or neoplasms at the base of the third ventricle. Dividing all of the axons at the optic chiasm, as may occur with bullet wounds following unsuccessful suicide attempts, results in **total blindness**.

CASE STUDIES

Ms. J. P.

HISTORY, JANUARY 20, 1989:   Two weeks prior to this examination, Ms. J. P., a 22-year-old white female, developed blurred vision in her right eye that she described as "spots in front of my eye." Her vision was severely disturbed and, worried about this, she saw an optometrist, who told her that it would probably clear up in a few weeks. The optometrist advised her to see a neurologist, which brought her to this examination. Presently the blurred vision is somewhat improved. She has no other complaints.

HISTORY: About 2 years ago, Ms. J. P. suffered from severe headaches. She was advised by her family physician to change jobs and leave her boyfriend. She did both and her headaches resolved.

NEUROLOGICAL EXAMINATION:

I.   MENTAL STATUS: Normal

II.   CRANIAL NERVES: All cranial nerve functions were tested and found to be normal except the following:

   A. Optic: Ophthalmoscopic examination of the two eyes revealed papillitis on the right. On confrontation, a scotoma in the left inferior quadrant of the right visual field was apparent. The left eye is normal.

   B. Oculomotor: There is a slight ptosis on

the right. The patient says that it has been present all her life.

III. MOTOR SYSTEMS: Strength was normal and symmetrical. Gait was steady and narrow based. Cerebellar functions were normal.

IV. REFLEXES: All MSRs were symmetrical and within normal limits.

V. SENSORY SYSTEMS: The patient reported no abnormalities to light touch, pin prick, or vibratory sensations.

Comment: Papillitis is a swelling and blurring of the papilla caused by an inflammation of the optic nerve close to its insertion into the eyeball. Papillitis produces disturbed vision and frequently is associated with scotomas. The eyeball may also be tender to touch and painful during normal eye movements. Usually no cause is determined, and two-thirds of patients have normal vision restored within a few weeks. The remainder are left with some permanent visual disturbance ranging from very slight blurring to total blindness. Although at the time of examination no cause for the papillitis can usually be determined, approximately 75% of these patients will develop multiple sclerosis (MS) *within 15 years.*

Multiple sclerosis is a **demyelinating disease of the central nervous system.** *Here, demyelinating is taken in its narrowest sense, referring to the principal pathological event, since nearly all neurological diseases eventually cause demyelination secondarily. There is no accepted etiology for MS, but there is growing evidence that it is caused by an infectious agent. Susceptibility to this agent is apparently limited to approximately the ages of 11 to 45. The infectious agent, 6 to 12 years later, appears to provoke an autoimmune reaction that is directed toward* **CNS myelin produced by oligodendrocytes.** *Peripheral myelin is never affected. During the immunological attack, inflammation, edema, and demyelination interfere with axonal function, producing* **symptoms restricted to the CNS.** *If the attack subsides, the inflammation and edema subside, but the demyelination remains.[6] The acute symptoms of disease diminish, but a residual, permanent deficit usually remains. For reasons not understood, after an indeterminate period of time, the patient may experience repeated autoimmune attacks, each attack followed by a period of remission. Each exacerbation leaves the*

---

[6]Recall the structure of the oligodendrocyte and its inability to regenerate its processes once lost (see Chapter 3).

patient more compromised. The immunological attacks appear to strike the CNS randomly, resulting in multifocal lesions. Therefore, multiple sclerosis is characterized by periods of exacerbation and remission, disseminated in space and time, and following a progressive course. There is considerable variability in the progression of MS among individuals. Some people experience one or two exacerbations followed by a lifelong remission. In others the disease follows a malignant course, bringing death within a few months without any remission.

There are no specific laboratory studies that permit an unequivocal diagnosis of MS; the diagnosis ultimately rests on the long-term picture of exacerbations, remissions, multifocal CNS lesions sparing the peripheral nervous system, and the exclusion of other possible diagnoses. The most useful procedures are (in order of their diagnostic value) MRI, evoked potential studies (BAER, VER, and SER),[7] demonstration of oligoclonal IgG bands on CSF electrophoresis, and the IgG index. The IgG index compares the ratios of IgG and albumin in the

CSF and serum.[8] A ratio greater than 1.7 and the demonstration of oligoclonal bands are found in about 90% of patients with clinically diagnosed MS. MRI can usually demonstrate multifocal white matter lesions (Fig. 11.27). There is no specific treatment of MS. Anti-inflammatory therapy using ACTH or methylprednisolone has been proved to hasten recovery of acute episodes, but there is no evidence that it alters the course of the disease or lessens the number or severity of subsequent relapses. Immunosuppressive therapy has not been proved to be beneficial at this time.

SUBSEQUENT COURSE: Although most physicians would suspect that Ms. J. P. has multiple sclerosis, at this time there is no history of exacerbations and remissions, nor is there evidence of multifocal CNS lesions. Therefore the diagnosis is not secure, so no such diagnosis was made. Ms. J. P. was told that her papillitis would probably resolve in a few weeks, and she was advised to return if she had any further trouble.

Almost 2 years later (November 1990) she was again seen by her neurologist. She was weak, her

---

[7]The visual evoked response (VER) and sensory evoked response (SER) are similar to the brainstem auditory evoked response (BAER), except that visual or sensory stimuli are used instead of auditory stimuli.

---

[8]
$$IgG\ index = \frac{CSF\ IgG/serum\ IgG}{CSF\ albumin/serum\ albumin}$$

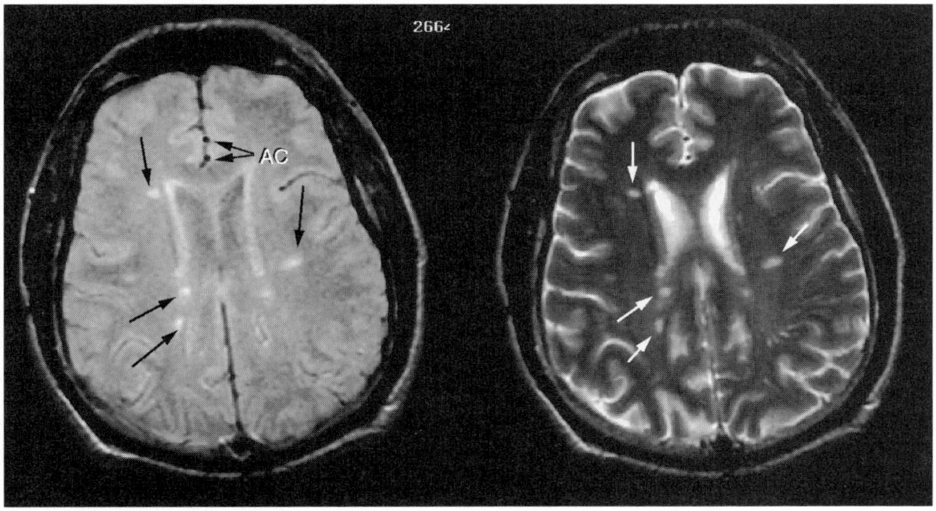

FIGURE 11.27.

T1-weighted (*left*) and T2-weighted (*right*) MR images in the same plane demonstrate several small white matter lesions (*arrows*) in Ms. J. P. Findings like these correlate closely with the clinical diagnosis of multiple sclerosis. The anterior cerebral arteries (*AC*) are easily seen in both images.

speech was dysarthric, and she was having difficulty swallowing liquids. She was very unsteady on her feet.

NEUROLOGICAL EXAMINATION:

I. MENTAL STATUS: Normal

II. CRANIAL NERVES:

A. **Facial:** When asked to grimace, there was a noticeable weakness of the lower half of the right side of the face. Her mouth deviated slightly to the right during conversation.

B. **Glossopharyngeal-vagus:** The gag reflex was diminished to stimulation on either the right or left. She had some difficulty swallowing liquids, choking almost always. Swallowing solids was difficult but did not cause her to choke.

III. MOTOR SYSTEMS:

A. There is marked weakness in her right arm and shoulder muscles as revealed by the drift test and direct strength testing. Her right hand grip is much weaker than the left.

B. Strength in her legs was symmetrical and within normal limits.

IV. REFLEXES:

A. MSRs from all five test points were 3+ bilaterally.

B. The Babinski sign was present bilaterally.

V. CEREBELLUM:

A. Her gait is wide based and ataxic; she nearly fell during the examination. She was unable to perform a tandem walk.

B. Finger-to-nose performance was ataxic with either arm. Heel-to-shin was ataxic with either leg.

C. There was marked dysdiadochokinesia in the right arm, much less on the left.

VI. SENSORY SYSTEMS: Her responses to pin prick, light touch, and vibration were diminished everywhere, but were nearly absent from the right arm and hand.

Comment: At this time there is objective evidence of CNS lesion(s); there is no evidence of involvement of the peripheral nervous system. The increased MSRs and the bilateral Babinski signs demonstrate the involvement of the corticospinal tract bilaterally, and the weakness of the facial and glossopharyngeal-vagus nerves points to the corticobulbar tract as well. Cerebellar involvement is indicated by the ataxia, dysmetria, and dysdiadochokinesia. Sensory involvement of the anterolateral system and the medial lemniscus is demonstrated by the sensory examination.

Ms. J. P.'s symptoms can be explained by a centrally placed lesion in the pons involving the corticobulbospinal tracts, the ALS, the medial lemniscus, and the deep pontine nuclei. Given her history of papillitis, multiple sclerosis is the most likely diagnosis, although her problems could be caused by any insult to the pons. A subsequent MRI study failed to demonstrate an intracranial mass or hemorrhage, but did demonstrate multiple periventricular abnormalities of the white matter and pons.

FURTHER APPLICATIONS

• When seeking a neurological diagnosis, one seeks to explain all the patient's signs and symptoms with as simple an explanation as possible. If a patient's situation can be explained by two widely separated lesions or a single lesion, the single lesion is the most likely correct diagnosis. Multiple sclerosis challenges this approach, at least in the long run. However, most exacerbations of MS are explained by a single lesion. Try to explain this patient's current difficulties by defining a single lesion, if possible, or by the smallest number of lesions. Make a drawing of the appropriate areas to illustrate your hypotheses.

• Explain the reason that MRI is very successful in demonstrating CNS demyelination while CT almost always fails to image white matter plaques caused by MS.

• With a friend or in a small group, debate the pros and cons of telling Ms J. P., after the first examination, that she probably has multiple sclerosis. Ask physicians you know what they do under such circumstances.

SUMMARY

Light is transformed into coded electrical signals by the eye. Passive optical elements of the eye, the cornea, and the lens focus images on the active transducer cells of the retina, the rods, and the cones. Current flow through the receptor cells is controlled by Na^+ ionophores that are gated by cGMP. Photons, absorbed by the visual pigments in the receptors, induce a series of G protein-mediated biochemical reactions that affect the availability of cGMP and consequently the flow of ionic current through the receptor. Current flow through the receptor modulates the release of neurotransmitter (probably glutamate) from the receptor's synapses.

Receptor cells synapse with bipolar and horizontal cells in the retina. The bipolar cells synapse with ganglion and amacrine cells. Although the synapses are chemical, only the ganglion cell supports action potentials; the intraretinal signaling is accomplished by graded local potentials. Visual information transduced into electrical form by the 6 million cones and 100 million rods converges on 1 million ganglion cells. In the process, the pixels of photic information detected by the receptor cells are combined into more complex center-surround patterns. Retinal processing enhances the detection of boundary conditions in the visual image.

Ganglion cell axons carry visual information to the lateral geniculate nucleus, the superior colliculus, and the pretectal area of the brainstem. The axons in the optic nerve are rearranged at the optic chiasm such that the visual image from one half of the visual field is carried to contralateral central structures, regardless of the eye of origin. The visual fields are mapped onto the receiving structures in a precise retinotopic organization. Visual information from the LGN is transmitted to the striate cortex, V_1, in the occipital lobe by the optic radiations. Visual information from the superior colliculus is transmitted to the pulvinar (among other places), from which it is sent to extrastriate visual cortical areas.

In V_1, simple cells are sensitive to bars of light oriented in a particular direction. Simple cells have relatively small receptive fields, being approximately the size of the bar of light. Complex cells are also sensitive to particularly oriented bars of light, only they have receptive fields much larger than the bar of light. The response of complex cells is an abstraction of the response of many simple cells with the same orientation. The response characteristics of simple and complex cells can be modeled by simple logical gating functions.

From the striate cortex, visual information is processed by at least two parallel cortical systems, each of which is composed of several serially linked, independently mapped visual processing areas. In very broad terms, the parallel paths abstract information about movement, color, shape, and spatial relationships from the visual image. Clinically, damage to these higher-order processing centers leads to specific visual agnosias that correspond with the visual quality abstracted by that center.

FOR FURTHER READING

Adams, R., and Victor, M. *Principles of Neurology*, 5th ed. New York: McGraw-Hill, 1993.

Brodal, A. *Neurological Anatomy in Relation to Clinical Medicine*, 3rd ed. New York: Oxford University Press, 1981.

Carpenter, M. B. *Core Text of Neuroanatomy*, 4th ed. Baltimore: Williams & Wilkins, 1991.

Detwiler, P. Sensory Transduction. In: *Textbook of Physiology*, 21st ed., edited by Patton, H., Fuchs, A., Hille, B., Scher, A., and Steiner, R. Philadelphia: W. B. Saunders, 1989, pp. 98–129.

Fawcett, D. *A Textbook of Histology*, 12th ed. Philadelphia: W. B. Saunders, 1986.

Fuchs, A. The Visual System: Optics, Psychophysics, and the Retina. In: *Textbook of Physiology*, 21st ed., edited by Patton, H., Fuchs, A., Hille, B., Scher, A., and Steiner, R. Philadelphia: W. B. Saunders, 1989, pp. 412–441.

Fuchs, A. The Visual System: Neural Processing Beyond the Retina. In: *Textbook of Physiology*, 21st ed., edited by Patton, H., Fuchs, A., Hille, B., Scher, A., and Steiner, R. Philadelphia: W. B. Saunders, 1989, pp. 442–474.

Gouras, P. Color Vision. In: *Principles of Neural Science*, 3rd ed., edited by Kandel, E., Schwartz, J., and Jessell, T. New York: Elsevier, 1991, pp. 467–480.

Haines, D. *Neuroanatomy: An Atlas of Structures, Sections, and Systems*, 3rd ed., Baltimore: Urban & Schwarzenberg, 1991.

Kandel, E. Perception of Motion, Depth, and Form. In: *Principles of Neural Science*, 3rd ed., edited by Kandel, E., Schwartz, J., and Jessell, T. New York: Elsevier, 1991, pp. 440–466.

Katz, L., and Callaway, E. Development of Local Circuits in Mammalian Visual Cortex. *Annu. Rev. Neurosci.* 15:31–56, 1992.

Kaupp, U. B., and Koch, K. W. Role of cGMP and Ca^{2+} in Vertebrate Photoreceptor Excitation and Adaptation. *Annu. Rev. Physiol.* 54:153–75, 1992.

Merigan, W. H., and Maunsell, J. H. R. How Parallel are the Primate Visual Pathways?. *Annu. Rev. Neurosci.* 16:369–402, 1993.

Miyashita, Y. Inferior Temporal Cortex: Where Visual Perception Meets Memory. *Annu. Rev. Neurosci.* 16:245–263, 1993.

Moore, K. L. *The Developing Human; Clinically Oriented Embryology*. Philadelphia: W. B. Saunders, 1988.

Nieuwenhuys, R., Voogd, J., and van Huijzen, C. *The Human Central Nervous System; A Synopsis and Atlas*, 3rd ed., Berlin: Springer-Verlag, 1988.

Weymouth, F. The Eye as an Optical Instrument. In: *Physiology and Biophysics*, 19th ed., edited by Ruch, T. C., and Patton, H. D. Philadelphia: W. B. Saunders, 1965.

Hypothalamus and Associated Systems

The **hypothalamus**, a subdivision of the diencephalon, lies on either side of the most ventral reaches of the third ventricle. It is separated from the thalamus by the **hypothalamic sulcus** (see Fig. 1.10). Most of its other boundaries are less distinct; the gray matter blends into the neighboring structures. At its rostral pole, the hypothalamus is bounded by the **lamina terminalis**, a thin sheet of nervous tissue that is the remains of the anterior pole of the neural tube and, in the adult, separates the diencephalon from the telencephalon (Figs. 12.1 and 12.2). The optic nerves and chiasm cradle the anterior portion of the hypothalamus and serve as important landmarks by which many of the hypothalamic structures are identified. The **mamillary bodies** mark the most posterior extent of the hypothalamus. Internally, its posterior boundary is marked by the subthalamus. The ventral portion of the hypothalamus narrows to a small neck, the **tuber cinereum** [L. *tuber,* a knob + L. *cinereus,* ashen (gray)], the ventral-most portion of which underlies the third ventricle, is slightly swollen, and is called the **median eminence**. The median eminence blends into the **infundibular stalk** [L. *infundibulum,* a funnel], which is continuous with the posterior lobe of the **hypophysis** [G. *hypo,* under + G. *phyo,* to grow] or **pituitary gland** [L. *pituita,* phlegm or mucus]. Together, the median eminence, infundibulum, and posterior pituitary gland are known as the **neurohypophysis**. The anterior portion of the pituitary gland is the **adenohypophysis** [G. *aden,* gland]. It is composed of a **pars tuberalis**, which has no endocrinological function, and a **pars distalis**, the secretory portion of the gland. Between these two major divisions of the hypophysis lies the **pars inter-media**, a portion of the gland that is poorly developed in the human.[1]

The cells of the hypothalamus have been organized into a number of named nuclei. Only a few of these are distinct, and most need concern only the specialist. Those that are important to the physician will be discussed in the appropriate sections that follow. In more general terms, the hypothalamus has been divided into a number of regions. The most rostral is the **anterior region**. It is composed of the area of the hypothalamus immediately superior to the optic tracts and chiasm. The **tuberal region** lies over the tuber cinereum, while the **posterior region** includes the mamillary bodies and that part of the hypothalamus immediately superior to them. The hypothalamus is also divided into a **medial** and a **lateral** part by an imaginary sagittal plane passing through the **fornix**. Consequently, each half of the hypothalamus is divided into six compartments. Most of the named nuclei lie in the medial compartments and are summarized in Table 12.1.

The constancy of the internal milieu is maintained by *local, nervous, hormonal,* and *behavioral* mechanisms. The interplay of these four mechanisms is essential to the long-range function and survival of the organism. The regulation of blood flow provides an excellent example of the interplay of these four systems. For example, blood flow in the capillary bed is regulated by the precapillary sphincter. This muscle is influenced by a number of *local* factors, including local O_2 and CO_2 tension and pH. Blood flow can be shunted between large capillary beds by the ac-

---

[1]The anatomy and physiology of the pituitary gland is more easily understood if one knows the embryological origin of its parts. Please see Appendix A.

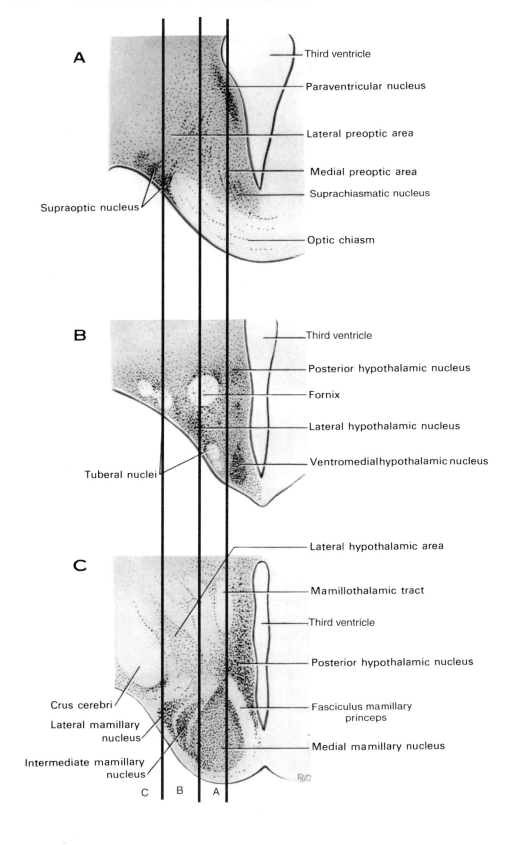

A

Third ventricle

Paraventricular nucleus

Lateral preoptic area

Medial preoptic area

Suprachiasmatic nucleus

Optic chiasm

Supraoptic nucleus

B

Third ventricle

Posterior hypothalamic nucleus

Fornix

Lateral hypothalamic nucleus

Ventromedial hypothalamic nucleus

Tuberal nuclei

C

Lateral hypothalamic area

Mamillothalamic tract

Third ventricle

Posterior hypothalamic nucleus

Fasciculus mamillary princeps

Medial mamillary nucleus

Crus cerebri

Lateral mamillary nucleus

Intermediate mamillary nucleus

C   B   A

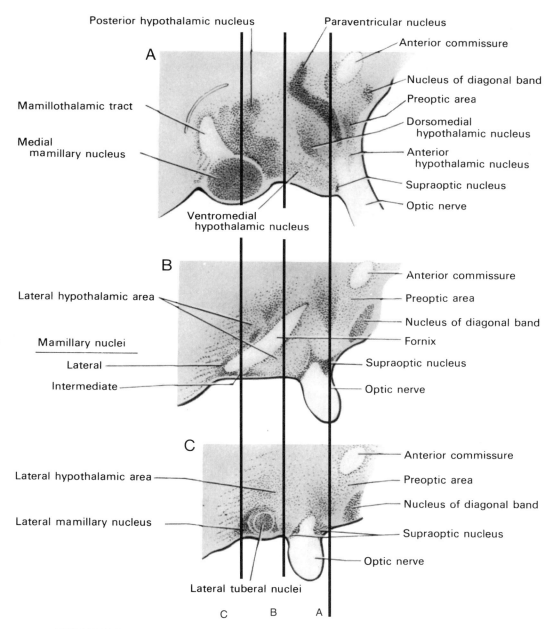

Posterior hypothalamic nucleus

Paraventricular nucleus

Anterior commissure

Mamillothalamic tract

Nucleus of diagonal band

Preoptic area

Medial mamillary nucleus

Dorsomedial hypothalamic nucleus

Anterior hypothalamic nucleus

Supraoptic nucleus

Optic nerve

A

Ventromedial hypothalamic nucleus

Lateral hypothalamic area

Anterior commissure

Preoptic area

Nucleus of diagonal band

Mamillary nuclei

Fornix

Lateral

Supraoptic nucleus

Intermediate

Optic nerve

B

Lateral hypothalamic area

Anterior commissure

Preoptic area

Nucleus of diagonal band

Lateral mamillary nucleus

Supraoptic nucleus

Optic nerve

C

Lateral tuberal nuclei

C   B   A

FIGURE 12.2.

Sagittal sections through the hypothalamus in the planes indicated in Figure 12.1.
Lines *A, B,* and *C* correspond to the planes shown in Figure 12.1.

FIGURE 12.1.

Frontal sections through the hypothalamus. **A,** At the level of the optic chiasm; **B,** At
the tuberal region; **C,** At the level of the mamillary bodies. Lines *A, B,* and *C* indicate
the planes of the sections illustrated in Figure 12.2.

Table 12.1. Source of Hypothalamic Releasing Hormones

Nucleus	Releasing Hormones
Supraoptic	Oxytocin, vasopressin
Paraventricular	Oxytocin, vasopressin, CRH
Preoptic	GnRH
Septal	GnRH
Arcuate	GnRH, GHRH, PRIH
Periventricular	TRH, GHIH

tion of the *nervous system,* which can greatly diminish flow to the digestive organs to favor the deep muscles and vice versa. Circulating *hormones,* such as angiotensin II, influence blood pressure in the entire organism by causing massive peripheral vasoconstriction. Finally, *behavioral responses,* such as one's assuming the prone position on feeling faint due to failing blood pressure, restores blood flow to the brain.

The hypothalamus organizes whole-body homeostasis through three distinct mechanisms. First, through descending connections with the brainstem and spinal cord, it *controls and regulates the autonomic nervous system.* Second, the hypothalamus itself *serves as an endocrine organ.* Secretions from hypothalamic neurons into the bloodstream control and regulate a number of physiological and endocrinological functions. Finally, through its connections with forebrain structures, it *plays an important, although very poorly understood, role in regulating the behavior of the organism.* In this chapter, each of the three functions and the associated anatomy of the hypothalamus will be considered separately.

Connections with the Autonomic Nervous System

Traditionally the nervous system is segregated into *voluntary* and *involuntary* divisions, the latter known as the **autonomic nervous system** or **ANS.** Such a dichotomy is too simple, but the concept of an *independent* or autonomous portion of the nervous system is useful. The autonomic nervous system is primarily an efferent system that controls and regulates several fundamental physiological functions. Among these are the regulation of the cardiovascular and digestive systems, temperature regulation, and reproductive functions. Some of these functions are entirely autonomous, such as the regulation of body temperature (over which we can exert no conscious control), while others are semiautonomous, such as sexual arousal (over which we have some conscious control—initiation of sexual activity, for example—but that has many elements, such as orgasm, that are autonomous).

ANATOMY OF THE AUTONOMIC NERVOUS SYSTEM

The ANS is composed of three divisions, the **sympathetic,** the **parasympathetic,** and the **enteric.** The sympathetic and parasympathetic divisions are composed of central and peripheral structures. The enteric division is exclusively peripheral.

Central neurons of the ANS reside in the brainstem and spinal cord and leave the CNS over cranial nerves or ventral roots. These motor axons synapse in peripheral **autonomic ganglia,** from which secondary axons arise that innervate the target organ. Therefore, the sympathetic and parasympathetic portions of the ANS are composed of **preganglionic** and **postganglionic** neurons. The cell bodies of preganglionic neurons lie in the CNS, while those of the postganglionic neurons reside in peripheral ganglia. The enteric division resides entirely within the layers of the gut and its closely associated organs, the gallbladder and the pancreas.

Parasympathetic Division

The preganglionic axons of the parasympathetic nervous system originate in brainstem nuclei or in the sacral spinal cord (Fig. 12.3). From the brainstem they travel with specific cranial nerves (III, VII, IX, or X); those originating in the spinal cord form the **pelvic splanchnic nerves** and are distributed to the pelvic structures via the inferior hypogastric plexus and the hypogastric nerve and its branches. These preganglionic parasympathetic axons are generally long, and synapse in ganglia that are near or in the target organ.

Sympathetic Division

The preganglionic neurons of the sympathetic nervous system originate in the **intermediolateral** nucleus of the thoracic and lumbar spinal cord. The sympathetic preganglionic axons are myelinated and generally short, synapsing in ganglia that are close to the CNS (Fig. 12.3). They leave the CNS in the ventral root, but soon separate from it as the **white ramus**. Most of these axons enter a chain of **paravertebral ganglia** that are close to and segmentally associated with all of the thoracic and the first two or three lumbar spinal segments. Three ganglia in the cervical region—the **inferior**,[2] **middle**, and **superior cervical ganglia**—are connected with the chain but have no direct segmental connections with the spinal cord. The chain of ganglia are interconnected, and preganglionic axons may synapse in more than one ganglion.

Postganglionic axons are unmyelinated and leave the chain ganglia as the **gray ramus** to rejoin the spinal nerves. They typically innervate the body wall and blood vessels. Sympathetic axons innervating the face and brain arise from the inferior and superior cervical ganglia and follow the carotid and vertebral arteries to their targets. There are also several **collateral sympathetic ganglia**, the **celiac**, the **superior mesenteric**, and the **inferior mesenteric** being the most prominent. They are not a part of the paravertebral chain, but exist independently. Axons leaving the collateral ganglia form nerves that innervate the viscera. Some of the larger organs—the heart, lungs, eyes and glands of the head—are innervated directly from the chain ganglia. An exception is the innervation of the adrenal medulla, which receives preganglionic axons directly from the spinal cord.

Enteric Division

The enteric division of the ANS lies entirely outside of the CNS It is composed of two nets of interconnected neurons, the **myenteric plexus** and the **submucosal plexus** (Fig. 12.4). Together they are composed of perhaps as many as 10^8 neurons. The myenteric plexus lies between the longitudinal and circular muscle layers of the gut, while the submucosal plexus lies between the circular muscles and the mucosa. Each consists of a multitude of ganglia that contain the neuronal cell bodies. The ganglia and the two plexuses are interconnected by axons that form an intricate network. Some of the neurons are sensory, registering the distention of the gut and the chemical environment of the mucosa. Other neurons affect the smooth muscle walls, the tone of the vasculature, and the secretory activity of the glands. The enteric system maintains the rhythmic and coordinated contractions of the intestinal tract. Although it can operate independent of the CNS, it is closely regulated by both sympathetic and parasympathetic innervation in the intact animal.

INNERVATION OF SPECIFIC ORGANS

Dysfunction of the ANS is a frequent sign of neurological involvement in the disease process. Therefore it is helpful for the physician to know the specific autonomic innervation of the major organs, since this helps one to understand the disease process and to reach appropriate diagnostic conclusions. Many of the following de-

[2]The inferior cervical ganglion is usually fused with the first thoracic ganglion, making a large, star-shaped structure commonly called the stellate ganglion.

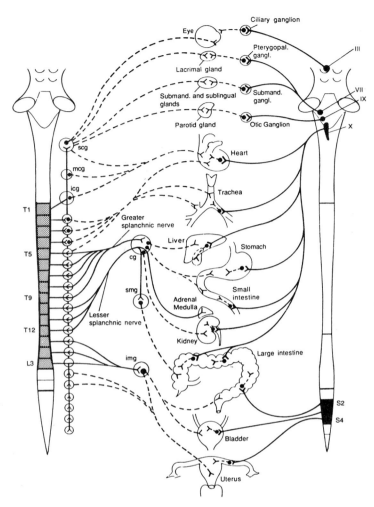

FIGURE 12.3.

Schematic depiction of the autonomic nervous system. Parasympathetic components are shown on the right, and sympathetic components on the left. Note that the preganglionic fibers (*solid lines*) are long for the parasympathetic division, terminating close to the target organ. Preganglionic sympathetic fibers are generally short, terminating either in the chain ganglia or in the remote collateral ganglia, all of which are a considerable distance from the target organ. Postganglionic fibers for both systems are shown as *dashed lines*.

scriptions depend on one's understanding of the autonomic components of the cranial nerves, which are presented in Chapter 9.

Structures of the Orbit

The parasympathetic innervation of the eye originates in the **Edinger-Westphal nucleus**. The preganglionic axons leave the brainstem with CN III, terminating in the **ciliary ganglion**.

Postganglionic fibers synapse on the circular muscles of the iris. Contraction of the circular muscle *causes the pupil to constrict*. Other fibers synapse on the ciliary muscle. Stimulation of the ciliary muscle allows the lens to thicken, *accommodating the eye for near vision*.

The sympathetic preganglionic innervation of the eye arises in the spinal cord (T1–T3) and reaches the **superior cervical ganglion**. Postganglionic fibers ascend into the head by fol-

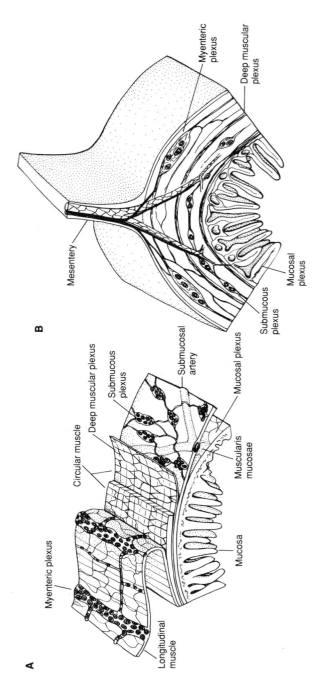

FIGURE 12.4.

A, The two layers of the enteric plexus are shown along with a dissection of the layers of the gut. **B,** A cross section of a segment of the gut showing the relationship of the enteric plexus to the mesentery.

lowing the carotid artery. These fibers join the tympanic division of CN IX and cross the middle ear cavity before joining the ophthalmic artery. On reaching the eye they synapse on the radial muscle of the iris. Contraction of this muscle *causes the pupil to dilate*. Sympathetic fibers also innervate the **tarsal muscle** and the **orbital muscle**. The former elevates the eyelid, while the latter cradles the eye in its socket.

Loss of sympathetic innervation to the eye, by destruction of either the preganglionic or postganglionic axons, produces a trilogy of signs — **miosis**, **ptosis**, and **enophthalmos** — known as **Horner's syndrome**. *Pupillary constriction (**miosis**) is caused by the unopposed action of the parasympathetic fibers on the circular muscle of the iris. The *drooping of the eyelid (**ptosis**) is caused by loss of the sympathetic innervation to the tarsal muscle. Finally, the *retraction of the eye into the socket (**enophthalmos**) is due to the loss of sympathetic innervation to the orbital muscle.[3] A fourth sign that is always present, *lack of sweating on the affected side,* although a result of sympathectomy, is not a named component of the syndrome only because Horner did not include it. Horner's syndrome is an important neurological event, even though its localizing value is somewhat limited. Interruption of the descending brainstem tracts that innervate the spinal cord (see below) or interruption of any part of the extensive peripheral path can produce the signs of Horner's syndrome. Common causes that may elude casual observation are mediastinal tumors, apical lung tumors, aneurysms of the aorta, and abscesses in the neck. Middle ear infections that interrupt the axons lining that cavity may also result in sympathectomy of the structures of the orbit.

The **lacrimal gland** receives both sympathetic and parasympathetic innervation. The **superior salivatory nucleus** provides the parasympathetic preganglionic innervation. Axons from this nucleus follow CN VII to innervate the **sphenopalatine ganglion**. The sympathetic innervation is similar to that already described for the orbital structures. Sympathetic fibers probably only innervate the blood vessels of the gland,

since sympathectomy does not interfere with lacrimation. In contrast, loss of the parasympathetic fibers can inhibit tearing to the point that the integrity of the cornea may be compromised (see Bell's palsy, Chapter 9).

The Heart

The preganglionic parasympathetic innervation of the **heart** arises from the **dorsal motor nucleus of X**. These axons follow the vagus nerve into the thoracic cavity. They separate from it as the **cardiac nerve** before innervating the heart. Within the heart they synapse on scattered neurons located at the base of the atria. From these cells arise the postganglionic axons that synapse in the **sinoatrial (SA)** and **atrioventricular (AV) nodes**. Their *activity slows the heart rate*.

The preganglionic sympathetic neurons affecting the heart reside at approximately T1 to T5 in the spinal cord. Their axons synapse in the corresponding thoracic ganglia as well as all of the cervical ganglia. Postganglionic axons from the cervical ganglia travel to the heart in the superior, middle, and inferior cardiac nerves, while the thoracic cardiac nerve carries the axons from the thoracic ganglia. The postganglionic fibers synapse on the SA node, the conduction system of the heart, the myocardium, and the coronary vessels. Activation of the sympathetic system *increases the heart rate and strength of contraction and dilates the coronary arterioles* of the heart muscles.

Peripheral Blood Vessels and Skin

The skin and peripheral vasculature receive only postganglionic sympathetic innervation. These axons arise from the chain ganglia and are distributed to the periphery with the somatic nerves, where they are distributed to **sweat glands**, **pilomotor muscles**, and the **smooth muscles of the arterioles** in skin and deep muscles. Stimulation of the sympathetic fibers *increases sweat production and causes cutaneous vasoconstriction and vasodilation of the deep muscle vasculature.*

The Respiratory System

The **nasal mucosa** receives its preganglionic sympathetic innervation by way of the superior

---

[3]One should note that there are few fibers in the orbital muscle, and the enophthalmos produced by its denervation is more apparent than real. Enophthalmos is nevertheless one of the classic signs of neurology and must be mentioned when discussing Horner's syndrome.

cervical ganglion with axons that follow the vascular supply into the mucosa. The parasympathetic supply arrives by way of the facial nerve and the **sphenopalatine ganglion** (see above for lacrimation). The **larynx** receives its parasympathetic innervation by way of the laryngeal nerve (a branch of the vagus nerve), the autonomic fibers of which synapse on scattered postganglionic cells in the walls of the larynx. The sympathetic innervation arrives from the middle cervical ganglion. The autonomic innervation of the nasal mucosa and the larynx is believed to be primarily vasomotor, with the *sympathetic innervation acting as vasoconstrictors, while the parasympathetic innervation effects vasodilatation.*

The **trachea**, **bronchi**, and **lungs** receive parasympathetic innervation from the vagus nerve. These axons synapse on postganglionic neurons in the trachea and bronchi that supply the glands and smooth muscles of the bronchial tree. The sympathetic innervation arrives from the inferior cervical ganglion and the first four thoracic ganglia. Postganglionic fibers are distributed to the smooth muscles and glands of the bronchi. *Parasympathetic stimulation produces bronchial constriction and increases secretion from the glands, while sympathetic stimulation produces dilation of the bronchi and decreases secretions.* The autonomic effect on pulmonary vessels is apparently slight.

The Digestive System

The preganglionic parasympathetic supply to the **submandibular** and **sublingual glands** comes by way of the facial nerve, while the **parotid gland** is supplied by the hypoglossal nerve. The postganglionic fibers arise from the **submandibular** and **otic ganglia** respectively. The sympathetic innervation to these glands originates in the superior cervical ganglion and reaches the glands by following the vasculature. The *parasympathetic fibers dilate the blood vessels in the glands and increase the production of serous saliva. Sympathetic stimulation constricts the blood vessels of the gland and produces an increase in the rate of mucous saliva production.*

The vagus nerve provides the parasympathetic preganglionic innervation to the **gut** from the lower, smooth muscle portion of the esoph-

agus to the junction between the transverse and descending colon. The remainder of the colon, the rectum, and the internal anal sphincter receive preganglionic innervation from neurons located in the sacral spinal cord (S2–S4). The preganglionic parasympathetic fibers synapse on neurons in the myenteric and submucosal plexus of the enteric nervous system. These plexuses act as the postganglionic neurons for the gut. Neurons in the myenteric plexus innervate the smooth muscle of the gut, while those in the submucosal plexus innervate the secretory glands. The preganglionic sympathetic innervation arises from the thoracic and lumbar spinal cord (T5–L2). These axons do not synapse in the chain ganglia, but innervate collateral ganglia (celiac, superior and inferior mesenteric) from which the postganglionic axons are distributed to the gut. *Digestion is aided by parasympathetic activity that increases motility and secretions and dilates the vasculature of the gut, while sympathetic activity reduces motility, secretions, and blood flow.* The extrinsic organs of digestion, the **liver** and the **pancreas**, are innervated in a manner similar to the gut with the exception that scattered postganglionic parasympathetic neurons substitute for the enteric plexus. Also, like the gut, *parasympathetic activity generally aids digestion, increasing pancreatic secretion (acinar and islets cells) and induces glycogenesis in the liver. Sympathetic activity decreases pancreatic islet secretion and induces glycogenolysis and gluconeogenesis in the liver.*

The Urogenital System

The kidneys are supplied with sympathetic innervation from the splanchnic nerve. There appears to be no parasympathetic innervation. Autonomic regulation of the kidney and urine production is minimal. The parasympathetic axons originating in the sacral spinal cord (S2–S4) form the **pelvic splanchnic nerve**. After joining the inferior hypogastric plexus, they are distributed to the pelvic viscera, including the **bladder** and **reproductive organs**. They synapse on neurons in the walls of the target organs. The preganglionic sympathetic innervation to these organs arises from neurons in T12 to L2 of the spinal cord. After forming the **lumbar splanchnic nerve** they synapse in the inferior mesenteric ganglion. The postganglionic fibers are distributed to the organs through the hypogastric nerve and plexus.

The neurological control and regulation of **micturition** [L. *micturio,* to urinate] is poorly understood (Fig. 12.5). Very briefly, signals conveying the bladder wall tension are carried by afferent fibers to the sacral cord, where they synapse with the parasympathetic preganglionic neurons, probably through interneurons. An increase in bladder wall tension is presumed to initiate a reflex contraction of the **detrusor muscle** in the wall of the bladder mediated by postganglionic parasympathetic fibers. This reflex also appears to bring about relaxation of the voluntary muscle of the external sphincter.[4] Urine flow can be stopped by voluntary constriction of the external sphincter. Once the bladder is empty, stretching of the bladder wall is minimal, reducing afferent stimulation of the parasympathetic fibers and allowing the detrusor to fully relax. This allows the bladder to fill and enlarge without increasing tension in the bladder wall. At some point, however, the bladder becomes so large that wall tension increases passively, causing the emptying process to be repeated. The

---

[4]The literature contains many references to an involuntary internal sphincter of the bladder. It is now recognized that this internal sphincter does not exist, but it is simply a continuation of the detrusor fibers. Contraction of the detrusor mechanically pulls these fibers away from the ureter. No separate innervation or inhibition is required.

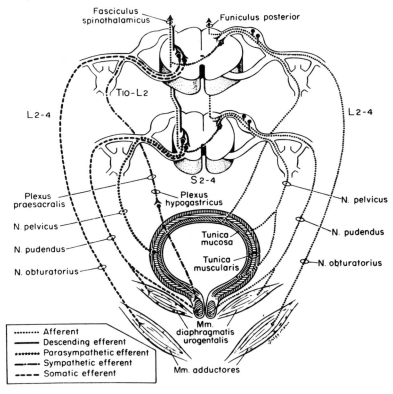

FIGURE 12.5.

A schematic representation of the spinal and peripheral pathways involved in micturition. Note that the principal innervation of the detrusor muscle of the bladder is from parasympathetic fibers via the pelvic nerves of the hypogastric plexus. Voluntary control is provided in part by the somatic innervation of the muscles of the urogenital diaphragm via the pudendal nerve.

role of the sympathetic fibers, if any, is not understood.

The previous discussion treats micturition as a simple reflex, which is approximately the case after spinal transection. Normally, however, there is considerable voluntary control over the process. The nature of this control is far from clear. Sensations of bladder fullness are brought to consciousness and one has considerable control over the timing of bladder emptying, implying that there must be supraspinal inhibition of the micturition reflex just described. Lesions to the spinal cord, pons, and frontal lobes can reduce or eliminate one's voluntary control. This is particularly common in people with multiple sclerosis and with certain types of dementia that affect the frontal lobes. However, the precise anatomical structures involved remain elusive.

Both sympathetic and parasympathetic innervation is required for copulation. The neurological events surrounding copulation are similar in the male and the female, if one takes into account the anatomical differences in the analogous structures. Therefore, only events as observed in the male will be presented here. Stimulation of the genitalia (sensations reach the spinal cord over the pudendal nerve) or erotic stimuli from supraspinal levels serve to increase parasympathetic outflow to the arterioles supplying the erectile tissue of the penis, causing vasodilatation and allowing its engorgement with blood and subsequent **erection**. Continued tactile and erotic stimuli bring about increased sympathetic activity that causes the contraction of the smooth muscles of the vas deferens and seminal vesicles. This brings **seminal fluid** into the urethra, a process know as **emission**. During this time, parasympathetic activity causes the prostate and bulbourethral glands to increase their secretions, which mix with the seminal fluid to form **semen**. Sympathetic activity, facilitated by the continued tactile stimulation of the shaft and glans penis together with other supraspinal erotic stimuli, culminates in abrupt, spasmodic contractions of the musculature of the urethra, prostate, bulbocavernosus and ischiocavernosus muscles, and the perineal musculature, causing **ejaculation** of semen from the penis. These muscles are said to be under both sympathetic and parasympathetic control. Detumescence follows as parasympathetic activity subsides and sympathetic activity continues, reducing penile vasodilatation.

The Adrenal Medulla

The autonomic innervation of the adrenal medulla deserves special consideration because the chromaffin cells of the adrenal medulla are derived from the neuroectoderm of the neural crest. These cells are in fact highly modified neurons that secrete catecholamines directly into the bloodstream. The gland is innervated by preganglionic sympathetic fibers that originate in the intermediolateral cell column at T8 to T11 and synapse directly on the chromaffin cells. Since the chromaffin cells of the adrenal medulla develop from the neural crest and migrate into the periphery, as do the cells that become the postganglionic neurons of the chain ganglia and collateral ganglia, *these secretory cells of the adrenal medulla are homologous to the postganglionic ganglion neurons*. This being the case, the secretory cells of the adrenal medulla mark an interface between the nervous system and the vasculature whereby the nervous system can secrete its transmitters directly into the bloodstream. Thus, the neurotransmitters epinephrine and norepinephrine act as hormones as well as neurotransmitters (Fig. 12.6).

AUTONOMIC AFFERENTS

Sensory fibers are also present in the autonomic nervous system and are the afferent fibers responsible for pain referred from the viscera (see Chapter 5). Another form of visceral pain that is often overlooked is caused by pain afferents innervating the blood vessels, especially the arteries. Vascular pain can be caused by ischemia (ligation of an artery or an arterial thrombosis is painful), or the injection of certain irritating substances. **Raynaud's syndrome** is characterized by a reflex vasoconstriction that results in severe ischemic pain in the affected extremities, usually the hands. The reflex vasoconstriction may be triggered by exposure to cold or by emotional stress, or it may be caused by systemic disease that affects the peripheral nerves.

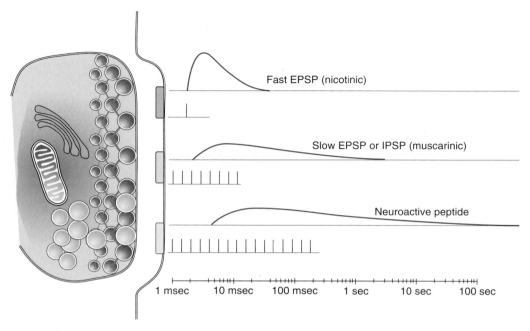

FIGURE 12.6.

The postsynaptic potentials (PSPs) evoked in a postganglionic cell vary greatly depending on the transmitter and the receptor available. The time course of the PSPs can vary over approximately five orders of magnitude. See text for details.

PHYSIOLOGY OF THE ANS

From the discussion of the autonomic innervation of the various organs, it is apparent that sympathetic and parasympathetic activity usually produces antagonistic physiological effects. For example, the sympathetic influences on the heart increase the heart rate and strength of contraction, while the parasympathetic influences decrease both. These effects are not always quite so predictable. Sympathetic activity reduces blood flow to the gut and skin but increases it to the deep muscles. Both parasympathetic and sympathetic activities increase salivation. Therefore, although it is tempting to make generalized statements about autonomic activity, few statements can be made that do not have significant exceptions. In spite of this it is fairly safe to generalize that the parasympathetic and sympathetic systems have contradictory actions and therefore the net physiological effect of autonomic activity rests on the balance of activity between them. As we have seen in previous chapters, the principal mode by which the nervous system functions is

through the essential competition of excitation and inhibition. In the "voluntary" nervous system this competition occurs at the synaptic interfaces of neurons. Ultimately, the ebb and flow of postsynaptic currents represents the inhibitory-excitatory competition. In the autonomic nervous system, this competitive interaction takes place at the effector organ, where the competitive struggle is resolved not in synaptic currents but in the final physiological effect on the organ itself.

The dichotomy of autonomic function is reflected in the types of neurotransmitters used by the sympathetic and parasympathetic systems. Although the preganglionic axons of each system use ACh and the postganglionic parasympathetic axons also liberate ACh, the postganglionic fibers of the sympathetic system secrete catecholamines. In addition to these "classic" transmitters, the preganglionic fibers co-release a number of neuroactive peptides that serve to modulate the postganglionic cell's response to ACh. To complicate matters further, there are a number of different receptors for ACh and NE, each producing quite different physiological effects. There are two major classes of cholinergic receptors known as **nicotinic** and **muscarinic**, because they were originally differentiated by their ability to bind these chemicals. Today subcategories of each of these receptors are now recognized. There are also two classes of norepinephrine receptors, known as **alpha** and **beta** receptors. As with the cholinergic receptors, subcategories are recognized.

In autonomic ganglia, the nicotinic cholinergic receptors are found on all of the postganglionic neurons. ACh bound to these receptors produce a fast (about 20 msec) excitatory postsynaptic potential (EPSP) in the postsynaptic cell (by opening Na$^+$ channels) that usually results in an action potential (Fig. 12.6). If this receptor is blocked with a competitive antagonist, transmission through the ganglion ceases. In addition to the fast nicotinic receptor, there are muscarinic receptors that produce slow (about 2 sec) postsynaptic potentials (PSPs). Since muscarinic receptors function through second-messenger systems (see Chapter 3), their effects are usually slower to develop and longer lasting. The PSPs produced by muscarinic receptors may be EPSPs or inhibitory postsynaptic potentials (IPSPs), depending on the type of receptor expressed on the postsynaptic cell. In both cases, the PSPs are produced only in response to trains of incoming action potentials; isolated action potentials have little effect. High-frequency trains are also required to cause the release of neuroactive peptides from the preganglionic terminals. These peptides produce very slow (about 4 min) EPSPs in the postsynaptic neuron. Peptides, used as neurotransmitters, are not inactivated immediately after their release from the presynaptic terminal. Consequently they diffuse within the interstitial space and affect other postsynaptic neurons. This is apparently a common feature of neuropeptide signaling in autonomic ganglia. Finally, within the autonomic ganglia, one finds chromaffin cells that appear to be dopamine-releasing interneurons. These cells are activated by incoming preganglionic axons and inhibit the postganglionic neurons.

In the target organs, a variety of receptors are seen, mostly among the adrenergic receptors. Some of the diverse effects of sympathetic acti-vation can be explained by the type of receptors expressed. For example, sympathetic neurons cause vasoconstriction in the visceral organs and in the skin because the smooth muscle of the arterioles express α receptors. The blood vessels in the deep muscles, however, express β receptors and sympathetic activity causes them to dilate. Heart muscle also expresses β receptors. The effects of sympathetic stimulation on heart rate and contractility can be selectively inhibited by pharmacologic antagonists that specifically bind to the β receptors. Finally, all tissue that expresses adrenergic receptors is capable of responding to circulating catecholamines. Since the ANS, through its ability to control secretions from the adrenal medulla, can function as an endocrine organ, it can globally and rapidly reinforce the neuronal sympathetic effect by enhancing adrenal medulla secretion.

SUPRASPINAL REGULATION OF THE ANS

In spite of its name and apparent autonomy, the autonomic nervous system is closely regulated by supraspinal structures. The **hypothalamus** serves as the principal integrating center for the ANS. It is connected with brainstem and spinal cord autonomic nuclei through three principal pathways, the **dorsal longitudinal fasciculus (DLF)**, the **mamillotegmental tract**, and the **medial forebrain bundle**.

The dorsal longitudinal fasciculus (see Appendix C) is the tract that serves to connect the hypothalamic nuclei with the sympathetic and parasympathetic preganglionic nuclei in the brainstem and spinal cord. This tract originates from cells located in the medial hypothalamus, adjacent to the third ventricle. The fibers descend into the brainstem and pass through the periaqueductal gray matter (PAG) and reticular formation to terminate in the dorsal motor nucleus of the vagus and probably also in the salivatory and Edinger-Westphal nuclei. The DLF also contributes to the reticular formation. It continues into the spinal cord, where it terminates in the intermediolateral cell column and the sacral autonomic nucleus.

The mamillotegmental tract originates in the mamillary body and descends into the brainstem, where it synapses in the mesencephalic and pontine reticular formation. A few axons also reach the reticular formation by way of the medial forebrain bundle.

Sensory information reaches the hypothalamus, but the pathways involved are poorly understood. Most sensory information reaching the hypothalamus seems to arrive by way of the brainstem reticular formation, originating from a variety of cell types that may express catecholamines, serotonin, or ACh. Somatosensory information from erogenous zones reaches the hypothalamus by several pathways. Some of these axons ascend in the dorsal longitudinal fasciculus and the medial forebrain bundle. Other diffuse, unnamed pathways are probably also involved.

The hypothalamus is a small, dense structure that serves many diverse functions. It has not been possible to closely correlate specific autonomic functions with individual hypothalamic nuclei. Regional specializations, however, do exist (Fig. 12.1). The **anterior regions** are closely associated with parasympathetic functions. For example, stimulation of the anterior region of the hypothalamus, particularly in or near the **anterior nucleus**, produces a slowing of the heart, lowering of blood pressure, peripheral vasodilation, salivation, increased peristalsis, and sweating. In contrast to this, the **posterior and lateral regions** are associated with sympathetic functions. Here, stimulation produces increased heart rate and blood pressure, dilation of the pupils, decrease or cessation of peristalsis, superficial vasoconstriction and deep vasodilation, and cessation of sweating.

Stimulation of the anterior or posterior regions of the hypothalamus usually produces a generalized autonomic response appropriate to the area of stimulation. From experimental studies in animals and clinical observations in humans, it appears that these regions of the hypothalamus effect an integrated total autonomic response. For example, body temperature is closely regulated by the hypothalamus. Thermosensitive cells exist in the anterior region (and perhaps in the posterior region as well) that monitor blood temperature. An increase in temperature, recorded in the anterior region, produces sweating, cutaneous vasodilation, and panting (in fur-bearing animals), actions that lower body temperature. Given the reciprocal nature of parasympathetic and sympathetic actions, these autonomic responses cannot be generated without simultaneous inhibition of the posterior regions of the hypothalamus. Lesions to the anterior regions eliminate these parasympathetic effects, resulting in an animal becoming hyperthermic because it cannot recruit the physiological apparatus to cool the body. In contrast to this, subnormal blood temperature produces the opposite effects: cutaneous vasoconstriction, shivering, and cessation of sweating. Lesions to the posterior region not only eliminate the sympathetic hypothalamic centers, but also the descending fibers from the more anterior parasympathetic centers. This produces a **poikilothermic** animal [G. *poikilos,* varied], since no thermoregulation is possible.

Connections with the Endocrine System

The distinctions that have traditionally been drawn between the endocrine and nervous systems have become increasingly blurred as our understanding of their respective physiological roles has broadened. Both systems are regulatory. The endocrine system generally regulates the internal milieu by means of blood-borne chemical signals that modify metabolism, reproductive states, digestion, and electrolyte balance. These signals reach every cell in the body, and the specificity of their action depends on the receptors expressed by individual cells. The nervous system regulates these same physiological systems by means of electrical signals that reach their targets by nerve axons. Although they are widely distributed, individual neurons do not reach every cell in the body. Nervous system specificity relies on the anatomical relationships between neurons and target cells. Both systems depend on delicately balanced negative feedback mechanisms to achieve their regulatory goals.

The primary criterion for distinguishing between the two systems—electrical versus chemical signaling—breaks down when one realizes that neurons are essentially secretory cells. Furthermore, some hormones are secreted and exert their effect locally without entering the bloodstream. All neurons secrete neurotransmitters as a means of signaling other neurons and of regulating secretory glands. Most neurons secrete neurotransmitters that affect only the cell on which they are released, although in a few cases they can diffuse beyond the synaptic cleft and affect propinquitous neurons, as we have just seen in the autonomic ganglia. A few neurons secrete neurotransmitters directly into the bloodstream, in which case the neurotransmitter can exert its influence in the manner of a hormone. We have just discussed one example of this type of neurosecretion, the secretion of norepinephrine from the chromaffin cells of the adrenal medulla. Other examples are discussed below.

One normally thinks of the CNS as being beyond the reach of the endocrine system because it is isolated by the blood-brain barrier so that blood-borne signals cannot affect it. More recently we have come to realize that this is not true. The circumventricular organs (CVOs) (see Chapter 1) represent specific areas in the brain where the blood-brain barrier is greatly modified or simply does not exist. Cells in these organs are endowed with special receptive properties so that they can serve as humoral-neuronal transducers. Furthermore, humoral secretions into the cerebral spinal fluid provide a means by which the brain can be affected by circulating chemical signals. The ability to secrete regulatory chemicals into the bloodstream and CSF, and the ability to respond to blood-borne and CSF-borne chemical signals, clearly makes the brain, at least in part, an endocrine organ.

It is convenient to define a **hormone** as a chemical signal that is conveyed to its target by the bloodstream or the CSF, while a **neurotransmitter** is a chemical signal that is restricted in its extracellular distribution to the interstitial fluid space. With these definitions, certain secretory products, such as norepinephrine and enkephalin, can be both a hormone and a neurotransmitter, depending on their mode of release.

The endocrinological role of the brain is an enormous topic that can be presented here only in the most abbreviated form. A more complete discussion of this subject can be found in the references. By means of introduction, the endocrinological role of the brain is most clearly represented in the relationship between the hypothalamus and the hypophysis, a brief discussion of which follows.

THE NEUROHYPOPHYSIS

Two nuclei in the hypothalamus, the **supraoptic** and **paraventricular nuclei**, contain large cells that send axons into the posterior lobe of the hypophysis (Fig. 12.7). Known as **magnocellular secreting neurons (MSNs)**, these cells produce the hormones **vasopressin (antidiuretic hormone [ADH])** and **oxytocin**. Synthesized as a prohormone that is packaged in large, 1200- to 2000-Å diameter vesicles, the precursor peptide is cleaved into the active hormone and a related peptide, **neurophysin**, within the vesicle as it is transported along the axon. Some of the magnocellular neurons produce **oxytocin** and its companion peptide **neurophysin I**. Others produce **vasopressin** and **neurophysin II**. None of the MSNs produce both sets of hormones simultaneously. The hormone-containing vesicles are transported along axons to the neurohypophysis. The magnocellular neurons and their axons also are electrically active and, in response to electrical activity, release hormones into the perivascular space of the neurohypophysis, where the peptides diffuse into the bloodstream by passing through fenestrated capillaries.

Oxytocin

The circulating hormone oxytocin has two principal physiological effects: it initiates milk secretion from the mammary glands and causes contraction of the uterine muscles during parturition. Oxytocin effects the release of milk by inducing the contraction of the myoepithelial cells of the mammary glands. Suckling by the infant produces tactile stimuli in the mother that reach the hypothalamus by neuronal pathways that have not been clearly delineated (Fig. 12.8). The suckling stimuli produce bursts of electrical activity in magnocellular neurons that cause the release of oxytocin into the neurohypophysis, where it enters the bloodstream. Approximately 13 seconds after release, intramammary pressure increases as milk enters the ducts of the gland.

The afferent path of this simple reflex is neural, making possible the interaction in the

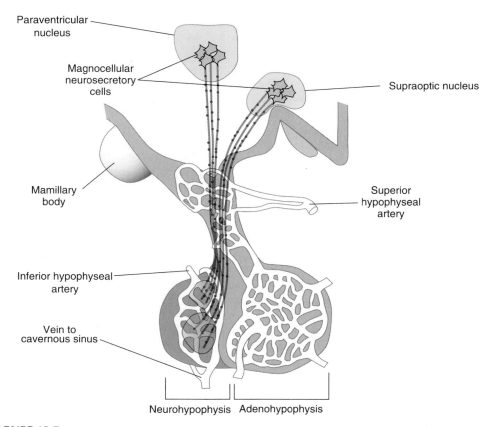

Paraventricular nucleus

Magnocellular neurosecretory cells

Supraoptic nucleus

Mamillary body

Superior hypophyseal artery

Inferior hypophyseal artery

Vein to cavernous sinus

Neurohypophysis   Adenohypophysis

FIGURE 12.7.

Schematic diagram of the location of magnocellular secreting neurons in the hypothalamus and their association with the neurohypophysis. Note that the neurohypophysis is not an endocrine gland, it is simply a specialized capillary bed that receives the hormones delivered to it from the hypothalamus and delivers them into the general circulation.

hypothalamus of other neural signals that can modify the reflex. For example, the reflex is facilitated by specific auditory and visual cues. Every lactating mother has experienced milk ejection in response to the cry of her hungry child. The reflex can also be inhibited by anxiety, a fact that sometimes interferes with initiation and full development of lactation.

Oxytocin also stimulates powerful contractions of the myometrium. Current evidence suggests that the initiation and progression of labor are closely related to biochemical events associated with the fetus, the placenta, and the chorionic membranes. Oxytocin itself does not initiate labor, but its release during parturition enhances the strength and frequency of uterine contractions. This release is signaled by a positive feedback loop. Pressure on the cervix is transmitted by neurons to the hypothalamus. This promotes the secretion of oxytocin, which is conveyed to the uterus by the bloodstream. The oxytocin increases the strength of uterine contractions, further pressuring the cervix.

Vasopressin

Vasopressin (ADH) is a circulating hormone that increases the permeability of the cell membranes of the collecting ducts in the kidney, allowing water and electrolytes to be reabsorbed in the kidney. Without this hormone, the kidney produces copious amounts of a very dilute urine, a condition known as **diabetes insipidus**. Vasopressin is only one of many factors that are im-

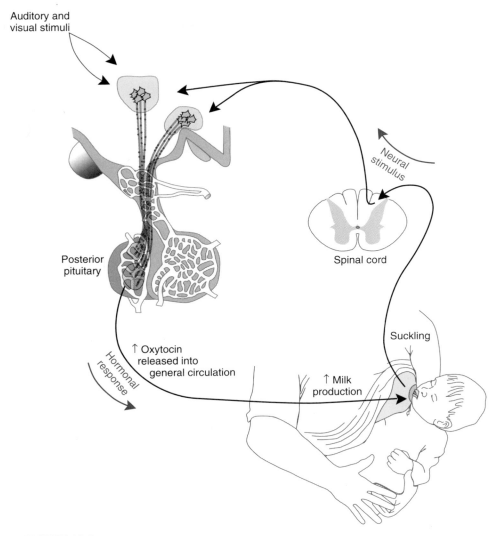

Auditory and
visual stimuli

Neural
stimulus

Spinal cord

Posterior
pituitary

↑ Oxytocin
released into
general circulation

Hormonal
response

Suckling

↑ Milk
production

FIGURE 12.8.

Diagrammatic representation of the lactation reflex. Note that this is an example of a mixed neuronal-endocrinological reflex. The afferent side of the loop is neural, the efferent side is endocrinological.

portant in the closely intertwined mechanisms of blood volume regulation, electrolyte balance, and blood pressure regulation. The magnocellular neurons that release vasopressin into the blood are influenced by several feedback systems that control the firing rate of these neurons and consequently the release of vasopressin. Three of these systems will be described here (Fig. 12.9).

First, although the magnocellular cells in the hypothalamus appear to be directly sensitive to plasma osmolarity, they also receive afferent connections from the **organum vasculosum (OV)**. This circumventricular organ (CVO) is exquisitely sensitive to plasma osmolarity because of its unusually permeable capillaries. Axons leaving the OV stimulate the secretion of vasopressin from MSNs. Increased plasma osmolarity also causes the sensation of thirst and induces drinking behavior. Experimental destruction of the OV in rats decreases their vasopressin secretion and renders them adipic [G. *dipsa,* thirst].

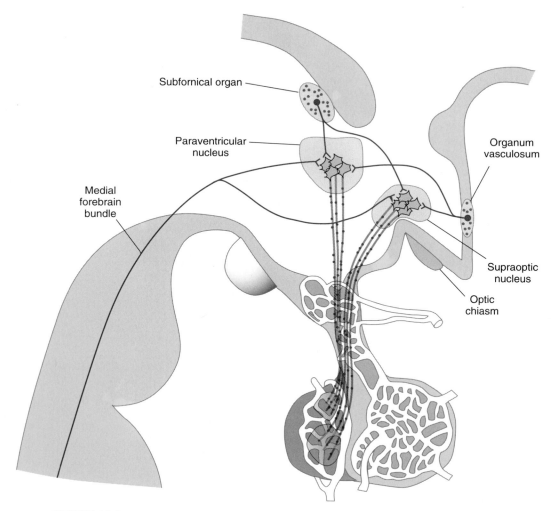

FIGURE 12.9.

The release of vasopressin is regulated by at least three mechanisms, illustrated here. The subfornical organ is sensitive to plasma osmolality, while the organum vasculosum detects circulating levels of angiotensin II. Both of these CVOs facilitate the release of vasopressin from the supraoptic and paraventricular nuclei. These nuclei are further regulated by ascending signals from the carotid bodies and aortic arch receptors. Therefore, vasopressin is under neural, endocrinological, and osmolar regulation.

The diabetes insipidus causes them to excrete a very dilute urine. Without compensatory drinking, the animals become chronically hypernatremic.

Second, another CVO, the **subfornical organ (SFO)**, is sensitive to circulating levels of **angiotensin II**, which, acting through connections between the SFO and the magnocellular neurons of the SO and PV neurons, increases the secretion of vasopressin. Angiotensin II plays a

key role in blood pressure regulation, electrolyte balance, and the maintenance of blood volume. It is the product of a chemical cascade that is initiated by the release of renin from the kidney in response to a decrease in perfusion pressure of the glomerulus, decreased sodium at the macula densa, or a decrease in extracellular fluid volume. In the periphery, angiotensin II causes peripheral vasoconstriction, which dramatically increases blood pressure; it is the most potent pressor hormone. It also causes the release of aldosterone, the mineralocorticoid that facilitates Na⁺ absorption by amiloride-sensitive sodium channels. In the hypothalamus it facilitates the release of vasopressin. Together, all of these mechanisms constitute an appropriate response to the decrease in blood pressure and blood volume that results from hemorrhage.

Third, blood pressure and oxygen tension also affect the release of vasopressin. Signals from the carotid bodies and aortic arch reach the hypothalamus from the nucleus of the solitary tract (see Chapter 9). These peripheral cardiovascular signals reach the magnocellular neurons of the anterior hypothalamus, where the information they convey is integrated with the other relevant signals just discussed.

THE ADENOHYPOPHYSIS

The adenohypophysis, unlike the neurohypophysis, is a true endocrine gland. Cells within the gland directly secrete a variety of hormones into the bloodstream. The release of these hypophysial hormones is regulated by hormones that are secreted by **parvocellular secreting neurons (PSNs)** of the hypothalamus. The adenohypophysis is a gland that not only secretes hormones but is under direct hormonal control as well.

All of the hypothalamic hormones that regulate the adenohypophysis are delivered to it by a specialized vascular system, the **hypophysioportal system** (Fig. 12.10). Arterial branches from the internal carotid arteries and from the circle of Willis form a **primary capillary network** in the median eminence of the hypothalamus. Emissary veins from this capillary network traverse the surface of the **pars tuberalis** to reach the **pars distalis**, where they form a **secondary venous sinus** within the gland. Emissary veins from this secondary sinus drain into the cavernous sinus. Parvocellular neurons of the hypothalamus send axons into the median

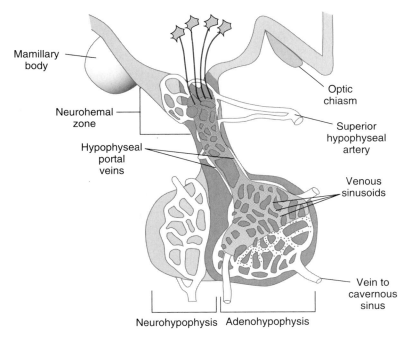

FIGURE 12.10.

The relationship of the hypothalamus with the adenohypophysis is depicted here. Note that the adenohypophysis is a true endocrine gland that contains secretory cells. These are under the control of hormones secreted into the blood by neurons in the hypothalamus and delivered to the adenohypophysis by a restricted portal circulation shown here.

eminence, where they *terminate on the primary capillaries of the hypophysioportal system.* Secretions from these neurons are carried by the blood from the median eminence into the pars distalis, where these **hypothalamic releasing hormones** diffuse from the secondary sinus bed to reach the secretory cells of the adenohypophysis. The hormones of hypothalamic origin may inhibit or facilitate the release of hormones from the adenohypophysial cells. Hormones secreted by the hypophysial cells enter the emissary venous sinus of the gland and are carried into the general circulation.

Adenohypophysial hormones are grouped into three classes based on their chemical structure: the glycoprotein hormones, the mammosomatotropic hormones, and the opiomelanocortin hormones. These hormones and their various releasing/inhibiting hormones are briefly considered below.

The Glycoprotein Hormones

The **glycoprotein hormones**—thyroid-stimulating hormone (TSH), luteinizing hormone (LH), and follicle-stimulating hormone (FSH)—consist of two protein subunits, each about 100 amino acids long. One subunit, the α chain, is essentially identical in all three hormones. The β chain is variable and accounts for the physiological specificity of the various hormones. These hormones also contain carbohydrate segments that are not only variable between hormones, but can vary within the same hormone. For example, there are at least eight different FSH molecules that differ only in the carbohydrate moieties. The significance of this variability is not known.

Gonadotropins (LH and FSH): The gonadotropins affect the reproductive organs in both males and females although their names reflect their action in the female. **Follicle-stimulating hormone (FSH)** promotes the growth of follicles in the ovary. After the follicle matures, a surge of FSH secretion, accompanied by an even greater surge of **luteinizing hormone (LH)** secretion, induces ovulation. Following ovulation, FSH and LH secretion returns to baseline levels. In addition to promoting the development of the follicle and causing ovulation, FSH

and LH induce the corpus luteum to produce steroids, particularly **progesterone** and **estrogen**. In the male, LH stimulates the Leydig cells[5] of the testes to produce testosterone, which is necessary for spermatogenesis and other male reproductive functions.

The secretion of LH and FSH is stimulated by **gonadotropin-releasing hormone (GnRH)**, a small 10–amino acid protein produced primarily in scattered parvocellular neurons of the **preoptic** and **septal** regions and in the **arcuate nucleus** of the hypothalamus (Fig. 12.11). Axons from these areas descend into the median eminence, where they liberate GnRH into the capillary bed of the hypophysoportal system.

Thyroid-Stimulating Hormone (TSH): Thyroid-stimulating hormone acts on the follicular cells of the thyroid gland. Its secretion from the pituitary gland is stimulated by **thyrotropin-releasing hormone (TRH)**. TRH is perhaps the smallest peptide hormone, being only three amino acids long. It is produced in the **periventricular nucleus**.

The Mammosomatotropic Hormones

The second class of hormones, the **mammosomatotropic hormones**—growth hormone (GH) and prolactin—consist of an α-helical chain about 200 amino acids long.

Growth Hormone (GH): Growth hormone acts throughout the body, stimulating the growth of bone structure and affecting the metabolism of cells. Its secretion is regulated by two hypothalamic hormones, one that facilitates the release of GH and the other that inhibits its release. **Growth hormone–releasing hormone (GHRH)** is produced in the **arcuate nucleus** (Fig. 12.12). It effects the release of growth hormone from the anterior pituitary. **Growth hormone–inhibiting hormone (GHIH)** (also known as **somatostatin**) is primarily produced in the **periventricular nucleus**, although it is also present in the supraoptic and paraventricular nuclei.

Prolactin: Prolactin promotes the development of breast tissue and milk production. Unique among the anterior pituitary hormones, prolactin

---

[5]The Leydig cells are also known as interstitial cells and in the male, LH is sometimes referred to as **interstitial cell stimulating hormone (ICSH)**.

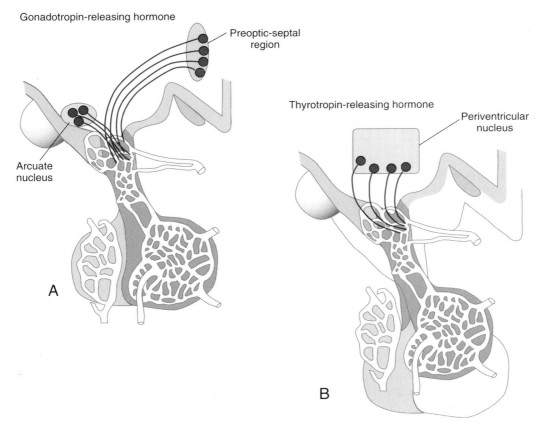

FIGURE 12.11.

Locations of parvocellular neurons that synthesize GnRH (**A**) and TRH (**B**), hypothalamic hormones that regulate the secretion of the glycoprotein hormones.

appears to be exclusively regulated by a release-inhibiting hormone, **prolactin release–inhibiting hormone (PRIH)**. PRIH is not a protein, but is probably **dopamine**. It is secreted into the portal system by diffuse neurons located in and near the **arcuate nucleus**.

The Opiomelanocortin Hormones

Finally, the **opiomelanocortin hormones**—adrenocorticotropic hormone (ACTH), melanocyte-stimulating hormone (MSH), β-lipotropin, and β-endorphin—are cleaved from a common prohormone, pro-opiomelanocortin. Of these hormones only ACTH is normally secreted by the adenohypophysis.

Adrenocorticotropic hormone (ACTH): The adrenal cortex is an endocrine gland that produces several types of steroids, including **androgens, estrogens, progesterones**, and the major classes of **mineralocorticoids** and **glucocorticoids**. The secretion of mineralocorticoids is controlled and regulated primarily by the renin-angiotensin system. The secretion of glucocorticoids falls under the regulation and control of the adenohypophysis. **Adrenocorticotropin (ACTH)** is the pituitary hormone that affects the adrenal cortex by stimulating steroid synthesis. The release of ACTH is controlled by **corticotropin-releasing hormone (CRH)**, a hormone that is produced in parvocellular neurons of the **paraventricular nucleus** and released in the capillary bed of the median eminence (Fig. 12.13). This nucleus also contains magnocellular neurons that secrete oxytocin and vasopressin. It is noteworthy that the ability of CRH to effect the

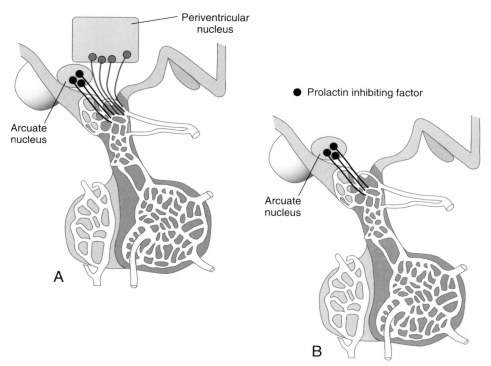

FIGURE 12.12.

Locations of parvocellular neurons that synthesize GHRH, GHIH (somatostatin) (**A**), and PRIH (probably dopamine) (**B**), hypothalamic hormones that regulate the secretion of the mammosomatotropic hormones.

release of ACTH is potentiated by vasopressin, and the co-release of these hormones from the parvocellular neurons has been reported.

REGULATION OF HORMONE SECRETION

The endocrine system, like the nervous system, is carefully regulated by negative feedback mechanisms that limit the effects of its actions within a very narrow range. Four types of negative feedback signals regulate the release of hormones from the adenohypophysis (Fig. 12.14).

Long-Loop Feedback

With the exception of prolactin, adenohypophysial hormones all stimulate the release of circulating hormones or metabolic products that can affect the hypothalamus. For example, es-

trogen inhibits the release of GnRH by acting on the hypothalamus. Somatomedins are secreted by the liver in response to GH. These metabolic products facilitate the release of GHIH, which inhibits the further release of GH. This indirect signaling system (*hypothalamic releasing hormone* $\overset{+}{\to}$ *hypophysial hormone* $\overset{+}{\to}$ *circulating hormone* $\overset{-}{\to}$ *hypothalamic releasing hormone*) is termed **indirect long-loop feedback** and appears to be the principal means of regulating the adenohypophysial hormones.

The secretion of hypophysial hormones can also be regulated by the circulating hormone directly (*hypothalamic releasing hormone* $\overset{+}{\to}$ *hypophysial hormone* $\overset{+}{\to}$ *circulating hormone* $\overset{-}{\to}$ *hypophysial hormone*) by **direct long-loop feedback** to the adenohypophysis. For example, estrogen acts not only at the hypothalamus, but also at the hypophysis to inhibit the release of

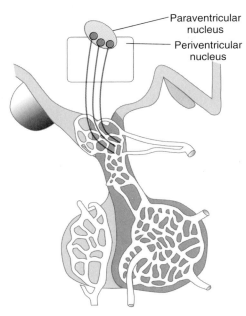

● Corticotropin-releasing hormone

Paraventricular
nucleus

Periventricular
nucleus

FIGURE 12.13.

Location of parvocellular neurons that synthe-size CRH, the hypothalamic hormone that regu-lates the secretion of ACTH, the only opiome-lanocortin hormone secreted by the anterior hypophysis.

LH and FSH. Direct feedback mechanisms ap-pear to reduce the ability of adenohypophysial cells to respond to releasing hormones, rather than limit their intrinsic secretory capabilities. In effect the circulating hormones "blind" the secretory cell to the releasing hormone signal provided by the hypothalamus.

Short-Loop Feedback

Blood flow in the hypophysoportal system can, under certain circumstances, reverse, and by doing so carries hypophysial hormones back to the hypothalamus, where they can inhibit the secretion of releasing hormones. Therefore, hy-pophysial hormones themselves are feedback signals that can affect the hypothalamus *(hypo-thalamic releasing hormone ⭢ hypophysial hormone ⭢ hypothalamic releasing hormone).* Such signaling is termed **short-loop feedback** because it does not involve the general circula-tion and part of the regulation of gonadotropins involves this short-loop inhibition of hypothala-mic neurons. Another form of regulation, **ultra-short-loop feedback**, exists. In this case the re-leasing hormones provide, by their very presence in the median eminence, negative feed-back regulation, since they can inhibit their own release from the hypothalamic neuron bouton.

Connections with Forebrain Structures

The hypothalamus lies at a crucial place in the neuraxis. It serves as a focal point where de-scending influences from cortical areas come to-gether to affect the most primitive survival func-tions of the organism: digestion, cardiovascular regulation, respiration, temperature regulation, defense mechanisms, and sex. It also serves as a focal point from which ascending information relating to these same survival functions can be distributed appropriately throughout the nervous system and to be integrated with the descending commands. Through these connections, behav-ioral mechanisms appear to become incorpo-rated into the physiological homeostatic mecha-nisms of survival. For example, feeding behavior is an essential component of digestion. Seeking shade or digging a hole to uncover cooler earth accompanied with panting is an essential facet of temperature regulation. Foreplay is essential to reproduction.

The role that the hypothalamus plays in the interface between sensory experience and be-havior is not understood. Both the anatomy and the physiology remain obscure, although much detail is known about each. Some of the more in-teresting experiments performed in this century have demonstrated how lesions to or stimulation of hypothalamic structures can trigger survival-oriented behavior. This is not to say that the hy-pothalamus is the "center" for behavioral moti-vation in the brain, but rather that hypothalamic structures are intimately associated with other neural structures that together can release stereo-typical behavior. The hypothalamus probably affects behavior through reciprocal connections with cortical structures. Research has focused

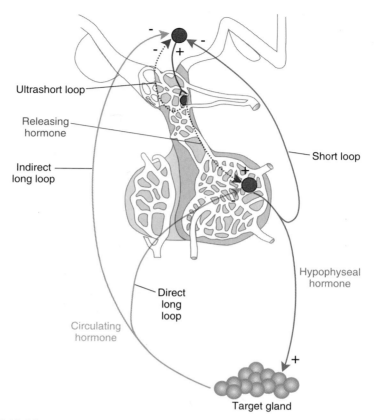

FIGURE 12.14.

This generalized diagram illustrates the various feedback mechanisms used by hypo-
thalamic-hypophyseal hormone systems.

on the relationships between the hypothalamus
and the associated archi-, paleo-, and neocortex.
The most important of these telencephalic structures are the **amygdaloid nuclei**, the **hippocampal formation**, the **septal nuclei**, the **cingulate
gyrus**, and the **prefrontal cortex**.

In the following paragraphs, the principal connections of the hypothalamus with telencephalic
areas are described, followed by a very brief discussion of some of the more important behavior
patterns associated with these structures.

THE AMYGDALOID NUCLEI

The amygdaloid nuclei [G. *amygdale*, al-
mond] lie in the temporal lobe, deep to the uncus [G. *onkos*, a hook], a swelling on its medial
aspect. The immediately overlying cortex, between the **rhinal sulcus** and the medial extent of
the temporal lobe, is called the **piriform cortex**

[L. *pirum*, pear + L. *forma*, form]. Caudal to it,
lying between the **collateral sulcus** and the **hippocampal sulcus**, and extending caudally to the
isthmus of the cingulate gyrus [L., a girdle,
from *cingo*, to surround] is the **entorhinal cortex** [G. *entos*, within, + G. *rhis*, nose] (Fig.
12.15). The amygdaloid nuclei are reciprocally
connected with the **olfactory bulb;** the **hypothalamus;** the **thalamus;** the **septal nuclei;** the
piriform, **entorhinal**, and **prefrontal** cortices;
the **cingulate gyrus;** and the **brainstem**. Since
the amygdala and its overlying cortex serve as
the principal cortical target for the central olfactory projections, the essential features of the olfactory system will be presented here.

Olfactory System

The olfactory sense is fundamentally important to many animals. Without it, the food gath-

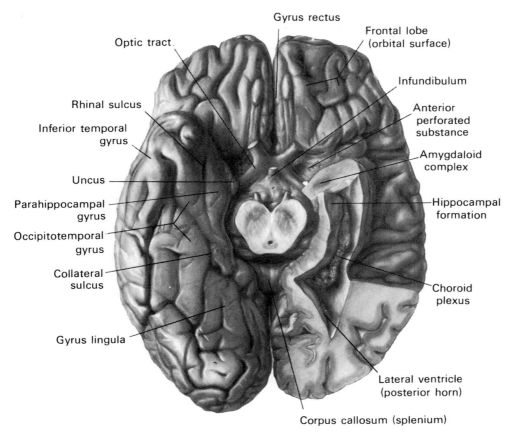

FIGURE 12.15.

Ventral view of the brain with part of the temporal lobe removed on the right to reveal the location of the amygdaloid complex and the hippocampal formation. The entorhinal area is a small portion of the parahippocampal gyrus that lies between the rhinal sulcus and the uncus. The piriform cortex lies on the medial aspect of the uncus.

ering and reproductive capacity of these animals would be severely limited. For example, in many species, specific odor-producing molecules (**pheromones**) are secreted that stimulate reproductive behavior in other members of the species. The ability to detect these molecules is remarkable. Pheromones are critical in triggering reproductive behavior in many mammals. Certain male insects can track a female for several miles by following her pheromone trail, which need be only 100 molecules per milliliter. Although not unimportant in the human, the role of olfaction is much diminished in human behavior. Reflecting this diminution of importance, the olfactory structures of the primate brain are relatively modest.

In the human, the **olfactory epithelium** lies in the olfactory cleft within the roof of the nasal cavity. The epithelium consists of **sustentacular cells**, **basal cells**, and **bipolar receptor cells** (Fig. 12.16). The surface of the epithelium is coated with a thick mucus that is secreted by the epithelial cells and from **Bowman's glands**. This mucus flows over the epithelial surface and is secreted continuously, cleansing the receptors of odoriferous molecules about every 10 minutes. The epithelium rests on a basement membrane that separates the superficial neuroepithelium from the deep lamina propria.

On the surface of the olfactory epithelium, the distal pole of the bipolar receptor cell terminates as a knob from which a variable number of

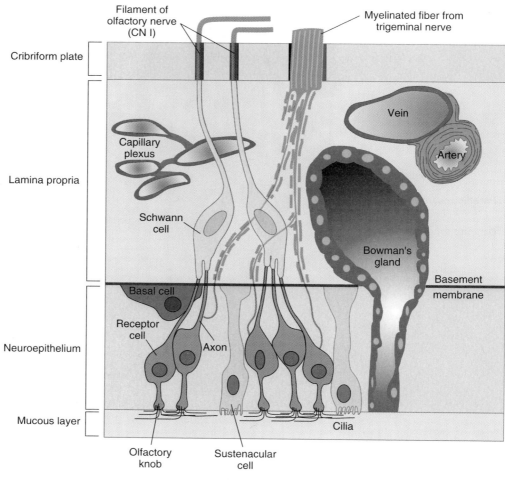

FIGURE 12.16.

A drawing illustrating the principal components of the olfactory epithelium.

cilia extend. Fully differentiated receptor cells exist for only about 30 to 60 days before they die and are sloughed. They are continuously replaced by differentiating basal cells. This is the only known case of neuron or receptor cell replacement in the adult vertebrate.

The cilia of the olfactory receptor cells, like those found in cochlear hair cells and phototransducers, contain the typical 9 + 2 arrangement of microtubules. Receptor sites, probably composed of membrane proteins, are located on the cilia. The binding of a single appropriate molecule to one receptor site causes a hypopolarization in the receptor cell that is sufficient to generate an action potential. It is hypothesized that different receptor sites on the cilia bind dif-

ferent classes of molecules and that each class has certain steric and/or charge characteristics that distinguish one class from another. Our perception of different odors could be explained on the basis of different receptor cells expressing different receptor molecules in a manner that is analogous to our perception of colors.

Unmyelinated axons emerge from the proximal pole of the receptor cell. The axons penetrate the basement membrane of the epithelium where they become myelinated by Schwann cells in the laminae propria. Bundles of axons pass through the **cribriform plate** [L. *cribrum,* a sieve, + *forma,* form] of the **ethmoid bone** [G. *ethmos,* sieve, + G. *eidos,* resemblance] as the **olfactory nerve (CN I)**. Branches of somatosensory nerves, derived from the trigeminal nerve, also pass through the cribriform plate and innervate the nasal and olfactory epithelia. These sensory fibers provide nociceptive sensitivity to the epithelium of the nasal cavity. The cribriform fenestrations in the ethmoid bone that are necessary for the passage of axons also provide a potential path by which infectious agents may gain entry into the cranium, where they can form abscesses in the frontal lobes.

After entering the cranium, the axons of the olfactory nerve immediately penetrate the **olfactory bulb**, within which they synapse with **mitral cells** and **tufted cells** in a complex structure

known as the **olfactory glomerulus** (Fig. 12.17). Several types of interneurons exist in the olfactory bulb. They provide for a rich lateral interaction between the primary neurons of the olfactory nerve and secondary mitral and tufted neurons. Axons from mitral and tufted neurons form the **olfactory tract**, which lies in the **olfactory sulcus** of the overlying frontal cortex. On reaching the **anterior perforated substance**, the olfactory tract divides into **lateral** and **medial olfactory striae**. The lateral olfactory stria, containing primarily axons from the olfactory bulb, passes lateral to the anterior perforated substance before entering the piriform cortex.

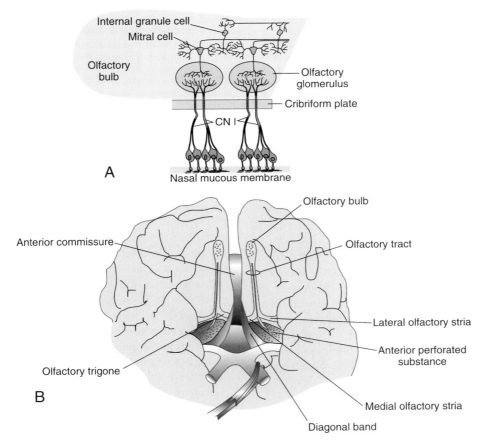

FIGURE 12.17.

A, Drawing of the synaptic arrangement of the cells in the olfactory bulb. Note that the short axons of the receptor cells of the olfactory epithelium form the olfactory nerve (CN I). They pass through the cribriform plate to synapse within the olfactory bulb. **B,** A ventral view of the brain showing the olfactory tract and striae. The optic chiasm and nerves have been displaced to reveal the olfactory trigone, the diagonal band of Broca, and the anterior commissure.

Here, the axons collateralize extensively, forming branches that terminate in the **piriform cortex**, the **entorhinal cortex**, and the **amygdaloid complex**. The medial olfactory stria contains mostly axons from the **anterior olfactory nucleus**, a collection of cells in the olfactory tract that receive mitral and tufted cell axon collaterals. The medial stria joins the anterior commissure to cross the midline. On the contralateral side, olfactory axons enter the olfactory tract and terminate in both the anterior olfactory nucleus and bulb.

Connections of the Amygdaloid Complex

As noted previously, in many animals, olfactory cues play a critical role in reproduction. Therefore, it should not be surprising that the amygdaloid complex has prominent connections with nuclei in the hypothalamus—nuclei that play a prominent role in reproduction.

Fibers leave the amygdala by two major pathways (Fig. 12.18). The most prominent path is the **stria terminalis**, a macroscopic fiber bundle that courses along the wall of the lateral ventricle. It follows the caudate nucleus to the rostral portion of the ventricle before entering the septal nuclei and the hypothalamus. Within the hypothalamus, most of the axons terminate in the preoptic and ventromedial areas. A second amygdalofugal pathway proceeds along the base of the brain and spreads out to innervate the dorsomedial nucleus of the thalamus, the prefrontal cortex, and the brainstem. In the brainstem, the major sties of termination are the **substantia nigra**, the **periaqueductal gray matter**, the **reticular formation**, the **nucleus of the solitary tract**, and the **dorsal motor nucleus of the vagus**. Afferent connections with the amygdala generally mirror the efferents.

Function of the Amygdala

The functional role of the amygdala is not well understood. Our knowledge is based on animal experiments and human lesions. These data suffer from a variety of methodological difficulties. Nevertheless, the evidence suggests that *the amygdala plays a role in determining one's affective perception of sensory stimuli.* This is most evident in evaluating one's *perception* of threat from environmental stimuli. In experimental animals, stimulation of the amygdala *produces an apparent behavioral state of anxiety,* a state that is signaled by autonomic cardiovascular, respiratory, and enteric responses as well as pupillary dilatation and piloerection. Humans who suffer from epileptic seizures that cause abnormal discharges in the amygdala often describe a feeling of intense fear or dread just before an epileptic attack. If not fear and dread, such patients may experience abnormal gastric sensations or olfactory hallucinations of a particularly disagreeable nature.

In contrast to this, in both animals and humans, lesions of the anterior temporal lobe (which includes the amygdala)[6] produce placidity, the opposite of fear. Individuals with such lesions do not respond to environmental stimuli as intensely as normal individuals. Humans with such lesions are said to have a "flat" affect. Previously aggressive animals (or humans) become docile or submissive when faced with threatening situations.

THE HIPPOCAMPAL FORMATION

The hippocampal formation [G. *hippocampos,* seahorse] is a complex structure that lines the medial aspect of the temporal lobe (Fig. 12.15). It extends from the amygdaloid complex to the isthmus of the cingulate gyrus. In cross section is resembles a seahorse, from which it gets its name (Fig. 12.19). The hippocampal formation consists of three divisions: the **dentate gyrus** (CA_4),[7] the **hippocampus proper** (CA_1, CA_2, and CA_3), and the **subiculum**. The dentate gyrus is a bulge of cortex bounded to the inferior by the **hippocampal sulcus** and to the superior by the **fimbria of the fornix**. Wrapping around the concave side of the dentate gyrus is the hippocampus proper. It is bounded on its lateral surface by the lateral ventricle. The hippocampus proper curves around the inferior aspect of the dentate gyrus, gradually becoming wider as it merges with the **entorhinal cortex (area 28)**.

---

[6]The anterior portion of the temporal lobe is occasionally excised in patients suffering from intractable epilepsy that originates from that structure.

[7]The designations CA_1 through CA_4 are based on histological differences that are unimportant to the physician. They are used here only for purposes of anatomical reference. CA is derived from *cornu ammonis,* an archaic eponym for the hippocampus.

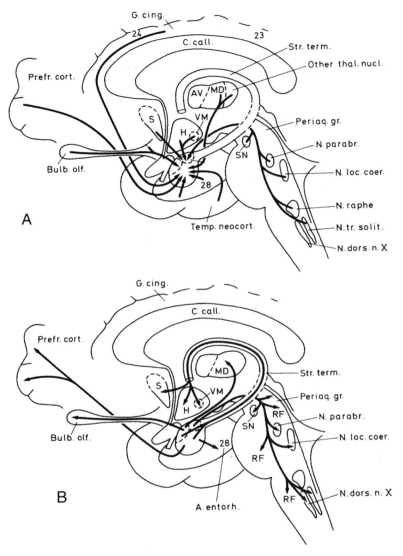

FIGURE 12.18.

  A, Principal afferent connections of the amygdaloid complex. **B**, Principal efferent connections. *Prefr. corta,* prefrontal cortex; *G. cing.,* cingulate gyrus; *C. call.,* corpus callosum; *Str. term.,* stria terminalis; *periaq. gr.,* periaqueductal gray matter; *N. parabr.,* parabrachial nerve; *N. loc. coer.,* locus ceruleus; *N. tr. solit.,* nucleus of the solitary tract; *N. dors. n. X,* dorsal motor nucleus of X; *SN,* substantia nigra; *S,* septal nuclei; *H,* hypothalamus; *VM,* ventromedial nucleus; *AV,* anteroventral nucleus; *MD,* medial dorsal nucleus. Numbers refer to the Brodmann areas.

This transition area is the **subiculum** [L. *subex,* a layer].

  The dentate gyrus and hippocampus are phylogenetically the oldest portion of the cerebral cortex, generally designated **archicortex**. They are composed of only three layers. The subiculum is of more recent phylogenetic origin and consists of three to five layers. The subiculum, the piriform cortex, and the cingulate gyrus are designated the **paleocortex**. The remainder of the cerebral cortex consists of six layers and is designated the **neocortex**. These designations

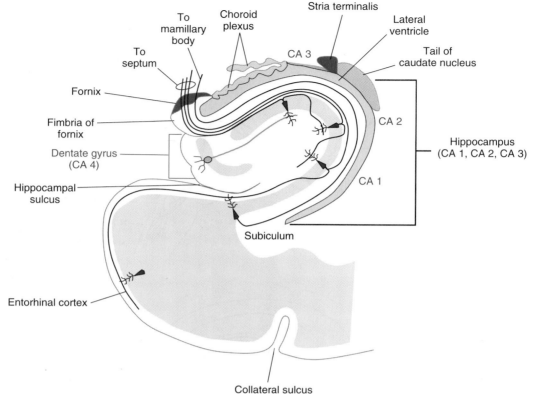

FIGURE 12.19.

   Diagram of the hippocampal formation. The designations CA_1 though CA_4 are based on histological differences. Note the synaptic relationships within the hippocampal system and the serial flow of information, step by step, from CA_4 to CA_1 and finally to the subiculum and entorhinal cortex. Note that the principal output from the hippocampal formation is to the cingulate gyrus. Fibers in the fornix arise from both the hippocampus proper and the subiculum, the former terminating mostly in the septal nuclei, while the latter terminate in the ipsilateral mamillary body.

are based on comparative anatomical studies and need concern the physician only as terms of reference.

Connections of the Hippocampal Formation

Pyramidal cells in the entorhinal cortex supply most of the fibers that enter the hippocampal formation. These axons pass through the subiculum to reach the dentate gyrus, where the majority terminate. This pathway is called the **perforant path** because of the way it penetrates the subiculum. Axons from cells of the dentate gyrus innervate the adjacent portion of the hippocampus proper (CA_3). Pyramidal cells leaving this area divide, with one branch entering the fornix while the other branch innervates the more distal portion of the hippocampus. Pyramidal cells from CA_1 innervate the subiculum. Axons leaving the subiculum innervate two areas, the adjacent entorhinal cortex and, by way of the fornix, the mamillary body of the hypothalamus. In addition to the principal afferents from the entorhinal cortex, a small number of axons from the septum innervate the hippocampal formation. These enter by way of the fornix, and most are cholinergic.

   The largest number of efferent connections of the hippocampal formation project from the subiculum to the adjacent entorhinal cortex (Fig. 12.20). There are also significant connections with the cingulate gyrus. Although they are numerous, the axons that make these connections are diffuse and carry no specific name. Nevertheless, one should not lose sight of the fact that the *principal efferent connections of the hippocampus are with cortical structures.*

The second efferent path from the hippocampal formation is the **fornix** [L. *fornix,* arch]. It is formed from axons of pyramidal cells of the hippocampal formation. They collect on the ventricular surface, forming a white structure known as the **alveus** [L. *alveus,* cavity, trough]. Tiny strands of axons leave the alveus as the **fimbria** [L. *fimbriae,* fringe], traverse the third ventricle, and collect as a compact bundle on the superior wall of the ventricle as the fornix. The fornix follows the ventricle, passing out of the temporal lobe, where it joins the inferior aspect of the septum near the midline. It proceeds anteriorly to the anterior commissure, where it divides into a **precommissural** and a **postcommissural** segment. Axons originating in the hippocampus proper follow the precommissural fornix and terminate in the septal nuclei. Axons originating

in the subiculum follow the postcommissural fornix into the hypothalamus, where they terminate in the ipsilateral mamillary body (Fig. 12.20).

The physiological role of the hippocampal formation is much debated. A great deal of evidence has accrued that shows it to be essential for certain types of memory tasks, but the hippocampal formation certainly is not a "memory bank" in the same sense that information is stored in specific locations as in computers or libraries. Human memory is much more complicated. An adequate discussion of neuron-based memory would fill a book (see For Further Reading). A brief introduction to the subject as it relates to the hypothalamus is presented here because memory disorders are an important part of neurology.

Memory

There are at least three types of memory systems in the human nervous system. The simplest system is the **reflex**. While we generally do not consider reflexes to be memories, in fact they do represent *"hard-wired" systems that are evoked by specific sensory stimuli* (for example, the

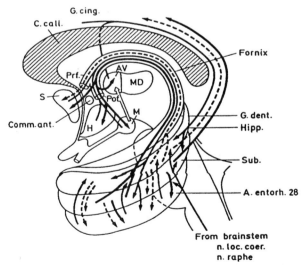

FIGURE 12.20.

Principal efferent connections of the hippocampal formation. *A. entorh.,* entorhinal area; *Sub.,* subiculum; *G. dent.,* dentate gyrus (CA$_4$); *Hipp.,* hippocampus (CA$_1$-CA$_3$); *Comm. ant.,* anterior commissure; *Prf.,* prefrontal cortex; *G. cing.,* cingulate gyrus; *C. call.,* corpus callosum; *n. loc. coer.,* locus ceruleus; *S,* septal nuclei; *H,* hypothalamus; *AV,* anteroventral nucleus; *MD,* medial dorsal nucleus.

muscle stretch reflex). To invoke a computer analogy, reflexes can be compared to read-only memory (ROM).

At the next level, humans have a type of memory that is usually called **procedural** memory. This type of memory refers to *learned skills* that are so well incorporated into our minds that we call on these remembered skill subconsciously. Walking, riding a bicycle, or using language are just a few examples of this type of memory. In our computer analogy, procedural memory is probably best represented by program subroutines or even entire programs that are called to execute a stereotyped algorithm.

Finally, there is **declarative** memory. This is the ordinary type of memory that we usually refer to when we say we are "remembering" things. In this case, declarative memory is the *skill to consciously recall experiences*, facts, or events. In the computer analogy, the recalled events are represented by the variable data operated on by the programs and stored in random-access memory (RAM) or magnetic memory (disks). The distinctions between the three types of memory are important because, as we are now learning, they each utilize different neurological mechanisms and brain structures. Here we are only concerned with declarative memory.

While it may seem obvious, it is worth noting that declarative memory mechanisms are composed of three functional elements: **acquisition**, **storage**, and **retrieval** (Fig. 12.21). The acquisition of sensory experiences and their internal representation as a neurological signal has been discussed in the previous chapters. We have already seen how various abstractions of sensory experiences are prepared at various parts of the cerebral cortex (see Chapter 11). Highly abstracted information from all parts of the cerebral cortex converges on the temporal cortex (see Chapter 13).

Human memory is bimodal because there are two types of storage mechanisms. The first phase is called **recent memory** because events can only be recalled for a few minutes after they occur. Unless there is some significance attached to the event, we soon forget it. This is equivalent to random-access memory in our computer analogy, since information in RAM, although relatively persistent, is easily lost. The inability to retain in memory most occurrences of daily life

is important, for without this gift of forgetting, our minds would be burdened with all the trivial happenings of a lifetime. Furthermore, selective forgetting is essential because it is doubtful that the mind has enough capacity to retain all the information we receive in our lifetimes.

The second phase is called **long-term memory**. Events to which we attach particular importance can be remembered, essentially indefinitely. After events become incorporated into recent memory, additional neuronal processing incorporates them into long-term memory. Once committed to long-term memory, past events can usually be recalled indefinitely and at will. Long-term memory is sometimes called permanent memory, but no memory is really permanent, as any student in the middle of an examination can attest! We gradually forget, over a period of years (Fig. 12.21), most events committed to long-term memory. To continue our computer analogy, long-term memory is like hard disk storage.

The permanence of memory depends on the context in which the events are presented. Strong emotional or silly associations help to sear experiences into long-term memory. Furthermore, long-term memory is an ongoing, active process. Permanence depends, to a large degree, on subsequent events. Frequent recollection aids retention. Also, making new associations with a remembered event can extend and enhance our ability to recall it.

This bimodal model of memory implies that a neurological mechanism exists that consolidates events from short-term to long-term memory. This mechanism is not understood. It is clear, however, that the **hippocampal formation** is absolutely required for long-term memory consolidation. The anterior and dorsomedial nuclei of the thalamus, the septal nuclei, and the nucleus accumbens have also been implicated, but their direct role in memory consolidation is less certain.

The role of the hippocampal formation in memory consolidation is illustrated by patients who have experienced bilateral lesions of the hippocampus, lesions that invariably include the inferior-medial temporal lobe and the amygdaloid complex. These people have no trouble remembering past events. They also have no trouble solving current problems or functioning

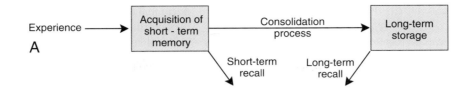

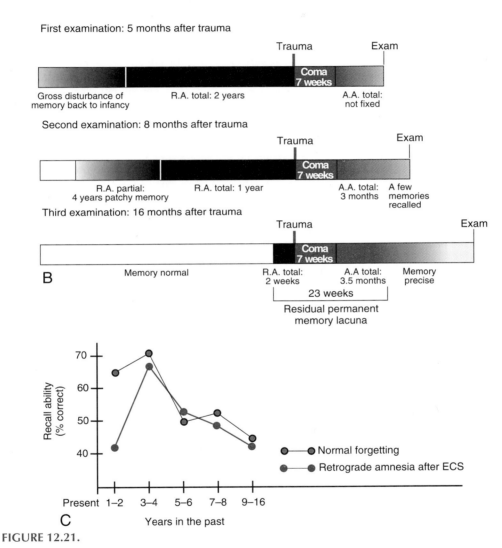

FIGURE 12.21.

A, Simple schematic drawing to show a "black box" relationship between long-term and short-term memory mechanisms and their readout systems. The Korsakoff amnesic state is brought about by a malfunction of the process of long-term memory consolidation. Previously stored long-term memory is present and available for recall. Similarly, recent events can be appreciated and remembered, but only for a limited time. **B,** Pattern of anterograde and retrograde memory changes following severe head injury. **C,** Alterations in memory following electroconvulsive shock (ECS). The *black line* represents the degree of "normal" forgetting over time. The *dashed line* represents memory losses after ECS. Note that, normally, although less is remembered from the distant past, those distant memories are more resistant to loss in the face of functional disturbances of the brain.

in everyday life. However, they *cannot consolidate current events into a permanent memory* that can be recalled at a later time. All of life's experiences between the time of their lesion and the present vanish.[8] Time ceases except for the immediate present and the remote past. This deficit carries no immediate emotional strain, since the problem cannot be recognized by the patient. With time, however, the patient becomes more and more baffled and confused. To the bewilderment of the patient, his environment changes instantly, the moment he leaves familiar surroundings. "How did these trees get so huge overnight?" "How could the Joneses' house vanish overnight?" As long as the patient can be maintained in a familiar time capsule he can function nearly normally. Thrust into the present world, he becomes disoriented and confused. An elegant description of such a patient has been provided by Oliver Sacks (see For Further Reading).

The inability to remember events that occurred prior to the onset of illness is called **retrograde amnesia**, while the inability to incorporate new information into memory is known as **anterograde amnesia**. The combination of limited retrograde amnesia with anterograde amnesia in an individual *who has otherwise normal cognitive abilities,* such as has just been described, is called the **Korsakoff amnesic state**, named after the Russian physician who first described the condition. This condition is much more common than one might suppose, since it is *produced by thiamin (vitamin B_1) deficiency.* It arises most frequently in people who abuse alcohol, since their thiamin reserves can be rapidly depleted. Often there are other neurological signs, the most important of which are ophthalmoplegia, ataxia and, occasionally, atrophy of the anterior lobe of the cerebellum. These signs may appear in the absence of Korsakoff's amnesic state. If there is no cerebellar atrophy, these ancillary symptoms are relieved by the administration of thiamin, usually within hours, but the amnesic state persists in spite of treatment.

Memory consolidation depends on the hip-pocampal formation. It receives information, via the entorhinal cortex, from all of the association areas of the neocortex, areas on which multi-modal forms of sensory information converge. This information passes serially through a three-neuron feed-forward path (dentate gyrus → CA_3 → CA_1 → subiculum) in the hippocampal formation and is then returned to these same association areas. The reinforcement of the neuronal signal by hippocampal efferents is presumed to be the process by which permanent memory traces are formed. The critical dependence of memory consolidation on the hippocampus was dramatically illustrated in a patient who developed a lesion that just involved CA_1, a lesion that was sufficient to cut the neocortex-hippocampus-neocortex loop. This patient exhibited all the manifestations of Korsakoff's amnesic state (Fig. 12.22).[9]

Another aspect of memory is familiarity. We all recognize places where we have been. This recognition is a form of memory recollection. Associated with this recognition is a sense of comfort that accompanies knowledge of one's immediate environment. Placed in a strange environment, animals and humans will explore their surroundings until a certain level of familiarity and comfort are achieved. Animal studies have suggested that the hippocampal formation plays a role in achieving this sense of surroundings. Individual hippocampal cells have been observed that fire only if the animal is in a specific place in a familiar environment. These "place cells" are very choosy about the environmental cues that cause them to fire. In this regard they are reminiscent of the "face-sensitive" cells discovered in the inferior temporal cortex (see Chapter 11). The importance of the "place cells" is suggested by the experience of humans who suffer epileptic seizures that originate in or near the hippocampus (see Chapter 13). Just before an attack, these patients often report a sense of

---

[8]Actually there is always some *retrograde amnesia;* that is, there is some amnesia of events before the time of the lesion. The amount of time lost varies among patients.

[9]Textbooks are filled with statements concerning the importance of the mamillary bodies in long-term memory consolidation. These statements are based on the connections between the subiculum and the mamillary bodies via the fornix and the frequent association of mamillary atrophy with Korsakoff's amnesic state. Recent animal studies, however, have clearly shown that transection of the fornix does not produce dramatic memory impairment. The role of the fornix and the mamillary bodies in memory consolidation has been greatly overstated.

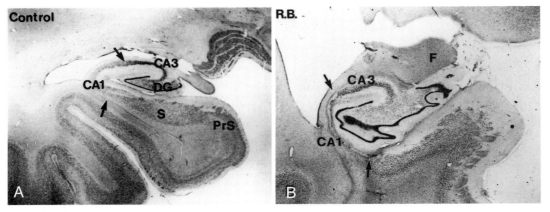

FIGURE 12.22.

A, Histology of a normal hippocampal formation. The CA$_1$ region is labeled. **B**, Histology of a similar view of the hippocampus from a patient with a pronounced Korsakoff amnesia (see text). Note that CA$_1$ (between the *arrowheads*) is nearly devoid of cells. This lesion extended the full length of the hippocampal formation and was the only significant neuron loss observed in this patient's brain.

unfamiliarity with their surroundings, a feeling often accompanied by great fear. This sense of unfamiliarity exists even if the patient objectively understands that he knows his surroundings and that they ought to be familiar to him.

THE SEPTAL NUCLEI

The lateral ventricles are divided at the midline by the septum pellucidum. In most mammals a few neurons occupy this sheet, but in the human it is essentially devoid of neurons. A small group of cells, the **septal nuclei**, exist at the rostral portion of the septum, near the midline and anterior to the anterior commissure and fornix (Fig. 12.23). This group is usually divided into three subnuclei: the medial, lateral, and dorsal septal nuclei. Several other collections of cells are variously included in the generic structure called the "septum" by some authors and excluded by others. The most prominent of these are the **nucleus of the diagonal band**, the **bed nucleus of the stria terminalis**, and the **bed nucleus of the anterior commissure**. The anatomy of this region is not well understood.

The septal nuclei receive afferent connections from the hippocampus proper via the precommissural fibers of the fornix (Fig. 12.23). The only other connection of the septal nuclei with the cortex is with the anterior portion of the

cingulate gyrus (area 24). The septal nuclei also receive fibers from the amygdala via the diagonal band. It receives afferents from the brainstem through the **medial forebrain bundle**. The most prominent of the brainstem nuclei that send axons to the septum are the **locus ceruleus**, the **dorsal raphe**, and the **periaqueductal gray matter**.

Efferents from the septal complex descend through the hypothalamus in the **medial forebrain bundle** and continue into the brainstem. Fibers in the medial forebrain bundle terminate throughout the hypothalamus, and particularly in the mamillary bodies. In the brainstem, they synapse in the locus ceruleus, the dorsal raphe, and the periaqueductal gray matter. Efferents from the septum also synapse in all parts of the hippocampal formation, entering it by way of the fornix. Similarly, septal efferents also innervate the amygdaloid complex. Septal efferents also project to area 24 of the cingulate gyrus.

The physiological significance of the septum is not clear. The most intriguing experiments relating to septal function have been performed in rats in which stimulating electrodes have been placed in the septum or the medial forebrain bundle. These animals can press a lever that will deliver an electrical stimulus to this bundle. The stimulus is apparently perceived to be of great significance to the rat, for it will press the lever

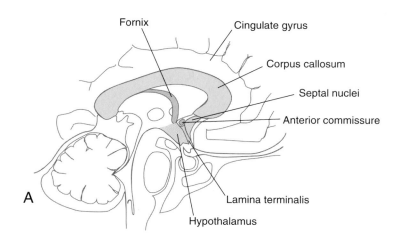

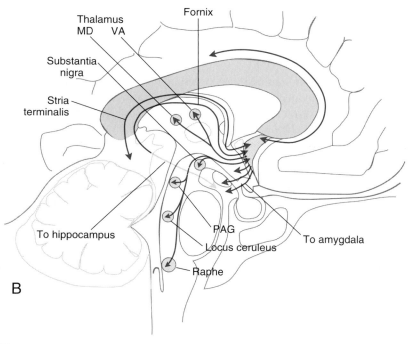

FIGURE 12.23.

A, Drawing of a midsagittal section showing the relation of the septal nuclei to the anterior commissure, the corpus callosum, the lamina terminalis, and the hypothalamus. **B**, Principal connections of the septal nuclei. *MD*, medial dorsal nucleus; *VA*, ventral anterior nucleus.

to the exclusion of all other activity, including feeding and sex. The results are more variable in similarly equipped humans, but many have reported a feeling of profound well-being. These experiments must be interpreted with caution, since stimulation of the medial forebrain bundle will affect many structures from the brainstem to the cingulate gyrus. They do nothing to elucidate septal function per se. However, the septal nuclei are strategically located and may very well serve as an important link between the hippocampal formation, amygdala, and hypothalamus and the cerebral cortex, in particular the cingulate gyrus.

THE CINGULATE GYRUS

The **cingulate gyrus** wraps around the external border of the corpus callosum (Fig. 1.3). Wrapping around its genu, the cingulate gyrus becomes the **subcallosal gyrus**. At the posterior portion of the corpus callosum, the **isthmus of the cingulate gyrus** blends into the parahippocampal gyrus. Deep to the cingulate gyrus is the **cingulum**, a large bundle of association fibers that connect the cingulate gyrus with many distant areas of the brain.

The subiculum of the hippocampal formation provides the most prominent connections with the cingulate gyrus (Fig. 12.24). There are two pathways by which the subiculum is connected with the cingulate gyrus, a direct and an indirect path. The direct path is by way of the cingulum. This path is reciprocal. The indirect path involves first the subicular fibers of the fornix that innervate the mamillary bodies in the posterior hypothalamus. From there, the **mamillothalamic tract** carries information to the **anterior nucleus** of the thalamus. The anterior thalamic nucleus, in turn, projects to most areas of the cingulate gyrus, but the heaviest innervation is

to the anterior region (area 24). There is a second indirect path from the hippocampal formation to the cingulate gyrus, originating not from the subiculum but from the hippocampus proper. These axons project to the septal nucleus by way of the fornix. From the septum, axons innervate area 24.

The cingulate gyrus, like all parts of the cerebral cortex, has important connections with specific thalamic nuclei. As previously mentioned, area 24 receives fibers from the anterior thalamic nucleus. Information from the cingulate gyrus is returned to the anterior nucleus, however, from the posterior region (area 23). The two cingulate regions, 23 and 24, are interconnected by numerous short association fibers in the cingulum. The anterior region (area 24) also projects, apparently unidirectionally, to the dorsomedial thalamic nucleus. This link is significant because the dorsomedial nucleus projects heavily to the prefrontal lobes of the cortex (see Chapter 13).

In addition to the well-documented reciprocal connections of the cingulate gyrus with the subiculum, it also projects to and receives information from the **temporal**, **parietal**, and **prefrontal lobes** of the cerebral cortex. These connections are with regions of the cerebral cortex that are commonly referred to as "asso-

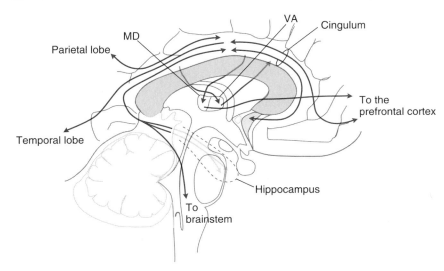

FIGURE 12.24.

Principal connections of the cingulate gyrus. *MD,* medial dorsal nucleus; *VA,* ventral anterior nucleus.

ciation areas." This term is imprecise and is often used differently by different authors. Generally, "association areas" of cerebral cortex are regions that are not primarily motor or sensory. They usually receive secondary or tertiary projections from primary motor or sensory regions. This concept will be discussed further in Chapter 13.

Finally, it must be mentioned that the anterior region of the cingulate gyrus projects to the amygdaloid nuclei. This connection provides two indirect paths by which the cingulate gyrus can influence the prefrontal cortex. First, the amygdala projects directly to the prefrontal lobes by way of the uncinate fasciculus. Second, it has reciprocal connections with the dorsomedial nucleus of the thalamus, the principal thalamic nucleus associated with the prefrontal lobes.

The function of the cingulate gyrus is not known. Stimulation in experimental animals seems to evoke the same sort of behavioral changes observed by stimulating the amygdala. Bilateral ablation of the anterior region has been said to decrease aggressiveness in animals, although the opposite has been reported too. Cingulotomies and ablation of portions of the cingulate cortex and cingulum have been performed in humans in an effort to modify socially unacceptable behavior associated with certain psychic disorders. This type of psychosurgery has met with variable results. Cingulotomy has been said to be effective in relieving chronic pain in some cases. These results have led some to suggest that the cingulate gyrus and/or those brain structures closely connected with it are important in attaching emotional quality or meaning to sensory signals, functions that have also been ascribed to the amygdala.

BEHAVIORAL CORRELATES

Knowledge of the kinds of behavior that can be affected by lesions to the hypothalamus and closely related structures is clinically useful, and therefore a very brief synopsis is presented here.

Feeding Behavior

Certain hypothalamic lesions alter the feeding behavior of the organism. Roughly speaking, ventromedial lesions produce hyperphagia [G. *hyper,* above, over + G. *phagein,* to eat], while lateral lesions produce aphagia. Experiments such as this need to be interpreted with caution, because the hypothalamus is a very small structure and fibers of passage are freely intermingled with the somas of neurons. Therefore, no lesion is limited to a single structure. Nevertheless, the clinical implications of these experiments are clear. Eating disorders commonly accompany hypothalamic tumors and lesions.

Rage

More dramatic, perhaps, are reactions that are commonly referred to as **sham rage**. Experimentally, removing the neocortex produces an animal that responds violently, with poorly directed attack behavior, in response to almost any stimulus. If the hypothalamus is removed, this rage reaction can no longer be provoked. In intact animals, rage can be evoked by stimulating various structures, including the septal nuclei and the anterior hypothalamus. Observations such as this suggest that neocortical structures suppress primitive behavior by acting through hypothalamic centers. When disconnected from neocortical elements, primitive behavior is released, or at least very poorly regulated. Humans with lesions affecting hypothalamic structures or their connections may exhibit bizarre attack behavior, and may even try to bite their physician (see the case of Ms. P. R.).

Sexual Behavior

The behavioral manifestations of reproduction are similarly affected by the hypothalamus. Stimulation of anterior regions often produces poorly directed mounting behavior. In one classic experiment, goats would mount and attempt to copulate with the nearest object when the anterior hypothalamus was stimulated electrically by an implanted electrode. Slightly different stimulus parameters or electrode placement would cause the goats to begin drinking instantly as soon as the stimulus began and to continue drinking until the stimulation ceased.

CASE STUDIES

Ms. P. R.[10]

HISTORY, NOVEMBER 12, 1962: Ms. P. R., a 20-year-old female bookkeeper, was admitted to the hospital complaining of polydipsia, polyuria, and bitemporal headaches. She has had no menses for the past 10 months. For her height, 5 ft 5 in., she is somewhat overweight (134 pounds), but otherwise in good health. A complete neurological examination, including an extensive visual field analysis, revealed no abnormalities.

LABORATORY STUDIES: After 10 hours of water deprivation, her urine specific gravity was 1.003. Urine output (volume unspecified) remained unchanged after a 45-minute infusion of 2.5% saline; however, urine production was inhibited by injection of 0.1 unit of vasopressin. Urine gonadotropins were 0 units in 24 hours. Cerebral spinal fluid examination revealed protein content of 43 mg/dL, glucose 47 mg/dL, and an opening pressure of 70 mm CSF. A pneumoencephalogram (radiological visualization of the ventricles by replacing the CSF with air) could not outline the third ventricle.

The diagnosis of diabetes insipidus was made based on the ability of vasopressin to inhibit urine production. The failure of the pneumoencephalogram to demonstrate the third ventricle suggested a hypothalamic tumor which subsequent craniotomy failed to confirm. Ms. P. R. recovered from surgery uneventfully. Her diabetes insipidus was controlled with vasopressin and she was discharged. Her headaches remained.

SUBSEQUENT COURSE, September 8, 1964: Ms. P. R. returned to the hospital in September of 1964. Since the previous July, her behavior had undergone a remarkable transformation. She became withdrawn, sometimes bursting out with unprovoked laughter, crying, or rage. At times she became confused and was found sleeping in the wrong bed. She occasionally conversed with imaginary people and had several times disrobed in public. Her physical examination was unremarkable except that she was still overweight and her pubic hair was sparse.

The neurological examination could not be performed because of the patient's aggressive and uncooperative behavior. She would, at times, strike out and attempt to hit, scratch, or bite the examiner. She was frequently disoriented as to place and time. Her memory was variable; sometimes she could not remember past events and at other times her memory seemed normal. Occasionally she was pleasant and cooperative, and expressed regret for her aggressive behavior. Her body temperature fluctuated, at times reaching 104°F. No source of infection was found. Laboratory studies at this time revealed decreased thyroid, adrenal cortical, and gonadal function in addition to the previously noted diabetes insipidus. CSF examination revealed a protein content of 73 mg/dL, glucose 82 mg/dL, and an opening pressure of 220 mm CSF.

A second pneumoencephalogram failed to outline the suprachiasmatic recess. Craniotomy revealed a tumor at the base of the third ventricle, but because of its critical location, no attempt was made to remove it. The patient recovered from surgery without complication, but her behavioral symptoms deteriorated. To quote from the original report:

"In spite of the sedation and the correction of her metabolic and hormonal deficits, the patient continued to display frequent outbursts of directed violence, consisting of hitting, biting, scratching, and throwing objects at attendants. Although these initially appeared to be unprovoked, we subsequently noted that the withholding of food, which she consumed in quantities of 8,000 to 10,000 calories per day, invariably evoked aggressive behavior. Near the end of her hospitalization, continuous feeding was found to be the only method which succeeded in maintaining the patient in a reasonably tractable state. . . . Her hyperphagia was unabated and by the end of her second month of hospitalization she had gained 24 kg (52.8 lb)."[11]

The patient died on December 23, 1964. Autopsy revealed a tumor located in the ventral medial portion of the hypothalamus that spared the dorsal and lateral regions and the mamillary bodies, but included the fornices and the median eminence (Fig. 12.25).

---

[10]This case was originally reported by Reeves and Plum (Reeves, A. G., and Plum, F. Hyperphagia, Rage, and Dementia Accompanying a Ventromedial Hypothalamic Neoplasm. *Arch. Neurol.* 20:616–624, 1969). Pedagogically useful cases of this nature are rare, and although this case is quite old and the technology available to the physicians at the time is considered primitive by today's standards, the correlation of symptoms with anatomy is so well illustrated and documented that this case warrants contemporary study.

[11]From Reeves, A. G. and Plum, F. Hyperphagia, Rage, and Dementia Accompanying a Ventromedial Hypothalamic Neoplasm; *Arch. Neurol.* 20:616–624, 1969). Copyright 1969, American Medical Association.

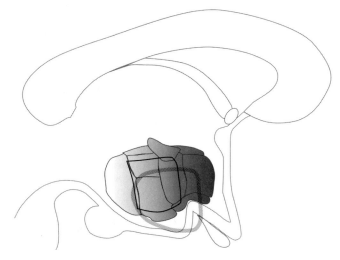

FIGURE 12.25.

Location of the tumor as revealed by autopsy of Ms. P. R.

Comment: In this case the tumor was probably limited initially to a small region in the posterior medial region of the hypothalamus that involved the arcuate nucleus and the magnocellular axons entering the neurohypophysis, the former being an important source of GnRH. As the tumor expanded to involve the median eminence, almost any endocrinological dysfunction would be expected, and this patient showed decreased function of several endocrine systems.

FURTHER APPLICATIONS

• Cases such as this, even though rare, are instructive because they reveal the close correlation between anatomical structures and their functions. Identify the lesioned structures that produced the following symptoms in this case: Diabetes insipidus; the unstable body temperature; the decreased endocrinological functions; the hyperphagia; and the rage reactions.

At the present time one cannot adequately explain the derangements in the patient's mental status. The waxing and waning of the patient's emotional state, including the incontinence of emotions, disrobing in public, lapses in memory, and at the end of her illness frank dementia are symptoms that have not been directly associated with hypothalamic function.

• At the time of this case (1962), modern imaging techniques were not available. What currently available mode of imaging would be most appropriate in this case, and, had it been available in 1962, how would it have affected the way this case was handled?

Ms. M. O.

HISTORY, OCTOBER 22, 1989:   Ms. M. O., an 80-year-old female, was in good health until she fell down a flight of stairs one evening, striking her head against the wall at the foot of the stairs. She arrived at the emergency room unconscious. Her daughter said that Ms. M. O. had been visiting her and had apparently gotten out of bed in the middle of the night, had become disoriented, and stepped into the stairwell, thinking that it was the bathroom. Before the accident Ms. M. O. had been living alone, driving her car, and leading an independent life in apparent good health. Fifteen years previously she had had a transient ischemic attack. After that incident, on the advice of her physician, she gave up smoking. She has had no recurrences.

X-rays taken in the emergency room revealed no broken bones. A CT scan revealed minimal high densities in the right parietal cortical area consistent with a right parietal contusion. A limited neurological examination at that time revealed conjugate doll's eye maneuvers, normal pupillary reflexes, normal MSRs from all test points, and flexor plantar responses. After about 2 hours she began to regain consciousness but was very disoriented. She was hospitalized.

NEUROLOGICAL EXAMINATION, OCTOBER 23, 1989:

I. MENTAL STATUS: Ms. M. O. was disoriented and confused. She communicated very little, but was basically cooperative.

II. CRANIAL NERVES:

A. **Olfactory:** Not tested

B. **Optic:** Unable to test visual fields because the patient could not cooperate.

C. **Eye movements:** Could not be tested directly, doll's eye maneuver showed grossly intact eye movements. Pupillary responses were present and consensual from both eyes. There was no ptosis.

D. **Trigeminal:** Not tested.

E. **Facial:** Face appeared symmetrical. Eyes seemed normally hydrated.

F. **Acoustic:** Not tested.

G. **Vagal-glossopharyngeal:** Not tested.

H. **Accessory:** Not tested.

I. **Hypoglossal:** Not tested.

III. MOTOR SYSTEMS: Could not test.

IV. REFLEXES: The MSRs from the left biceps, brachioradialis, and quadriceps muscles were more brisk than from the right. The other test points were not evaluated. The Babinski sign was present on the left, absent from the right.

V. SENSORY SYSTEMS: Could not test.

SUBSEQUENT COURSE: In the hospital Ms. M. O. became febrile, which was soon discovered to be caused by pneumonia, probably caused by aspiration. The infection responded rapidly to antibiotic therapy. Although her general confusion and disorientation improved during the first few days of hospitalization, it never fully resolved. On the fourth day, she became delirious and then unconscious. Plasma sodium levels were found to be 108 mEq/L; urine osmolality was 523 mosm/kg. Her fluid intake was restricted to 500 mL/day. Over the next 24 hours her plasma sodium increased to 119 mEq/L. She regained consciousness and her confusion and disorientation gradually improved. During the following 2 weeks, her plasma sodium levels fluctuated between 105 mEq/L to 129 mEq/L, and her level of consciousness, confusion, and disorientation fluctuated in concert with the sodium levels. Eventually she stabilized and was discharged to a rehabilitation unit, where she remained for 4 weeks. At that time she was discharged to a nursing home, where she lived until her death from unrelated causes on the 28th of September, 1991.

Comment: Blood volume and osmolality are carefully regulated within very narrow limits by humoral and behavioral mechanisms that are coordinated by the nervous system. It follows, therefore, that insult to the nervous system can potentially disrupt this homeostatic mechanism. Closed head injury, brain surgery, meningitis, encephalitis, stroke, subarachnoid hemorrhage, and neoplasms can all interfere with hypothalamic function and lead to the syndrome of inappropriate secretion of ADH (SIADH). *The mechanisms by which this occurs are not known, but the syndrome is common and well recognized.*

The present case is typical. Following a closed head injury, ADH (vasopressin) secretion continues inappropriately, decreasing water loss through the kidney, resulting in expansion of the blood volume and consequent dilution of the plasma sodium (hyponatremia) and reduction in blood osmolality. Hyponatremia and decreased blood osmolality normally suppress drinking and ADH secretion but fail to do so in these cases. As plasma sodium levels fall below about 120 mEq/L, higher cortical functions become increasingly compromised, as manifested in loss of alertness, drowsiness, coma, convulsions, and death. Sodium levels are usually restored by restriction of water intake rather than by infusing sodium, since the latter course frequently results in central pontine myelinolysis.

FURTHER APPLICATIONS

- Considering what you know about the ionic mechanisms of the action potential and of synaptic transmission, speculate about the possible mechanisms involved in SIADH that compromise cortical functions.

- What parts of the nervous system are involved in maintaining osmolality of the blood? How could the diverse set of causal factors listed above affect these structures?

Mr. B. T.

HISTORY, JULY 4, 1989: Mr. B. T. was brought to the hospital by his family after he became delirious and confused at the family Fourth of July picnic. Although it was a hot July day—the temperature was 87°—Mr. B. T. felt cold and asked for a blanket. The family, thinking he was joking, ignored his request. Later they found him wrapped in a blanket, shivering and delirious.

In the emergency room, Mr. B. T. was no longer delirious. He was shivering violently and wrapped in the three blankets he had asked for from the nurses. Blood pressure was 132/82, pulse was 97, and temperature was 103°F. He was not perspiring.

NEUROLOGICAL EXAMINATION:

I. MENTAL STATUS: Mr. B. T. was awake, alert, and oriented as to date, time, and place. He stated that his family thought he was delirious because he had asked for a blanket. He said that he just was very cold. He was aware that it was a hot July day. He said he didn't know why he was cold.

II. CRANIAL NERVES:

A. **Olfactory:** He could detect and correctly identify methyl salicylate from both nostrils.

B. **Optic:** There was a dense bitemporal hemianopsia on confrontation testing.

C. **Eye movements:** He could track the examiner's moving finger in all directions. There was no obvious weakness of any of the muscles of ocular motion. There was no nystagmus. Pupillary responses were present and consensual from both eyes. There was no ptosis.

D. **Trigeminal:** Sensations to cotton touch and pin prick were present and symmetrical from all divisions. Corneal reflex was present. Masseter muscles were equal.

E. **Facial:** Grimace was symmetrical and full. Eyes appeared normally hydrated.

F. **Acoustic:** Hearing was grossly normal to finger rubbing.

G. **Vagal-glossopharyngeal:** Voice was normal. Pharyngeal arches were high and symmetrical. Uvula was midline. Gag reflex was present.

H. **Accessory:** Sternocleidomastoid and trapezius muscles seemed to be of normal strength and were symmetrical.

I. **Hypoglossal:** Tongue protruded in the midline. There were no fasciculations. There was no atrophy.

III. MOTOR SYSTEMS: Strength was everywhere normal and symmetrical. Drift test was normal. Gait was normal. Tandem walk, hopping, Romberg, heel-to-shin, and finger-to-nose were all performed without difficulty.

IV. REFLEXES: All MSRs were within normal limits and symmetrical. Plantar signs were flexor.

V. SENSORY SYSTEMS: Sensations to pin prick and cotton touch were everywhere present and symmetrical.

Comment: Finding Mr. B. T. shivering and wrapped in three blankets on a hot July day might suggest that he is delusional or hysterical. A psychiatric illness, however, is less likely in light of the bitemporal hemianopia revealed during the neurological examination and the lack of sweating in spite of his elevated body temperature. A CT scan of the head was ordered, which revealed a large mass at the level of the optic chiasm.

SUBSEQUENT COURSE: A craniotomy was performed and the mass successfully removed. Mr. B. T. recovered from surgery without complication. He subsequently moved from the area and was lost to follow-up.

FURTHER APPLICATIONS

• Explain how the major findings in this case—the bitemporal hemianopia and the lack of sweating—can be explained by a single-locus lesion. Where is the lesion and what structures are involved?

• Explain how the patient could be shivering and wrapping himself in a blanket in spite of his abnormally high body temperature. What non-neurological cause for increased body temperature ought to be considered?

SUMMARY

The hypothalamus is a small mass of neural tissue located at the base of the diencephalon. Through its neural connections it is linked with the three principal regulatory systems of the organism: the autonomic nervous system, the endocrine system, and behavior.

Nuclei in the hypothalamus are the source for

the main descending signals that control the parasympathetic and the sympathetic divisions of the autonomic nervous system. These signals descend in the dorsal longitudinal fasciculus, the mamillotegmental tract, and the medial forebrain bundle. The anterior region of the hypothalamus is closely associated with parasympathetic functions (slowing of the heart, decrease in blood pressure, peripheral vasodilatation, increased gastric motility, salivation, and sweating), while the posterior and lateral hypothalamic regions are associated with sympathetic functions (increased heart rate and blood pressure, peripheral vasoconstriction, decreased gastric motility, lack of salivation, and cessation of sweating).

The peripheral components of the parasympathetic system are composed of a long efferent preganglionic axon that originates in certain cranial nerve nuclei or in a small cell column in the sacral spinal cord. The preganglionic axon synapses in or near the target organ, from which a short postganglionic axon innervates the target tissue. The peripheral components of the sympathetic system originate in the spinal cord as short preganglionic efferent neurons. They synapse either in the spinal chain ganglia or collateral ganglia, from which a long postganglionic axon emerges to innervate the target tissue.

The principal efferent influence of the hypothalamus on the endocrine system is by means of its connections with the hypophysis. Endocrinological signals can reach the hypothalamus by means of the circumventricular organs. Neurosecretory granules containing prohormones for vasopressin or oxytocin are produced by magnocellular secreting neurons in the anterior hypothalamus. Axons from these neurons pass into the neurohypophysis and release the active form of these hormones into the perivascular space, from which they diffuse into the bloodstream.

The adenohypophysis, a true endocrine gland, is partially controlled by hormones produced in the hypothalamus. These regulatory hormones are synthesized in parvocellular secreting hypothalamic neurons, transported along axons, and released into the hypophysoportal system, which transports them to the adenohypophysis. There they regulate the release of the circulating hormones TSH, LH, FSH, GH, prolactin, and ACTH. The release of these hormones as well as the hypothalamic releasing hormones is partially regulated by long-loop and short-loop negative feedback systems.

The hypothalamus affects behavior through its connections with forebrain structures, principally the amygdaloid nuclei, the hippocampal formation, the septal nuclei, the cingulate gyrus, and the prefrontal cortex. These structures are interconnected by complex reciprocal pathways. Fibers from the amygdaloid nuclei enter the hypothalamus via the stria terminalis. The principal efferent connections of the hippocampal formation are with the cingulate gyrus via fibers originating from the subiculum and adjacent entorhinal cortex. Axons also leave the hippocampal formation via the fornix and synapse in the mamillary bodies (from the subiculum) and the septal nuclei (hippocampus proper). The septal nuclei send axons into the hypothalamus via the medial forebrain bundle. Hypothalamic connections with the cingulate gyrus are indirect, being mediated by the anterior and dorsomedial nuclei of the thalamus, the hippocampal formation, the amygdaloid nuclei, and the septal nuclei. These connections are important because the cingulate gyrus is connected with the prefrontal, temporal, and parietal lobes of the cerebral cortex.

Stimulation of or lesions to various parts of the hypothalamus directly affect certain kinds of primitive behavior. Ventromedial lesions to the hypothalamus disrupt feeding behavior to the extent that the animal (or human) becomes hyperphagic. Lesions to lateral areas produce the opposite effect, aphagia. Unprovoked rage reactions can be produced by stimulation to the septal nuclei or the anterior hypothalamus. Depending on electrode placement, anterior hypothalamic stimulation can also produce inappropriately directed sexual mounting behavior or unquenchable drinking.

FOR FURTHER READING

Adams, R., and Victor, M. *Principles of Neurology.* New York: McGraw-Hill, 1993.

Brodal, A. *Neurological Anatomy In Relation to Clinical Medicine.* New York: Oxford University Press, 1981.

Clifton, D. K. The Anterior Pituitary. In *Textbook of Physiology,* Ch. 63, edited by Patton, H. D., Fuchs, A. F., Hille, B., Scher, A. M., and Steiner, R. Philadelphia: W. B. Saunders, 1989, pp. 1202–1214.

Cornett, L. E. The Adrenal Medulla. In *Textbook of Physiology,* Ch. 65, edited by Patton, H. D., Fuchs, A. F., Hille, B., Scher, A. M., and Steiner, R. Philadelphia: W. B. Saunders, 1989, pp. 1229–1238.

Davis, M. The Role of the Amygdala in Fear and Anxiety. *Annu. Rev. Neurosci.* 15:353–375, 1992.

Dodd, J., and Role, L. The Autonomic Nervous System. In *Principles of Neural Science,* Ch. 49, edited by Kandel, E., Schwartz, J., and Jessell, T. New York: Elsevier, 1991, pp. 761–776.

Dorsa, D. M. Neurohypophyseal Hormones. In *Textbook of Physiology,* Ch. 61, edited by Patton, H. D., Fuchs, A. F., Hille, B., Scher, A. M., and Steiner, R. Philadelphia: W. B. Saunders, 1989, pp. 1173–1183.

Ito, M., and Nishizuka, Y. *Brain Signal Transduction and Memory*. London: Academic Press, 1989.

Knopp, R. H., and Magee, M. S. Pregnancy and Parturition. In *Textbook of Physiology,* Ch. 70, edited by Patton, H. D., Fuchs, A. F., Hille, B., Scher, A. M., and Steiner, R. Philadelphia: W. B. Saunders, 1989, pp. 1380–1407.

Kupfermann, I. Hypothalamus and Limbic System: Peptidergic Neurons, Homeostasis, and Emotional Behavior. In *Principles of Neuroscience,* Ch. 47, edited by Kandel, E., Schwartz, J., and Jessell, T. New York: Elsevier, 1991a, pp. 735–749.

Kupfermann, I. Hypothalamus and Limbic System: Motivation. In *Principles of Neuroscience,* Ch. 48, edited by Kandel, E., Schwartz, J., and Jessell, T. New York: Elsevier, 1991b, pp. 750–760.

Rowland, L. *Merritt's Textbook of Neurology*. Philadelphia: Lea & Febiger, 1989.

Sacks, O. The Lost Mariner. In *The Man Who Mistook his Wife for a Hat,* Ch. 2. New York: Harper & Row, 1987, pp. 23–42.

Simpson, J. B. The Circumventricular Organs and Brain Barrier Systems. In *Textbook of Physiology,* Ch. 64, edited by Patton, H. D., Fuchs, A. F., Hille, B., Scher, A. M,. and Steiner, R. Philadelphia: W. B. Saunders, 1989, pp. 1215–1228.

Smith, M. S. Lactation. In *Textbook of Physiology,* Ch. 71, edited by Patton, H. D., Fuchs, A. F., Hille, B., Scher, A. M., and Steiner, R. Philadelphia: W. B. Saunders, 1989, pp. 1408–1422.

Squire, L. R. *Memory and Brain*. New York: Oxford University Press, 1987.

Squire, L. R., Knowlton, B., and Musen, G. The Structure and Organization of Memory. *Annu. Rev. Psychol.* 44:1993.

Steiner, R., and Cameron, J. L. Endocrine Control of Reproduction. In *Textbook of Physiology,* Ch. 68, edited by Patton, H. D., Fuchs, A. F., Hille, B., Scher, A. M., and Steiner, R. Philadelphia: W. B. Saunders, 1989, pp. 1239–1262.

Zola-Morgan, S., and Squire, L. R. Neuroanatomy of Memory. *Annu. Rev. Neurosci.* 16:547–564, 1993.

13 The Cerebral Cortex

The cerebral cortex is, perhaps, the most enigmatic part of the mammalian nervous system. Its essential structure is quite simple yet it seems to embody the most complex functions performed by any living tissue. It is to the cerebral cortex that we attribute learning, perception, self-awareness, free will, and the most enigmatic of all neuronal functions, consciousness. How these complex cortical functions are achieved is not known. The simple structure of the cortical column seems to be an essential factor enabling the complex information transformations performed by the cerebral cortex. In this chapter we will briefly examine (a) the EEG, (b) altered states of consciousness, (c) the properties of cortical columns acting in concert, and (d) some of the clinically relevant functions associated with the cerebral cortex.

The Electroencephalogram

A cortical column is a self-contained information processing unit. The connections the columns make with other columns and subcortical structures means, however, that they do not function in isolation. As already noted, they are connected with adjacent columns, adjacent gyri, lobes, hemispheres, and subcortical structures such as the thalamus and brainstem. These numerous connections ensure that interconnected cortical columns function as an **ensemble**. Working together, the columns have *ensemble properties* that are characteristic of the group, in much the same way as a symphony orchestra has properties much different from its individual

instruments. The whole is much more powerful than the sum of its parts.

Although there are very good ways of monitoring the activity of single neurons with intra- and extracellular electrodes, there are no satisfactory methods for measuring the activity of neuron ensembles. A crude measure of the electrical activity of ensembles of columns are the electrical potentials that can be recorded from the cortical surface (electrocorticogram or ECoG) or from the surface of the scalp (**electroencephalogram** or **EEG**). The EEG is clinically useful.

A typical EEG is recorded from 16 electrodes that are attached to the scalp (Fig. 13.1). Amplifiers detect the electrical difference between pairs of electrodes and that potential is amplified and displayed on a polygraph (typically 16 to 20 individual channels) as a voltage versus time graph. Because each channel of the recording represents the electrical difference between pairs of electrodes, the activity displayed in each channel of the polygraph recording is derived from a relatively isolated portion of the cerebral cortex. Before the advent of modern imaging techniques, the EEG was the only non-invasive way of localizing brain dysfunction. It is still useful in that regard, when the underlying pathology cannot be visualized by CT or MRI.

The EEG is not a particularly good measure of cortical activity. It is limited by the fact that it reflects primarily the activity of pyramidal cells. Pyramidal cells are large and oriented vertically with respect to the cortical surface. When active they create large, electrical dipoles that are aligned, like the cell bodies, perpendicular to the cortical surface. This parallel arrange-

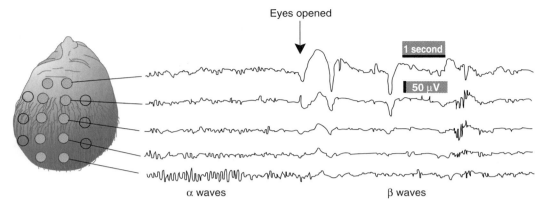

FIGURE 13.1.

The standard placement of electrodes for recording the EEG. Pairs of electrodes are led to differential amplifiers that amplify the potential difference between the electrodes. The electrical difference for each pair of electrodes is displayed simultaneously as a separate channel on a polygraph. A typical EEG from an awake, but relaxed individual. Note the waxing and waning of the alpha waves seen in the occipital trace. The large waves in the anterior traces were caused by an eye blink (*arrow*). Opening the eyes suppresses alpha waves. The remainder of the recording illustrates beta wave activity.

ment permits the extracellular electrical fields of pyramidal cells to interact, magnifying their effect. In contrast, most of the cortical interneurons are small and symmetrical. Being symmetrical, they do not produce strong electrical dipoles. Their electrical fields tend to be oriented randomly, which causes them to cancel each other.

Another limitation of the EEG is its narrow frequency range. For a number of technical reasons, the EEG is restricted to a frequency range from about 0.5 Hz to about 30 Hz. Therefore it represents primarily the activity of postsynaptic potentials, not action potentials. An EEG recording (Figs. 13.1 and 13.2) is divided into four frequency bands for ease of description and analysis. These are the **beta** (13 to 30 Hz), **alpha** (8 to 13 Hz), **theta** (4 to 8 Hz), and **delta** (0.5 to 4 Hz) bands. A normal, awake but quiet individual will produce primarily beta and alpha waves. The beta waves, characterized by high frequency and low amplitude, are produced by multiple **asynchronous** postsynaptic potentials. Because they are asynchronous, the PSPs interfere randomly and are unable to summate. **Synchronization** of the PSPs magnifies the recorded amplitude of the waveform because the extracellular currents can summate. During periods of relaxed wakefulness, transient periods of synchronization develop. They are signaled in the

EEG as a series of waxing, then waning, low-frequency alpha waves. This represents a sequence of increasing then decreasing synchrony among PSPs occurring in pyramidal cells. Since alpha waves are associated with a relaxed but awake state, sensory stimuli, such as opening the eyes or hearing a novel sound, will extinguish the alpha waves and re-establish an EEG dominated by beta waves. Hence beta waves are associated with an alert state.

Altered States of Consciousness

The EEG is a useful clinical and research tool. With it one can determine the overall activity of the cerebral cortex and, to some extent, localize abnormal activity to relatively small cortical areas. Because of this, *the EEG is an essential ancillary method for differentiating coma from cerebral death, and for diagnosing disorders of sleep and epilepsy.*

SLEEP

The human brain oscillates between periods of wakefulness and periods of **sleep**. If deprived

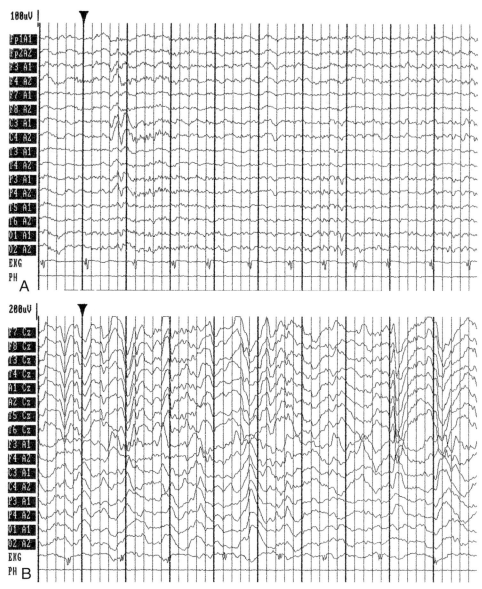

FIGURE 13.2.

The state of sleep is commonly divided into four stages based on the EEG record. Illustrated here are EEG records taken while the patient was awake and alert, and during various stages of sleep. **A**, Stage 1 sleep, often called quite wakefulness, is characterized by a low amplitude, high frequency record. **B**, Stage 2 is signaled by the development of high amplitude "sleep spindles," 13–16 Hz waves superimposed on a beta wave background. **C**, Stage 3 is similar to stage 2, but with the addition of delta waves. **D**, Stage 4 is dominated by delta waves with no spindling. The REM stage is indistinguishable from Stage 1 except for the presence of rapid eye movements.

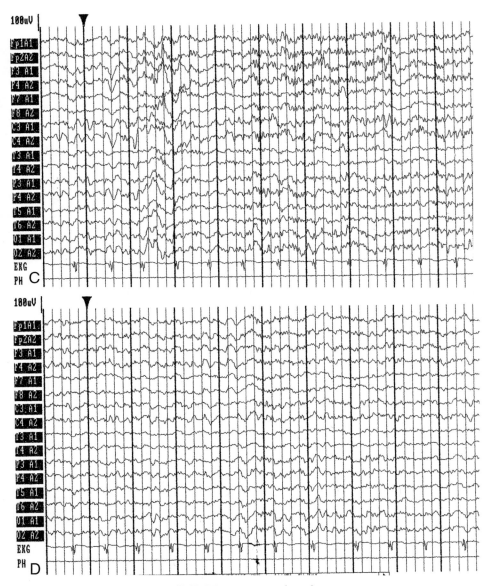

FIGURE 13.2.—*continued*

of sleep, many animals inexplicably die within weeks. This observation has not been confirmed in the human, but the human brain deprived of sleep does become increasingly dysfunctional. Sleep-deprived humans first become irritable, fatigued, and have difficulty concentrating. Motor skills then deteriorate and sustained thoughtful activity becomes impossible. Eventually sensory disorientation develops and the individual becomes hallucinatory. The need for sleep is absolute. The reasons we need sleep are unknown.

Stages of Sleep

The EEG of individuals undergoes characteristic transitions during the course of sleep (Fig.

13.2). The sleep EEG can be categorized into various stages that differ primarily in the degree of synchrony, represented by increasing dominance of the EEG record by theta and delta waves. As one enters sleep, the EEG becomes more and more synchronized. In the brainstem and spinal cord, there is moderate postsynaptic inhibition of the α-motor neurons that produces skeletal muscle relaxation. The autonomic nervous system activity becomes dominated by the parasympathetic division; gastric motility increases, the heart rate and blood pressure decrease, and temperature regulation becomes erratic. During the most synchronized phase of sleep, the EEG suddenly takes on a highly desynchronized pattern that is indistinguishable from the wakeful state. During this desynchronized period of sleep there is intense postsynaptic inhibition of the α-motor neurons and at the same time presynaptic inhibition of all the sensory systems. This massive inhibition of motor and sensory systems makes the individual very difficult to arouse. In effect the nervous system has disconnected itself from the exterior world and internally generated signals dominate cortical activity. This is impressed on consciousness as dreaming. Dreaming is signaled externally by characteristic rapid eye movements (REM) and the twitching of the muscles of the face and upper extremity.

Sleep Cycles

Clinicians and researchers divide sleep into five categories: Stages 1–4 and REM sleep. The first level, Stage 1, is a "light" sleep dominated by a desynchronized EEG. Stages 2, 3, and 4 are characterized by increasing synchronization of the EEG; each stage is considered to be "deeper" that the previous one. The stages progress in order, stage 1 being the first level which is reached approximately 10 minutes after retiring. It is followed by stages 2, 3, and 4. The order then reverses until stage 1 is reached, followed by REM. This completes a **sleep cycle**. Typically, one completes between three and five cycles during the night, each cycle taking approximately 90 minutes (Fig. 13.3). Early in the night, sleep is dominated by a slow wave sleep (stages 2, 3, and 4). REM bouts are brief and widely spaced. By morning, there is little slow wave sleep and the REM bouts become more frequent and longer in duration. This is only a generalization, for these sleep patterns vary with age. Children acquire much more of the highly synchronized sleep than the elderly, but the total amount of REM acquired every night remains fairly constant throughout life, accounting for about 20% of the total sleep time. From EEG records it is apparent that sleep is not a period of

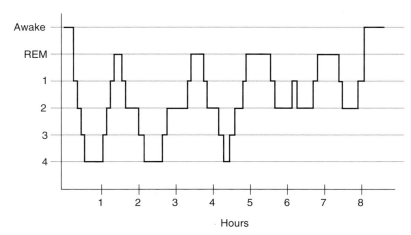

FIGURE 13.3.

This histograms represents the amount of time an adult spends in each stage of sleep during one period of sleep. Note that several cycles occur during the night and that more time is spent in REM sleep toward the end of the sleep period, at the expense of time spent in stages 3 and 4.

brain quiescence. Quite the contrary. As recorded by EEG, neuronal activity is prominent during sleep and this is reflected by metabolic studies that show brain metabolism to be as high during sleep as it is during wakefulness.

Sleep is not just a manifestation of altered cortical activity. Sleep involves the entire brain. There is a great deal of evidence that supports the hypothesis that the thalamus, the principal subcortical source of cortical afferents to the brain, establishes the major states of brain activity. Thalamic activity, in turn, is driven by a number of brainstem nuclei; among these, the reticular formation, the raphe nuclei, and the periaqueductal gray (PAG) are particularly important. It now seems certain that *brainstem nuclei regulate the sleep-wakefulness cycle* through their action on the thalamus. In animals, if the diencephalon and telencephalon are isolated from the brainstem, activity in the cerebral cortex becomes highly synchronized and the EEG assumes a pattern that is identical with the synchronized phase of sleep. However, the desynchronized phase of sleep and REM episodes fail to develop. In humans, lesions involving primarily the PAG and some tumors of the diencephalon have a similar effect. Nuclei in the brainstem associated with specific neurotransmitters, most notable the locus ceruleus (norepinephrine), the raphe nuclei (serotonin), and the ventral tegmental area (dopamine) also project widely to parts of the cerebral cortex and may play a role in regulating its major activity states (wakefulness or stages of sleep). Their exact role remains elusive.

Disorders of Sleep

Disturbances of sleep fall into four categories: disorders of initiating and maintaining sleep during normal sleeping periods (**insomnia**), disorders of excessive sleep during normal waking periods (**hypersomnia**), disorders of the sleep cycle, and other nonsleep dysfunctions associated with sleep (**parasomnia**). A full description of the various sleep disorders can be found in the references. Only the most prominent examples of each class will be discussed here.

Difficulty in initiating or maintaining sleep is very common, affecting perhaps as many as 40% of the population although few people seek pro-

fessional help. Most forms of insomnia are psychophysiologic and transient, reflecting periods of stress or emotional disturbances in one's life. Some people are awakened from sleep by repetitive muscle twitches (**nocturnal myoclonus**). A third problem, **nonobstructive sleep apnea**, is characterized by hypopnea due to a diminished central respiratory drive that may be severe enough to result in diaphragmatic arrest. The resulting hypoxemia and hypercapnia arouses the patient repeatedly during the night. All forms of insomnia result in excessive sleepiness during normally wakeful periods.

Excessive sleepiness during normal waking periods can be caused by sleep disorders other than insomnia. One of the most common problems, **obstructive sleep apnea**, is caused by an excessive relaxation of the pharyngeal muscles. This results in snoring and restriction or collapse of the upper airway during sleep. The problem is more prevalent in those over the age of 40 and is 20 times more common in men than women. It is exacerbated by obesity. The airway obstruction can continue for some time, but even though oxygen saturation may drop as low as 50% the patient usually is not awakened from sleep. Stages 3 and 4 are, however, greatly diminished and the resulting sleep deprivation causes the patient to experience excessive daytime somnolence. The problem can be relieved by tracheostomy, but continuous positive airway pressure maintained by a nasal mask is a nonsurgical alternative that is also very effective.

Narcolepsy is a fairly common disorder, affecting about 4/10,000 and is inherited as an autosomal recessive trait. It is a form of hypersomnia that is characterized by uncontrollable napping periods during the day, cataplexy, hallucinations while falling asleep, and sleep paralysis. Cataplexy is a brief loss of muscle tone without loss of consciousness. It often occurs during periods of emotional excitement. Hypnagogic [G. *hypnos,* sleep + G. *agogos,* leading into] hallucinations occur during the onset of sleep and may take the form of a dream. Sleep paralysis is a profound paralysis that overtakes the patient while falling asleep or upon awakening. The patient is fully aware of the condition and will remember it in detail. Any sensory stimulus, such as a light touch from a bed partner is sufficient to terminate the condition, or it will spontaneously subside after a few minutes. Narcolepsy is thought to be a fragment of REM sleep that inappropriately penetrates the wakeful state. On retiring for the evening, people with narcolepsy fall asleep almost immediately and enter REM sleep without first passing through the other stages of sleep. This rapid REM onset is considered diagnostic for the disease.

Disturbances of the normal sleep-wakefulness cycle are thought to be problems associated with the human's biological clock. In humans, certain biological functions such as body temperature, hormone secretions, and psychological performance vary during the day. However, if kept in isolation without external clues, these biological functions free-run on a cycle that is about 25 hours. Various triggers, especially light cycles, entrain these biological clocks to the 24-hour period of the earth's rotation. The inability to entrain these systems to the rhythm of the earth leads to sleep disturbances. For most people, these disturbances are caused by sociological events such as jet travel across several time zones or work shift changes. After a day or two one's internal biological clocks become re-synchronized with the earth. Some people appear to have an error in the entrainment mechanism. Their clocks free-run and are only occasionally synchronized with the earth. Their lives can be seriously disrupted since they are also not synchronized with the daily life pulse of society.

A variety of other disturbances are associated with sleep, but are not actual disturbances of the sleep cycle itself (parasomnia). For example, sleepwalking (somnambulism) is common in children (incidence about 15%). Sleepwalking occurs during stages 3 and 4, not during REM, and the sleepwalker seldom recalls the event when awakened. Therefore, sleepwalking is not an acting out of a dream, as commonly supposed. Nor is sleepwalking a manifestation of epilepsy (discussed later in this chapter). Children usually outgrow it and it is considered a benign condition. The onset of sleepwalking in the adult may be a sign of psychiatric disease that should be investigated. In contrast to sleepwalking, which is almost always a childhood phenomenon, a REM-associated sleep disorder is manifested in adults. It occurs exclusively during REM and appears to be an acting out of threatening dreams. The sleeping patients become violent and dangerous, often injuring themselves and their bed partners. The condition responds well to anticonvulsant drugs.

BRAIN INJURY AND CONSCIOUSNESS

Injury to the brain frequently results in alterations in consciousness. A strong blow to the head may jar the brain enough to cause a temporary loss of consciousness. This may be a simple **concussion** [L. *concussio*, to shake violently] or, if part of the brain is bruised, a **contusion** [L. *contusio*, a bruising]. The loss of consciousness is brief, never lasting more than a few minutes. Longer term alterations of consciousness that result from more serious brain insult are called **confusion**, **stupor**, **coma**, and **cerebral death**.

Confusion

A clouding of consciousness, known as confusion, is the least disturbed alteration of the conscious state. In this condition a patient's thinking processes are slowed and the patient has difficulty integrating experiences into the thought processes. The patient may be disoriented as to time and place, be inattentive, and have difficulty carrying out simple commands. Speech may be slow, sentences incomplete, and vocabulary limited.

Stupor

A more profound disturbance of consciousness is known as stupor. Patients are minimally conscious and can only be brought to awareness by strong sensory stimuli, at which time they may not be able to respond to simple commands even though they seem conscious and aware of the examiner. They will easily lapse back into stupor. A patient that can no longer be aroused is in coma.

Coma

Coma is an unconscious state in which *the metabolic activity of the brain is reduced from normal levels*. The comatose individual cannot be aroused from coma and is unresponsive to painful stimuli. In the recovered patient, the unconsciousness of coma is also characterized by amnesia. This amnesia is generally complete, with no sense of time having passed, and in that regard, is much like the anesthetic state.

Coma is unlike sleep. During sleep, the brain's metabolic activity is equal to or may even exceed the waking state. Furthermore, some level of awareness seems to be present during sleep since dreaming reflects, at the very least, a distorted form of consciousness. Consciousness is also suggested by the preservation of the sense of time during sleep. One is always aware of the passage of time even after a period of the soundest slumber. Finally, sensory stimuli can terminate sleep and bring one into a fully conscious, awake state. Sleep is a natural facet of normal brain activity; coma is a depressed, pathological state.

Alterations in neuronal function can also produce what is usually called **metabolic coma**. In these cases, coma is brought about by a generalized reduction in neuronal function. This can be expressed as an impaired ability to produce action potentials due to ionic imbalances, impaired production or recognition of neurotransmitters due to disturbances in protein metabolism, or even widespread but incomplete loss of cortical neurons due to exposure to toxins. A few of the more common causes of metabolic coma are **hypoxia**, **hypoglycemia**, disturbance in cerebral or blood **pH**, **electrolyte imbalances**, particularly those affecting potassium and sodium, and **toxins**. Toxins may be from industrial sources (CO_2 poisoning), of felonious intent (cyanide), self-induced (barbiturates), or iatrogenic [G. *iatros,* physician + G. *gen,* produce] (from prescribed drugs). Determining the cause of coma can be a major challenge.

The most extreme form of coma results when the electrical function of the neurons ceases, but the cells remain metabolically alive. If the brainstem is involved, somatic death rapidly follows due to the collapse of the respiratory functions.

Cerebral Death

Advances in medicine have improved methods of resuscitation to the point that the body may be alive and functional while, at the same time, the brain, although alive, is not functional. Under such circumstances, the body can remain viable for as long as proper nursing care is pro-vided. The coma, however, if profound and irreversible, has been termed cerebral death or brain death. In the most perplexing cases, the cerebral cortex may be partially or totally nonfunctional, while the brainstem functions normally. Whether or not a person in such a condition is living or dead is a legal and ethical question (see discussion later in this chapter). The relevance of the issue to the practice of medicine has become more pressing as the need for donor organs and issues related to allocation of medical resources have become more acute.

BRAINSTEM INJURY AND CONSCIOUSNESS

Alterations in the state of consciousness can be produced by lesions to the cerebral cortex, thalamus, and reticular formation. However, lesions to the cerebral cortex do not produce alterations of consciousness unless they are massive and, usually, bilateral. By contrast, small, highly localized lesions to the thalamus or mesencephalon can produce profound and irreversible coma. These alterations in consciousness are thought to be caused by eliminating the ability of the thalamus to control cortical activity. This can be brought about either by direct thalamic damage or by disconnecting the thalamus from the brainstem structures that regulate its function.

The tentorium separates the superior from the inferior cranial fossa. The tentorial notch, the rim of dura that circles the brainstem at approximately the level between the mesencephalon and the diencephalon, is a site where certain brain structures can be trapped and injured. Because of this, expanding lesions superior to the tentorium have quite different symptoms from those that develop inferior to the tentorium. Therefore, it is convenient to divide processes that affect consciousness into **supratentorial** and **infratentorial** lesions to reflect the location of the primary pathology. Differentiating supratentorial from infratentorial causes of altered consciousness is essential since many supratentorial causes can be successfully treated while the prognosis for infratentorial lesions is more ominous.

Supratentorial Lesions

Supratentorial space occupying lesions frequently affect consciousness slowly, in a step-

wise fashion. This is due, in part, to the fact that the brain, CSF, and blood of the cerebral circulation are enclosed by a rigid structure, the skull. Any expansion of one of these elements is borne at the expense of the others and eventually causes an increase in intracranial pressure. For example, an expanding tumor will displace brain tissue and cause the ventricles to diminish in size. As the mass expands, further collapse of the ventricles eventually becomes impossible, so intracranial pressure increases.

When intracranial pressure exceeds diastolic blood pressure, cerebral perfusion is diminished. The first effect on consciousness is confusion.

Stupor follows as the mass expands and intracranial pressure continues to increases. Pressure on the brainstem frequently stimulates emetic centers, causing vomiting. This can be dramatic, in which case it is described as **projectile vomiting**. If a supratentorial mass becomes large enough, it will displace supratentorial structures through the tentorial notch. This displacement shifts the position of the brainstem and pulls the contralateral oculomotor nerve over the posterior clinoid process, stretching it and creating a third nerve palsy with loss of the light reflex from the eye (Fig. 13.4).

The displacement of supratentorial structures

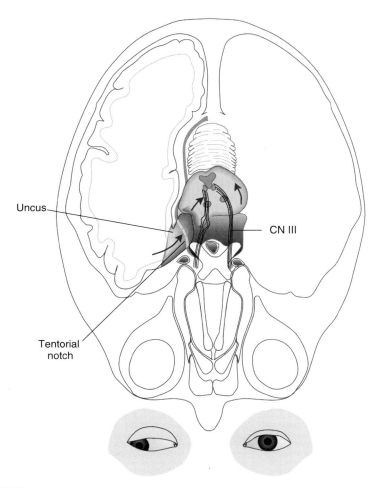

FIGURE 13.4.

An expanding supratentorial mass can cause herniation of the uncus through the tentorial notch, which displaces the upper brainstem and causes the third nerve to be stretched over the posterior clinoid process. This ominous condition is signaled by confusion, stupor, or coma combined with a unilateral third nerve palsy.

through the tentorial notch can also constrict the upper brainstem, which results in local ischemia. The subsequent loss of perfusion renders the upper brainstem structures dysfunctional, which effectively separates the diencephalon from the brainstem. Coma ensues since the thalamus is deprived of the essential brainstem activation necessary for consciousness.

Further expansion of the mass may affect the lower brainstem directly which will affect the respiratory centers of the medulla. Cheyne-Stokes respiration develops, soon followed by the total collapse of the respiratory centers and, ultimately, death.

Infratentorial Lesions

In contrast to this scenario, *expanding infratentorial lesions first produce symptoms associated with the cranial nerves.* Vertigo, nausea, deafness, facial paralysis, nuclear ophthalmoplegia, and projectile vomiting may be early signs, depending on the placement of the lesion. If the lesion involves the upper pons before affecting the medullary respiratory centers, coma may immediately develop without the preceding signs of confusion and stupor. The EEG in such cases is not isoelectric, since the neurons in the cerebral cortex are still alive and electrically functional. In such cases, the EEG displays the patterns associated with slow-wave sleep. If the lesion is reversible, recovery is possible. It the lesion remains stable, the patient may remain in a coma-sleep state indefinitely.

SEIZURES

Under certain circumstances, the ensemble of cortical neurons fails to function in a coordinated and purposeful manner. The activity is described as a **seizure** when the neuronal activity is highly synchronized and results in a loss of consciousness. Therefore, *seizures are characterized by a prolonged synchronous discharge of large numbers of cortical neurons.* This neuronal state brings about disturbances of perception, inappropriate motor functions, and alterations in awareness and consciousness. **Epilepsy** is an illness affecting about 1% of the American population. It is characterized by repeated

seizures that continue intermittently throughout life unless successfully treated. Seizures are classified as **general** or **partial**, based on the initial location of the brain involved and the manner in which the seizure spreads to other areas. Partial seizures account for about two thirds of all epilepsy. **Febrile seizures** are a specialized class outside the general illness of epilepsy.

Partial Seizures

Seizures that are initiated from an identified brain structure are termed partial (sometimes focal seizures), in order to suggest the partial involvement of the brain. The initial symptoms of partial seizures are appropriate to the location of the epileptiform locus. The patient experiences these symptoms as an **aura**. For example, if the locus were in the somatosensory cortex (areas 3, 1, 2), the patient would experience paresthesias in the corresponding body area. Similarly, a locus in the motor cortex (area 4) would produce myoclonic twitching from the appropriate muscles. If the temporal lobes is the locus, particularly in the region of the uncus, the aura may be an unpleasant odor. Psychic auras are common from loci in the temporal lobe. For example, feelings of great fear or dread are common, probably because of the involvement of the amygdaloid nuclei (see Chapter 12). Abnormal experiences of premonition, *déjà vu,* or unfamiliarity, *jamais vu,* are common. These may represent hippocampal activity. Autonomic signs, such as pupillary dilation, pallor, flushing, or palpitations, may also accompany the onset of partial seizures originating in the temporal lobe, auras that probably reflect the intimate connections of the temporal lobe with the hypothalamus. Partial seizures frequently progress, becoming tonic-clonic seizures after a few seconds (see discussion later).

General consciousness may be unaffected during partial seizures in which case they are termed **simple partial seizures**. In these cases the seizure remains confined to the locus and symptoms are limited to the aura. About 75% of partial seizures affect consciousness, in which case they are termed **complex partial seizures**. In such cases, immediately following the aura, the patient usually engages in some type of automatistic activity of which, after the seizure has abated, there is no recollection. The activity is usually quite simple, consisting of arranging available objects (clothing, objects on one's desk, dishes), or simply walking aimlessly about the immediate environment. More rarely, the patient may undress, leave the immediate environment, and walk outdoors into traffic. Simpler automatisms, such as lip smacking, chewing, and swallowing, are very common.

General Seizures

In contrast to partial seizures, general seizures are believed to immediately involve the entire brain. There is no aura or other indication that a restricted part of the brain may be involved. There are two major types, **absence seizures** and **tonic-clonic seizures**.

Absence seizures[1] are characterized by a loss of conscious awareness without loss of muscle tone. The individual suddenly loses awareness for several seconds. They occur repeatedly, having a frequency from several to several hundred a day. There are no obvious motor signs. In some individuals there may be a fluttering of the eyelids or minor chewing movements. Immediately after the seizure, the patient is fully aware, mentally alert and has no recollection of the passage of time or of events that may have occurred during the seizure. Absence seizures usually begin between the ages of 2½ and 20 years. About half of the individuals have occasional tonic-clonic seizures. Few have other neurological problems. Absence seizures are readily recognized by the 3-Hz spike-wave complex recorded by the EEG.

Tonic-clonic seizures[2] are dramatic events. The patient loses consciousness and falls. At the onset of the seizure, there is a generalized stiffening of the body as all the muscles contract (tonic phase). The initiation of the seizure is frequently heralded by a vocalization as air is involuntarily expelled through the larynx. The tonic phase lasts about a minute, during which time breathing ceases. The patient may become cyanotic. Following the tonic phase, rhythmic symmetrical muscle jerking occurs (clonic phase). This continues for about a minute, during which time the tongue may be lacerated, salivation may increase (causing frothing at the mouth), and there may be urinary incontinence. Following the clonic phase there is several minutes of postictal stupor and confusion. Following full recovery, the patient has no recollection of the seizure.

Febrile seizures occur in about 4% of children between the ages of 6 months and 4 years. These are almost always general seizures and occur in association with a rapidly rising fever. A probability of a second attack is much greater if the first attack occurs at an early age; the probability drops rapidly with maturity. True febrile seizures are considered a benign consequence of infancy. They are not a predictor of childhood or adult epilepsy. The principal difficulty for the physician is separating febrile seizures from true epilepsy or from seizures having a treatable cause. Consequently, the diagnosis of all seizures in infants must be vigorously pursued.

Causes of Seizures

Seizures may be initiated by a number of factors, not all of which are understood. They may be caused by isolated events such as concussions, hypoglycemia, meningitis, and encephalitis. Such seizures are usually one-time occurrences related to the insult. Seizures can also be the unwelcome herald of brain tumors, brain abscesses, intracranial vascular lesions, and metabolic disease. Seizures can also be caused by certain drugs, withdrawal from other drugs (especially barbiturates and alcohol), and neurotoxins (particularly carbon monoxide).

Perhaps the best documented cause of seizures is "scarred" brain tissue. After brain injury, neurons and glia die. Their remains are phagocytized and replaced by proliferating astrocytes. The axons of the damaged and dying neurons, separated from the soma, degenerate, a process that deafferents other neurons. Undamaged axons in the vicinity of deafferented neurons sprout new processes, many of which form new synapses to replace those lost. This may upset the balance between excitation and inhibition in the deafferented neurons. Some may become chronically

---

[1]Formerly known as *petit mal* seizures, a term that is falling into disuse because it has been applied to so many unrelated types of seizures as to have become medically meaningless.

[2]The term *grand mal* is no longer applied to tonic-clonic seizures.

hypopolarized to the extent that they spontaneously fire in the manner of a cardiac pacemaker, forming an **epileptic focus**.

An epileptic focus by itself is not a seizure, nor is it a necessary sign of epilepsy. Many people have an identifiable focus and never have seizures. In other cases, for reasons that are not understood, the epileptic focus transiently expands from time to time, recruiting additional neurons in a wider and wider area of the cerebral cortex. This produces a generalized seizure. Consciousness is usually lost when both hemispheres become involved. If the epileptic focus is near the central sulcus, one can observe the orderly recruitment of neurons by observing the advance of rhythmic motor contractions that corresponds with the somatotopic organization of the precentral gyrus. This was first noted by the 19th century neurologist Hughlings Jackson and is known today as the "Jacksonian march." There are many forms of epilepsy for which no focus is identifiable. In many of these cases an epileptiform focus probably does exists, but it does not register on an EEG. In other cases there may be no focus and the initiating mechanism is unknown.

Hemispheric Specializations

The most obvious fact relating to the gross anatomy of the brain is that it is a bilaterally symmetrical paired organ. There are only minor structural differences between the left hemisphere and the right; for example, there is slightly more brain tissue at the left temporoparietal junction and the surface of the left temporal lobe posterior to Heschl's gyrus is somewhat larger in some individuals. These findings are subtle and would hardly lead one to believe that there were major differences in function between the left and right hemispheres.

Animal studies and clinical evidence, however, suggests that there are important hemispheric differences. For example, about 90% of the human population prefers to use the right hand for most delicate motor activities. Strokes affecting the right hemisphere produce a left hemiparesis and left homonymous hemianopia,

but usually leave the essential personality, intellect, and consciousness of the patient intact. They do not usually affect speech, reading, or writing skills. Left hemispheric strokes, on the other hand, produce a right hemiparesis and right homonymous hemianopia, but also can render the individual incapable of using language. Large lesions to the left hemisphere can destroy an individual's essence as a human being and render one incapable of normal consciousness.

The association of right handedness and left hemispheric language function is almost universal. Only about 4% of the population has language representation in the right hemisphere. The reasons for the prevalence of language skills in the left hemisphere are not known. Many people with language skills in the right hemisphere suffered neo- or perinatal brain damage to the left hemisphere. It is presumed that language development transferred to the opposite hemisphere as a compensation for the injury, proving that the right hemisphere is capable of language function. However, brain injury does not explain all instances of right hemispheric language development. For example, language representation has been found to be bilateral in bilingual interpreters (people who verbally translate language as it is being spoken), which suggests that if the linguistic pressures are great, the right hemisphere can be recruited. There are many exceptions to these observations, exceptions that render generalizations difficult to justify.

Left-right hemisphere differences in areas other than language have only recently been systematically studied in the human. This is possible when the corpus callosum is divided as treatment for cases of intractable epilepsy. These patients appear neurologically normal; it was many years before specific test situations were developed that elucidated the effects of hemispheric isolation. These test situations take advantage of the fact that visual images can be presented to the two hemispheres independently (Fig. 13.5A). An image presented to the right visual hemifield is perceived by the left hemisphere and accurately reported verbally. An image presented only to the left visual hemifield is perceived by the right hemisphere, but cannot be accurately reported verbally. In this case, either

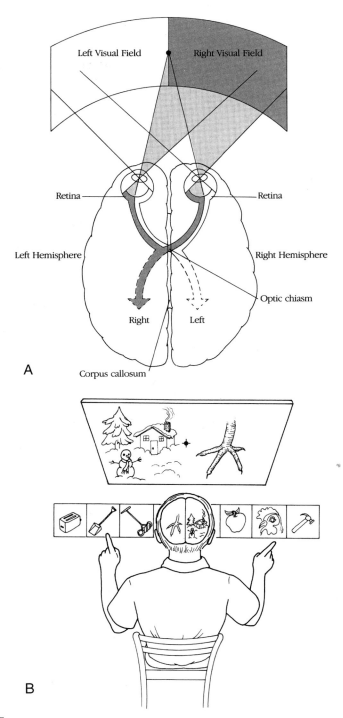

FIGURE 13.5.

In individuals with isolated hemispheres, the two hemispheres can be separately pre-sented with information by taking advantage of the contralateral mapping of the visual hemifields on the primary visual cortex. The patient must fixate on the center spot and the stimulus must be brief (less than 100 msec) to ensure that the stimulus reaches only one hemisphere.

the verbal report is a confabulation (made up without regard to facts) or the patient says that nothing was presented. The left hand, however, which is controlled by the right hemisphere, can correctly convey what was seen by pointing to a picture or picking up an appropriate object.

In addition to documenting in a rigorous manner the almost total lateralization of language in most individuals, the "split-brain" experiments have given us insights into the lateralization of functions to the right hemisphere that could not be uncovered by other means. In general, functions relating to spatial problem solving are most often lateralized to the right. These are problems that are solved holistically and in parallel, by what seems to be an intuitive approach. Examples of such problems are face recognition, three-dimensional puzzle solving, map reading, and melodic recognition. The left hemisphere is more adept in solving problems that require a linear step-wise, logical solution. Spoken language and mathematics are examples of such problems.

The left hemisphere also has a particular need to bring all observations into a logical unity. If the hemispheres are divided, this often leads the left hemisphere to invent a rationalization of the behavior of the left hand over which it has no control. One famous example of this occurred when an individual with separated hemispheres was shown two pictures (Fig. 13.5*B*). The right hemisphere saw a house in a snowstorm while the left hemisphere saw a chicken claw. The patient was asked to select with his hands items that were logically related to the picture. The right hand picked out a chicken head while the left hand picked out a shovel. When asked why he made those particular choices, the patient replied that the chicken head obviously went with the chicken claw while the shovel was to clean up after the chicken!

The need for communication between the hemispheres is dramatically revealed by the puzzle-solving paradigm. One patient with isolated hemispheres, when asked to solve the puzzle with both hands, demonstrated an amazing display of interhemispheric competition. The right hand grabbed all of the pieces of the puzzle and placed them at the far right portion of the table. It then attempted to put them together and failed. The left hand furtively grabbed two

or three pieces and placed them at the extreme left side of the table where it proceeded to assemble them correctly. The left hand tried to grab another piece, but when the right hand (meaning the left hemisphere) became aware that some of the pieces were at the left side of the table, it grabbed them and returned them to the right side, disassembling the partially completed structure in the process. The puzzle was never solved.

Aphasia

Language is the use of complex abstract symbols to represent one's perception of the world to another individual using an agreed upon set of syntactical rules that help to convey and to clarify the underlying semantics. It is a function unique to the human cerebral cortex and language is the quintessential feature of the human mind.[3] Language is expressed outside the brain in multiple forms. In human society, one form, the verbal-written language, is universal. It is commonly expressed in a written or spoken form but can also be represented in other ways, such as by gestures in the American Sign Language or in a tactile form in Braille. **Aphasia** is the inability to transform the internal, mental representation of verbal language into its external spoken and written forms and *vice versa*, the inability to transform the external spoken or written forms into a meaningful internal verbal form.[4] Aphasia must be differentiated from **dysarthria**, speech loss due to a motor disorder, **dysphonia**, disorders of the larynx, or speech disorders secondary to **dementia** (see later discussion).

ANATOMICAL CORRELATES OF LANGUAGE

The verbal-written language areas of the brain are restricted to a narrow band of tissue, the op-

---

[3] A few chimpanzees and gorillas have been taught to sign with their hands or to use computer interfaces in an effort to determine if these animals are capable of linguistic expression. The interpretation of these experiments is hotly debated. Regardless of these experiments, humans remain the only animal that developed the use of language through the natural course of evolution.

[4] The definition of aphasia includes all forms of language, including American Sign Language. Documentation of these cases is sparse, but see Bellugi et al. for a review of language deficits in the deaf following cortical lesions.

erculum on the lateral surface of the left[5] hemisphere, surrounding the lateral sulcus (Fig. 13.6). The primary auditory cortex (areas 41 and 42) lies deep within the lateral fissure. Surrounding the primary auditory cortex is area 22, an auditory association area. Area 39, in the parietal cortex, and its surrounding cortex marks the posterior boundary of the higher order language region of the brain. The frontoparietal operculum extends along the lateral fissure from area 39 to its anterior limit at area 44. Areas 44 and 45 mark the anterior limit of the superior opercular structures associated with higher language function. The arcuate fasciculus connects the posterior (area 39) region with the anterior (areas 44 and 45).

The entire language area is strategically located. The temporal operculum is connected by many association fibers to the primary auditory cortex on the one side and to the lateral temporal lobe on the other side, a relationship that is presumed to link spoken language with memory (see Chapter 12). To the posterior, area 39 is connected by association fibers with the visual association cortex, a connection that is presumed to link language with its visual representation in the cortex. The anterior region (areas 44 and 45) is linked by short association fibers with the motor areas of the face, tongue, and larynx. Lesions to each of these areas produces specific deficiencies in language.

---

[5] As already discussed, in about 4% of the human population, spoken language is on the right, but in order to simplify this discussion, spoken language will be discussed as if it were the exclusive province of the left hemisphere.

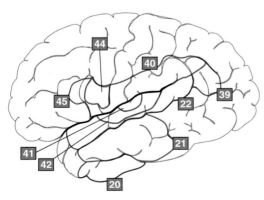

FIGURE 13.6.

The areas of the left hemisphere that are associated with spoken and visual forms of language.

EXPRESSIVE APHASIA

The inability to transform one's mental linguistic thoughts into spoken words is known as **expressive aphasia**.[6] Patients with this condition have a largely unimpaired understanding of both spoken and written language. The frustration of understanding but being denied the ability of expression usually causes the patients to attempt to verbalize. Such efforts produce only the most rudimentary forms of vocal expression. Unfortunately, the ability to write is usually impaired as well. Expressive aphasics generally display great frustration at their inability to speak or write effectively.

The nature of the condition is made very apparent to the informed observer when the patient demonstrates good control over the muscles of the face and mouth. The patient can chew food, swallow it, and vocalize nonverbally. The patient can hum melodies (and sometimes sing the words). Depending on the size of the lesion, motor deficits of the right arm and face can be seen, but these are not paralyzing and are not the cause of the language impairment. The true affliction of expressive aphasia is *the inability to initiate verbal motor commands in response to the internal will to do so.*

As one might expect, there is a very wide range of disability shown by patients with expressive aphasia. The least affected may be able to speak slowly and with difficulty. In favorable cases, this may improve rapidly with a nearly complete recovery. In these cases, most evidence suggests that only areas 44 and 45 are affected. In the more seriously affected, the frontoparietal operculum extending nearly the entire length of the lateral sulcus may be involved. CT studies have also shown that the insula is rarely spared, although this may be an incidental finding. Always spared in cases of expressive aphasia is area 39 and the inferior temporal operculum.

RECEPTIVE APHASIA

Another language disturbance, **receptive aphasia**[7] interferes with one's ability to understand externally presented language. Such a patient has no apparent understanding of either

---

[6]Formerly Broca's aphasia.
[7]Formerly Wernicke's aphasia.

spoken or written language and cannot extract meaning from linguistic symbols. In spite of this deficiency, they verbalize effortlessly. The speech they produce is generally understandable in the sense that most of the words can be recognized. However, one is soon impressed by the fact that *there is no syntactical or semantic meaning to the vocalizations*. It is as if the word generating part of the language system were connected to a random phrase generator. Since the patient cannot communicate effectively, there is no way to determine if the internal representation of language is intact or also has been deranged. The only clue, perhaps, is the fact that such individuals seem to show no particular anxiety or distress at their condition, unlike the people with expressive aphasia who are considerably affected.

Patients with receptive aphasia usually have a lesion that involves area 39. Frequently there is a quadrant- or hemianopia due to the involvement of the indirect optic radiations that pass through the temporal lobe and lie deep to this area. Like expressive aphasia, there are many variations of receptive aphasia, depending on the extent and the exact location of the lesion. For example, reading may be preserved in some patients who, at the same time, have no comprehension of spoken language. As with most insults to the cerebral cortex, receptive aphasia usually improves with time, although full recovery is not common.

CORTICAL LANGUAGE PROCESSING

Observation of aphasic patients has led to certain theories about the representation of verbal language processing in the human brain. These theories recognize that *language is not a single process and that various aspects are relegated to different parts of the brain*. Auditory and visual association areas are thought to parse the sensory input into generic linguistic abstractions specific to each modality. These abstractions are integrated into a universal language representation that endows it with meaning. This probably occurs in area 39. The frontoparietal operculum transforms the universal representation into generic motor commands that are apparently specified more explicitly into vocal motor patterns by areas 44 and 45. There is some evidence that a similar premotor area, adjacent

to the hand area in the motor cortex, exists for writing.

The parceling of linguistic functions suggests that specific lesions ought to produce linguistic dissociations. For example, if only visual or auditory association areas are lost, the complementary language skill should be preserved if area 39 remains intact. Loss of the arcuate fasciculus, an association tract that connects the parietal with the frontal lobe, ought to leave comprehension of spoken and written language intact while disrupting the ability to speak intelligibly. Loss of only area 39 should devastate all language skills. While this schema is overly simplified, to some degree these dissociations have been observed in patients who have suffered infarctions restricted to these anatomical sties. The interpretation of these "accidental experiments" is open to considerable debate since the exact lesion is seldom ascertained and the few patients who are systematically observed are not subjected to controlled protocols.

OTHER LANGUAGE REPRESENTATIONS

Language is an abstraction and representation of the world in a symbolic form. The symbols are interrelated to each other by specific rules of *syntax* that help to organize and clarify the underlying *semantics*. Symbols and syntax need not be limited to the spoken-written form that is so dominant in human society. For example, mathematics is a symbolic representation of the world with its own specific syntax. It is a highly developed and profoundly expressive language Those who are skilled in its use can express and develop ideas that cannot be adequately expressed in other forms. The language of mathematics does not, however, have a unique phonetic structure. This it shares with the spoken-written language. Cortical lesions to the left parietal association areas, especially those close to area 39, severely disrupt mathematical functions.

Leonard Bernstein, in a particularly insightful series of lectures, proposed that music is a language in the same sense that English, German, or Chinese are languages. He argues persuasively that music has a *phonetic, syntactic,* and *semantic* structure that is entirely analogous with verbal language. Music, of course, does not have the same ability that verbal language has to literally transcribe the external world into symbolic expression. Rather, music can powerfully express nonliteral symbolic meaning, much as poetry does. Like poetry, Bernstein argues, the power of musical expression depends on its *ambiguity,* a property that endows both poetry and music with multiple parallel meanings.

Unlike mathematics, music has an independent phonology that makes possible the dissociation of musical expression from verbal expression. Most aspects of musical cortical function seem to be associated with the right hemisphere and consequently lesions to the right parietal-temporal lobes produce severe disruptions in the perception, appreciation, and expression of music while sparing written-vocal language. By contrast, as we have seen, expressive verbal aphasia, being associated with left hemispheric lesions, usually leaves musical expression relatively intact. If Bernstein's hypothesis is correct, our usual concept of language as exclusively a verbal (written-oral) abstraction of the world is too narrow. A more complete concept of language would acknowledge that it is verbal, mathematical, and musical.

Agnosia

In order to function as a successful organism, humans have the capacity and the need to attach "meaning" to sensory stimuli. The concept of "meaning" is defined as that which has significance to the mind. For humans, significance is usually brought about by relating sensory experiences to one's mental reference system, a reference system that includes a *personal schema* of one's self, *memories* of past similar experiences, *analogies* with stimuli derived from other sensory modalities, and *expectations* derived from past experience. **Agnosia** [G. *a* + *gnosis,* without knowledge] is the clinical term applied to those situations where a patient is unable to attach meaning or significance to sensory stimuli. *Agnosia is caused by a lesion that leaves the primary sensory pathways intact.* The sensory stimuli reach the brain and are perceived as stimuli, but they cannot be integrated into the individual's mental system of reference.

ASOMATOGNOSIA

Patients who suffer parietal lesions of the non-speaking hemisphere frequently manifest a dramatic inability to attach meaning to stimuli coming from the contralateral half (usually left) of their world space. They may pay little or no attention to visual stimuli presented from the left side in spite of the fact that vision is intact. The condition, **asomatognosia** [G. *a* + *soma* + *gnosis,* without body knowledge], also leaves them unaware of tactile stimuli from the left side of the body. Furthermore, when asked to move a left extremity, they may move the right, or make no movement at all. When shown their own left extremity they may seem surprised. Asked to identify it, they may deny that it is their own arm or leg and assign its possession to the examiner or to God, or simply express total bewilderment.

The inability to incorporate left-sided stimuli into the mental frame of reference extends to self-directed activity also. Patients with right parietal lesions, when asked to bisect a line, will invariably mark the midpoint far to the right of center. If asked to draw the petals on a daisy or to number the face of a clock, they will place all of the objects on the right half of the figure (see Fig. 7.11). These patients frequently have a very difficult time dressing themselves for they often fail to clothe the left side of their bodies. Lipstick may be applied only to the right half of the mouth or men may shave only the right half of their face.

Right parietal lesions produce more than left-sided neglect. Usually there is an inability to solve spatial problems. Clothing presents additional difficulties, for these patients may be unable to figure out the orientation of a shirt, and, for example, place their arm into the neck opening. With such a bad start, further attempts at dressing result in hopeless entanglement. These individuals will also have enormous difficulty reading maps and relating maps to the real world. They cannot follow routes and frequently get lost.

While asomatognosia is usually discussed in relation to the right (nonspeaking) hemisphere, it has also been observed following lesions in the parietal lobe of the left (speaking) hemisphere. Many cases of left hemispheric asomatognosia are undoubtedly obscured by the associated aphasia that makes effective communication with the patient impossible. When observable, however, the neglect syndrome is quite prominent when the left parietal lobe is affected.

VISUAL AGNOSIAS

There are a number of visual agnosias, all of which are fairly rare, but instructive. Two have already been discussed, **prosopagnosia**, the inability to recognize previously familiar faces and visual motion agnosia (see Chapter 11). **Object agnosia** is the inability to identify familiar objects by visual inspection even though ordinary vision is intact and there is no aphasia. The object can be immediately recognized by touch, sound, or smell. **Achromatopsia** is a visual agnosia limited to color perception. Patients cannot distinguish objects on the basis of hue although other aspects of vision are relatively normal. This must be differentiated from **color blindness**, an inherited disorder related to the loss of one or more of the visual pigments (see Chapter 11). The visual agnosias are related, more or less, to bilateral lesions of the inferior medial portion of the occipital lobe where it blends with the temporal lobe. They may appear independently or in combination.

The occurrence of visual agnosias that are independently related to different aspects of visual perception, motion, color, and object recognition, implies that there are physically separate areas of the cerebral cortex devoted to extracting extremely specific information from the visual stimulus. Furthermore, information from these separate sensory processing centers must then be "integrated" at other remote areas of the brain (usually called "association cortex") before **perception** is realized. This is difficult to verify from human lesions because few patients fulfill all the theoretical criteria for agnosia (loss of perception while all sensory, linguistic, and mental faculties are intact). The behavioral deficits of humans with various agnosias do seem to lend support to the notion that multimodal integration is an essential component to the attachment of meaning to sensory phenomena; animal studies offer some validation for this idea. As discussed in Chapter 11, there is considerable physiological data to support the hypothesis that the extraction of specific features of sensory signals is parceled to different physical regions of

the cerebral cortex. However, animal studies cannot elucidate the nature of perception.

Disorders in Personality and Motivation

In 1848, Phineas P. Gage, a shrewd, patient, and energetic railroad foreman, was tamping dynamite into a drill-hole in rock. The dynamite exploded, driving the tamping rod through his left orbit. The rod, 3.5 feet long and 1.25 inches in diameter, emerged from the superior surface of his skull, anterior to the coronal suture near the midline, destroying most of the frontal lobes in the process. Dr. J. M. Harlow attended this man, daily cleansing the wound by scraping away pustulant and necrotic brain tissue until the defect closed and healed. Dr. Harlow followed this patient for a number of years and published his findings in 1868. The most remarkable feature of the case of Mr. Gage, other than his initial recovery,[8] was the fact that he had few neurological deficits. In spite of the relatively minor motor and sensory deficits, Mr. Gage suffered a marked change in personality.

While "previous to his injury . . . he possessed a well-balanced mind, and was looked upon by those who knew him as a shrewd, smart business man, very energetic and persistent in executing all his plans of operation." After his recovery, "the equilibrium . . . between his intellectual faculties and animal propensities seems to have been destroyed. He is fitful, irreverent, indulging at times in the grossest profanity (which was not previously his custom), manifesting but little deference for his fellows, impatient of restraint or advice when it conflicts with his desires, at times pertinaciously obstinate yet capricious and vacillating, devising many plans of future operation, which are no sooner arranged than they are abandoned in turn

---

[8]Such a recovery is not unprecedented. An elderly, retired physician told this author that soon after he left medical school he was called to a farm where there had been a terrible accident. A 2-year-old child had been kicked in the head by a horse. Arriving at the scene, the young physician noted that the right hemisphere had been almost entirely removed by the horse's hoof. The wound had clotted, but was contaminated with straw, dirt, and manure. He advised the family to prepare themselves for the child's imminent death but much to his surprise, the boy survived, grew to manhood, graduated from college, and led a full and productive life.

for others appearing more feasible. A child in his intellectual capacity and manifestations, he has the animal passions of a strong man. . . . In this regard his mind was radically changed, so decidedly that his friends and acquaintances said he was 'no longer Gage'." (Harlow, 1868)

Gage lived for many years, yet never recovered his original personal qualities. The changes observed by Harlow have now become recognized as the **prefrontal lobe syndrome** which results from lesions to that portion of the frontal lobes anterior to the premotor area (area 6). The most important features of the syndrome are (*a*) **personality changes**, (*b*) deficits in **strategic planning**, (*c*) **perseveration**, (*d*) the **release of "primitive" reflexes**, and (*e*) **abulia**.

Personality changes accompany the apathetic state. Patients with prefrontal lobe lesions frequently exhibit a lack of compassion and sensitivity to social conditions. They are usually uninhibited and they may make inappropriate jokes or comments that are hurtful to others. They seem to have no comprehension of the concept of social restraint.

Persons with lesions to the prefrontal lobes also exhibit certain mental deficiencies that are usually thought to be associated with strategic thinking or planning. When given a problem that requires planning for future action, they may not be able to solve it. This is not a memory defect for they can describe the problem. They cannot apply that knowledge or abstract the problem in ways that suggest options for a solution.

Another very characteristic feature of frontal lobe lesions is perseveration, the constant repetition of an action in response to a situation. This may be manifested in a number of ways. For example, a patient, when asked to name objects, may correctly identify a pen that is presented for identification. The next object, a set of keys, will also be identified as a pen; likewise a handkerchief. In other words, once the idea of "pen" has been established, the patient gets stuck on that one concept and cannot free the thought processes in a way that allows a new idea to be entertained. In a different, more subtle context, the patient may devise a correct solution to a problem. If, however, the problem is changed, requiring a new solution, the patient will doggedly stick to the original strategy in spite of its ineffectiveness.

Abulia [G. *a* + *boule,* without will] refers to a general slowing of the intellectual faculties. The patient appears to be apathetic. His speech is slowed and his spontaneous participation in social intercourse is decreased almost to the point of nonparticipation. The profoundly abulic patient may lie motionless and mute for weeks, in spite of having the ability to speak and move and to have all sensory modalities intact. This latter condition is sometimes referred to as akinetic mutism.

There also are objective signs of prefrontal lobe lesions. These signs are usually "primitive" reflexes associated with infancy that are suppressed with development. A good example is the suckling reflex evoked by touching the cheek. The head turns reflexively to the side of the stimulus and the mouth attempts to locate and suckle the nipple. Scratching the palmar surface of the hand evokes a reflex closure of the fingers, allowing the infant to grasp whatever touches his hand. In the adult, these reflexes, normally suppressed, are released with loss of prefrontal lobe function. The elicitation of primitive reflexes is an important part of the neurological examination as it may provide the only objective evidence of frontal lobe involvement.

Dementia

Dementia is a general term describing the global loss of higher cortical functions, especially memory, personality, and language. Dementia can be caused by any process that interferes with cortical function, including disease, hydrocephalus, toxins, and multiple cerebral infarcts. The most common cause of dementia, however, is **Alzheimer's disease**, a degenerative disease that comprises about half of all diagnosed dementias. Dementia secondary to vascular disease accounts for another 25% of presentations.

Alzheimer's disease is noted for the loss of the neurons of the telencephalon, particularly the large pyramidal cells of the cerebral cortex. The remaining cells contain less RNA and are smaller than normal. The loss of neurons is accompanied by a proliferation of astrocytes and a thinning of the cortex that results in a widening of the gyri, a marker that can be visualized with CT or MRI. **Senile** or **neuritic plaques** are found scattered throughout the cortical grey matter at autopsy. These plaques contain an **amyloid** [G. *amylon* starch + *eidos,* like] protein, and fragments of degenerated dendrites, terminals, and lysosomes. **Neurofibrillary bodies** composed of twisted pairs of tubules are found within many neurons. The loss of neurons, the numbers of plaques, and neurofibrillary bodies correlates quite closely with the degree of dementia. These pathological changes occur throughout the cerebral cortex, but the temporal, frontal, and parietal lobes seem to be more severely affected in most cases. The most severely damaged part of the brain is CA_1 of the hippocampus, and the subiculum. Loss of these areas may be presumed to cause the distinctive memory deficits that are associated with this disease.

Alzheimer's disease usually begins after the sixth decade of life, but can be manifested sooner. Its incidence increases with age (Fig. 13.7). The incidence of onset for the ages between 30 and 59 is 0.02%, which rises about tenfold to 0.3% between 60 and 69. Another tenfold increase occurs between the ages of 70 and 79, when the rate is 3.2%. The 80- to 89-year-old population runs a risk of 10.8%. The disease is, in part, hereditary, being an autosomal dominant gene carried on chromosome 21. The expression of the gene is not, however, determined by hereditary influences alone. First-degree descendants of someone with Alzheimer's disease have a risk about 4 to 5 times greater than nonrelated individuals. The risk factors determining penetrance are not known. It is interesting to note that nearly all individuals with Down's syndrome (trisomy 21) develop amyloid plaques and neurofibrillary bodies after the age of 40.

The clinical picture of Alzheimer's disease is one of relentlessly progressing dementia with relative sparing of other neural functions until the later stages of the disease. For example, motor skills remain relatively intact. The muscle stretch reflexes remain normal and the plantar signs are flexor. The most characteristic symptoms involve memory, spatial orientation, and personality changes.

Memory deficits are frequently the first symptoms that come to light. Initially this is manifested as simple forgetfulness; the inability to recall a name or think of a word. Later, Alzheimer's patients will have difficulty incorporating present experiences into long-term memory, a situation that is similar to the Korsakoff amnesic state (see Chapter 12). Patients will also begin to have difficulties with speech, particularly aspects that require memory. For example, a patient will have increasing difficulty finding the right word to use and will hesitate, causing the speech to become interspersed with pauses. Eventually, the patient becomes mute. As the deterioration of the cerebral cortex becomes more pronounced and more profoundly affects the language areas, spoken and written language comprehension also becomes defective.

When the loss of neurons involves the parietal lobes, particularly on the right side, the pa-

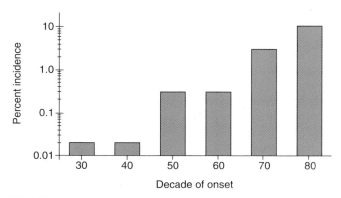

FIGURE 13.7.

The onset of Alzheimer's disease increases dramatically with age.

tient will become increasingly disoriented (see earlier discussion of agnosia). They get lost easily and cannot follow even a simple map. They have trouble dressing for they cannot correlate the spatial arrangement of the clothing with their bodies. They may display any of the agnosias associated with the parietal lobe. When the frontal lobes are involved, personality changes associated with the prefrontal lobe syndrome become manifest. In the latter stages of the disease, the primitive reflexes appear, followed by incontinence of bowel and bladder, an incontinence that appears to be more related to apathy than to a true lack of motor control.

New drugs appear to slow the progression of the disease in some cases, but the disease process itself remains refractory to medical intervention. It is important, however, to accurately diagnose Alzheimer's disease in order to differentiate it from treatable causes of dementia. The most common treatable causes of dementia are normal-pressure hydrocephalus, operable intracranial masses, nutritional deficiencies, endocrinological disorders that affect general metabolism, and drug intoxication. In addition, among the elderly, the symptoms of depression can mimic those of dementia.

Ethical Issues of Death

Mrs. C. B., a 25-year-old female, entered the hospital for a routine prenatal ultrasound examination. Unfortunately, the examination revealed an anencephalic fetus. The parents were counseled and advised of their options, which included carrying the infant to term (with a 65% probability of stillbirth), or termination of the pregnancy. The parent's religious beliefs precluded abortion so, after considerable deliberation, they asked that, if the baby were to be born alive, it be declared brain dead at birth so that the organs could be donated for transplant. The health care providers could not grant the parent's request because anencephalic babies are born with a functional brainstem and do not meet the current legal requirements for "brain death." The parents petitioned the courts and were denied their request. The baby was carried to term, but

by the time she died of natural causes, her internal organs had deteriorated to the point that organ donation was no longer possible.

One of the unintended consequences of modern medical advances has been the blurring of the lines between life and death. Before the second half of the twentieth century, the death of an individual was marked by the rapid and nearly simultaneous collapse of all organ systems. Now, with medical intervention, all organ systems do not necessarily cease to function simultaneously. The need for a formal definition of death, compatible with a partly functioning body, gave rise the concept of "brain death" in the 1960s. The *Uniform Determination of Death Act* (1980) and the *Report to the President's Commission* (1981) have set a legal and practical foundation for establishing death in the United States. Unfortunately, there is little agreement among physicians, philosophers, and the legal profession about the presuppositions those authors brought to the discussion and, consequently, with the conclusions that they have drawn.

At present the legal questions related to determining death are not settled. Criteria for determining death vary from nation to nation and, within the United States, vary from state to state. Individual hospitals set their own standards for determining death in accordance with the ethical-legal environment within which they operate; it is not possible to give, in these few pages, a description of a protocol for establishing death. In general, no single criterion can be used to determine death. For example, electrocerebral silence, as determined by EEG, is one criterion. It is an imperfect tool, however, since drug-induced metabolic coma frequently produces electrocerebral silence, but is usually reversible. In order to provide a reasonable assurance that death is certain, most protocols require some evidence of *brainstem death* as well as *cerebral death*. The former is often determined by documenting sustained apnea or by the demonstration of an isoelectric BAER.

What constitutes "brain death"? There is the general notion that someone without a functioning brain is "dead," even if other bodily functions are present. The original, practical approach to brain death was that loss of brain function served as a reliable *predictor* of collapse of all other organ system functions. That collapse was particularly certain if the brainstem had ceased to function, since respiration ceases. A more recent approach to brain death rests on the presumption that human life may be divided into a cognitive (human consciousness and cognition) form and a biological form (all other organic functions such as digestion, breathing, etc.). Assuming that division is valid, the loss of cognitive capability would signal the loss of human capacity and therefore the *extinction of the person* as a human being. Since consciousness and cognition appears to reside exclusively in the cerebral cortex, the loss of cortical function alone would be sufficient to signal the end of human capacity even if the brainstem remained viable.

The anencephalic infant brings these issues into sharp focus. These babies that are born alive have no cerebral cortex, but do have a functioning brainstem. They clearly have biological life. They respond to touch, sound, and light. Many of them have eye movements and may even follow an object with their eyes. All of these responses can be associated with brainstem function alone and, as long as any part of the brainstem is functional, they cannot be legally declared "brain dead" in most countries. The ethical debate centers on how much cognitive capability they have. Those who would define death in cognitive terms argue that anencephalic infants have no human cognitive capacity, and are therefore "brain dead." They argue that the laws are not consistent with our understanding of brain function. Others argue that since the neurological capacity of an anencephalic infant is comparable to that of intact normal infants for at least the first two months of life, they should be accorded the same rights and projections that are afforded normal infants. They argue that the laws should remain conservative, not withstanding the need for organ donors and other practical considerations.

The ambiguity of death is forcing us, as a society, to re-examine what we believe to be the essence of our humanity. What really separates humans from other life forms? Is it our consciousness? If so, what defines and differentiates human consciousness from animal awareness? Is human consciousness a unique consequence of human brain function? Without conscious-ness have we lost our human qualities? Can we be alive as *human* beings if we have no conscious awareness? Of course, none of these questions cannot be answered definitively. Knowledge of the structure and function of the brain does have a bearing on how we think about these questions. Struggling with these questions is necessary if one is to find a morally acceptable, as well as practical way of facing the problems that the ambiguity of death has thrust upon modern society.

CASE STUDIES

Mr. G. P.[9]

HISTORY, NOVEMBER 30, 1993: Mr. G.P., a 36-year-old landscaper, was seen in the emergency room because a large tree branch fell on his head as he was cutting down a tree with a chain saw. The branch pushed his chest onto the handle of the chain saw, the blade of which became stuck in the ground. He complained of chest and back pain as well as a headache. The 8 cm laceration on his occipital scalp was repaired. A cervical spine x-ray and neurologic examination were normal. He was discharged.

Forty-eight hours later Mr. G. P. took approximately twelve 200 mg ibuprofen tablets over a brief period of time because of worsening interscapular pain. Four hours later, 52 hours after the initial injury, he complained of sudden worsening of the back pain which radiated to the anterior chest and shoulders. He fell to the floor unable to move his legs, and complained of numbness of his legs. He was able to move his arms and his left toe. As he was brought to the emergency room by ambulance, he lost all motor activity of his lower extremities and he noted tingling of his arms. Over a 2-hour period in the emergency room and radiologic suite, he developed weakness of hand grip.

NEUROLOGICAL EXAMINATION, DECEMBER 2, 1993:

1. MENTAL STATUS: normal

2. CRANIAL NERVES: II to XII normal

3. MOTOR SYSTEMS: Strength 0/5 in both lower extremities; strength of hand grip and wrist flexors, 4/5.

---

[9] This case was kindly provided to me by Dr. Mark Walsh.

4. REFLEXES: Muscle stretch reflexes from the brachioradial, biceps, and triceps are 2/4 bilaterally. The gastrocnemius and quadriceps MSRs are 0/4 bilaterally. There is no response to plantar stimulation. Cremasteric and abdominal reflexes are absent bilaterally. There is paradoxical abdominal motion with breathing. There is no rectal tone and the anal wink reflex is absent.

5. CEREBELLUM: Not tested.

6. SENSORY SYSTEMS: There is a complete sensory loss from the lower extremities, abdomen and chest approximately from the level of C_7 to T_1.

7. ANCILLARY TESTS: A thoracic spine x-ray was normal. A CT scan of the thoracic spine was also normal. A CT scan of the cervical spine suggested an epidural hematoma at C_6–C_7 . A myelogram revealed a long circumferential epidural hematoma extending from C_2 to T_3 with a nearly complete block at C_6–C_7 (Fig. 13.8).

SUBSEQUENT COURSE: The patient was immediately taken to surgery where a laminectomy from C_3 to T_3 was performed in order to relieve the pressure on the spinal cord and evacuate the hematoma. In the recovery room, Mr. G. P. noted normal movement of his arms and fingers. Over the next two weeks he regained complete recovery of his motor and sensory functions in the reverse order of the ascending paralysis.

Comment: In Mr. G.P.'s case, the extreme and rapid flexion of the spine over the handle of the impaled chain saw probably stretched the spinal cord over the anterior bodies of the vertebrae, contusing the venous plexus from approximately C_6–C_7, causing some relatively minor bleeding. The large quantities of ibuprofen subsequently taken by the patient caused platelet dysfunction which resulted in significant bleeding from the already compromised venous plexus. The bleeding that resulted from this strangulated the spinal cord. The sudden weakness of the lower extremities followed by the

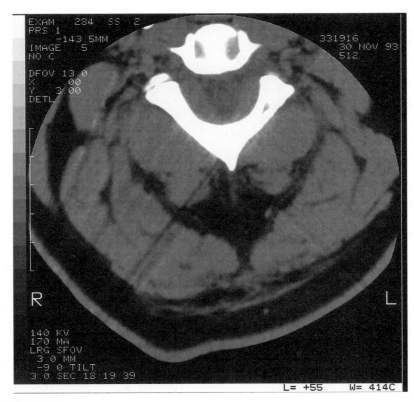

FIGURE 13.8.

This CT from Mr. G. P. was taken several days after his injury and after signs of progressing paralysis had developed. Note the extent of the epidural hematoma.

gradual onset of partial motor and sensory loss of the upper extremities dramatically demonstrated the sequential compression of the corticospinal tract fibers. The corticospinal fibers that extend to the lumbar cord are superficial to those serving the thoracic region, while the fibers that terminate in the cervical spinal cord lie deepest in the tract (see Chapter 7). After the pressure on the spinal cord was relieved, the return of motor function mirrored the pattern of motor losses with function returning first to the deepest lying corticospinal fibers serving the upper extremities, followed in turn by the increasingly more superficial fibers serving the chest, abdomen, and lower extremities. The excellent recovery obtained in this case is attributed to the prompt diagnosis and rapid decompression. Prolonged compression of the spinal cord normally results in permanent paralysis.

FURTHER APPLICATIONS

Head trauma followed by delayed paralysis might lead one to consider an intracranial epidural or subarachnoid hematoma.

- How do you suppose this case might have presented had there been an intracranial bleed? Describe the pattern of paralysis you would expect to observe.

- How would you differentiate between an epidural and a subarachnoid intracranial hematoma? Is that distinction important? Would bleeding from the external auditory meatus be significant in making this distinction? Explain.

- Under what circumstances would you expect alterations in consciousness? How would they present?

- What pattern of sensory losses, if any, would you expect to observe?

- Would you expect cranial nerve involvement? If so, which nerve(s)? What would you see? Explain.

- In the present case, how do you explain the absence of plantar signs?

Mr. C. V.[10]

HISTORY, NOVEMBER 5, 1989: Mr. C.V. is a 70-year-old Baptist minister with a history of hypertension, congestive heart failure, and insulin-dependent diabetes mellitus. His wife brought him

[10] This case was given to me by Drs. Mark Walsh and Janet Peterson.

to the emergency department with the complaint, "My husband's eyes be buggy and he be sleepy."

The patient was in good health until 1 hour prior to admission, when he suddenly was noted to have a waxing and waning level of consciousness associated with widening of the palpebral fissures and extension of the neck when attempting to visually fix objects.

PHYSICAL EXAMINATION: Vital Signs: Temperature, 97°; heart rate 80, blood pressure 160/78, respirations 20.

NEUROLOGICAL EXAMINATION: (taken soon after admission to the emergency room)

I.   MENTAL STATUS: There was a progressive obtundation noted in the emergency department. At first, he could be aroused and was able to give his name and knew he was in the hospital, although there was definite confusion. Within 1 hour the patient was snoring and arousable only with a sternal rub (stupor).

II.   CRANIAL NERVES:

  A. **Olfaction**: not noted.

  B. **Vision**: full visual fields.

  C. **Oculomotion**: paralysis of upward gaze on volition, but eyes would track vertically during the doll's eye maneuver.

  D. **Trigeminal**: normal.

  E. **Facial**: normal.

  F. **Audition**: not tested.

  G. **Vagal-glossopharyngeal**: Normal gag and voice.

  H. **Accessory**: not noted.

  I. **Hypoglossal**: tongue protruded in the midline.

III. MOTOR SYSTEMS: No weakness noted. Standing, walking, and hopping not tested because of mental status.

IV. REFLEXES: Bilaterally, the MSRs were 1/4 from the quadriceps and 2/4 from the brachioradialis biceps and triceps. There was no MSR from the gastrocnemius bilaterally.

V. SENSORY SYSTEMS: Response to pinprick was symmetrical and seemed within normal limits, given the patient's impaired mental status.

VI. ANCILLARY TESTS: A CT scan was read as normal. Because the clinical picture of paralysis of upward gaze and alteration of levels of consciousness are classic signs of a lesion in the mesencephalon, and because an embolism or thrombosis (not visible by CT) seemed unlikely, the normal CT was completely unexpected. Finally, after considerable deliberation, additional CT images were obtained and sent to the radiologist, who immediately amended his previous normal reading. A 1.5-cm lesion in the right side of the quadrigeminal plate was noted and believed to represent a mesencephalic hematoma in the region of the aqueduct (Fig. 13.9).

SUBSEQUENT COURSE: The patient remained intermittently obtunded with persistent paralysis of the upward gaze. Each day the patient's wife would express concern over his running nose. This seemed trivial in light of the fact that her husband's survival was still in doubt. Mrs C. V. continued to badger the resident, and she demanded that the young physician address the issue of the running nose. He noted that, in fact, there was considerable nasal discharge. He then probed the matter further by sitting down face-to-face with the patient's wife to discuss what he felt was the insignificant problem of a running nose. Once seated, the wife told him that there was a temporal relationship between the nasal symptoms and the neurologic symptoms. She asked if there might be a causal association between the two. The resident indulgently replied that it was unlikely. However, she was insistent, at which point the resident fixed her gaze and, with exasperation, asked why she was so insistent on linking the running nose with his cerebral hemorrhage. She repled, "Well, doctor, I just wonder if that cocaine he was taking before he got a runny nose had anything to do with his eyes bein' buggy and he bein' so sleepy." This fixed the resident's attention.

Mr. C. V. eventually made a full recovery. He stopped using cocaine and the other recreational drugs to which he had become accustomed and returned to his ministry. He died a few years later of unrelated causes.

*Comment: **The important clinical signs of supranuclear ophthalmoplegia of vertical gaze combined with the rapid deterioration of consciousness are classic symptoms of a mesencephalic lesion. It is***

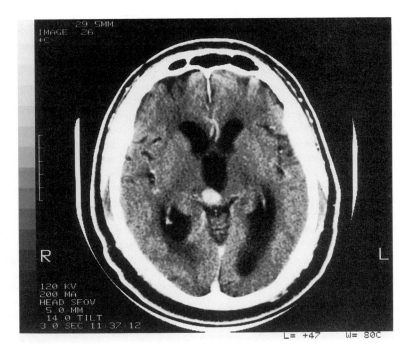

FIGURE 13.9.

CT of Mr. C. V. The abnormality in the region of the quadrigeminal plate is small, but evident on this CT image. Slices on either side of this image do not show the lesion.

clear that the patient's obtundation was caused by dysfunction of the mesencephalic reticular formation and that his loss of vertical gaze was due to a lesion involving the superior colliculus or pretectal area. These small structures are close together only at the rostral pole of the mesencephalon. Given the rapid onset of symptoms, this lesion is probably due to a stroke.

Two principles of history taking and diagnosis are demonstrated by this case. First, always determine the location of the lesion before seeking ancillary data. If the ancillary data are not congruent with the lesion as you have placed it, figure out how the discrepancy occurred. Do not simply take another opinion at face value. In this case, the discrepancy between the clinical picture seen in the emergency room and the opinion of the radiologist arose because the bleed was so small it did not appear on the initial CT. The difference of opinion was resolved by making subsequent CT images cut at different levels in the mesencephalon, a process that eventually revealed the lesion. This was of great diagnostic importance because during the early stages of a stroke, CT can reveal only fresh blood, not ischemic infarction. In the absence of a demonstrable lesion by CT, an ischemic event is assumed and anticoagulation with heparin is normally indicated. This would have had disastrous results in this case.

Second, the wife's constant badgering had a point. It was only after the young doctor took the appropriate amount of time to carefully listen to her that she confided in him what she knew all along to be the cause of her husband's illness: cocaine. The cocaine had raised his blood pressure and caused a small hemorrhage from one of his brittle diabetic vessels. Had the patient's use of cocaine been known on the night of admission, the probability of a hemorrhagic stroke would have been more certain and re-evaluation of the CT results would have been much easier to justify.

The moral of the story is to listen carefully to the history, believe your patient, and trust your observations.

FURTHER APPLICATIONS

- How do you know that this is supranuclear ophthalmoplegia, rather than nuclear ophthalmoplegia?

- An ischemic stroke is not revealed by CT during the first 48 hours. The initial read of the CT is consistent with an ischemic stroke. What, do you suppose, are the elements in this case that made the emergency room physicians uneasy about the normal CT?

- Describe the clinical observations that clearly place this lesion below the tentorium? Particularly discuss the importance of the changing levels of consciousness.

- Described the expected sequence of events had the patient experienced a hemorrhagic stroke involving one of the left thalamostriate arteries.

- Review the risk factors for stroke. How are they pertinent in this case?

SUMMARY

The cerebral cortex derives its higher order, complex functions from the interactions between the cortical columns, the collections of cells that form the fundamental computational unit of the cerebral cortex. Together these columns function as ensembles creating a functional complex more powerful than the individual components. There are no satisfactory methods available for monitoring the properties of ensembles of neurons.

The EEG is a crude measure of the electrical events of the cerebral cortex. In spite of its limitations, the EEG is a useful clinical tool. The EEG reflects PSP activity in pyramidal cells, activity that is divided into four frequency bands: the beta, alpha, theta, and delta. The lower frequencies signal synchronization among pyramidal cells while higher frequencies reflect a lack of synchronization. High frequencies are associated with states of mental alertness.

The EEG is useful in determining various states of consciousness. The most common state of altered consciousness is sleep. Sleep is a period of intense brain activity that is somehow necessary for normal brain function. Deprived of sleep, the brain becomes dysfunctional, incapable of sustained mental activity during wakefulness. Based on EEG recordings it is divided into five stages labeled 1 through 4 and REM. During an evening of sleep, one passes through the stages in sequence is a series of sleep cycles, each culminating with a period of REM. Dreaming occurs only during REM.

The various stages of brain activity are determined by the brainstem that drives thalamic nuclei. The thalamus, in turn, sets cortical activity. Lesions to the brainstem, particularly to the PAG, locus ceruleus, and the raphe nuclei, can produce an irreversible cortical state that is similar, if not indistinguishable, from sleep.

Sleep disorders are characterized as insomnia,

hypersomnia, or parasomnia. Most types of insomnia are related to stress and are temporary. Non-obstructive sleep apnea, a disorder in which hypopnea is caused by a depressed central respiratory drive. The resulting hypoxemia results in dozens of awakenings during the night and a disruption of the normal sleep cycle. The most common hypersomnia, obstructive sleep apnea, produces excessive daytime sleepiness although it does not usually awaken the patient during the night. Narcolepsy, another hypersomnia, is thought to be a REM-like state that intrudes during normally wakeful periods. Disturbances of the normal circadian rhythm can also produce hypersomnia during inappropriate times. Motor disturbances during sleep, such as sleepwalking in children, are considered parasomnia.

Brain injury can alter the conscious state. In increasing order of severity, the altered states of consciousness are confusion, stupor, coma, and cerebral death. Altered states of consciousness can be brought about by a number of insults including concussion or contusion, increased intracranial pressure, disruptions in electrochemical balance or pH, hypoxia, hypoglycemia, and drugs or toxins. Supratentorial lesions usually produce altered stages of consciousness that progress from confusion to coma in a orderly manner. Infratentorial lesions can produce coma and death immediately, without progressing through all the less severe stages if the brainstem centers that control cortical activity and the respiratory centers are affected early in the course of events.

Seizures are periods of abnormal hyperactive synchronized neuronal activity in the cerebral cortex. Seizures may be classified as partial if they originate in a restricted part of the brain or general if they appear to originate in the entire cerebral cortex. Partial seizures are associated with motor or sensory auras, appropriate to the locus of the seizure. Partial seizures are considered simple if consciousness is not affected, or complex if mental awareness is altered. In the latter case, the post-aura period is characterized by automatisms. General seizures always affect consciousness. Absence seizures consist only of periods of unawareness of one's surroundings. They are brief (seconds) and frequent (several to several hundred times a day). Clonic-tonic general seizures are dramatic events that involve both loss of consciousness and brief periods of altered muscular activity. Febrile seizures are general seizures of infancy associated with a fever of rapid onset. Epilepsy is a condition of repeated seizures (unless successfully treated) that continues throughout life.

The cerebral cortex is not functionally homogeneous. Language operations are performed in the left hemisphere in all but 4% of the population. Mathematical problems, which are solved in a linear, step-wise manner, are solved in the left hemisphere. Spatial or pattern problems, such as face recognition and musical skills, which are solved in a parallel, holistic manner, are generally performed in the right hemisphere. Specific cortical lesions in humans produce functional deficits that are related to the location of the lesion. The aphasias are produced by lesions to the left hemisphere in the region of the lateral fissure. Lesions to the posterior area produce receptive aphasia while those more anterior and involving the frontal operculum produce expressive aphasia. Agnosia results from lesions to the transition areas between the parietal, occipital, and temporal cortical lobes. Asomatognosia is particularly pronounced if the right posterior parietal cortex is lost while visual agnosias are more readily produced by lesions to the inferior occipital-temporal cortex. Lesions to the prefrontal lobes produce personality changes, strategic planning deficits, perseveration, abulia, and the release of primitive reflexes. Dementia is not related to losses of a specific portion of the cerebral cortex, but a generalized loss of cortical neurons, particularly pyramidal cells. Alzheimer's disease is the most common dementia.

The technical ability of modern medicine to maintain the viability of major organ systems in spite of the loss of function of the cerebral cortex and brainstem has raised moral and ethical issues concerning the determination of death. These issues deal with such questions as the nature of humanity itself and its relationship to the brain and brain functions.

FOR FURTHER READING

Adams, R. D., and Victor, M. *Principles of Neurology*. New York: McGraw-Hill, 1993.

Bellugi, U., Poizner, H., and Klima, E. S. Language, Modality and the Brain. *Trends Neurosci.* 12:380–388, 1989.

Bernstein, L. *The Unanswered Question: Six Talks at Harvard*. Cambridge, MA: Harvard University Press, 1976.

Churchland, P. S., and Sejnowski, T. J. *The Computational Brain*. Cambridge, MA: MIT Press, 1992.

Denny-Brown, D. The Frontal Lobes and Their Functions. In: Feiling, A., ed. *Modern Trends in Neurology*. New York: Hoeber, 1951, p. 65

Edelman, G. M. *Neural Darwinism: The Theory of Neuronal Group Selection*. New York: Basic Books, 1987.

Edelman, G. M. *The Remembered Present: A Biological Theory of Consciousness*. New York: Basic Books, 1989.

Edelman, G. M., and Mountcastle, V. B. *The Mindful Brain*. Cambridge, MA: MIT Press, 1978.

Harlow, J. M. Recovery from the Passage of an Iron Bar through the Head. *Mass. Med. Soc. Publ.* 2:327–346, 1868.

Jaynes, J. *The Origin of Consciousness in the Breakdown of the Bicameral Mind.* Boston: Houghton Mifflin, 1976.

Lechtenberg, R. *Seizure Recognition and Treatment.* New York: Churchill Livingstone, 1990.

McCullagh, P. *Brain Dead, Brain Absent, Brain Donors, Human Subjects or Human Objects:* New York: John Wiley & Sons, 1993.

Penrose, R. *The Emperor's New Mind: Concerning Computers, Minds, and the Laws of Physics.* New York: Oxford University Press, 1989.

Plum, F., and Posner, J. B. *The Diagnosis of Stupor and Coma.* Philadelphia: F. A. Davis, 1980.

Searle, J. Minds and Brains Without Programs. In: Blakemore, C., and Greenfield, S., eds. *Mindwaves.* Oxford: Basil Blackwell, 1987, pp. 209–233.

Springer, S. P., and Deutsch, G. *Left Brain, Right Brain.* New York: W. H. Freeman, 1985.

APPENDIX A
NEUROEMBRYOLOGY

T. R. Kingsley

A bit of historical background often goes a long way toward understanding the status quo. In the same way, knowledge of the developmental history of the nervous system, or neuroembryology, can help in understanding the relationships between parts of the system, both anatomic and functional. The introduction to the basic embryology that follows is intended to serve these functions. Although recognizable components of the nervous system do not form until the third week of development, a brief review of pre-neurological embryology will be presented to introduce terms and concepts that will set the stage for the neuroembryology that follows.

Early Development

Human development begins with fertilization, the union of a sperm and egg to form a single-celled **zygote**. The first week following fertilization is marked by rapid cell proliferation via **cleavage**, a type of mitotic division that increases cell number without increasing cytoplasmic mass. Following cleavage, cells begin to **differentiate;** that is, they become committed to a specific developmental state.

The cells formed as a result of cleavage are termed **blastomeres** [G. *blastos,* germ + G. *meros,* part] (Fig. A.1). By the fourth day following fertilization, there are 12 to 16 blastomeres that form a **morula** [L. *morus,* mulberry], a solid spherical mass. The morula is transformed into a **blastocyst** when it acquires a fluid-filled central cavity (**blastocyst cavity**). Two recognizable cell populations are present at this stage, the **trophoblast** [G. *trophe,* nutrition], which forms the surface of the sphere, and the **embryoblast** or **inner cell mass**, a knot of centrally located cells that becomes attached to the inner surface of the trophoblast. In general, the trophoblast will give rise to the **fetal membranes** (**amnion, chorion,** fetal components of the **placenta,** and the **primary yolk sac**), while the embryoblast will give rise to the embryo itself.

During the second week of development, the process of implantation is completed and the process of forming of the three primary germ layers, or **gastrulation**, begins. The embryoblast is first transformed into the **embryonic disc**, a flattened, nearly circular plate of cells. Soon two distinct layers of cells can be discerned within the embryoblast. One layer, which consists of tall columnar cells, is known as the **epiblast** [G. *epi,* upon]. The other, a layer of cuboidal cells, constitutes the **hypoblast** [G. *hypo,* under]. At this point, the embryoblastic tissue may be described as a **bilaminar disc**.

Completion of gastrulation, the formation of a **trilaminar disc**, is signaled by a rapid proliferation of cells in the epiblast. This results in the formation of a thickened strip of tissue that extends from the border of the embryonic disc to a point near its mid-region. The strip is known as the **primitive streak**. The origin of the primitive streak marks the future caudal end of the embryo, and thus its location defines the craniocaudal axis. It also permits the identification of the dorsal and ventral surfaces of the embryo. Epiblastic cells migrate from the deep surface of the primitive streak and form a layer of tissue that intervenes between the epiblast and the hypoblast. This layer, known as **embryonic mesoderm**, progresses to the margins of the embryonic disc. It eventually occupies all areas except two small axial regions: the **prochordal plate**, near the cranial end of the embryo, and the **cloacal membrane**, near the caudal end of the embryo.

With the formation of embryonic mesoderm, the embryonic disc becomes trilaminar. The layer originally designated as epiblast is now called **ectoderm** [G. *ekdos,* outside + G. *derma,* skin]; the layer originally designated as hypoblast is now called **endoderm** [G. *endo,* within]. The three primary germ layers—ectoderm, mesoderm, and endoderm—give rise to all of the more highly differentiated structures of the body. Ectoderm produces the epidermis and the nervous system. Endoderm gives rise to the epithelial linings of the respiratory tract and digestive tract, and portions of organs associated with the digestive tract. Mesoderm is the source of the remainder of the organism, including all varieties of connective tissue and muscle, vascular structures, blood cells, bone marrow, and portions of internal organs

Following the development of embryonic meso-

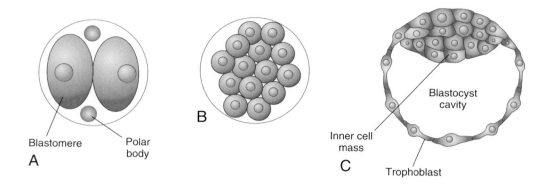

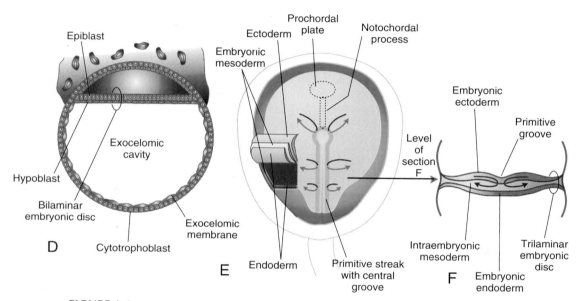

FIGURE A.1.

Early embryological development. Following fertilization, the cell mass subdivides into two blastomeres within 30 hours (**A**). The morula, a solid mass of 12 to 16 blastomeres, is formed by the third day (**B**). A fluid-filled blastocyst is produced by the fourth day (**C**). During the second week following fertilization, cells of the embryoblast form a bilaminar disc. The upper layer is designated epiblast, the lower, hypoblast (**D**). During the third week, cells of the epiblast (ectoderm) proliferate to form a primitive streak. Epiblastic cells migrate down from the region of the primitive streak to form a layer of cells (mesoderm) that intervenes between epiblast and hypoblast. This transforms the embryo into a trilaminar disc (**E**, dorsal view; **F**, transverse section).

derm, the primitive streak regresses toward the caudal end of the embryo. Normally, it degenerates completely, but occasionally remnants persist, giving rise to a tumor generally located in the region of the lower spinal column. A tumor that originates from remnants of the primitive streak is known as a **teratoma** [G. *teras*, monster + G. *oma*, tumor]. Since they develop from pluripotential tissue, such tumors may exhibit a variety of component tissue types.

The first structure to develop within the newly forming embryonic mesoderm is the **notochord** [G. *motos*, back + G. *chorde*, string]. This is a rodlike structure that begins to form at the cranial terminus of the primitive streak, and extends cranially as far as the prochordal plate. As the primitive streak regresses caudally, the notochord follows. The lengthened notochord now defines the craniocaudal axis of the embryo and lends some rigidity to the embryonic structure. It also has other important functions that are of particular interest to students of neuroscience. First, the notochord induces the differentiation of a portion of overlying ectoderm to form the primordium of the nervous system. Second, the notochord is the center around which the developing vertebrae form. While the notochord eventually degenerates in the regions where the vertebral bone develops, it persists in the intervertebral spaces, forming the **nucleus pulposus** [L. *pulpa,* flesh] of the **intervertebral disc.**

Development of the Nervous System

The term **neurulation** describes the sequence of events that characterizes the formation of the neural plate and neural folds (Fig. A.2). As the notochord begins to develop (approximately day 16), it induces the ectoderm dorsal to it to proliferate, forming the **neural plate**. By day 18, an invagination, the **neural groove**, develops along the midline of the neural plate. At the lateral edges of the groove are the **neural folds**.

By day 21, the neural folds have begun to approach each other in the midline, eventually meeting and forming a closed **neural tube**. As they do so, a portion of the tissue that was found in the area of the neural fold fails to become incorporated into the neural tube. This tissue, the **neural crest**, loses its connection with the overlying surface ectoderm and comes to occupy a midline position, interposed between the neural tube and the surface ectoderm. Within a short time, the neural crest divides into left and right portions that migrate to positions dorsolateral to the neural tube.

Neural crest cells are important to the formation of the peripheral nervous system, since they give rise to all sensory cells of peripheral ganglia. They are also the source of Schwann cells, and cells of the chromaffin system, including paraganglia, adrenal medulla, and cells within the carotid and aortic bodies. Neural crest cells can migrate throughout the embryo and are reported to give rise to nonneuronal cell populations as diverse as the melanocytes in the skin, the stria vascularis of the inner ear, and enteroendocrine cells.

Early Development of the Central Nervous System

The central nervous system, which includes the brain and spinal cord, is derived from the neural tube.

DEVELOPMENT OF THE NEURAL TUBE

Conversion of the neural groove into a closed tube occurs over a period of several days early in week 4.

Closure begins in the region of the fourth to sixth pairs of somites and proceeds in both the cranial and caudal directions. The tube is temporarily open at both ends. The cranial opening, called the **rostral neuropore**, closes by about day 25. The caudal opening, the **caudal neuropore**, closes by about day 27. The region of the fourth pair of somites marks a dividing point for the central nervous system. Rostral to this point, the neural tube will form the brain, while caudal to it, the tube will develop into the spinal cord.

The neural tube is initially a thin-walled structure having a large central lumen, or **neural canal**. The walls consist of pseudostratified neuroepithelium, the cells of which reach from the lumen to the margin of the tube (Fig. A.3 *A* and *B*). Beginning in the fourth week, the neuroepithelial cells proliferate rapidly and differentiate into the precursors of categories of cells that will make up the tissue of the central nervous system.

As cell replication progresses, the walls of the neural tube thicken. Simultaneously, the neural canal diminishes to form a smaller channel that will ultimately form the **ventricular system** of the brain and the **central canal** of the spinal cord. Gradually, layers (or zones) of cells become apparent within the tissue of the tube (Fig. A.3*C*). Remnants of the original neuroepithelial layer remain adjacent to the lumen and form the **ventricular zone (ependymal layer)**. This layer will form the epithelial lining of the ventricles and central canal and the epithelium of the choroid plexus. Cells that migrate away from the lumen form the **mantle (intermediate) layer**. All neurons of the central nervous system and most of the supporting cells, or **macroglia** [G. *makros,* large + G. *glia,* glue], are derived from this layer. A third region, the marginal layer, lies farthest from the lumen. It is composed primarily of axons that grow out from cells of the mantle layer. In the mature nervous system, highly cellular regions that are derived from the intermediate zone will constitute **gray matter**, while regions that are composed primarily of axons will constitute **white matter**.

ORIGIN OF CELL TYPES

Neurons develop from precursor cells that have migrated to the intermediate zone. Evidence suggests that the primitive neurons, or **neuroblasts**, are assisted in their migration by a population of cells called **radial glial cells**. This cell group differentiates from cells in the parent neuroepithelial cell layer, and continues to span the entire width of the neural tube as it thickens. They serve as guides along which the neuroblasts migrate as they are formed.

Initially the neuroblast has no processes, and it is therefore called an **apolar neuroblast**. Soon two processes, a primitive axon and a primitive dendrite, develop and the cell is termed a **bipolar neuroblast**. The dendritic process then degenerates, leaving a **unipolar neuroblast**. The dendritic process is subsequently replaced with multiple dendrites, transforming the cell into a **multipolar neuroblast**.

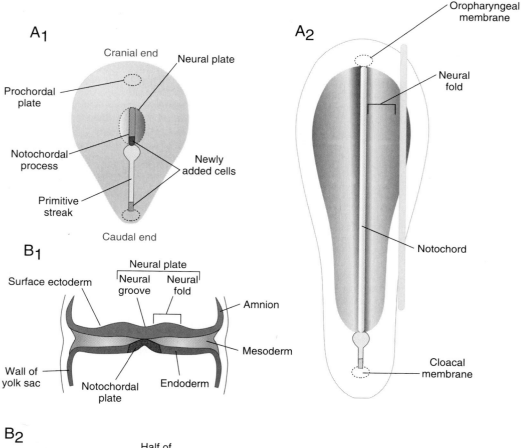

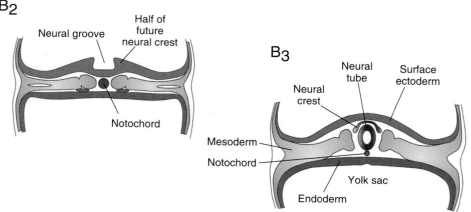

FIGURE A.2.

Neurulation. Development of the notochord induces the overlying ectoderm to transform into neuroectoderm (**A₁** at 17 days; **A₂**, at 21 days, dorsal view). This tissue develops as a flattened neural plate. Viewed in cross-sections, a central neural groove, with lateral neural folds, begins to form by day 18 (**B₁).** The groove deepens (**B₂**), then folds together to form a neural tube by day 21. Cells from the region of the neural fold separate from the developing tube and the overlying ectoderm, forming by day 28 a band of tissue called the neural crest (**B₃**). This will later separate into left and right portions, located dorsolateral to the neural tube.

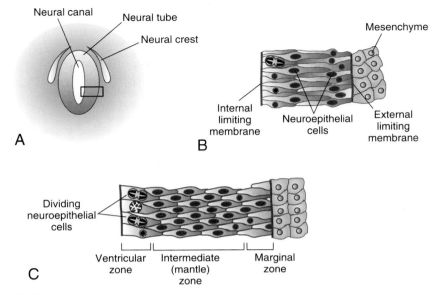

Neural canal

Neural tube

Neural crest

Mesenchyme

A

Internal limiting membrane

Neuroepithelial cells

External limiting membrane

B

Dividing neuroepithelial cells

Ventricular zone

Intermediate (mantle) zone

Marginal zone

C

FIGURE A.3.

Development of the cell layers of the primitive central nervous system. During the fourth week the wall of the neural tube (**A**) consists of neuroepithelial cells that span the width of the tube from the internal limiting membrane to the external limiting membrane (**B**). Continued proliferation of neuroepithelial cells causes the tube to thicken, and gradually three zones can be defined (**C**).

After most of the neuroblasts have developed, other cells of neuroepithelial origin enter the intermediate zone and differentiate into supporting cells of the central nervous system called macroglia. Primitive forms of these cells are called **glioblasts** (or spongioblasts). Some of these become **astroblasts** [G. *astro*, star], that develop into **astrocytes**. Others become **oligodendroblasts** [G. *oligo*, few, + *dendron*, tree], that mature into **oligodendrocytes**. An additional type of supporting cell found within the central nervous system, the **microglia**, is not derived from the neuroepithelium. These cells arise from surrounding primitive mesoderm and subsequently migrate into the CNS to function in a phagocytic capacity.

DEVELOPMENT OF AFFERENT AND EFFERENT REGIONS

Continued proliferation of cells along the extent of the neural tube causes the walls to become thicker. The pattern of cell proliferation produces a shallow groove along the inner, lateral surface of the walls on each side. This groove, called the **sulcus limitans** [L. *sulcus*, groove + L. *limes*, limit], extends along the entire length of the future spinal cord and as far rostral as the future midbrain. It divides the neural tube into a dorsal region, called the **alar plate** [L. *ala*, wing] (Fig. A.4), and a ventral region called the **basal plate**. Structures that develop in the alar plate region mediate **afferent functions** [L. *af-ferens*, to bring to] (i.e., information

being conveyed into the central nervous system), while structures found in the basal plate region mediate **efferent functions** [L. *ef-ferens*, to bring out] (i.e., information being conveyed from the central nervous system toward the periphery). Thus, the sulcus limitans forms a handy landmark, neatly dividing the spinal cord and part of the brain into functional regions.

The developing neural tube is surrounded by primitive mesoderm that soon differentiates into connective tissue wrappings called **meninges**. The meninges begin development as a single thick membrane called the **primitive meninx** [G. *meninx*, membrane]. This membrane differentiates secondarily into an outer layer, the **pachymeninx** [G. *pachys*, thick], equivalent to the **dura mater**, and a thinner inner zone that forms the **leptomeninges** [G. *lepto*, delicate], which includes the **pia mater** [L. *pius*, tender + L. *mater*, mother] and the **arachnoid** [G. *arachne*, spider] layer. The pia mater and arachnoid layer are initially apposed, but they become separated by the **subarachnoid space**. This space later is filled with cerebrospinal fluid.

Further Development of the Spinal Cord

By the sixth week of development, the spinal cord begins to assume the morphology that will characterize

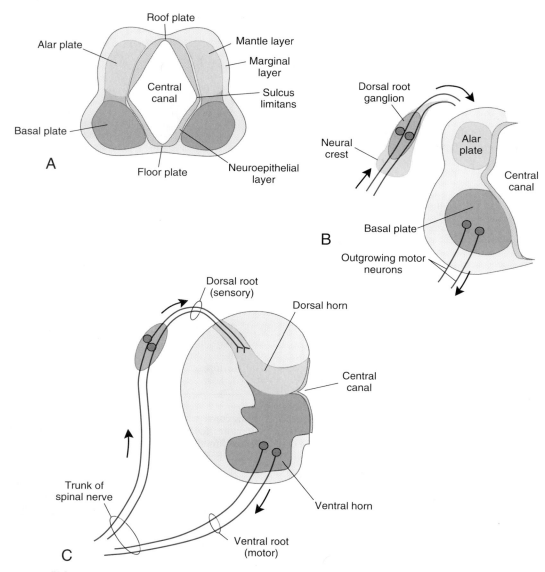

FIGURE A.4.

The alar and basal plates evolve into sensory and motor regions respectively. The line that divides the two areas, the sulcus limitans, is an important referential landmark that extends from the spinal cord through the mesencephalon (**A**, week 6). By the sixth week the sensory ganglion, developing from the neural crest, sends axons into the alar plate region of the CNS while, at the same time, motor neurons in the basal plate extend axons out of the CNS (**B**). The sensory ganglion extends peripheral axons that meet motor axons leaving the CNS. Joined, they form a spinal nerve that contains both afferent (sensory) and efferent (motor) components (**C**).

its adult form. The cell bodies of the mantle layer have produced an area of gray matter that occupies the region near the central canal, while their axons in the marginal layer have begun to form a well-defined layer of white matter. **Dorsal root ganglia** cells, peripheral ganglia derived from neural crest, appear as clumps of cells ad-

jacent to the dorsolateral portion of the spinal cord (Fig. A.4). Nerve processes growing out from them can be observed growing into the alar plate region of the developing cord, thus forming the **dorsal root** of the spinal nerve. **Ventral nerve** rootlets can also be observed emerging from the basal plate region. Organization of cells in the central gray matter to form functional associated groups of cells, or **nuclei**, is accomplished by the fourteenth week of development.

From the time of neural tube formation until the beginning of the third month of development, the spinal cord extends the entire length of the embryo, into the developing coccygeal region. Spinal nerves exit through intervertebral foramina at approximately their level of origin (Fig. A.5A). However, as development progresses, the rate of growth of the vertebral column exceeds that of the spinal cord. This causes the caudal end of the spinal cord to lie at relatively more rostral regions of the vertebral column (Fig. A.5 B and C). In the adult, the termination of the spinal cord lies in the upper lumber region (T12 to L3). Spinal nerves from the lower lumbar, sacral, and coccygeal regions retain their original points of exit between the vertebrae, but they lengthen as the rate of growth of the vertebral column outstrips that of the spinal cord. As a result, the lower end of the vertebral column is occupied by a sheaf of nerve fibers known as the **cauda equina** [L. *cauda,* tail + L. *eqine,* horse]. The dura mater and arachnoid continue to line the entire length of the vertebral column in the adult. The pia mater terminates at the caudal end of the spinal cord, known as the **conus medullaris** [G. *konos,* cone + L. *medulla,* marrow]. Only a thin thread of pia mater, the **filum terminale** [L. *filum,* thread] extends from the conus medullaris to the first coccygeal vertebra, defining the discrepancy between the initial and final extent of the spinal cord. Because of this anatomical orientation, the subarachnoid space is quite large below the level of the conus medullaris. Consequently, the vertebral column below the level of L3 is a suitable site to insert a needle into the subarachnoid space to obtain a sample of cerebrospinal fluid, a procedure known as a **spinal lumbar puncture**. A needle inserted in this region has little probability of damaging the spinal cord.

Congenital Malformations of the Spinal Cord

Failure of the rostral or caudal neuropore to close results in a variety of congenital malformations of the brain and spinal cord respectively. The most common form, lack of closure of the caudal neuropore, can produce defects in the spinal cord itself and in structures that lie dorsal to it, including meninges, vertebral arch, dorsal musculature, and skin. **Spina bifida** [L. *bifidus,* cleft] is a generic term that describes spinal malformations, and **crania bifida** describes cranial malformations. Both conditions always include some form of nonfusion of the overlying bone, but the conditions encompass a hierarchy of severity, ranging from asymptomatic to fatal.

Spina bifida occulta [L. *occulo,* to cover or hide] (Fig. A.6A) represents the minimal degree of defect,

in which the halves of the vertebral arch fail to grow together in the medial plane. This produces a dimple often marked by the presence of a tuft of hair. The defect is often apparent on the surface and does not involve the underlying meninges or neural tissue. The result is generally asymptomatic and represents a curiosity found at the L5 or S1 vertebra in approximately 10% of the population.

More severe degrees of malformation are called **spina bifida cystica** [G. *kystis,* bladder], and are characterized by a sac-like protrusion into the cleft between the unfused halves of the vertebral arch. In **spina bifida with meningocele** (Fig. A.6B), the sac contains only meninges and cerebrospinal fluid; in **spina bifida with meningomyelocele** (Fig. A.6C), the sac also contains tissues of the spinal cord or nerve roots. Both of these forms of defect can occur at any level of the vertebral column, but are most common in the lumbar region. The most severe and very rare form is **spina bifida with myeloschisis** [G. *myelo,* marrow + G. *schisis,* cleaving] (Fig. A.6D). This condition results from a failure of the neural tube to close over an extensive region of the spinal cord, producing an exposed flat mass of neural tissue.

The various forms of spina bifida cystica produce different degrees of neural deficit at levels appropriate to the level of the lesion. Generally there is a loss of skin sensation for regions innervated by nerves at that level, along with some degree of muscle paralysis. Meningomyeloceles at the lumbosacral level often confer paralysis of bladder or anal sphincters. A diagnosis of spina bifida cystica in the developing fetus can be made by measuring the level of **α-fetoprotein** in the amniotic fluid. This protein, normally present in small amounts, is greatly elevated in the presence of spina bifida cystica. The condition may also be visualized via an ultrasound examination after the tenth week of gestation.

Further Development of the Brain

The region of the neural tube that will form the brain is initially similar in appearance to the region destined to form the spinal cord. However, beginning in the fourth week the portion of the neural tube rostral to the fourth pair of somites undergoes a series of dilations and foldings that rearrange the tissue into a compact spherical mass and define the functional regions of the adult brain (Fig. A.7). This process is initiated when the tube dilates to form three **primary brain vesicles**, the **prosencephalon** [G. *proso,* forward + G. *enkephalos,* brain] (or forebrain), the **mesencephalon** [G. *mesos,* middle] (or midbrain), and the **rhombencephalon** [G. *rhombos,* a rhombus] (or hindbrain).

During the fifth week the prosencephalon subdivides to form the **telencephalon** [G. *telos,* end] and

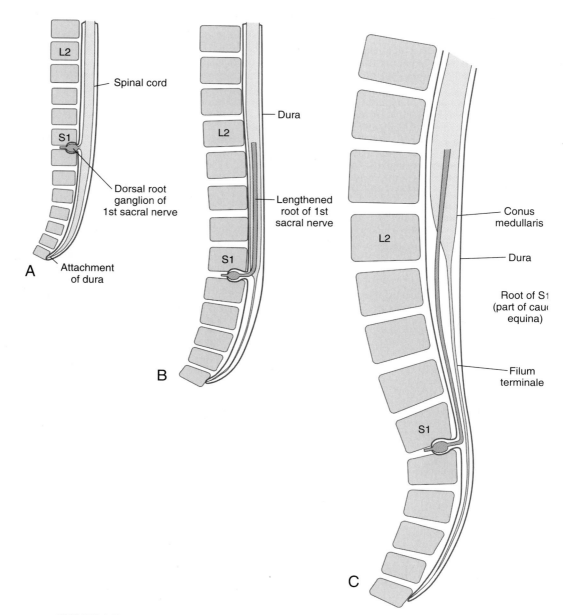

FIGURE A.5.

The development of the cauda equina results from the continued elongation of the vertebral column and dural sac after the spinal cord has reached its final length. The spinal nerves are drawn away from the spinal cord until, in their adult form, they extend several centimeters within the dural sac before leaving the CNS and entering the peripheral space. (**A**, third month; **B**, 5–6 months; **C**, neonate). The portion of the dural sac between the conus medullaris and its termination at the first coccygeal vertebra forms the lumbar cistern, from which cerebral spinal fluid can safely and conveniently be sampled.

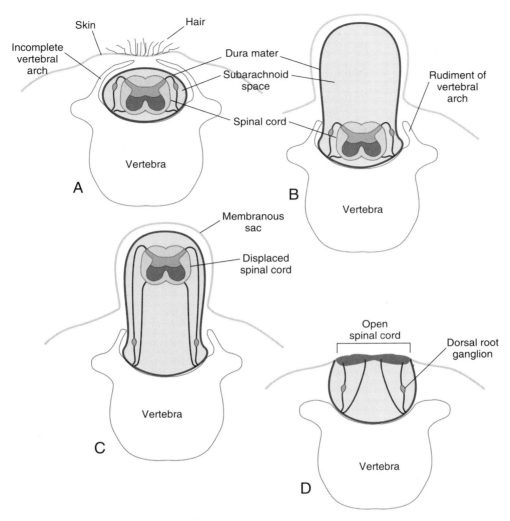

FIGURE A.6.

Common malformations of the spinal cord due to failure of the neural groove to close completely. **A**, **Spina bifida occulta**. The dural sac and spinal cord are normal. The dorsal aspect of the vertebra is not closed, and a hairy patch is frequently seen over the area. **B**, Spina bifida with **meningocele**. The bony deficit is larger than in spina bifida occulta, resulting in a ballooning of the dural sac. The neural elements, however, remain in their proper place and there are usually no neurological deficits. **C**, Spina bifida with **meningomyelocele**. This is similar to spina bifida with meningocele, except that the spinal cord and its associated spinal nerves are displaced into the expanded dural cavity. There are always neurological deficits associated with this malformation. **D**, Spina bifida with **myeloschisis**. In this case the failure of the neural groove to close is extreme, resulting in a failure of the spinal canal to form. The exposed neural tissue is little more than a malformed neural plate.

---

the **diencephalon** [G. *dia,* through]. The rhombencephalon subdivides to form the **metencephalon** [G. *meta,* after] and the **myelencephalon** [G. *myelo,* marrow]. The mesencephalon does not divide. Each of the resulting **five secondary brain vesicles** consists of a wall of neural tissue bordering a fluid-filled cavity. The cavities persist in the adult as the **ventricular system** of the brain. The sulcus limitans continues rostral from the spinal cord as far as the junction between the mesencephalon and the diencephalon, extending the regional division into alar and basal plates up to that level.

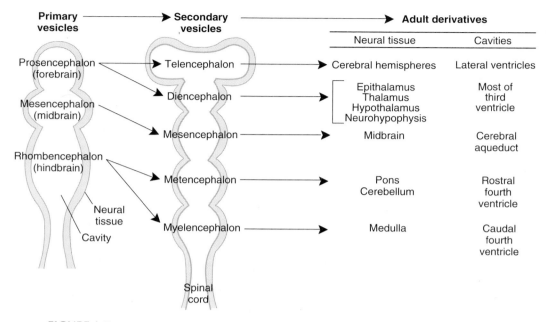

FIGURE A.7.

Following closure of the rostral neuropore, the neural tube rostral to the level of the fourth somites forms three swellings, the primary vesicles (about day 28). By day 35, the prosencephalon and rhombencephalon divide. This produces the five secondary brain vesicles from which the definitive brain structures—neural tissue and cavities—develop.

The folding of the brain occurs in two phases (Fig. A.8). During the fourth week, the prosencephalon bends ventrally, forming a ventrally directed concavity in the mesencephalic region; a similar ventrally directed concavity is formed by a ventral bending at the juncture of the rhombencephalon with the spinal cord. In this manner, the **midbrain flexure** and the **cervical flexures** are formed. A third flexure, having a dorsal concavity, forms at the juncture of the metencephalon and myelencephalon during the sixth week. This **pontine flexure** exerts forces that result in the thinning of the roof of the metencephalon and myelencephalon.

MYELENCEPHALON

The tissue of the myelencephalon develops into the **medulla oblongata** of the adult brain. The rostral portion of the myelencephalic cavity becomes part of the **fourth ventricle**, while the caudal portion of the cavity narrows to form the **central canal**, which is continuous with the central canal of the spinal cord.

The caudal portion of the myelencephalon retains some of the anatomical characteristics of the adjacent spinal cord (Fig. A.9A). The lumen remains a small **central canal** marked by the sulcus limitans, and the division into alar and basal plates persists. There is, however, a departure from the arrangement of components of the mantle and marginal layers. In the cau-

dal myelencephalon, groups of neuroblasts in the mantle layer (gray matter) of the alar plate migrate into the marginal layer (white matter) to become the **inferior olivary nuclei**. At the dorsal border of the myelencephalon neuroblasts form nuclei on either side of the midline. These, named from medial to lateral, become the **gracile nucleus**, the **cuneate nucleus**, and the **nucleus of the spinal tract of cranial nerve V**.

In the rostral portion of the myelencephalon, the arrangement of components diverges further from the pattern found in the spinal cord (Fig. A.9B). As the result of the development of the pontine flexure, neural tissue originally located in the dorsomedial portion of the neural tube is pushed laterally, as if on a hinge positioned at the sulcus limitans. The fluid-filled canal enlarges to form the broad, flattened, rhomboid-shaped **fourth ventricle**. The lateral movement of the dorsal neural tissue leaves behind a thin ventricular roof composed only of meninges apposed to ependyma. Located along the floor of the fourth ventricle are neuroblasts that form a series of cell columns. The columns that lie dorsolateral to the sulcus limitans represent tissue of the alar plate and are associated with sensory function. Those that lie ventromedial to the sulcus limitans represent tissue of the basal plate and are associated with motor function. Each column is concerned with processing neural information of a specific type; that is, its function is associated with a

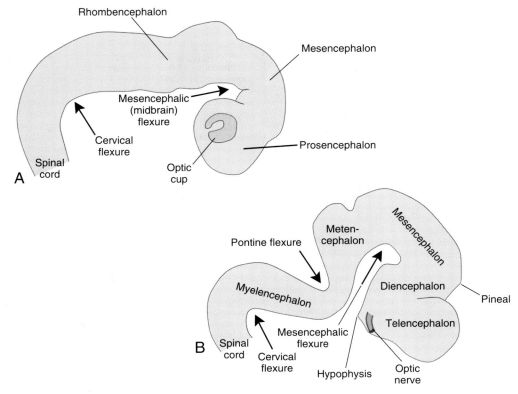

FIGURE A.8.

Folding of the rostral neural tube. During the fourth week, the rostral neural tube begins to flex at the junction of the spinal cord and rhombencephalon (forming the cervical flexure) and at the mesencephalon (forming the mesencephalic flexure) (**A**). During the sixth week, a flexure (the pontine flexure) develops between the metencephalon and the myelencephalon (**B**). This folding produces a compact mass of tissue that approximates the shape of the definitive brain.

single neural **modality** (Box A.1). The relative positions of cell columns associated with specific modalities remain consistent along the extent of the rostral myelencephalon (Fig. A.9*B*).

Nerve fibers that emerge from the neuroblasts that form these cell columns contribute to structures called **cranial nerves** that exit from the brainstem and innervate peripheral structures. Twelve pairs of cranial nerves (designated by roman numerals I through XII) extend out from various portions of the brainstem. Some of them mediate a single modality (e.g., special somatic sensory information), but others mediate a wide assortment of modalities. Cell columns of the myelencephalon develop into nuclei that are associated with cranial nerves XII through IX.

METENCEPHALON

The neural tissue of the metencephalon gives rise to the **pons** [L. *pons,* bridge] and **cerebellum** [L. dim. of *cerebrum,* little brain]. Its cavity develops into the

rostral portion of the fourth ventricle. Prior to the eighth week of development, the metencephalon resembles the myelencephalon. It has a broad, rhomboid ventricle with a thin roof. Components of the alar plate are located in the dorsolateral region, while basal plate components are found in the ventromedial region.

Pons

The pons develops in the region lateral and ventral to the fourth ventricle. The alar plate components form columns of cells located in the extreme dorsolateral region of the pons. Some neuroblasts from the alar plate region migrate into the ventral portion of the pontine area, where they join with neuroblasts that have migrated from the alar plate region of the myelencephalon, to form the **basal pontine nuclei** (Fig. A.10*A*).

The area that borders the floor of the fourth ventricle constitutes the **tegmentum of the pons** [L. *tego,* to cover]. This region develops from the basal plate,

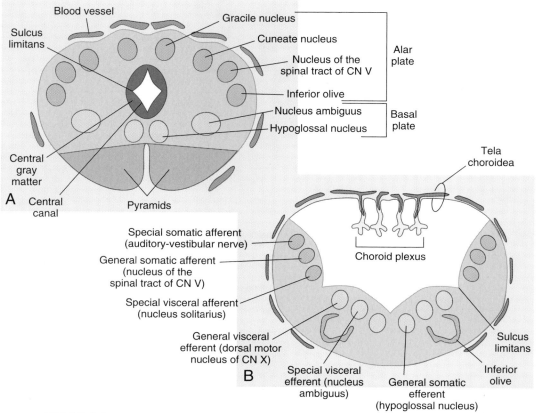

FIGURE A.9.

A, Transverse section through the caudal part of the myelencephalon at 5 weeks. At this stage it is very similar to the spinal cord, having a small central canal and a nearly spherical cross section. The cells of the alar and basal mantle regions are beginning to migrate and become organized into individual nuclei in the marginal layer. **B**, Transverse section through the rostral myelencephalon at 5 weeks. The lateral movement of the alar plate has transformed the fourth ventricle into a flattened thin-roofed structure. A series of cell columns is found adjacent to the floor of the ventricle; dorsolateral columns are alar plate derivatives, while ventromedial columns are derived from the basal plate.

and contains columns of cells that are associated with motor function.

Ventral to the tegmentum is the **basis of the pons.** This region is initially small, occupied by the basal pontine nuclei. As development proceeds, large numbers of axons arise from the basal pontine nuclei and cross the midline as they grow toward the contralateral cerebellar hemisphere (Fig. A.10*B*). These fibers form the **middle cerebellar peduncles (brachium pontis** [L. *brachium,* arm]), the development of which greatly enlarges the basis of the pons and is responsible for the predominantly fibrous appearance of the ventral pons in the adult.

The cell columns located in the dorsolateral and tegmental portions of the pons give rise to nuclei that

are associated with cranial nerves VIII through V, which emerge from the pons.

Cerebellum

The cerebellum begins to develop from the alar plate of the metencephalon. Cells of the alar plate that border the dorsolateral margins of the fourth ventricle form the **rhombic lip** (Fig. A.10*A*). Cells of the rostral end of rhombic lip proliferate, forming a projection into the fourth ventricle. They also grow toward the midline in the area dorsal to the roof of the fourth ventricle. These two projections form the **cerebellar plate.** By the third month, the intraventricular projection has regressed, leaving only the plate of tissue

BOX A.1.

Neural Modalities

The term "neural modality" describes the nature of the information carried by a given axon. In classical neural terminology there are a number of modalities defined.

Afferent (Sensory): Refers to information transmitted from peripheral sensors, or **receptors,** toward the central nervous system.

Efferent (Motor): Refers to information transmitted from the central nervous system toward peripheral **effectors,** such as muscle cells or secretory cells that respond to neural stimuli.

Sensory and motor information can be further defined, based on the anatomic distribution of the innervation.

Visceral: Refers to information transmitted to or from internal organs (viscera) or regions derived from the branchial arches.

Somatic: Refers to information transmitted to or from the remainder of the body (nonvisceral structures); i.e., skin, muscles, etc.

A final, rather arbitrary distinction is made, based on the embryological origin of a structure.

Special: Refers to information transmitted to or from a specified subgroup of structures within the "visceral" or "somatic" categories. **Special visceral** refers to information transmitted to or from structures derived from the branchial arch region of the embryo. **Special somatic** is a term used only to qualify sensory or afferent information, and describes information transmitted from the organs of special sense (e.g., retina, cochlear organ, taste buds, olfactory organ, vestibular organ).

General: Refers to information transmitted to or from all structures within the "visceral" or "somatic" categories other than those included in the "special" category.

A given *axon* is limited to transmitting a single modality; that is, it might function as a special somatic afferent or as a general visceral efferent, etc. *Peripheral nerves* are composed of a mixed population of axons, and consequently a given peripheral nerve might transmit a number of modalities.

---

covering the roof of the ventricle (Fig. A.11*A*). This tissue persists, forming the **vermis** [L. *vermis, worm*] of the cerebellum.

During the fourth month, the lateral portions of the cerebellar primordium increase in size, forming the **lateral lobes** of the cerebellum (Fig. A.11*B*). By the end of the fourth month, fissures divide the region (Fig. A.11*C*). One fissure segregates tissue of the caudal portion of the cerebellar primordium, defining the caudal **flocculonodular lobe** [L. *floccus,* tuft of wool + L. *nodulus,* knot]. The **nodule** forms the medial portion of this lobe, and **flocculus** forms the lateral portion. Tissue of the lateral lobes lies rostral to the flocculus, and the vermis lies rostral to the nodule. At

approximately this time, the lateral lobes are subdivided into **anterior** and **posterior lobes**.

Within the substance of the developing cerebellum, neuroblasts of the mantle layer form the **dentate** and **deep cerebellar nuclei**. Other neuroblasts migrate toward the surface of the developing lobes to form the highly cellular **cerebellar cortex**.

The embryological development of the structure of the cerebellum mirrors its phylogenetic development. The lobes of the cerebellum that are phylogenetically the oldest are concerned with more basic, primitive function. Thus the **archicerebellum** [G. *archos,* chief, first] is composed of the flocculonodular lobe and is the oldest phylogenetically. It has connec-

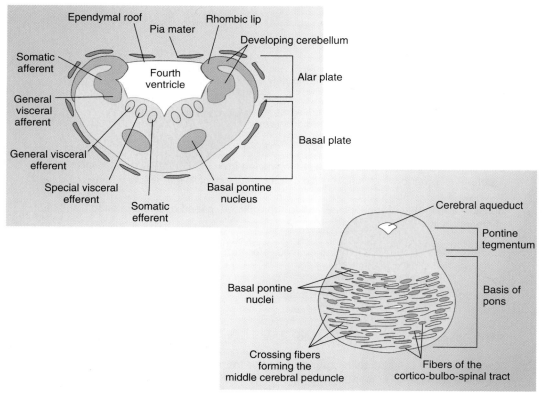

FIGURE A.10.

A, Transverse section through the metencephalon at about the end of the fifth week of development. The organization of nuclei across the floor of the fourth ventricle is similar to that found in the rostral myelencephalon. In the metencephalon, the alar plate further differentiates into the structures associated with the cerebellum. A portion of the alar plate also migrates ventrally to form the basal pontine nuclei. These develop into the principal source of afferent axons to the cerebellar hemispheres. **B**, Transverse section through the adult metencephalon. The greatly expanded basis is formed principally of fibers from the basal pontine nuclei that cross the midline en route to the cerebellum.

tions with the vestibular apparatus. The **paleocerebellum** [G. *paleos,* ancient], the vermis and anterior lobes, evolved later. It has reciprocal neurological connections with the spinal cord. The **neocerebellum**, the posterior lobes, is the most recent phylogenetic addition. It has reciprocal neurological connections to the cerebral cortex.

MESENCEPHALON (MIDBRAIN)

The midbrain undergoes little change during its development (Fig. A.12). The cavity of the mesencephalon is a narrow channel, the **cerebral aqueduct**, that connects the third and fourth ventricles. The dorsal portion of the mesencephalon forms the **tectum** [L. *tectum,* roof]. Neuroblasts of the alar plate proliferate and form paired sets of large nuclei, the **superior colliculi** [L. *collis,* hill] (rostral), and the **inferior colliculi** (caudal).

The mesencephalon ventral to the sulcus limitans develops primarily from the basal plate and may be divided into a more dorsal tegmentum and a ventral basis. Nuclei associated with the **third and fourth cranial nerves**, the **red nuclei** and the nuclei of the reticular formation, are found here. The marginal zone at the ventral extreme of the mesencephalon becomes quite prominent as fibers from the developing cerebral cortex pass through the midbrain. These fiber bundles become a prominent landmark, known as the **cerebral peduncles** [L. *pedunculus,* dim. of *pes,* foot]. Cranial nerves III and IV emerge from the mesencephalic region of the brain.

DIENCEPHALON

The diencephalon represents the thickened walls of the caudal portion of the original prosencephalon.

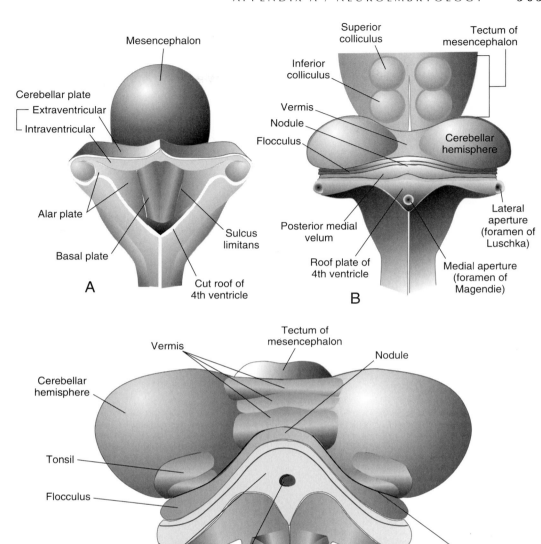

FIGURE A.11.

A, Dorsal view of the metencephalon at 2 months. The rhombic lip of the metencephalon expands medially, merging at the midline and forming the cerebellar plate in the process. **B**, Dorsal view of the metencephalon at 4 months. As development of the cerebellum continues, the three principal divisions of the adult cerebellum become evident: the flocculus-nodulus, the vermis, and the hemispheres. **C**, Dorsal view of the metencephalon at 5 months. The cerebellum continues to expand dorsally and laterally. The posterolateral fissure separates the floccular-nodular lobe from the remainder of the cerebellum. With further development, the puckering seen on the surface of the vermis continues and leads to the formation of the folia of the adult structure.

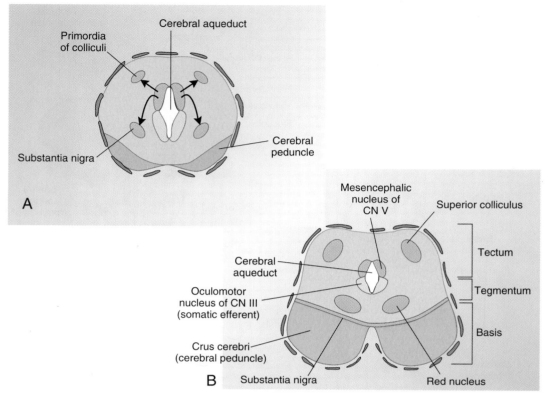

FIGURE A.12.

A, Transverse section through the mesencephalon at 5 weeks. The primordia of the superior and inferior colliculi begin to separate from the alar plate and migrate toward the dorsal margin of the neural tube. Similarly, cells that are destined to become the substantia nigra, red nucleus, and parts of the reticular formation begin to migrate away from the basal plate. **B**, Transverse section through the mesencephalon at 11 weeks. Fibers originating in the cerebral cortex become layered onto the surface of the basis of the mesencephalon, forming the prominent cerebral peduncles. The substantia nigra and red nucleus are recognizable.

Its cavity forms the **third ventricle**. At the junction of the mesencephalon with the diencephalon, the sulcus limitans terminates, and the distinction between alar and basal plates does not extend into the diencephalon and telencephalon. The diencephalon is divided instead by two sulci, the **epithalamic sulcus** and the **hypothalamic sulcus**. These separate the tissue of the walls of the diencephalon into three regions that become the **epithalamus**, **thalamus**, and **hypothalamus** (Fig. A.13). Major elements within these regions are recognizable by the eighth week of development.

The epithalamus forms from the roof and dorsolateral wall of the diencephalon. This area regresses as development proceeds. Definitive structures that arise from this area include the **pineal gland (pineal body)** [L. *pineus,* pine], a solid cone of cells that projects into the roof of the third ventricle, and the **habenular nuclei** [L. *habena,* strap]. Two bands of

fibers form transverse commissures, linking the right and left halves of the diencephalon. These are the **habenular** and **posterior commissures**.

The thalamus develops in the midlateral region of the walls of the diencephalon. Neuroblasts proliferate extensively and form groups of thalamic nuclei (the **anterior**, **ventral**, **medial**, and **lateral groups of thalamic nuclei**, and the **medial and lateral geniculate bodies**). These expand laterally, and also protrude into the cavity of the third ventricle, reducing it to a narrow slit. The thalamic tissue frequently fuses in the midline, forming the **interthalamic adhesion (massa intermedia)**.

The hypothalamus develops from the tissue ventral to the hypothalamic sulcus. In the hypothalamic region, neuroblasts of the mantle layer give rise to a number of nuclei that are associated with autonomic or endocrine function. The base of the hypothalamus

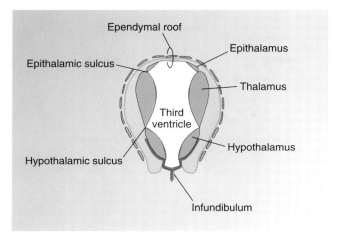

FIGURE A.13.

Transverse section through the diencephalon at 8 weeks. The division of the region into epithalamus, thalamus, and hypothalamus by the epithalamic and hypothalamic sulci is apparent. Note that the sulcus limitans does not exist at this level and the distinction between alar and basal plates loses its meaning.

and a ventral extension of tissue in this region develop into an endocrine gland, the **neurohypophysis**.

ORIGIN OF THE PITUITARY GLAND

Extending from the basis of the hypothalamic region is a glandular structure, called the **pituitary gland** or **hypophysis cerebri** [G. *hypophysis,* an undergrowth]. Although grossly it appears as a single structure, its embryological development and adult function reveal the hypophysis cerebri to be two autonomous glands, linked anatomically. The pituitary gland begins to develop during the fourth week from two separate origins, neuroectoderm and oral ectoderm (Fig. A.14*A*). The neuroectodermal component begins as a downgrowth from the ventral surface of the diencephalon, termed the **infundibulum**. The oral ectodermal component originates as a diverticulum (**"Rathke's pouch"**) that projects dorsally from the primitive mouth cavity (stomodeum). Rathke's pouch continues to grow dorsally toward the infundibulum, and by the eighth week has reached a position adjacent to it, and has lost its connection with the stomodeum.

During further development, the infundibulum expands to form the neural portion of the hypophysis cerebri, generally referred to as the neurohypophysis (Fig. A.14 *B* and *C*). Neurohypophyseal components include the part of the base of the hypothalamus known as the **median eminence**, a thin stalk (the original infundibulum), and a distended tip, the **pars nervosa**, from which hormones are secreted.

The parts of the hypophysis cerebri that are derived from Rathke's pouch (oral ectoderm) are jointly referred to as the **adenohypophysis** (Fig. A.14 *B* and

C). A small component, the **pars tuberalis**, lies adjacent to the infundibular stalk and wraps around it. The distal portion of the sac expands to form the **pars distalis**, the source of a number of hormones. The posterior wall of the original pouch that lies adjacent to the pars nervosa does not develop extensively in humans. It persists as a small population of cells that make up the **pars intermedia**. During development, the lumen of Rathke's pouch is frequently obliterated. However, it may persist in the adult gland as a narrow cleft, sometimes occupied by colloidal cysts.

ORIGIN OF THE OPTIC TRACT AND RETINA

During the fourth week of development, an **optic vesicle** develops as an evagination from the lateral aspect of each side of the prosencephalon, prior to its division into the telencephalon and diencephalon. These vesicles are the primordia of the **optic tract** and **retina**. After the subdivision of the prosencephalon, the fibers that make up the optic tract (axons of the retinal ganglion cells) retain their connection to the diencephalic portion.

TELENCEPHALON

The telencephalon forms the rostral end of the original neural tube. Its rostral wall is the **lamina terminalis** [L. *laminae,* plate + L. *terminus,* boundary].

Large lateral evaginations of the telencephalon form dorsal and rostral to the optic vesicles during the seventh week. These **cerebral vesicles** are the primordia of the cerebral hemispheres. Each has a cavity, a **lateral ventricle**, that is an extension of the cen-

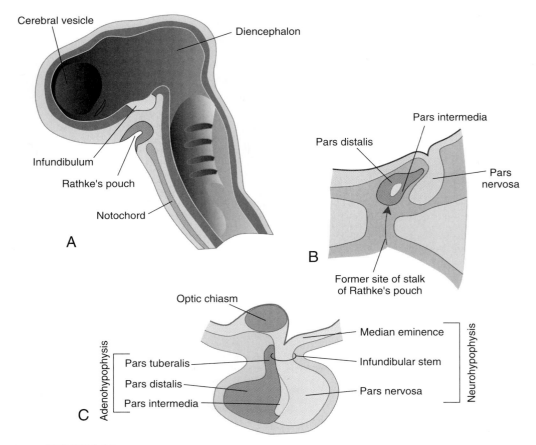

FIGURE A.14.

A, Sagittal section of the cranial end of a 6-week embryo. Tissue of the infundibulum forms on the ventral surface of the diencephalon. Rathke's pouch can be seen extending dorsally from the roof of the stomodeum. **B**, Sagittal section through the region of the developing hypophysis cerebri at approximately 10 weeks. Rathke's pouch has reached a position adjacent to the developing infundibulum, having lost contact with the oral cavity. The anterior surface of the pouch proliferates, forming the pars distalis of the hypophysis. **C**, The definitive structure.

tral cavity and is continuous with the third ventricle via the **intraventricular foramen** (Fig. A.15). The cerebral vesicles grow rapidly upward, forward, and backward, overgrowing the diencephalon, mesencephalon, and hindbrain. Above and in front of the diencephalon the medial surfaces of the two vesicles contact one another, forming the flattened medial surfaces of the cerebral hemispheres.

The inferior portion of the cerebral vesicles grows less rapidly, and develops into a thick layer of tissue that will become the **corpus striatum** [L. *corpus,* body + L. *striatus,* furrowed] (Fig. A.15). Later, nerve fibers leaving the developing cerebral cortex, en route to the brainstem and spinal cord, form a wide fiber tract called the **internal capsule** that divides the corpus striatum into the **caudate nucleus** (superior) [L.

cauda, tail], the **putamen** [L. *puto,* to prune] (inferolateral), and the **globus pallidus** [L. *pallidus,* pale] (inferomedial). The internal capsule and the caudate nucleus elongate, conforming to the contour of the lateral ventricle as it expands in the rostral-caudal plane.

The non-striatal portion of the wall of the vesicle, the **pallium** [L. *pallium,* cloak, mantle], remains thinner, and is the primordium of the cerebral cortex (Fig. A.15*B*). The medial pallial wall remains quite thin, and invaginates into the lateral ventricle, forming the **tela choroidea** [L. *tela,* web + L. *choroeides,* membrane-like]. The line of invagination, apparent on the medial surfaces of the hemispheres, is termed the **choroidal fissure.**

The pallial walls initially have the same layers as the rest of the developing neural tube (ventricular,

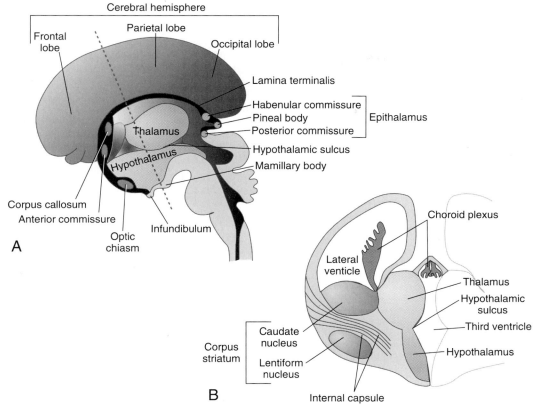

FIGURE A.15.

A, Sagittal section through the developing cerebral hemispheres at approximately 10 weeks, showing the association with the diencephalon. The various commissures in the region of the lamina terminalis are apparent at this stage. **B**, Transverse section through the plane shown (*dashed line*) in **A**. As the internal capsule develops, it divides the corpus striatum into a superior caudate nucleus and an inferior lentiform nucleus that later becomes the putamen and globus pallidus.

mantle, and marginal). Later, cells of the marginal zone migrate outward, forming a highly cellular surface. The surface of the cerebral cortex is therefore composed of gray matter, and the nerve fibers that extend from these cells extend centrally, forming a large mass of white matter.

The area of the lamina terminalis, the original rostral extent of the neural tube, develops into a site in which fiber tracts form that connect corresponding areas of the cerebral hemispheres with one another (Fig. A.15A). Fiber tracts that serve this function are called **commissures** [L. *committo,* to join]. The **anterior commissure** is a small fiber bundle that develops early. The largest commissure is the **corpus callosum**. The corpus callosum initially occupies the dorsal region of the lamina terminalis, but eventually it becomes so large that it extends over the roof of the diencephalon. In the ventral portion of the lamina terminalis, half of the fibers from the retina form the **optic chiasm** [G.

chiasma, crossing of two lines], crossing the midline to join the optic tract of the opposite side.

THE CHOROID PLEXUS

In several regions of the ventricular system of the brain there are areas in which little neural tissue intervenes between the ependyma that lines the ventricle and the overlying meninges. This means that the highly vascular pia mater lies adjacent to the ependyma. At points within these regions the pia mater undergoes extensive proliferation, forming projections into the ventricles that resemble clusters of grapes. These projections retain their ependymal covering. The modified pia mater (vascular connective tissue) together with its ependymal covering form the **chorioid plexus**. The pia mater component is referred to as the tela choroidea, while the ependymal component is referred to as the **lamina epithelialis** [G. *epi,* upon + G. *thele,* nipple, term originally applied to the skin covering the nipple]. The choroid plexus is a secretory tissue responsible for the production of **cerebrospinal fluid (CSF)**. Areas of choroid plexus are located in the roof of the fourth and third ventricles and in the medial walls of the lateral ventricles (Fig. A.15*B*).

Congenital Malformations of the Brain

Congenital abnormalities of the brain are relatively common and can result from errors in development of the neural tissue itself or of associated structures (e.g., mesodermal derivatives). Causes include a wide range of **genetic** and **environmental factors**. The brain is particularly susceptible to environmental **teratogens** (agents that can increase the incidence of congenital malformations) during the first 16 weeks of development.

ANENCEPHALY

Anencephaly [G. *an,* without] (or meroanencephaly) is a condition that results from failure of the rostral neuropore to close. The forebrain primordia and the overlying skull do not develop normally, and the brain is exposed, a condition known as **exencephaly** [G. *ex,* out]. Generally components of the brainstem (midbrain, pons, medulla) are present; consequently the term **meroanencephaly** [G. *meros,* part] is technically more correct, but is not in common use. Anencephaly is a common defect, occurring approximately once per 1000 births, and is a condition that inevitably precludes extrauterine life. The causes are not well understood, but genetic factors undoubtedly play a role, since there is a well-established familial incidence. Teratogenic agents may also be involved, as suggested by experimental evidence in rats. The existence of anencephaly can be detected by ultrasonography, fetoscopy, and radiography as early as 14 weeks, and, like spina bifida, is usually associated with an elevated level of α-fetoprotein in the amniotic fluid. (See Case Discussion in Chapter 13.)

MICROCEPHALY

Microcephaly [G. *mikros,* small + G. *kephale,* head] is a condition in which the calvaria and brain are abnormally small. The etiology is believed to involve failure of the brain to develop, since pressure from the developing brain normally is responsible for enlargement of the calvaria. The cause of this abnormality is unclear and may involve genetic or environmental agents. Infants afflicted with this disorder are severely mentally retarded.

HYDROCEPHALUS

Hydrocephalus [G. *hydor,* water] is a term that describes any abnormality involving expansion of the ventricular system of the brain. Generally this results from blockage of CSF circulation that causes an excess production of CSF relative to its reabsorption. The concomitant increase in intracranial pressure causes the volume of the ventricles to increase. In infants the sutures of the calvaria have not yet fused. Consequently, this increased pressure results in expansion and thinning of the brain and enlargement of the calvaria. This also leads to atrophy of the cerebral cortex and white matter, and compression of the basal ganglia and diencephalon. The most common cause of this phenomenon is fetal viral infection (**cytomegalovirus** and *Toxoplasma gondii*).

The Peripheral Nervous System

The peripheral nervous system is a network of **nerves**—groups of functionally associated axons—and **ganglia**—groups of functionally associated neuron cell bodies (Box A.1). All sensory neurons of the peripheral nervous system are derived from neural crest cells. The neurons that will make up the ganglia of the peripheral nervous system begin to develop during the fifth week, shortly after neural crest tissue is segregated from the neural ectoderm of the neural tube. Further neuronal proliferation and differentiation continue throughout the course of fetal development and into early childhood. The cell processes that make up peripheral nerves can originate from neurons located in the peripheral nervous system. They can also grow out from neurons within the brain (efferent fibers of cranial nerves) or in the spinal cord (from motor neurons in the ventral horn or from the lateral horn). Development of these axons begins during the fourth week and also continues throughout fetal life and early childhood.

Supporting cells of the peripheral system (**Schwann cells**) are also derived from neural crest tissue. Within ganglia, Schwann cells border the cell bodies of the neurons, forming a capsule-like structure. The capsule continues into a **neurilemmal sheath**, also formed by Schwann cells, that invests the axon.

ELEMENTS OF THE PERIPHERAL NERVOUS SYSTEM

Various types of ganglia and nerves are organized to form functional units within the peripheral nervous system.

Peripheral Sensory Ganglia

Peripheral sensory ganglia include the thirty-one pairs of **dorsal root ganglia (or spinal ganglia)**

that lie adjacent to each segment of the spinal cord, as well as the peripheral sensory ganglia of **cranial nerve (CN) V (trigeminal), CN VII (facial), CN VIII (vestibulocochlear), CN IX (glossopharyngeal), and CN X (vagus)**. These ganglia (with the exception of CN VIII) are composed of **pseudounipolar neurons**. These are cells in which the embryonic bipolar status has been modified to form a single process. This process splits in the vicinity of the neuron, forming a peripheral branch that terminates in a sensory ending, and a central branch that enters the dorsal horn of the spinal cord. The neurons of the spiral ganglion of the cochlea and the vestibular ganglia of CN VIII (vestibulocochlear nerve) are an exception, and remain bipolar.

Autonomic Ganglia

Autonomic ganglia are clusters of small multipolar neurons. They include the **sympathetic paravertebral ganglia**, which lie adjacent to the vertebral bodies in the thoracolumbar region of the spinal column, and the **sympathetic collateral ganglia** (e.g., the cardiac, celiac, and mesenteric plexus) found in the thorax and abdomen. They also include the **parasympathetic ganglia**, such as the submucosal and myenteric plexuses, and other plexuses located within the tissue of viscera.

Spinal Nerves

A **mixed spinal nerve** contains **motor elements** that are derived from neuroblasts in the basal plate of the developing spinal cord, and **sensory fibers** that develop from neuroblasts found in the dorsal root ganglia.

Late in the fourth week of development, axons grow out from neuroblasts in the basal plate of the spinal cord. Axons emerging at a given level form a group of fibers called a **ventral nerve rootlet** (Fig. A.4B). A series of rootlets emerge along the extent of the spinal cord.

Neuroblasts in the dorsal root ganglion transform from bipolar to pseudounipolar (see Fig. A.4B). The central branch of their process grows toward the alar plate region of the spinal cord. Groups of these fibers from a given dorsal root ganglion form a **dorsal root**. The distal branch of the process grows toward the periphery, eventually meeting and joining with the axons of the ventral rootlet to form a mixed spinal nerve (Fig. A.4C).

Each spinal nerve divides immediately into **dorsal and ventral primary rami** [L. *rami,* branches]. The dorsal primary ramus innervates the dorsal axial musculature, the vertebrae, the posterior intervertebral joints, and part of the skin of the back. The larger ventral primary ramus innervates the limbs and the ventrolateral parts of the body wall, and forms the major nerve plexus. As the limbs develop from **limb buds**, the ventral rami of the corresponding segments grow into the bud and innervate the muscles and skin of the developing limb in a segmental manner. The segmental distribution of the spinal nerves is reflected in the pattern of **dermatomes** that can be mapped on the body surface. Each dermatome represents the region innervated by the dorsal root ganglion associated with one specific level of the spinal cord (see Chapter 5).

Cranial Nerves

The cranial nerves are groups of axons that have their origin within the brain. With the exception of **CN I (olfactory nerve)** and **CN II (optic nerve)**, the function of the cranial nerves is associated with nuclei that form from cell columns in the alar and basal plates of the mesencephalon and rhombencephalon.

There are twelve pair of cranial nerves. They are generally divided into three groups, based on their embryological origin (see Box A.1). The **somatic efferent cranial nerves** include **CN IV (trochlear), CN VI (abducens), CN XII (hypoglossal)**, and most of **CN III (oculomotor)**. The cells of origin of these nerves are located in the basal plate region of the brainstem, in the region that mediates somatic efferent function. Their axons are distributed to muscles of somite origin.

The **nerves of the branchial arches** include **CN V (trigeminal), VII (facial), IX (glossopharyngeal), and X (vagus)**. They serve structures that are derived from embryonic branchial arches. Motor fibers of these nerves originate in the basal plate region, in the cell columns concerned with general and special visceral efferent function. Their sensory fibers terminate in the alar plate region in cell columns concerned with general and special visceral afferent and somatic afferent function.

The **nerves of the special senses** include **CN I (optic), II (olfactory), and VIII (vestibulocochlear)**. CN VIII is a highly specialized somatic afferent nerve, the cells of which terminate in the alar plate of the brainstem. CN I develops as an evagination of the wall of the brain, and consequently actually represents a fiber tract of the brain.

SUBSYSTEMS WITHIN THE PERIPHERAL NERVOUS SYSTEM

The elements of the peripheral nervous system can be functionally grouped into two systems, one that innervates visceral structures, the **autonomic nervous system**, and one that innervates the remainder of the body (skin, muscles, etc.), the **somatic nervous system**.

The Autonomic Nervous System

This system can be further subdivided into the **sympathetic nervous system** and the **parasympathetic nervous system**.

The sympathetic nervous system includes the sympathetic trunks (paravertebral ganglia) and the collateral ganglia. **Preganglionic fibers** arise from neurons in the central nervous system and synapse within the peripheral sympathetic ganglia; **postganglionic fibers** emerge from the ganglion cells and continue to the target organ. Postganglionic fibers emerging from the sympathetic trunk travel with spinal nerves and mixed peripheral nerves, while those emerging from collateral ganglia form purely autonomic **splanchnic nerves**.

The ganglia of the parasympathetic nervous system are small clusters of cells located in or near the organ they innervate. Preganglionic fibers arise from neurons in nuclei of the brainstem (associated with CN III, VII, IX, and X) and in the sacral region of the spinal cord. Because of the location of the parasympathetic ganglia, preganglionic fibers are long, while postganglionic fibers have only a short distance to travel before reaching their target cells.

The Somatic Nervous System

This portion of the peripheral nervous system has a well-developed sensory component as well as motor components. The sensory portion includes peripheral sensory receptors (specialized or simple nerve endings) that form a unit with afferent nerve fibers. Most of these fibers are processes of neurons located in the dorsal root ganglia; others are afferent components of cranial nerves. Motor components of the somatic system consist of fibers that emerge from the ventral horn of the spinal cord or fibers that make up the efferent components of cranial nerves.

FOR FURTHER READING

Hamilton, W. J., Boyd, J. D., and Mossman, H. W. *Human Embryology*. Baltimore: Williams & Wilkins, 1962.

Jacobson, M. *Developmental Neurobiology*. New York: Plenum Press, 1991.

Moore, K. L. *The Developing Human; Clinically Oriented Embryology*. Philadelphia: W. B. Saunders, 1988.

Rowland, L. P. *Merritt's Textbook of Neurology*. Philadelphia: Lea & Febiger, 1989.

Sadler, T. W. *Langman's Medical Embryology*, 7th ed. Baltimore: Williams & Wilkins, 1995.

Shepherd, G. M. Developmental Neurobiology. In: *Neurobiology*, Ch. 9. New York: Oxford University Press, 1994, pp. 192–225.

THE NEUROLOGICAL EXAMINATION

R. E. Kingsley and Steven R. Gable

The neurological examination allows a rapid and accurate assessment of the function of the nervous system. It is elegant in its simplicity. It does not rely on complex or technologically advanced instruments; its power arises from the knowledge and observational skills of the examiner. Only a brief description of the neurological examination can be given here—one that is appropriate for use in every primary care setting. The standard neurological examination as presented here is quite simple, and it is an essential skill for every physician to master. More detailed explanations of the examination can be found in the references.

For purposes of explanation, the neurological examination may be divided into seven categories. However, the examination is never performed in this way. Rather, the elements are combined in a way that is efficient for the physician and that disturbs the patient as little as possible. Therefore, in this appendix, the examination will first be described in detail, followed by an outline of the examination as it is practiced.

I. History

The most important part of the neurological examination is having patients give a narration of their difficulties. Since the history is taken early in the physician's encounter with the patient, it is the best time to begin to establish a bond of trust. It is essential to *listen* to the patient. Only by truly listening with an open mind can one be objective and not impose one's preconceived evaluation of the patient on the history. Although the patient must not be coached or prompted, discreet questions are necessary from time to time to draw out important clinical points that, to the patient, may not seem important. Almost invariably, patients will spontaneously describe their condition in a way that leads the knowledgeable physician toward the correct diagnosis. Toward this end, the physician must be sure to make note of the **chief complaint**, the **time course** of the patient's difficulties, and the **distribution of symptoms**.

Some important aspects of the history may need gentle prompting. For example, the patient may not realize that illnesses of related family members may be significant, and most patients will have a certain reluctance to describe sexual dysfunction, incontinence, seizures, or drug use (alcohol, tobacco, prescribed, and recreational).

2. Mental Status

Evaluating the state of mind of the patient is essential when evaluating the nervous system. The patient may be suffering from a psychiatric disorder, or an organic disease process may be interfering with the patient's ability to communicate. The mental status of the patient is first assessed as he is giving the history.

Separating psychiatric from organic disorders can sometimes be difficult. Generally speaking, the more specific the patient's complaints, the more likely is organic disease. For example, neurological illness is suggested if the patient says something like "Six months ago I was able to lift the salt bags to fill my water softener, but now I can't" rather than "I feel weak and tired all the time. Some days I just can't seem to get out of bed."

The mental status is frequently described in terms of levels of consciousness and state of memory. Is the patient **awake?** If awake, is he **alert**, or confused or stuporous? Is the patient **orientated** to time and place? Ask the patient where he is; what is the time; the day of the week? Finally, give the patient three things to be remembered. Make it clear that he will be asked to recall those things later in the examination. The things to remember should be a number, a color, and an object or place. After naming these three items, immediately ask the patient to repeat them to ascertain that the patient heard and understood the request.

3. Cranial Nerves

During the neurological examination, every component of every cranial nerve should be tested.

OLFACTORY NERVE

The common habit of dismissing the olfactory nerve as unimportant is to be strongly discouraged. Unilateral anosmia is an important clinical sign and may be the only early objective sign of a frontal lobe abscess or tumor (see Chapter 12). Having the patient smell coffee or smelling salts is an extremely poor method of testing olfaction. In the case of coffee, the patient knows by visual association what odor is being presented, and therefore the physician has no independent means of assessing the reliability of the patient's report. Also, some patients may confuse the warmth of the vapors for the olfactory sensation the physician is testing. Smelling salts are a poor choice, because they are irritating and will stimulate the pain receptors of the nasal membranes. The best olfactory stimulus is **methyl salicylate**, a colorless aromatic liquid that has a nonirritating, pleasant odor similar to wintergreen. The patient has no clue as to the nature of the odor being presented, yet will invariably report in some unambiguous way having experienced the odor, describing it as "spearmint," "Ben-Gay", etc.

To perform the examination, place a vial containing cotton saturated with methyl salicylate under one nostril while occluding the other with a finger. Ask the patient to sniff and report the odor. Repeat for the other nostril.

OPTIC NERVE

There are many aspects of the sense of vision; for example, acuity, color, scotopic (day) versus photopic (night) vision, etc. The neurologist is not interested in ophthalmological problems, so tests of refraction, glaucoma, or color blindness etc. are not a normal part of the neurological examination. The neurologist is most concerned with signs of **papilledema** and **patterns of visual field losses** (see Chapter 11).

Papilledema can only be visualized by a fundoscopic examination using an ophthalmoscope, and this visualization is an essential part of every neurological examination. Visual field disturbances can be crudely determined by standing in front of the patient while the patient fixates on the examiner's nose. While the patient covers one eye with a hand, the examiner places his two hands in various quadrants of the visual fields. By twitching the fingers of one or both hands, one can determine if the patient can see from the corresponding quadrant simply by asking him to point to the fingers that are moving. One tests each of the four quadrants in both eyes.

NERVES OF OCULAR MOTION

The thee pairs of nerves that innervate the extraocular eye muscles—the oculomotor, the trochlear, and the abducens—can be tested independently by having the patient gaze approximately 25 to 30 degrees to one side (see Chapter 9). With the eyes in this position, elevation will be accomplished by the supe-

rior rectus muscle in the abducted eye and the inferior oblique muscle in the adducted eye. Depression is controlled by the inferior rectus muscle in the abducted eye and the superior oblique muscle in the adducted eye. The action of the medial and lateral rectus muscles (abduction and adduction of the eye) are tested simply by asking the patient to maintain lateral gaze while testing the other muscles. While testing the eye movements, one notes the presence or absence of nystagmus (see below).

The oculomotor nerve also has a parasympathetic component that must be tested (see Chapter 9). This is done by shining a light into one eye while observing the two pupils. They should both constrict together (consensual light reflex). This test is not valid if the patient is in a brightly lit room, for the eyes may already be constricted from the ambient light.

It is not easy to observe both pupils simultaneously, so it is suggested that each eye be tested twice in the following manner. Aim the light from the ophthalmoscope at the forehead and draw it down to shine on the eye while the patient is gazing at the wall behind the examiner and not at the light. Pause for a moment while watching the pupil of the stimulated eye, then raise the light to the forehead. Repeat the process, only now observe the contralateral pupil. Do the same for the other eye. Note that the light is moved in a vertical direction, not horizontally from eye to eye. This vertical motion gives the pupils time to dilate between stimuli. In cases where the consensual light reflex is weak or absent from one eye, shining the light directly from the reactive eye to the nonreactive eye will cause the nonreactive eye to dilate paradoxically if the pupil had previously constricted consensually. This sign is called the "Marcus Gunn pupil."

Pupillary constriction also occurs during accommodation and convergence. To test this, have the patient fix his gaze on a distant object and then look at a near object. As the eyes converge, both pupils should constrict. Because of the linkage between accommodation and pupillary constriction, one must be careful when testing the light reflex to avoid having the patient fix his gaze on a near object, such as the testing lamp.

TRIGEMINAL NERVE

The sensory portion of the trigeminal nerve is tested by touching the skin of the face with a wisp of cotton and the cold surface of the tuning fork. One must test the separate dermatomes for all three divisions (see Chapter 9). Pin prick testing of the face is usually deferred, unless there is a particular reason to do so, because of the obvious discomfort to the patient.

The corneal reflex is always tested. This reflex involves two nerves: the trigeminal, which carries the afferent signal to the brain, and the facial, which carries the efferent component. It is evoked by touching the cornea with the cotton wisp. Since most people will blink if they see an object coming directly at their eye, the cotton is brought to the cornea from the side while the patient is looking straight ahead.

The motor component of the trigeminal nerve can be tested by having the patient isometrically bite. Paralysis of a masseter muscle can usually be seen and flaccidity can be palpated. As an additional test, the jaw will deviate to the paralyzed side on opening the mouth due to the paralysis of the external pterygoid muscle.

FACIAL NERVE

The motor component of the facial nerve innervates the mimetic muscles of the face and the orbicularis oculi (the latter is tested as part of the corneal reflex). The critical observation to make with respect to the function of the facial nerve is the pattern of paresis or paralysis of the mimetic muscles (see Chapter 9). Paralysis of these muscles is usually quite apparent as facial asymmetry when one asks the patient to grimace. Also observe the nasolabial fold, which will be flattened on the affected side. Especially note whether or not the muscles that raise the eyebrows are involved.

The parasympathetic components of the facial nerve affect salivation and lacrimation. These functions are not specifically tested during the neurological examination, but sometimes can be inferred from the history if the patient complains of a dry mouth or eye. Corneal ulceration may be secondary to poor lacrimation.

AUDITORY-VESTIBULAR NERVE

The principal goals in testing the auditory division of this nerve are first to determine if hearing is grossly normal and, if not, to differentiate between conductive and sensorineural hearing losses. Rubbing the fingers together over the external auditory meatus is a common practice that can usually determine if hearing is grossly present but, unless the patient is profoundly deaf, provides very little useful information. Rinne's test accurately distinguishes between conductive and sensorineural hearing losses and is much preferred to the finger test. Weber's test can only be interpreted in light of the results of Rinne's test (see Chapter 10). Weber's test provides a comparison of hearing efficiency between the two ears. If either of these tests indicates a conductive hearing loss, use the otoscope to examine the external auditory meatus for wax or other types of simple obstructions. Central hearing losses can only be determined by special audiological methods and cannot be determined by the routine neurological examination.

The vestibular system is usually not specifically tested. Spontaneous nystagmus, however, may be a sign of vestibular end-organ or nerve disease, particularly if it is accompanied by vertigo, nausea, and unsteadiness with a tendency to fall (see Chapter 9). Nystagmus can also be caused by cerebellar or brainstem lesions. Therefore, if nystagmus is present these structures must be carefully examined for corroborative evidence. When vestibular end-organ or nerve dysfunction is suspected, an electronystagmogram (caloric testing) is indicated.

GLOSSOPHARYNGEAL-VAGUS NERVE COMPLEX

The sensory components of theses nerves supply the inside of the mouth and pharynx. They can be tested by evoking the gag reflex, which is simply done with a tongue depressor placed in the back of the pharynx. Test both sides independently. The entire oral cavity can be explored with the tongue depressor to check general sensations, if warranted. While eliciting the gag reflex, one should observe the uvula and pharynx for asymmetries (Chapter 9). The quality of the voice should be noted for any signs of stridor or hoarseness.

ACCESSORY NERVE

This nerve is tested by checking the general strength of the sternocleidomastoid and trapezius muscles (see Chapter 9). The sternocleidomastoid muscle is tested by having the patient rotate the head while the examiner is offering resistance against the chin. The muscle being tested is opposite the direction of rotation. In addition to checking the strength of the trapezius muscles, one should also note the symmetry of the two muscles.

HYPOGLOSSAL NERVE

This nerve innervates the intrinsic muscles of the tongue and is tested by observing the tongue for signs of atrophy or fasciculations and having the patient stick out the tongue in the midline. Deviation from the midline position indicates weakness on the side of the deviation (Chapter 9).

---

4. Motor Systems

Testing the motor systems is possibly the most illuminating portion of the neurological examination, since few diseases fail to affect motor functions in some characteristic way. The motor system testing can be divided into four categories: observing **gait**, testing **strength** and **tone**, observing **abnormal movements**, and checking for **ataxia**.

GAIT

To observe the gait, have the patient walk away, turn around in place, and return. Normal gait is smooth, on a narrow base, with both arms gently swinging. Weakness may be revealed by circumduction of the leg during the return swing, or by a limp. A shuffling gait, with the arms stiffly held in one place without swinging, suggests parkinsonism. Ataxia suggests cerebellar disease. If the patient cannot follow a straight line, but meanders back and forth in an almost dance-like manner, this may suggest basal ganglion disease.

Next have the patient do a deep knee bend, hop on each foot in turn, and then walk on heels and then toes. This tests for strength, balance, and coordination. Finally, as a further test of coordination, have the patient tandem walk heel-to-toe.

STRENGTH AND TONE

A very sensitive test of strength is to have the patient stand with the arms held straight ahead and supinated, palms flat and horizontal, with the fingers held together. Any drift (either downward or laterally), pronation, or separation of the fingers indicates weakness in that extremity.

All major muscle groups should be directly tested for strength by having the patient perform an isometric contraction against resistance offered by the examiner. Since strength varies considerably among individuals, one is not looking for absolute strength per se, but is comparing the relative strength between the bilateral groups. While performing this **direct strength testing**, one observes the various muscle groups for **atrophy** and **fasciculations** (see Chapter 6), **spasticity** (see Chapter 7), or **rigidity** (Chapter 8).

Spasticity and rigidity are directly ascertained by passive joint movement. The examiner rapidly lengthens a muscle group while the patient is relaxed; for example, by pronating and supinating the arm. Spasticity reveals itself by offering a rapidly developing resistance, or "catch," during the passive movement, a "catch" that is almost immediately released. Rigidity feels like bending a lead pipe. Cogwheel rigidity feels like a series of small "catches" during passive movement.

ABNORMAL MOVEMENTS

Abnormal, involuntary movements suggest basal ganglion disease (see Chapter 8). Look for a resting tremor that relents during voluntary movement. This is a classic sign of parkinsonism. Also look for unusual fidgetiness, which may be an early sign of chorea. Athetosis, dystonia, and tics are usually self-evident.

ATAXIA

Cerebellar disease is usually quite evident by observing the gait, hopping, and the tandem walk. To further document suspected cerebellar disease, ask the patient to touch his nose with a finger and then to touch your finger which is held in front as a target (see Chapter 8). This movement should be direct, smooth, and coordinated. If in doubt, move your finger as the patient is reaching for it to see if there is a smooth redirection of the intended movement. Test the lower extremity by having the patient run the heel along the shin of the opposite leg. Finally, have the patient slap the thighs alternately with the palms and the backs of the hands.

5. Reflexes

The reflexes provide important, objective signs of neurological status. Always test the five basic **muscle stretch reflexes**—biceps, triceps, brachioradialis, quadriceps and gastrocnemius—on each side (see Chapters 6 and 7). Also, always check the **plantar signs** (Babinski) and, if there is any doubt, reinforce this with the Bing sign. The other **FRA reflexes**, the cremasteric and abdominal, are occasionally useful. Finally, the atavistic reflexes (rooting, snout, glabellar, palmomental, and grasp) are useful in evaluating the cerebral cortex (see Chapter 13).

6. Sensory Systems

Testing the sensory systems can be difficult because it requires the cooperation and understanding of the patient. Only the patient can experience sensations, and those experiences are inescapably colored by the patient's emotional state. Sensory experiences cannot be objectively reported to the physician. On the other hand, many of the sensory tests, like the methyl salicylate test mentioned above, are designed to objectively draw out any inconsistencies in the patient's report (see Chapter 5). Because of the subjective nature of the sensory examination, many physicians perform only a cursory examination. This is unfortunate, for the sensory examination is usually the best indicator of the level of a discrete lesion. Furthermore, learning that the patient is unable to provide a consistent report of sensory perceptions, or that the patient is deliberately malingering, is valuable information.

The **Romberg test** is an objective sensory test. It is performed by having the patient stand on a narrow base with eyes open. If the patient is stable, ask him to close the eyes. A slight increase in instability is normal, but gross instability or falling suggests that proprioceptive sensations from the lower extremities are defective. Instability with the eyes open points to cerebellar disease.

Proprioception can be further tested by holding the tip of the patient's finger or toe with one hand and asking the patient to report whether the digit has been moved up or down. Normal patients can detect the slightest movement. One must be careful to shield the patient's vision from the extremity.

The dorsal column system is further tested with a wisp of cotton applied very lightly to the skin, or just to a single hair. One asks the patient if it can be felt. One checks the dermatome, peripheral nerve, or stocking-glove patterns as appropriate by asking the patient if there is a substantial difference between consecutive touches, strategically placed on opposite sides of suspected boundaries.

Finally, the dorsal column system is tested with the 128-Hz tuning fork. Place the stem of the vibrating fork on a distal bony prominence of an extremity. When the patient can no longer feel the vibration, immediately place the stem on another, more proximal prominence. It should not be felt. If vibrations are perceived at the more proximal point after distal extinction, suspect peripheral neuropathy.

The anterolateral system is tested by pin prick. Use a new clean pin for every patient to avoid transmitting blood-borne diseases. As with the dorsal column system, draw out any patterns of sensory loss.

7. Higher Cortical Functions

Most higher cortical functions (as opposed to primary sensory or motor functions) are tested during the history portion of the examination. The various tests are designed to draw out functional differences as they are related to the various lobes and, where appropriate, to the left-right differences between the lobes (see Chapter 13). These tests are not to be confused with tests for general dementia (which, more or less, affects the cerebral cortex globally).

Lesions involving primarily the frontal lobes are marked by a general deterioration in the personality. The patient may be more belligerent or less socially respectful than usual. This is best determined by interviewing a close relative or spouse to determine any deterioration over time. Perseveration is an important frontal lobe sign that can be specifically tested during the neurological examination by presenting a series of objects to the patient for identification. If the frontal eye fields are involved, there may be **enhanced fixation** of gaze (see Chapter 8).

Aphasia, either expressive or receptive, is usually apparent. The left-sided neglect syndrome, associated with right parietal lobe lesions, can be specifically tested by having the patient copy drawings provided by the examiner (see Chapter 13). Drawing petals on a daisy or drawing a clock or a square are excellent choices. Mathematical functions are more closely associated with the left parietal lobe. This is traditionally tested by having the patient subtract 7 serially, beginning with 100. More subtle parietal lobe lesions can be determined by evoking the **optokinetic reflex** (see Chapter 9).

Practical Application of the Neurological Examination

The following outline is a suggested method for performing the neurological examination efficiently. While everyone must develop one's own method, it is important to do the examination the same way every time. This ensures that nothing will be overlooked. Obtaining data in the same order, at every presentation, also helps to organize the information systematically.

I. Instruments

  A. Reflex hammer—Reflexes; MSRs

  B. Clean unused safety pins (use a new one for each patient)—Sensory; pain

  C. Wisp of cotton—Sensory; light touch

  D. Oto-ophthalmoscope—Sensory; retina, external auditory meatus, light reflex

  E. Methyl salicylate—Sensory; olfaction

  F. Tuning forks; 128 Hz and 256 Hz—Sensory; vibration, hearing

  G. Tongue depressors—Gag reflex, plantar signs

  H. Optokinetic tape—Parietal function; optokinetic nystagmus

  I. Pad of paper and pencil—Note taking and for patient's drawings

II. Patient seated or in bed if in hospital setting.

  A. History—Emphasize *chief complaint, time course,* and *distribution of symptoms.* Do not lead the patient, to avoid imposing preconceived notions on the patient's narrative. Let the story flow naturally.

  B. Mental Status—Evaluate patient's mental status while holding the history-taking conversation. Is the patient awake, alert, and oriented to space and time? At this time, tell the patient three things to remember (a number, a color, and a place are frequently used).

  C. During this period, note any abnormal movements such as tics, tremor, or fidgetiness. Do these movements cease with volition?

III. Have the patient get up and walk.

  A. Evaluate the gait. Follow this with deep knee bends, heel & toe walking, tandem walking, and hopping.

  B. Feel the head, checking for trauma, old and new. Use methyl salicylate to check olfaction, and cotton to test corneal reflex and light touch over the face.

  C. While facing the patient, combine the Romberg, drift, and finger-to-nose tests.

  D. Use the ophthalmoscope to inspect the fundus and test the light reflex.

  E. Test eye movements and visual fields.

  F. Have the patient grimace, phonate, and protrude the tongue in the midline. While the mouth is open, elicit the gag reflex using the tongue depressor. Observe the muscle tone of the pharynx. Palpate the masseter muscles while the patient is biting.

IV. Have patient lie on the examining table or bed.

A. Have the patient run a heel along the shin of the opposite leg.

B. Use the safety pin and then the cotton wisp to test sensations over the arms, the legs, and (if warranted) the body. Either establish or rule out dermatome, peripheral nerve, or stocking-glove patterns to any abnormalities.

C. If warranted, check the abdominal and cremasteric reflexes.

V. Have the patient sit up.

A. Use the 128-Hz tuning fork to test temperature and vibratory sensations. Vibration should extinguish at the ankle and knee and at the wrist and elbow at approximately the same time.

B. Use the 256-Hz tuning fork for the Rinne and Weber tests. If there is a conductive hearing loss, use the otoscope to examine the external meatus for wax and check the tympanic membrane.

C. Do general strength testing.

D. Check proprioception by testing movement perception of the fingers and toes.

E. Test reflexes and signs

1. MSRs—Use the reflex hammer to test these five: Biceps, C5, C6; Triceps, C6, C7; Brachioradialis, C5, C6; Quadriceps, L3 to L5; Gastrocnemius, S1.

2. FRAs—Use the tongue depressor to check for the Babinski sign; use the safety pin to check the Bing sign.

3. Atavistic—Do your favorites, but these are good: Glabellar, grasp, and palmomental.

F. Test higher cortical functions

1. Use the optokinetic tape to elicit optokinetic nystagmus.

2. If indicated, use the pad of paper and pencil to have the patient make appropriate drawings.

3. Ask the patient to recall the three memory test items.

4. Ask the patient to do serial 7 subtractions.

5. Hold up simple objects for identification. Not only determine if the patient can make the appropriate identifications, but note any tendency to perseverate.

VI. Things to note:

A. All of the cranial nerves are tested, and for most of them, all components are tested.

B. Important systems are tested in more than one way.

C. Only very simple instruments are needed. Everything is easily carried in a small physician's bag.

D. With the exception of sensory tests, all results are independent of the patient's consciousness understanding. The results then become **signs** as opposed to **symptoms**.

E. The patient is disturbed as little as possible. There are only two major changes in patient positioning.

F. With practice, this examination takes very little time, usually no more than 5 minutes.

FURTHER APPLICATIONS

- Make a list of all the cranial nerves and their components. Identify where in the neurological examination each of these is tested. Are any missed? Which?

- Identify which parts of the neurological examination test the cerebellum, the basal ganglia, the upper motor neuron system, the lower motor neuron system, the anterolateral system, and the dorsal column systems.

- Practice this examination on a friend until you have it memorized. It is important that you do it exactly the same way every time. Leave nothing out. Get into the habit of taking notes (use the pad of paper and pencil) while you do the examination.

FOR FURTHER READING

Adams, R. D., and Victor, M. *Principles of Neurology.* New York: McGraw-Hill, 1993.
Bickerstaff, E. R., and Spillane, J. A. *Neurological Examination in Clinical Practice.* Boston: Oxford, 1989.
DeMeyer, W. *Technique of the Neurologic Examination.* New York: McGraw-Hill, 1980.
Geraint, F. *Neurological Examination Made Easy.* New York: Churchill Livingstone, 1993.
Goldberg, S. *The Four Minute Neurological Exam.* Miami: MedMaster, 1992.
Joynt, R. J. *Clinical Neurology.* Philadelphia: J. B. Lippincott, 1993.
Rowland, L. P. *Merritt's Textbook of Neurology.* Philadelphia: Lea & Febiger, 1989.

SECTIONS OF THE BRAIN

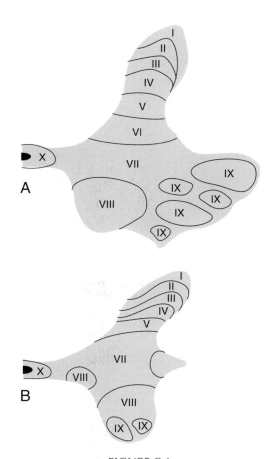

FIGURE C.1.

Laminae of Rexed. **A**, Cervical; **B**, Thoracic; **C**, Lumbar.

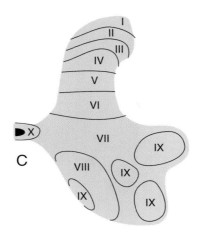

FIGURE C.1.—*continued*

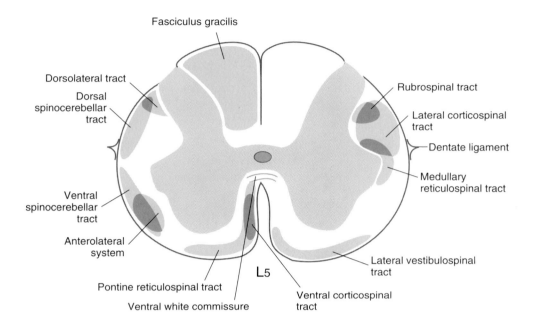

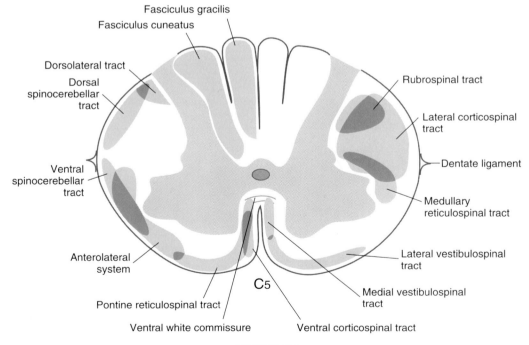

FIGURE C.2.

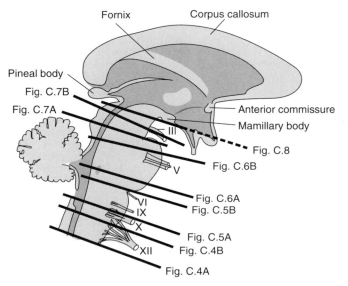

FIGURE C.3.

The levels of the cross sections of the brainstem shown in Figures C.4 through C.8 are indicated here.

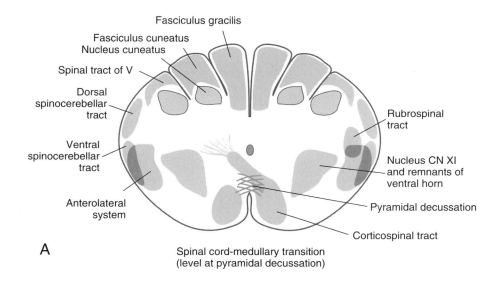

Fasciculus gracilis
Fasciculus cuneatus
Nucleus cuneatus
Spinal tract of V
Dorsal spinocerebellar tract
Ventral spinocerebellar tract
Anterolateral system

Rubrospinal tract
Nucleus CN XI and remnants of ventral horn
Pyramidal decussation
Corticospinal tract

A

Spinal cord-medullary transition
(level at pyramidal decussation)

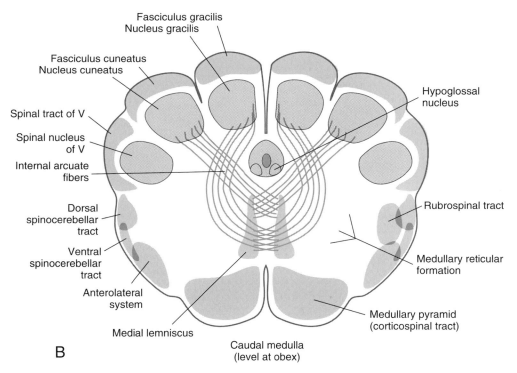

Fasciculus gracilis
Nucleus gracilis
Fasciculus cuneatus
Nucleus cuneatus
Spinal tract of V
Spinal nucleus of V
Internal arcuate fibers
Dorsal spinocerebellar tract
Ventral spinocerebellar tract
Anterolateral system
Medial lemniscus

Hypoglossal nucleus
Rubrospinal tract
Medullary reticular formation
Medullary pyramid (corticospinal tract)

B

Caudal medulla
(level at obex)

FIGURE C.4.

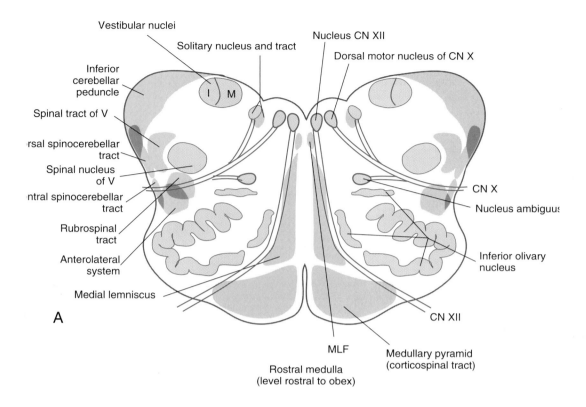

Rostral medulla
(level rostral to obex)

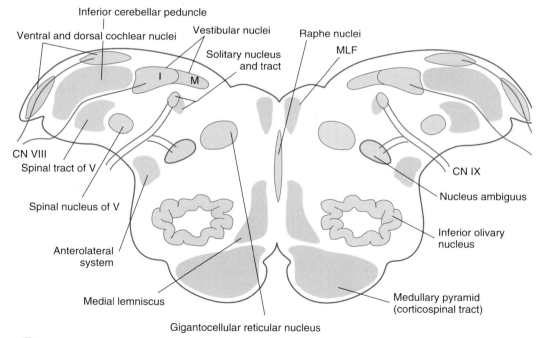

Rostral medulla
(level at pontomedullary junction)

FIGURE C.5.

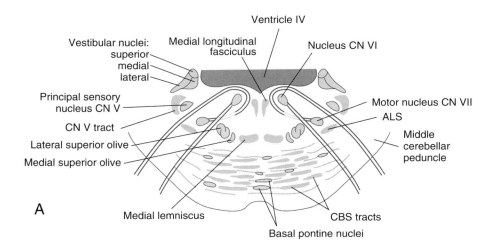

A

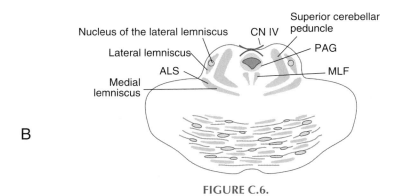

B

FIGURE C.6.

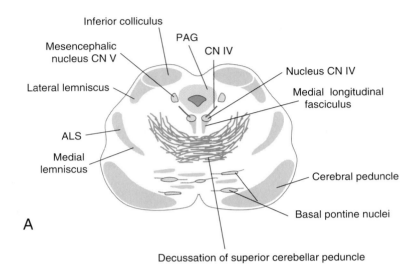

A

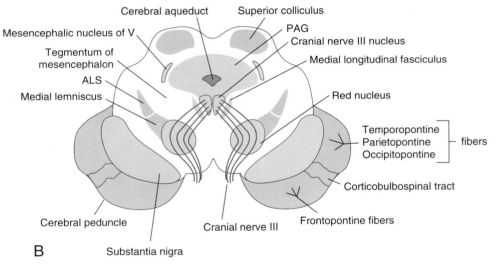

B

FIGURE C.7.

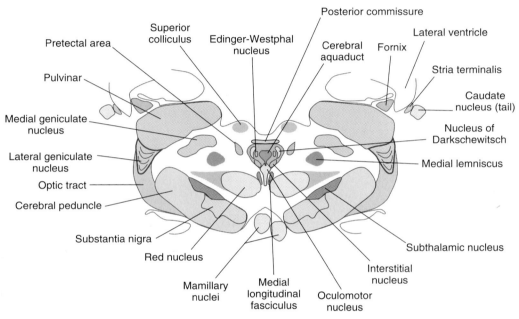

Posterior commissure

Superior
colliculus

Pretectal area

Edinger-Westphal
nucleus

Cerebral
aquaduct

Fornix

Lateral ventricle

Pulvinar

Stria terminalis

Medial geniculate
nucleus

Caudate
nucleus (tail)

Nucleus of
Darkschewitsch

Lateral geniculate
nucleus

Medial lemniscus

Optic tract

Cerebral peduncle

Substantia nigra

Subthalamic nucleus

Red nucleus

Interstitial
nucleus

Mamillary
nuclei

Medial
longitudinal
fasciculus

Oculomotor
nucleus

FIGURE C.8.

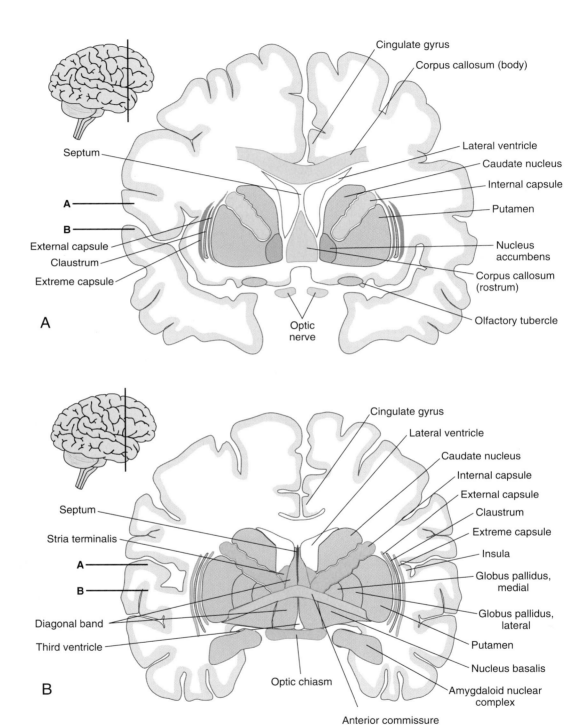

A

Cingulate gyrus
Corpus callosum (body)
Lateral ventricle
Caudate nucleus
Internal capsule
Putamen
Nucleus accumbens
Corpus callosum (rostrum)
Olfactory tubercle
Optic nerve
Septum
A
B
External capsule
Claustrum
Extreme capsule

B

Cingulate gyrus
Lateral ventricle
Caudate nucleus
Internal capsule
External capsule
Claustrum
Extreme capsule
Insula
Globus pallidus, medial
Globus pallidus, lateral
Putamen
Nucleus basalis
Amygdaloid nuclear complex
Anterior commissure
Optic chiasm
Third ventricle
Diagonal band
A
B
Stria terminalis
Septum

FIGURE C.9.

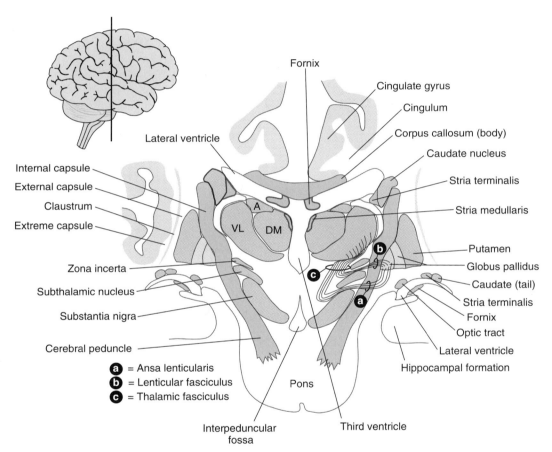

Fornix

Cingulate gyrus

Cingulum

Corpus callosum (body)

Caudate nucleus

Lateral ventricle

Stria terminalis

Internal capsule

Stria medullaris

External capsule

Claustrum

A

Extreme capsule

VL   DM

Putamen

Globus pallidus

Zona incerta

Caudate (tail)

Subthalamic nucleus

Stria terminalis

Substantia nigra

Fornix

Optic tract

Cerebral peduncle

Lateral ventricle

ⓐ = Ansa lenticularis

Hippocampal formation

ⓑ = Lenticular fasciculus

ⓒ = Thalamic fasciculus

Pons

Interpeduncular
fossa

Third ventricle

FIGURE C.10.

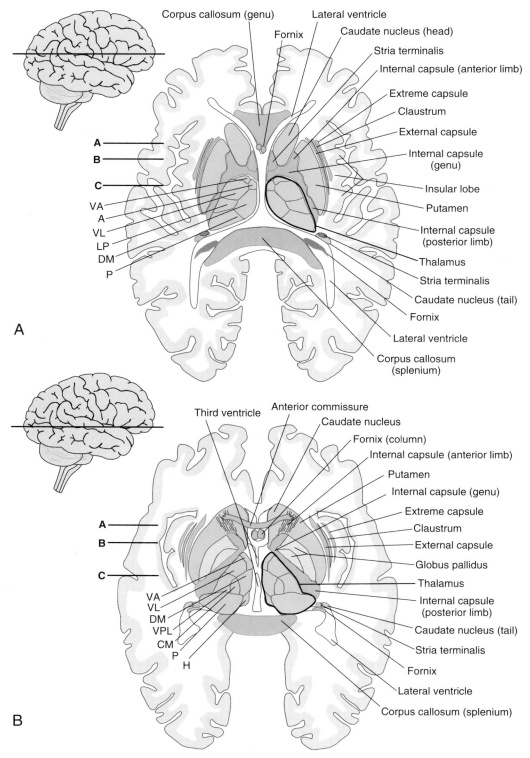

A

Corpus callosum (genu)
Fornix
Lateral ventricle
Caudate nucleus (head)
Stria terminalis
Internal capsule (anterior limb)
Extreme capsule
Claustrum
External capsule
Internal capsule (genu)
Insular lobe
Putamen
Internal capsule (posterior limb)
Thalamus
Stria terminalis
Caudate nucleus (tail)
Fornix
Lateral ventricle
Corpus callosum (splenium)

A
B
C
VA
A
VL
LP
DM
P

B

Third ventricle
Anterior commissure
Caudate nucleus
Fornix (column)
Internal capsule (anterior limb)
Putamen
Internal capsule (genu)
Extreme capsule
Claustrum
External capsule
Globus pallidus
Thalamus
Internal capsule (posterior limb)
Caudate nucleus (tail)
Stria terminalis
Fornix
Lateral ventricle
Corpus callosum (splenium)

A
B
C
VA
VL
DM
VPL
CM
P
H

FIGURE C.11.

GLOSSARY

The selection of terms included in this glossary is necessarily arbitrary. I have skewed the selection toward medical terms that may be unfamiliar or that have a specific medical connotation that is different from the general use of the term. I have not included anatomical words unless they are common synonyms for the currently accepted term. Almost all of the items in this glossary appear in boldface type in the text. Reference to the chapter(s) in which they appear are provided. For those whose interest in medical terminology extends beyond the scope of this glossary, the resources listed at the end are recommended.

abulia [G. *a* + G. *boule,* without will] (Chapter 13)— Lack of willpower. Refers to a condition associated with frontal lobe lesions in which the affected individual is apathetic and slowed in speech and other motor actions. The profoundly affected may lie motionless for weeks.

accommodation (Chapter 9)—In ophthalmology, refers to the ability of the lens to change its shape to allow a nearby object to be brought into focus on the retina.

action potential (Chapters 3, 4)—The rapid reversal of the neuron's membrane potential and its subsequent recovery. The action potential is a signal that conveys information within the nervous system.

adequate stimulus (Chapter 4)—The notion that a stimulus of a unique energy form is required to activate a specific receptor modified especially to receive that form of energy. In fact, receptors can be activated by a number of different energy forms but usually have the lowest threshold for only one type of energy.

Aesculapius—The romanized version of Asclepios, the Greek god of medicine. The staff of Aesculapius—a stick, entwined by a single snake—is the symbol of the medical profession. *See* caduceus.

afferent [L. *af-ferens,* to bring to] (Chapters 1, 9)—In reference to the central nervous system, afferent signals are those that *approach* the CNS. Afferent fibers are axons that convey information to the CNS. *See* efferent.

alar plate—Region of the embryonic nervous system, dorsal to the sulcus limitans, from which sensory neurons intrinsic to the CNS develop.

AM—*See* amplitude modulated.

aminoglycoside antibiotics (Chapters 9, 10)—A class of antibiotic that interferes with protein synthesis in prokaryotic cells.

amplitude modulated (AM) (Chapter 4)—A type of information coding. Amplitude-modulated signals convey information by coding the intensity of the carrier signal in proportion to the information signal. *See* frequency modulated; analog; digital.

analog [G. *analogos,* according to ratio, proportionate] (Chapter 4)—With reference to signaling systems, analog signals vary in proportion to the energy content of the stimulus. *See* digital; amplitude modulated; frequency modulated.

aneurysm [G. *aneurysma,* wide] (Chapters 1, 7)—A weakness in the arterial wall that causes the artery to widen. A dissecting aneurysm is one in which the layers of the artery separate. A berry aneurysm is a small, balloon-like dilatation in the artery that communicates with the main branch by means of a narrow passage, making the aneurysm look like a berry during angiography.

anterograde [L. *antero,* in front + L. *gradior,* to go] (Chapter 2)—To advance in the forward direction. *See* orthograde; retrograde.

aphasia (Chapter 13)—Inability to communicate by language, in any of its forms—reading, writing, speaking, listening or signing—due to loss of cerebral function independent of motor or sensory losses.

arteriovenous malformation (AVM) (Chapters 1, 7)—An inappropriate anastomosis between an artery and vein without an intervening capillary bed.

atavism [L. *atavus,* a remote ancestor] (Appendix B)—Literally, the appearance of disease that can be attributable to a remote ancestor. In neurology, this refers to certain reflexes that are present in infancy but not thereafter unless there is deterioration of cerebral cortical function.

ataxia [G. *a,* without, + G. *taxis,* order] (Chapter 8, Appendix B)—In neurology, a lack of coordination due to diminution or loss of cerebellar function. Most commonly seen in inebriated individuals.

atherothrombotic infarction (Chapter 7)—An infarct caused by an arterial thrombosis. *See* infarct; thrombosis.

athetosis [G. *athetos,* without position or place] (Chapter 8)—A type of involuntary movement characterized by continuous, slow, writhing motor activity in one or more parts of the body.

atrophy [G. *a*, without + G. *trophe*, nourishment] (Appendix B)—In neurology, a wasting away of muscle mass. Atrophy of disuse follows immobilization and is relatively minor. Denervation atrophy is profound and is an important clinical sign of lower motor neuron disease.

AVM—*See* arteriovenous malformation.

Babinski sign (Chapter 7)—The reflex extension and flaring of the toes in response to scraping the lateral margin of the sole of the foot. A sign of upper motor neuron disease. The normal response is to flex the toes.

BAER—*See* brainstem auditory evoked response.

basal plate (Appendix A)—Portion of the embryonic nervous system, ventral to the sulcus limitans, from which efferent neurons develop. *See* alar plate.

Beevor's sign (Chapter 6)—Rostral movement of the umbilicus during contraction of the rectus abdominis muscles when there is paresis or paralysis of the caudal portions of this muscle. In normal individuals there is no movement of the umbilicus.

bilaminar disc (Appendix A)—In embryology, refers to the embryoblast during the early phase of gastrulation when it consists of two layers, the epiblast and the hypoblast.

Bing sign (Chapter 7)—The reflex flexion of the foot in response to stabbing a pin into the dorsum of the foot. A sign of upper motor neuron disease. The normal response is to extend the foot away from the pin.

brachium conjunctivum (Chapters 1, 5, 8)—Synonym for the superior cerebellar peduncle.

brachium pontis (Chapters 1, 5, 8)—Synonym for the middle cerebellar peduncle.

brainstem auditory evoked response (BAER) (Chapter 10)—A sensory evoked potential elicited by auditory stimuli. The response provides information about the integrity of the auditory brainstem nuclei.

brainstem coma (Chapter 13)—Coma induced by a lesion in the brainstem that deprives the cerebral cortex of essential activation by subcortical structures.

Brown-Séquard syndrome (Chapters 5, 7)—The characteristic set of signs and symptoms produced by hemisection of the spinal cord.

caduceus [L. *caduceus*, the staff of Mercury]—A staff entwined by two snakes and with two wings at the top. It is the symbol of the U.S. Army Medical Corps. *See* Aesculapius.

carpal tunnel syndrome (Chapter 5)—Paresthesias in the hand, particularly within the distribution of the median nerve, caused by compression of that nerve by the flexor retinaculum.

CBS tract—*See* corticobulbospinal tract.

central hearing loss (Chapter 10)—Hearing loss due to damage to retrocochlear structures. *See* conductive hearing loss; sensorineural hearing loss.

cerebral death (Chapter 13)—Usually is used to describe death of the cerebral cortex while lower areas of the CNS, particularly brainstem structures, remain viable.

cerebral embolism (Chapter 7)—The passage of a blood clot from a remote structure such as the heart into the vasculature of the brain, causing infarction.

cerebral hemorrhage (Chapter 7)—Arterial bleeding into cerebral tissue or the subarachnoid space. Not to be confused with a contusion or bleeding into the epidural space.

chorea [G. *choros*, a dance] (Chapter 8)—An involuntary movement that appears to be fragments of purposeful movement.

chromatolysis [G. *chroma*, color + G. *lysis*, dissolution] (Chapter 6)—The dispersion of the Nissl substance (rough endoplasmic reticulum) that occurs following the separation of the axon from the soma.

circumventricular organs (CVOs) (Chapter 1)—Small areas within the brain that contain modified capillaries that allow blood-borne signals (hormones) to reach and affect neurons. These brain structures lie near the midline and at the ventricular surface, giving them their generic name.

Clarke's column (Chapter 8)—Synonym for nucleus dorsalis of the spinal cord.

clasp knife reflex (Chapters 6, 7)—This reflex is evoked by stimulating the Golgi tendon organs (GTOs), which in turn cause inhibition of the homonymous motor neuron pools.

clonus [G. *klonos*, a tumult] (Chapter 7)—The rapid (5- to 10-Hz) beating of an extremity such as a hand or foot, when one of the muscles controlling it is placed under rapid and sustained tension. Clonus occurs only after injury to the CBS tract innervating the motor neuron pool controlling the muscle. *See* spasticity.

cogwheel rigidity (Chapter 8)—Patients with certain basal ganglion diseases, particularly parkinsonism, display a rigidity to passive movement that is not always constant, but may be released in small steps, giving the impression of a cogged wheel being turned. *See* lead pipe rigidity; spasticity.

coma (Chapter 13)—State of unconsciousness from which the patient cannot be aroused by sensory stimuli, but cerebral cortical activity remains.

concussion [L. *concussio*, to shake violently] (Chapter 13)—State of brief cerebral cortical dysfunction resulting from a violent blow to the head.

conductive hearing loss—Hearing loss caused by the inability of sound pressure waves to reach the cochlea. *See* central hearing loss; sensorineural hearing loss.

confusion (Chapter 13)—In medicine, the state of decreased cerebral cortical function in which the patient is not entirely oriented to time and place, even though he is conscious and aware.

contusion [L. *contusio*, a bruising] (Chapter 13)—A bruise of the brain tissue resulting from a severe blow to the head, but that has not caused cerebral hemorrhage.

corona radiata (Chapters 1, 5)—The portion of the cerebral white matter that lies superior to the internal capsule and radiates, like a crown, into the cerebral cortex.

corticobulbospinal tract (CBS) (Chapter 7)—The

motor tract that originates in the cerebral cortex and terminates on the brainstem and spinal interneurons and motor neurons affecting voluntary movement.

decussation [L. *decusso,* to make in the form of an X] (Chapter 5)—In neuroanatomy, any paired bundle of axons that cross the midline.

déjà vu (Chapter 13)—Literally, seen before. In neurology, refers to an epileptic aura characterized by a sensation that one's surroundings are unusually familiar or have been seen previously, even if one knows objectively that the familiarity is false.

dementia (Chapters 8, 13)—The loss of cognitive capabilities without objective loss of sensory or motor functions.

digital—With reference to signaling systems, digital signals code the stimulus in the time domain; their energy content remains constant. *See* analog; frequency modulation.

doll's eye maneuver (Chapter 9)—While the unconscious patient is lying in the supine position, the head is rotated. Normally, the eyes will rotate in the opposite direction, giving the impression that gaze is fixed straight ahead, much like the weighted glass eyes in a china doll.

dysarthria (Chapters 3, 13)—The inability to articulate speech due to lack of coordination, spasticity, or weakness of the muscles of speech.

dysdiadochocinesia [G. *dys,* difficult; + G. *diadochos,* working in turn; + G. *kinesis,* movement] (also spelled **dysdiadochocinesia**) (Chapter 8)—The inability to perform smoothly, rapid alternating movements requiring the coordination of antagonistic muscle groups.

dyskinesia [G. *dys,* bad + G. *kinesis,* movement] (Chapter 8)—A general term describing any difficulty in performing voluntary motor activity.

dystonia [G. *dys,* bad + G. *tonos,* tension] (Chapter 8)—A condition in which a set of antagonistic muscles are involuntarily kept in prolonged isometric contraction.

Eaton-Lambert syndrome (Chapter 3)—An autoimmune disease affecting the voltage-gated Ca^{2+} channels, characterized by motor weakness and an incrementing electromyelographic response to repetitive stimulation.

efferent [L. *ef-ferens,* to bring out] (Chapters 1, 9, Appendix A)—In reference to the central nervous system, efferent refers to signals that *evade* (i.e., leave) the CNS. Efferent fibers are axons that convey information away from the CNS. *See* afferent.

embolism [G. *embolos,* a wedge or stopper] (Chapter 7)—A detached blood clot (or other foreign object) that has occluded an artery.

endogenous opioids (Chapter 5)—Neuroactive peptides produced by the body that bind to the same receptors in the brain that bind morphine.

epilepsy (Chapter 13)—A chronic disorder of the central nervous system characterized by seizures that occur periodically.

EPSP—*See* excitatory postsynaptic potential.

excitatory postsynaptic potential (EPSP) (Chapter 3)—A hypopolarizing postsynaptic potential.

fasciculations [L. *fascis,* bundle] (Chapter 6, Appendix B)—An involuntary twitching of a motor unit, frequently but not always the result of muscle denervation.

Fourier analysis—The mathematical analysis of an arbitrarily complex waveform from which a series of pure sinusoidal waveforms are extracted, the sum of which reconstructs the original.

FRA (Chapters 6, 7, Appendix B)—Flexion reflex afferent; receptors that elicit a polysynaptic reflex facilitating flexor motor neuron pools.

frequency modulated (FM) (Chapter 4)—A type of information coding. Frequency-modulated signals convey information by varying the timing of elements of the carrier signal in proportion to the information signal. *See* amplitude modulated; analog; digital.

FM—*See* frequency modulated.

Gilles de la Tourette syndrome (Chapter 8)—A neurological disorder characterized by involuntary tics and repetitive compulsive behavior, often including involuntary cursing (coprolalia). Thought to be a hereditary disorder carried by an autosomal dominant gene with variable penetrance, it is associated with abnormally low levels of dynorphin and defective dopamine uptake mechanisms.

glabrous skin [L. *glaber,* smooth] (Chapter 4)—Smooth, hairless skin.

Guillain-Barré syndrome (Chapter 2) (Also known as Landry-Guillain-Barré syndrome, acute inflammatory polyneuropathy, acute autoimmune neuropathy, and postinfectious polyneuritis)—An autoimmune disease directed against peripheral myelin. *See* multiple sclerosis.

hemiplegia—Paralysis of half of the body divided along the midsagittal plane.

Horner's syndrome (Chapter 12)—Classically, the unilateral constellation of three symptoms—miosis, ptosis, and enophthalmos—resulting from sympathectomy. Lack of sweating over the affected side of the face is an additional valuable clinical sign.

hypotonia (Chapter 8)—Decrease in resting muscle tone.

hypoxia (Chapter 13)—Decrease in arterial O_2 saturation.

hysteresis [G. *hysteresis,* a coming later] (Chapter 7)—A delay in effect on the application of a force.

idiopathic [G. *idios,* individual. + G. *pathos,* suffering] (Chapter 6)—Without known cause.

infarct [L. *in-farcio,* pp. *-fartus,* to stuff into] (Chapter 7)—Death of tissue due to loss of vascular perfusion. The word *infarction* is commonly misused to refer to an atherothrombotic infarction. *See* atherothrombotic infarction.

inhibitory postsynaptic potential (IPSP) (Chapter 3)—An alteration in the membrane permeability that buffers the membrane potential of neurons against subsequent hypopolarization. Generally, but not always, a hyperpolarizing postsynaptic potential.

IPSP—*See* inhibitory postsynaptic potential.

isometric (Chapter 6)—In physiology, the contraction of a muscle held at constant length.

isotonic (Chapter 6)—In physiology, the contraction of a muscle under constant tension.

jamais vu (Chapter 13)—Literally, never seen before. In neurology, refers to an epileptic aura characterized by a sensation that one's surroundings are unfamiliar or have never been seen previously, even if one objectively understands that the sensation is false.

keratoconus [G. *keras,* horn + G. *konos,* cone] (Chapter 11)—Deformation of the normally spherical cornea into a cone shape due to thinning of the cornea at the margin.

Korsakoff amnesic state (Chapter 12)—A type of anterograde amnesia characterized by the inability to consolidate short-term memory into long-term memory. Correlated with specific damage to the CA region of the hippocampal formation.

lateral corticospinal tract (Chapter 7)—A subset of the CBS tract that is located in the lateral white columns of the spinal cord.

lead pipe rigidity (Chapter 8)—Patients with certain basal ganglion diseases, particularly parkinsonism, display a rigidity to passive movement. This rigidity gives slowly to constant pressure, giving the impression of bending a lead pipe. *See* cogwheel rigidity; spasticity.

leukodystrophy (Chapter 2)—A generic term describing any degeneration of the white matter of the brain.

Lhermitte's sign (Chapter 6)—A sensation of electric shocks descending the spine upon flexion of the neck. A sign of meningeal irritation.

ligand-gated ion channels (Chapter 3)—A membrane protein that adjusts its permeability to one or more ions when a specific ligand is bound to a specific receptor on the protein.

line of Gennari (Chapter 1)—An unusually prominent band of white matter in layer IV of the primary visual cortex (Brodmann area 17) that is visible to the naked eye and gives this area its name, the striate cortex.

Lissauer's tract (Chapter 5)—Synonym for the dorsolateral tract.

Marcus Gunn pupil (Appendix B)—In neurology, this term applies to a pupil that paradoxically dilates when light is shone into the eye.

Ménière disease (Chapter 9)—A neurological disorder characterized by extreme vertigo due to inappropriate unilateral stimulation of the labyrinth.

metabolic coma (Chapter 13)—Coma of chemical rather than structural origin, such as hypoxia, the presence of hypnotic drugs, or hypoglycemia.

miosis (Chapter 12)—Constriction of the pupil.

monoplegic [G. *monos,* single, + G. *plege,* stroke] (Chapter 7)—Paralysis of a single extremity.

monosynaptic reflex (Chapter 6)—Synonym for the muscle stretch reflex.

MSR—*See* muscle stretch reflex.

multiple sclerosis (MS) (Chapters 2, 7, 11)—An autoimmune disease directed against central myelin. *See* Guillain-Barré syndrome.

muscle stretch reflex (MSR) (Chapter 6, Appendix B)—A reflex initiated by the rapid stretching of a muscle that stimulates the primary receptors of the muscle spindle apparatus and results in a facilitation of the motor neuron pools serving the homonymous muscle.

myasthenia gravis (Chapter 6)—An autoimmune disease directed against the ACh receptors in the motor endplate of striated muscle.

myelogram (Chapter 1)—A radiographic procedure in which the subarachnoid space of the spinal cord is filled with a radiopaque dye. Subsequent spinal x-rays reveal the outline of the dural sack.

neurofibrillary bodies (Chapter 13)—Collections of tubules found within neurons of patients suffering from Alzheimer's disease. These tubules are similar to but not identical with the microtubules that are normally found in neurons and other cells.

Nissl substance (Chapter 2)—The historical name given to the rough endoplasmic reticulum of neurons.

nociceptors [L. *noceo,* to injure + L. *capio,* to take] (Chapter 4)—Receptors that are most sensitive to noxious stimuli, including tissue-killing stimuli such as heat and the byproducts of tissue destruction. These receptors should not be called pain receptors because pain is a perception, not a stimulus.

nystagmus [G. *nystagmos,* a nodding] (Chapter 8, 9)—The rhythmic oscillation of the eyes, the oscillation being fast in one direction and slow in the opposite direction. The nystagmus is named after the direction of the fast phase.

operculum [L. *operculum,* a cover or lid] (Chapter 1)—The part of the cerebral cortex that covers the insula, thus obscuring and forming a cover over it.

optokinetic nystagmus (OKN) (Chapter 9)—Nystagmus produced by a series of images passing before the eyes.

orthograde [G. *orthos,* correct, straight + L. *gressus,* to walk]—Moving in the normal direction. In neuroscience, moving along an axon beginning at the soma, as in orthograde conduction. *See* anterograde; retrograde.

ototoxic (Chapter 10)—Toxic to the hair cells of the vestibular apparatus or cochlea.

papilledema (Chapter 11, Appendix B)—Edema of the papilla, the point of entry of the optic nerve into the eye. This is caused by an increase in intracranial pressure. Not to be confused with papillitis.

papillitis (Chapter 11)—Inflammation of the head of the optic nerve where it enters the eye, usually caused by an exacerbation of multiple sclerosis. Not to be confused with papilledema.

paraplegia [G. *para,* beside + G. *plege,* stroke] (Chapter 7)—Paralysis of the two lower extremities.

paresis [G. *paritemi,* to let go, slackening] (Chapter 7)—A weakness or partial paralysis.

Parkinson's disease (Chapter 8)—The disease described by James Parkinson, a progressive neurological disorder characterized by bradykinesia, a 4- to 7-Hz tremor of rest, and lead-pipe or cogwheel rigidity.

parkinsonism (Chapter 8) — A generic term used to describe all diseases of the basal ganglia that have the common features of resting tremor, bradykinesia, and rigidity. This includes Parkinson's disease as well as a number of other significantly different illnesses such as supraoptic ophthalmoplegia.

paroxysmal — The sharp or sudden onset and recurrence of the manifestations of disease.

Pelizaeus-Merzbacher disease (Chapter 2) — A neurological disease characterized by lack of development of CNS myelin with normal peripheral myelin, due to a genetic error that prevents the correct assembly of PLP.

peritrichal receptor [G. *peri*, around + G. *thrix*, hair] (Chapter 4) — Sensory receptor located at the base of the hair shafts and sensitive to their movement.

perseveration (Chapter 13) — In neurology, the repeated and inappropriate naming of each object in a series by the name given to the first object, even though they are all different.

plegia [G. *plege*, stroke] — Profound weakness or paralysis.

poikilothermal [G. *poikilos*, varied + G. *therme*, heat] (Chapter 12) — Varying temperature in accordance with the environment.

poliomyelitis [G. *polios*, gray + G. *myelos*, marrow + G. *itis*, inflammation] (Chapter 6) — Literally, an inflammation of the gray matter of the spinal cord. A viral disease that, among other things, kills α motor neurons, particularly those of the lumbosacral spinal cord.

postsynaptic potential (PSP) (Chapter 3) — An alteration in membrane potential produced by electrochemical events originating at synapses terminating on that neuron.

presynaptic inhibition (Chapter 3) — An electrochemical alteration in the bouton membrane that reduces the efficiency of transmitter release from that bouton, thereby making its effect less pronounced on the postsynaptic neuron.

presbyacusis [G. *presby*, old + G. *akouo*, to hear] (Chapter 10) — The decrease in high-frequency sensitivity to sound that occurs with the advancement of age.

presbyopia [G. *presbys*, old + G. *ops*, eye] (Chapter 11) — The decreased elasticity of the lens that results in the inability to bring near objects into focus, a process that occurs with advancing age.

primitive reflexes — *See* atavism.

ptosis (Chapters 3, 12) — Drooping of the eyelid.

quadriplegia [L. *quadri-*, four + G. *plege*, stroke] (Chapter 7) — Paralysis of all four extremities.

radicula [L. dim. of *radix*, root] — A spinal root.

Raynaud's syndrome (Chapter 12) — Paroxysms of pain and cyanosis in the distal portions of the extremities, brought on by emotion or cold. Commonly associated with sympathectomy of the extremities, as in peripheral neuropathy or peripheral nerve compression.

restiform body (inferior cerebellar peduncle) (Chapters 5, 8) — Synonym for the inferior cerebellar peduncle.

retrograde [L. *retro*, in back + L. *gradior*, to go] (Chapter 2) — To advance in the backward direction. *See* anterograde; orthograde.

satellite cells (Chapter 2) — Supporting cells that surround the neuron soma in ganglia. As these cells are now recognized to be identical to Schwann cells, this term is falling out of favor.

sensorineural hearing loss — Hearing loss due to damage to cochlear structures or the auditory nerve. *See* conductive hearing loss; central hearing loss.

sign (Appendix B) — In medicine, an abnormality observed by the physician and independent of the observation of the patient. *See* symptom.

sound pressure level (SPL) (Chapter 10) — Sound pressure of 0.0002 dyne/cm^2 used to establish a 0-dB reference for audiological testing.

spasticity — An increase in muscle tone resulting from an increased sensitivity of the muscle stretch reflex (MSR). This is the cardinal sign of an upper motor neuron lesion. *See* cogwheel rigidity; lead pipe rigidity.

spina bifida [L. *bifidus*, cleft] (Appendix A) — The generic term used to describe any of a set of developmental disorders in which the structures dorsal to or derived from the neural tube do not close during development.

spinal disk syndrome (Chapter 6) — The constellation of signs of unilateral (rarely bilateral) radicular pain and weakness of muscles innervated by the same spinal root, caused by compression of the spinal nerve by a prolapsed spinal disk.

strabismus [G. *strabismos*, a squinting] (Chapter 11) — Lack of parallel tracking of the two eyes, resulting in diplopia.

stroke — The rapid loss of neural function that can be explained by infarction of a portion of the brain located within the territory of a single artery, and that does not resolve within 24 hours.

stupor (Chapter 13) — A state of unconsciousness from which the patient can be aroused, but not brought to full awareness.

subiculum [L. *subex*, a layer] (Chapter 12) — An area of transitional cerebral cortex between the hippocampus proper and the entorhinal cortex.

symptom (Appendix B) — In medicine, an abnormality observed by the patient and reported to the physician. Symptoms are necessarily subjective. *See* sign.

thrombosis [G. *thrombosis*, a curdling] (Chapter 7) — The formation of a clot, particularly a blood clot.

thrombotic stroke (Chapter 7) — The occlusion of a cerebral artery by the formation of a thrombus within that artery as opposed to an occlusion by an embolism.

TIA — *See* transient ischemic attack.

tic (Chapter 8) — An involuntary rapid and repeated contraction of a small group of muscles that results in spasmodic movement.

tic douloureux [Fr. *tic*, spasmodic movement + Fr.

douloureux, painful] (Chapter 9)—Synonym for trigeminal neuralgia.

tinnitus [L. *tinnio,* a jingling] (Chapter 10)—The non-hallucinatory perception of sound where none exists.

transcutaneous electrical neural stimulation (TENS) (Chapter 5)—Stimulation of a nerve by passing electrical current noninvasively through the skin.

transient ischemic attack (TIA) (Chapter 7)—The rapid loss of neural function that can be attributable to dysfunction of a portion of the brain located within the territory of a single artery, and that resolves within 24 hours.

vestibulo-ocular reflex (VOR) (Chapter 9)—The reflex movement of the eyes in the opposite direction to the movement of the head. *See* doll's eye maneuver.

VOR—*See* vestibulo-ocular reflex.

Waardenburg's syndrome (Chapter 10)—The association of abnormal pigmentation and congenital deafness and or retinitis pigmentosa.

wallerian degeneration (Chapter 6)—Synonym for orthograde axonal degeneration.

ADDITIONAL RESOURCES

Stedman's Medical Dictionary, 26th ed. Baltimore: Williams & Wilkins, 1995.

Dorland's Illustrated Medical Dictionary, 28th ed. Philadelphia: W. B. Saunders, 1994.

Lockard, I. *Desk Reference for Neuroscience.* New York: Springer-Verlag, 1992.

FIGURE AND TABLE CREDITS

Cover design based on an original figure by Caryl Erickson.
Inside covers from Hayman, L. A., Berman, S. A., and Hinck, V. C. *Correlation of Cerebral Vascular Territories with Cerebral Function by Computed Tomography.*
Reprinted courtesy of Eastman Kodak Company, © Eastman Kodak Co.
Etymologies from *Stedman's Medical Dictionary*, 26th ed. Baltimore: Williams & Wilkins, 1995.

Figure 1.2. From Mettler, F. A. *Neuroanatomy,* 2nd ed. St. Louis: C.V. Mosby, 1948.

Figure 1.3. From Mettler, F. A. *Neuroanatomy,* 2nd ed. St. Louis: C.V. Mosby, 1948.

Figure 1.5. From Mettler, F. A. *Neuroanatomy,* 2nd ed. St. Louis: C.V. Mosby, 1948.

Figure 1.6. From Mettler, F. A. *Neuroanatomy,* 2nd ed. St. Louis: C.V. Mosby, 1948.

Figure 1.7. After Brodmann, K. *Vergleichende Lokalisation lehre der Grosshirnrinde in ihren Prinzipien dargestellt auf Grund des Zellenbaues.* Leipzig: J. A. Barth, 1909.

Figure 1.9. From Mettler, F. A. *Neuroanatomy,* 2nd ed. St. Louis: C.V. Mosby, 1948.

Figure 1.13. From Lewis, *Gray's Anatomy.* Philadelphia: Lea & Febiger, 1924.

Figure 1.21. From Mettler, F. A. *Neuroanatomy,* 2nd ed. St. Louis: C.V. Mosby, 1948.

Figure 1.28. Magnetic resonance image courtesy of the Magnetic Resonance Imaging Center, South Bend, IN.

Figure 1.30. After Bailey, *Intracranial Tumors,* 2nd ed. Springfield, IL: Charles C Thomas, 1948.

Figure 1.33. From Shaver et al. Progress in Brain Research. *J. Comp. Neurol.* 306:1981.

Figure 1.34. From Shaver et al. Progress in Brain Research. *J. Comp. Neurol.* 293:1990.

Figure 2.1. From Peters, A., Palay, S. L., Webster, H. D. *The Fine Structure of the Nervous System: Neurons and Their Supporting Cells,* 3rd ed. New York: Oxford University Press, 1991.

Figure 2.9. From Peters, A., Palay, S. L., Webster, H. D. *The Fine Structure of the Nervous System: Neurons and Their Supporting Cells,* 3rd ed. New York: Oxford University Press, 1990.

Figure 2.10. Courtesy of Dr. Pasko Rakic.

Figure 2.12. From Brown, A. G. *Organization of the Spinal Cord.* New York: Springer-Verlag, 1991.

Figure 2.13. From Peters, A., Palay, S. L., Webster, H. D. Synapses. In: *The Fine Structure of the Nervous System: Neurons and Their Supporting Cells,* 3rd ed. New York: Oxford University Press, 1990.

Figure 2.16. From Fawcett, D., and Raviola, E. *A Textbook of Histology,* 12th ed. New York: Chapman and Hall, 1994.

Figure 2.17. From Peters, A., Palay, S. L., Webster, H. D. *The Fine Structure of the Nervous System: Neurons and Their Supporting Cells,* 3rd ed. New York: Oxford University Press, 1990.

Figure 2.18. From Peters, A., Palay, S. L., Webster, H. D. *The Fine Structure of the Nervous System: Neurons and Their Supporting Cells,* 3rd ed. New York: Oxford University Press, 1990.

Figure 2.19. From Peters, A., Palay, S. L., Webster, H. D. *The Fine Structure of the Nervous System: Neurons and Their Supporting Cells,* 3rd ed. New York: Oxford University Press, 1990.

Figure 2.20. From Rakic, P. *J. Comp. Neurol.* 145: 61–84, 1972.

Figure 2.21. From Peters, A., Palay, S. L., Webster, H. D. *The Fine Structure of the Nervous System: Neurons and Their Supporting Cells,* 3rd ed. New York: Oxford University Press, 1990.

Figure 2.22. Courtesy of Dr. D. Kent Morest.

Figure 2.24. From Coggeshall, R. E. A fine structural analysis of the myelin sheath in rat spinal roots. *Anat Rec* 194: 1979.

Figure 2.26. From Peters, A., Palay, S. L., Webster, H. D. *The Fine Structure of the Nervous System: Neurons and Their Supporting Cells,* 3rd ed. New York: Oxford University Press, 1990.

Figure 2.27. After Hall, Z. *An Introduction to Molecular Neurobiology.* Sunderland, MA: Sinauer Associates, 1992.

Figure 2.28. After Hall, Z. *An Introduction to Mole-*

cular Neurobiology. Sunderland, MA: Sinauer Associates, 1992.

Figure 3.14A and **B.** Patch clamp data kindly supplied by Dr. David Colquhoun.

Figure 3.14C. After Hall, Z. *An Introduction to Molecular Biology.* Sunderland, MA: Sinauer Associates, 1972, Fig. 5, p. 93.

Figure 3.16. After Shepherd, G. *The Synaptic Organization of the Brain.* New York: Oxford University Press, 1990.

Figure 3.17. After Shepherd, G. *The Synaptic Organization of the Brain.* New York: Oxford University Press, 1990.

Figure 3.19. After data in Rall, W. In: *Neural Theory and Modelling,* edited by R. F. Reiss. Stanford, CA: Stanford University Press, 1964.

Figure 3.25. After data in Harvey, A. M., et al. Observations on the nature of myasthenia gravis: the phenomena of facilitation and depression of neuromuscular transmission. *Bull. Johns Hopkins Hosp.* 69:547–565, 1941.

Table 3.1. After data in Smith, C. U. M. *Elements of Molecular Neurobiology.* New York: John Wiley & Sons, 1989; and Partridge, L., and Partridge, L. D. *The Nervous System: Its Function and Its Interaction with the World.* Cambridge, MA: The MIT Press, 1993.

Figure 5.12. Based on data in Brown, A. G. *Organization in the Spinal Cord: The Anatomy and Physiology of Identified Neurons.* Berlin: Springer-Verlag, 1981.

Figure 5.15. After Penfield, W., and Rasmussen, T. *The Cerebral Cortex of Man: A Clinical Study of Localization of Function.* New York: Macmillan, 1950.

Figure 5.19. From Curtis, B., et al. *An Introduction to the Neurosciences.* Philadelphia: W. B. Saunders, 1972.

Figure 5.24. Magnetic resonance image courtesy of the Magnetic Resonance Imaging Center, South Bend, IN.

Figure 6.4. From Crowe, A., and Matthews, P. B. C. Further studies of static and dynamic fusimotor fibres. *J. Physiol. (Lond.)* 174:132–151, 1964.

Figure 6.6. Based on data in Brown, A. G. *Organization in the Spinal Cord: The Anatomy and Physiology of Identified Neurons.* Berlin: Springer-Verlag, 1981.

Figure 6.9. From Hunt, C. C., and Kuffler, S. W. Further study of efferent small-nerve fibres to mammalian muscle spindles. Multiple spindle innervation and activity during contraction. *J. Physiol. (Lond.)* 113:283–297, 1951.

Figure 6.11. After Monster, A. W., and Chan, H. Isometric force production by motor units of extensor digitorum communis muscle in man. *J. Neurophysiol.* 40:1432–1443, 1977.

Figure 6.14. Magnetic resonance image courtesy of the Magnetic Resonance Imaging Center, South Bend, IN.

Figure 7.3. After Penfield, W., and Rasmussen, T. *The Cerebral Cortex of Man: A Clinical Study of Localization of Function.* New York: Macmillan, 1950.

Figure 7.9. From Weinrich, M., and Wise, S. P. The premotor cortex of the monkey. *J. Neurosci. 2:* 1329–1345, 1982.

Figure 7.10. After data in Roland, P. E., et al. Supplementary motor area and other cortical areas in organization of voluntary movements in man. *J. Neurophysiol.* 43:118–136, 1980.

Figure 7.11. From Curtis, B. A., et al. *An Introduction to the Neurosciences.* Philadelphia: W. B. Saunders, 1972.

Figure 7.12. From Georgopoulos, P., et al. On the relations between the direction of two-dimensional arm movements and cell discharge in primate motor cortex. *J. Neurosci.* 2:1527–1537, 1982.

Figure 7.15. From Denny-Brown, D. *Proc. R. Soc. Lond.* B104:252–301, 1929.

Figure 7.22. From Rowland, L. P. *Merritt's Textbook of Neurology,* 8th ed. Philadelphia: Lea & Febiger, 1989, Fig. 123-1.

Figure 8.10. After Albin et al. *Trends Neurosci.* 12: 1989.

Figure 8.11. Magnetic resonance image courtesy of the Magnetic Resonance Imaging Center, South Bend, IN.

Figure 8.12. After Albin et. al. *Trends Neurosci.* 12: 1989.

Figure 8.13. Based on Hallett, M., and Khoshbin, S. A physiological mechanism of bradykinesia. *Brain* 103:301–314, 1980.

Figure 8.14. After data in Lee et al. In: Desmedt, ed, *Adv. Neurol. 39:* 1983.

Figure 8.15. From Rowland, *Merritt's Textbook of Neurology,* 8th ed. Philadelphia: Lea & Febiger, p. 659, Fig. 118-1.

Figure 8.16. After data in Albin et al. *Trends Neurosci.* 12: 1989.

Figure 8.17. After data in Rowland, *Merritt's Textbook of Neurology,* 8th ed. Philadelphia: Lea & Febiger, 1989.

Figure 8.19. From Thatch, W. T. Coordination and Learning. *Annu. Rev. Neurosci.* 15:403–442, 1992.

Figure 8.25. From Thach, W. T. Coordination and Learning. *Annu. Rev. Neurosci.* 15:403–442, 1992.

Figure 8.26. From Eccles, J. C., Ito, M., and Szentagothai, J. *The Cerebellum as a Neuronal Machine.* New York: Springer-Verlag, 1967.

Figure 9.5A. From Fawcett, D., *A Textbook of Histology.* 12th ed. New York: Chapman and Hall, 1994.

Figure 9.5B. From Friedman, I., and Ballantyne., J. *Ultrastructural Atlas of the Inner Ear.* London: Butterworth-Heinemann, 1984.

Figure 9.6A. From Iurato, S. *Submicroscopic Structure of the Inner Ear.* New York: Pergamon Press, 1967.

Figure 9.6B. From Paparella, M., and Shumrick, D., eds. *Textbook of Otolaryngology,* vol. 11. Philadelphia: W. B. Saunders, 1991.

Figure 9.7. Original micrographs courtesy of Dr. C. Gary Wright.

Figure 9.8. From Iurato, S. *Submicroscopic Structure of the Inner Ear.* New York: Pergamon Press, 1967.

Figure 9.9. From Kandel, E., et al. *Principles of Neural*

Science, 3rd ed. New York: Appleton & Lange, 1991, pp 506–507.

Figure 9.14A–E. From Brodal, A. *Neurological Anatomy.* New York: Oxford University Press, 1981.

Figure 9.14F–G. From Jefferson. *Br. J. Surg.* 26:267–302,1938.

Figure 9.20A. Magnetic resonance image courtesy of the Northwest Center for Medical Education, Indiana University School of Medicine.

Figure 9.20A. Arteriogram courtesy of the Northwest Center for Medical Education, Indiana University School of Medicine.

Figure 9.21. From Rowland, L. *Merritt's Textbook of Neurology,* 8th ed. Philadelphia: Lea & Febiger, 1989, Fig. 119-1.

Figure 10.1B. From Brobeck, J., ed. *Best & Taylor's Physiological Basis of Medical Practice,* 9th ed. Baltimore: Williams & Wilkins, 1973.

Figure 10.2. Electron micrograph courtesy of Dr. C. Gary Wright.

Figure 10.3. From Northern, J. L., and Downs, M. P. *Hearing in Children,* 4th ed. Baltimore: Williams & Wilkins, 1991.

Figure 10.4A–C. From Friedmann, I., and Ballantyne, J. *Ultrastructural Atlas of the Inner Ear.* London: Butterworths, 1984.

Figure 10.5B. From Wenstrup, J., and Suthers, R. *J. Comp. Physiol. A.,* 155:75–89, 1984.

Figure 10.6B, 1. After von Békésy, G. Wave motion in the cochlea. In: Wever, E. G., ed. *Experiments in Hearing.* New York: McGraw-Hill, 1960, Fig. 12-17.

Figure 10.6B, 2. After Tonndorf. In Rassmussen, G., and Windle, W, eds. *Neural Mechanisms of the Auditory and Vestibular Systems.* Springfield, IL: Charles C Thomas, 1960.

Figure 10.6B, 3. After von Békésy, G. The patterns of vibrations in the cochlea. In: Wever, E. G., ed. *Experiments in Hearing.* New York: McGraw-Hill, 1960, Fig. 11-58.

Figure 10.8A. After Kiang, N. Stimulus coding in the auditory nerve and cochlear nucleus. *Acta Otolaryngol.* 59:186–200, 1965.

Figure 10.8B. Threshold data after Neff, W., and Hind, J. Auditory thresholds of the cat. *JASA* 27:480–483, 1955. Characteristic frequency data after Buser, P., and Imbert, M. *Audition.* Translated by Kay, R. Cambridge, MA: MIT Press, 1992.

Figure 10.9A. After data in Rose, J., et al. Some effects of stimulus intensity on the response of auditory nerve fibers in the squirrel monkey. *J. Neurophysiol.* 34:685–699, 1971.

Figure 10.9B. After data in Rose, J., et al. Patterns of activity in single auditory fibers of the squirrel monkey. In: de Reuck, A., and Knight, J., eds. *Hearing Mechanisms in Vertebrates.* London: Churchill, 1968, pp. 144–157.

Figure 10.15. Audiograms courtesy of Mary Donigan.

Figure 10.17. From Meyerhoff, W., ed. *Diagnosis and Management of Hearing Loss.* Philadelphia: W. B. Saunders, 1984.

Figure 10.18. Courtesy of Saint Joseph's Medical Center, South Bend, IN.

Figure 10.19. Magnetic resonance image courtesy of the Magnetic Resonance Imaging Center, South Bend, IN.

Figure 11.4. After data in Duane, A. *Am. J. Ophthalmol.* 5:865–877, 1922.

Figure 11.8. From Fawcett, D. *A Textbook of Histology,* 12th ed. New York: Chapman & Hall, 1994.

Figure 11.10. Electron micrograph courtesy of S. Flieslor and J. Dant.

Figure 11.11. From Fine, B., and Yanoff, M. *Ocular Histology,* 2nd ed. New York: Harper & Row, 1979.

Figure 11.12. After data in Østerberg, *Acta Ophthalmol.* Suppl. 6, 1935.

Figure 11.15. From Patton, H., et al. *Textbook of Physiology,* 21st ed. Philadelphia: W. B. Saunders, 1989.

Figure 11.26. From Curtis, B., et al. *An Introduction to the Neurosciences.* Philadelphia: W. B. Saunders, 1972.

Figure 12.1. From Carpenter, M. B., and Sutin, J. *Human Neuroanatomy.* Baltimore: Williams & Wilkins, 1983.

Figure 12.2. From Carpenter, M. B., and Sutin, J. *Human Neuroanatomy.* Baltimore: Williams & Wilkins, 1983.

Figure 12.3. From Shepherd, *Neurobiology.* New York: Oxford University Press, 1994.

Figure 12.4. From Kandel, E., et al. *Principles of Neural Science,* 3rd ed. New York: Appleton & Lange, 1991.

Figure 12.5. From Bors, E. and Porter, R. W. Neurosurgical considerations in bladder dysfunction. *Urol. Int.* 24:114–133, 1970.

Figure 12.15. From Mettler, F. A. *Neuroanatomy,* 2nd ed. St. Louis: C.V. Mosby, 1948.

Figure 12.18. From Brodal, A. *Neurological Anatomy.* New York: Oxford University Press, 1981.

Figure 12.20. From Brodal, A. *Neurological Anatomy.* New York: Oxford University Press, 1981.

Figure 12.21B. After Barbizet, J. *Memory and Its Pathology.* San Francisco: W. H. Freeman, 1970, Fig. 9, p. 126.

Figure 12.21C. After data in Marler, P., and Terrace, H. S., eds. *The Biology of Learning.* New York: Springer-Verlag, 1984.

Figure 12.22. From Squire, L. *Memory & Brain.* New York: Oxford University Press, 1987.

Figure 13.1. Electroencephalogram courtesy of South Bend Neurology, South Bend, IN.

Figure 13.2. Electroencephalogram courtesy of South Bend Neurology, South Bend, IN.

Figure 13.5A. From Springer, S. P., and Deutsch, G. *Left Brain, Right Brain.* New York, W. H. Freeman, 1985.

Figure 13.5B. From Gazzaniga, M. S., and LeDoux, J. E. *The Integrated Mind.* New York: Plenum Press.

Figure 13.8. Magnetic resonance image courtesy of the Saint Joseph Medical Center, South Bend, Indiana.

Figure 13.9. Magnetic resonance image courtesy of the Saint Joseph Medical Center, South Bend, Indiana.

INDEX

Page numbers in *italics* denote figures; those followed by "t" denote tables.

Posterior Cerebral Territory in Axial

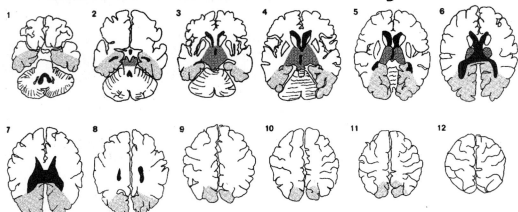

Cortical Branches

Cortical Functions

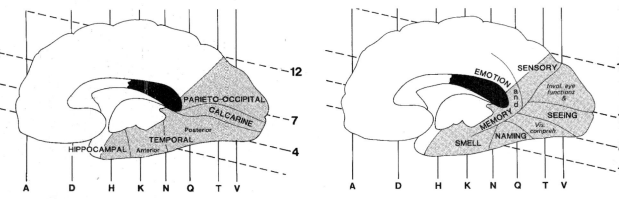

Cortical Branches labels: PARIETO-OCCIPITAL, CALCARINE, Posterior, TEMPORAL, Anterior, HIPPOCAMPAL — 12, 7, 4 — A D H K N Q T V

Cortical Functions labels: EMOTION, SENSORY, Invol. eye functions &, MEMORY, SEEING, Vis. compreh., SMELL, NAMING — 12, 7, 4, 1 — A D H K N Q T V

Posterior Cerebral Territory in Coronal

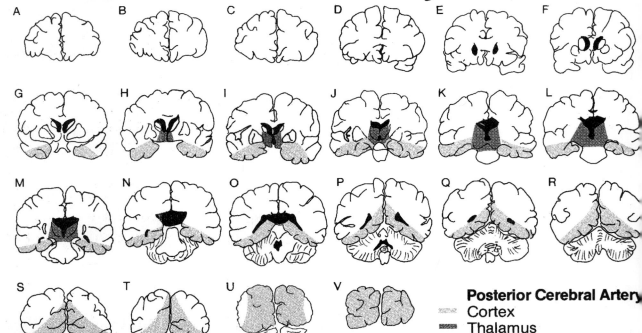

Posterior Cerebral Artery
Cortex
Thalamus
Corpus callosum